MINING AND ENERGY LAW

Mining and Energy Law is a comprehensive introduction to the laws affecting the resources sector in Australia.

It is the only textbook available that encompasses a discussion of both the law and the policy of mining and resource regulation in existing and emergent areas, for the Commonwealth as well as for the states and territories.

The book begins by examining the ownership framework underpinning natural resources within Australia and reviews the proprietary status and scope of mining tenements, and the regulation of offshore petroleum extraction. It goes on to explore the legal regimes that have emerged in response to more recent developments, such as coal seam gas, renewable energy and geo-sequestration. Other chapters consider the relevance for the mining and energy sector of environmental protection and management laws in addition to climate change regulation. The book concludes with a discussion of the commercial and contractual arrangements commonly used by transacting parties operating in the sector.

Each chapter includes:

- review questions to reinforce key concepts
- legislative extracts
- case discussion
- lists of further reading for those wishing to investigate particular topics in more detail.

Mining and Energy Law is the ideal starting point for anyone seeking to understand the regulatory regimes and policy challenges relevant to one of Australia's most important industry sectors.

Samantha Hepburn is Professor in the School of Law at Deakin University.

MINING AND ENERGY LAW

Samantha Hepburn

CAMBRIDGE
UNIVERSITY PRESS

CAMBRIDGE
UNIVERSITY PRESS

477 Williamstown Road, Port Melbourne, VIC 3207, Australia

Cambridge University Press is part of the University of Cambridge.

It furthers the University's mission by disseminating knowledge in the pursuit of
education, learning and research at the highest international levels of excellence.

www.cambridge.org
Information on this title: www.cambridge.org/9781107663237

© Cambridge University Press 2015

This publication is copyright. Subject to statutory exception
and to the provisions of relevant collective licensing agreements,
no reproduction of any part may take place without the written
permission of Cambridge University Press.

First published 2015

Cover designed by Kerry Cooke
Cartography by Tony Fankhauser
Typeset by Aptara Corp.
Printed in Singapore by C.O.S. Printers Pte Ltd

A catalogue record for this publication is available from the British Library

*A Cataloguing-in-Publication entry is available from the catalogue
of the National Library of Australia at www.nla.gov.au*

ISBN 978-1-107-66328-7 Paperback

Reproduction and communication for educational purposes

The Australian *Copyright Act 1968* (the Act) allows a maximum of
one chapter or 10% of the pages of this work, whichever is the greater,
to be reproduced and/or communicated by any educational institution
for its educational purposes provided that the educational institution
(or the body that administers it) has given a remuneration notice to
Copyright Agency Limited (CAL) under the Act.

For details of the CAL licence for educational institutions contact:

Copyright Agency Limited
Level 15, 233 Castlereagh Street
Sydney NSW 2000
Telephone: (02) 9394 7600
Facsimile: (02) 9394 7601
E-mail: info@copyright.com.au

Cambridge University Press has no responsibility for the persistence
or accuracy of URLs for external or third-party internet websites
referred to in this publication and does not guarantee that any content
on such websites is, or will remain, accurate or appropriate.

Permissions: CLR extracts reproduced with permission
of Thomson Reuters (Professional) Australia Limited,
www.thomsonreuters.com.au. Extracts from the
National Gas (South Australia) Act 2008 (SA) reproduced
with permission of the Government of South Australia.

FOREWORD

Minerals and energy law ranges across many issues and has always been complex. Australia's *Constitution* makes it especially complex, with laws in the six states and two territories being different from each other and sometimes in conflict with Commonwealth law. It has become even more complex over recent years with new technologies and concern for climate change generating new policy, law and regulation – and with conflicts over policy leading to major changes over short periods.

Hepburn manages to provide an intelligible, comprehensive account of Australian minerals and energy policy and law without losing the reader in the detail. The greatest strength of the book is to make a bewildering reality comprehensible to the reader who comes new to the subject.

This book takes the reader through the major issues, explains how these issues have led to the adoption of laws, reviews comprehensively the more difficult issues to ensure that the reader has not been left behind, presents detail from important laws, discusses cases to demonstrate how the law has worked in practice, and suggests many avenues for further reading for those who want to dig deeper into the subject.

It is an excellent book for the law student wanting to understand the broad context of her or his subject. It is just as useful for students seeking to understand minerals and energy from other disciplines and wanting to see how their subject is affected by law.

The treatment of the conventional resources issues is thorough and reliable. The early chapters – 2 to 4 – deal authoritatively with old and familiar minerals and energy issues. Resource titles differentiated at the exploration and mining stages of resources development have their own history in Australia, with the history and surviving law varying across states and territories. The resource production leases may be old, but have new overlays of complexity with the recent introduction of international legal issues – with the globally unique joint development zone with Timor, with the law of the sea, with World Heritage listing of some prospective areas, and with new international obligations of varying kinds under the *United Nations Framework Convention on Climate Change*. Natural gas operations introduce new areas of complexity related to pipeline access. Chapter 10 discusses the important question of taxation of minerals rents and again manages to make comprehensive treatment of a complex subject readable.

The most impressive and exciting innovation in this book is the treatment of the issues arising out of new technologies and products (unconventional gas) and new policy objectives (climate change, leading into policy on renewable energy and carbon capture and storage). The discussion of unconventional gas introduces us to new issues arising out of new technologies and to the polarised political response to them – varying across Australia – and to the interaction of new laws covering the new issues, with laws on biodiversity and conservation that themselves have been subject to recent change.

Renewable energy has become much more important in Australia over the past decade as a result of the federal and state Governments' interest in having Australia contribute to the global effort to reduce the impact of climate change. Renewable energy raises different issues in access to land and resources than the exploitation of fossil fuel resources, and has generated new law and regulations. Chapter 6 manages to relate the evolution of regulation designed to promote growth in renewable energy to the international discussion and negotiation of emissions reduction targets. Chapter 7 takes us into deep exploration of another set of laws and regulations that have their origins in mitigation of climate change – the capture and storage of carbon emissions from fossil fuel combustion – and then the many ways in which climate change concerns have interacted with energy policy. The discussion of international climate change mitigation notes the discordance today between Australian and other countries' approaches.

The reality in which Australian minerals and energy law operates in the years immediately ahead will continue to change rapidly, forcing more change in policy, law and regulation. Better understanding of the realities, including the realities of other countries' action on climate change, will also be a source of pressure for change. Students and others who have established a strong foundation in Australian minerals and energy policy and law through this book will have an advantage in understanding complex new realities as they reveal themselves to us.

Ross Garnaut
University of Melbourne

CONTENTS

Foreword by Ross Garnaut   *v*
Acknowledgement   *xiii*
Acronyms   *xiv*
Table of cases   *xviii*
Table of statutes   *xx*

1 Ownership of minerals and natural resources   1

    1.1 Introduction   2
        1.1.1 The nature and scope of energy resources in Australia   2
    1.2 Ownership of the subsurface strata at common law   5
    1.3 Public ownership for minerals and petroleum   10
    1.4 The proprietary status of mining tenements   14
    1.5 Royal minerals: Gold and silver   18
    1.6 Ownership of renewable resources: Hydro-electricity, geothermal, solar, and wind   21
        1.6.1 Hydro-electrical power   22
        1.6.2 Geothermal energy   22
        1.6.3 Ownership of water in Australia   23
        1.6.4 Access entitlements for wind and solar   28
        1.6.5 Market progression   30
    1.7 Division of land and resources: Overlapping tenures   31
    1.8 Land access and compensation   34
    1.9 Native title, cultural heritage, and mining rights   40
    1.10 Review questions   46
    1.11 Further reading   47

2 Resource titles: Permits, licenses, and leases   51

    2.1 Introduction   52
    2.2 Mining approval process: Exploration, assessment, and extraction phases   52
    2.3 Exploration licences and permits   54
        2.3.1 Approval process   54

	2.3.2	The character of an exploration licence/permit	57
	2.3.3	Relevant legislative provisions	59
		2.3.3.1 New South Wales	59
		2.3.3.2 Western Australia	62
		2.3.3.3 Queensland	64
	2.3.4	Proprietary status of the exploration licence	69
	2.3.5	Legal status of offshore exploration permits: s 51(xxxi) *Constitution*	70
2.4	Retention licences and assessment leases		75
2.5	Mining and production leases		80
	2.5.1	Statutory character of a mining lease	80
	2.5.2	General terms and conditions of a mining lease	83
	2.5.3	General entitlements of a mining lease	84
2.6	Review questions		90
2.7	Further reading		92

3 Australian offshore petroleum and minerals regulation — 94

3.1 Introduction — 95

3.2 Constitutional arrangements for offshore regulation — 97
- 3.2.1 United Nations Convention on the Law of the Sea — 103
- 3.2.2 Territorial waters — 105
- 3.2.3 State jurisdiction in the territorial sea — 106
- 3.2.4 Contiguous zone, exclusive economic zone, and the continental shelf — 107
- 3.2.5 The high seas — 113
- 3.2.6 *Offshore Petroleum and Greenhouse Gas Storage Act 2006* (Cth) — 117
- 3.2.7 Overview of the *OGGSA* title framework — 119
  - 3.2.7.1 Petroleum exploration permit — 120
  - 3.2.7.2 Petroleum retention lease — 120
  - 3.2.7.3 Petroleum production licence — 121
  - 3.2.7.4 Infrastructure licence — 123
  - 3.2.7.5 Pipeline licence — 124
- 3.2.8 Sea installations — 124
- 3.2.9 Offshore petroleum safety: NOPSEMA — 125
- 3.2.10 Joint petroleum area and Greater Sunrise: Australia and Timor-Leste — 127
- 3.2.11 *Offshore Minerals Act 1994* (Cth) — 130

3.3 Review questions — 131

3.4 Further reading — 132

4 Natural gas regulation — 136

4.1 Introduction — 137

4.2	What is natural gas?		139
4.3	The Australian gas market		140
4.4	The regulatory framework for natural gas		143
	4.4.1	The former natural gas access code	144
	4.4.2	The National Gas Law: Functions of the AER and the AEMC	147
	4.4.3	The National Gas Law: An overview	158
	4.4.4	Pipeline classification under the National Gas Law	159
	4.4.5	Light and fully regulated pipelines	164
	4.4.6	Access determination disputes	166
	4.4.7	Access arrangement information for light and full regulation pipelines	167
4.5	Review questions		171
4.6	Further reading		172

5 Unconventional gas regulation — 175

5.1	Introduction		176
5.2	What is unconventional gas?		178
	5.2.1	Shale gas	179
	5.2.2	Tight gas	179
	5.2.3	Coal seam gas	179
5.3	How is unconventional gas extracted?		180
	5.3.1	Hydraulic fracturing and horizontal drilling	181
	5.3.2	Water pumping for coal seam gas extraction	182
5.4	Environmental and social issues associated with unconventional gas extraction		184
5.5	Regulatory frameworks for unconventional gas: Queensland and New South Wales		188
	5.5.1	Queensland: *Petroleum and Gas (Production and Safety) Act 2004*	188
		5.5.1.1 Regulatory requirements for resource titles	188
		5.5.1.2 Access and compensation framework	196
	5.5.2	New South Wales	202
		5.5.2.1 The ownership framework under the *Petroleum (Onshore) Act 1991* (NSW)	207
		5.5.2.2 Petroleum licences and conditions under the *Petroleum (Onshore) Act 1991* (NSW)	208
		5.5.2.3 Land access, compensation and access disputes	212
		5.5.2.4 Codes of practice for CSG: Fracture stimulation and well integrity	216
		5.5.2.5 The gateway process	219

		5.5.2.6	*Environmental Planning and Assessment Act 1979* (NSW); *Protection of the Environment Operations Act 1997* (NSW)	223
		5.5.2.7	The *Water Management Act 2000* (NSW) and the Aquifer Interference Policy	227

    5.6 Regulatory framework: *Environment Protection and Biodiversity Conservation Act 1999* (Cth)   230

    5.7 Review questions   232

    5.8 Further reading   233

6 Renewable energy: Regulation, the RET, wind energy, and the market framework   **239**

    6.1 Introduction   240

    6.2 What is renewable energy?   241

    6.3 The renewable energy market   244

    6.4 Statutory regulation: *Renewable Energy (Electricity) Act 2000* (Cth)   246

    6.5 Background to the RET   250

    6.6 Australian Renewable Energy Agency (ARENA), the Clean Energy Regulator (CER), and the Clean Energy Finance Corporation (CEFC)   252

    6.7 The Large-scale Renewable Energy Target   254

    6.8 The SRES and the Solar Credits Scheme   257

    6.9 The economics of renewable energy   258

    6.10 National electricity market   260

    6.11 Wind energy: Regulatory and practical issues   262

    6.12 Review questions   273

    6.13 Further reading   274

7 Carbon capture sequestration   **279**

    7.1 Introduction   280

    7.2 What is carbon capture and storage?   282

    7.3 Why do we need CCS?   286

    7.4 Capturing CO_2 288

        7.4.1 Pre-combustion technology   289

        7.4.2 Post-combustion technology   289

        7.4.3 Oxyfuel combustion   289

		7.4.3.1	Enhanced oil recovery operations	290
		7.4.3.2	CCS operations in the cement industry	290
	7.5	Transporting CO_2		291
	7.6	Storing CO_2		292
	7.7	International CCS projects		292
	7.8	CCS in Australia		294
	7.9	Regulating the storage of CO_2 in Australia		296
		7.9.1	Victoria: *Greenhouse Gas Geological Sequestration Act 2008*	297
		7.9.2	Queensland: *Greenhouse Gas Storage Act 2009*	301
		7.9.3	The Australian offshore statutory regime	304
			7.9.3.1 Commonwealth: *Offshore Petroleum and Greenhouse Gas Storage Act 2006*	305
			7.9.3.2 Victoria: *Offshore Petroleum and Greenhouse Gas Storage Act 2010*	307
		7.9.4	Legal liability for carbon capture in Australia	307
			7.9.4.1 Environmental concerns	307
			7.9.4.2 Tortious actions	308
			7.9.4.3 Regulatory standards	309
	7.10	Review questions		310
	7.11	Further reading		311
8	**Climate change and mining and energy policy**			**315**
	8.1	Introduction		316
	8.2	Changes in the climate system: Atmosphere, ocean, cryosphere, and sea level		318
	8.3	The legal framework		320
		8.3.1	The IPCC	321
		8.3.2	The *United Nations Framework Convention on Climate Change*	321
		8.3.3	The *Kyoto Protocol*	324
		8.3.4	The *Energy Charter* and the *Energy Charter Treaty*	329
		8.3.5	The Harvard Project: a new proposed framework for climate change	333
	8.4	The economics of climate change		335
	8.5	The impact of climate change on Australian mining and energy industries		342
	8.6	Review questions		348
	8.7	Further reading		349

9	**Environmental regulation**	**354**
9.1	Introduction	355
9.2	Jurisdictional framework	360
9.3	Bilateral agreements	362
9.4	Environmental impact assessment	365
9.5	Environmental assessment of onshore mining projects in Western Australia	368
9.6	Environmental assessment of onshore mining projects in Queensland	374
9.7	Environmental assessment of onshore mining projects in New South Wales	381
9.8	Commonwealth environmental legislation: *Environment Protection and Biodiversity Conservation Act 1999*	387
9.9	Review questions	398
9.10	Further reading	399
10	**Mining agreements and revenue frameworks**	**405**
10.1	Introduction	406
10.2	Mining agreements	407
	10.2.1 Concession agreements	409
	10.2.2 Profit sharing contracts	410
	10.2.3 Risk service contracts	414
	10.2.4 Joint venture agreements	414
10.3	The core elements of a mining agreement	417
10.4	Revenue frameworks	422
	10.4.1 Royalties	422
	10.4.2 Royalty rates and mining taxes in the Australian context	425
	10.4.2.1 Coal royalty rates	428
	10.4.2.2 Iron ore royalty rates	428
	10.4.2.3 Petroleum royalty rates	428
	10.4.2.4 Privately owned minerals	429
	10.4.2.5 Meaning of a well-head	430
	10.4.2.6 Mining taxes in Australia	434
	10.4.2.7 Minerals resource rent tax	435
	10.4.2.8 Petroleum resource rent tax	436
10.5	Review questions	437
10.6	Further reading	438
Index		*444*

ACKNOWLEDGEMENT

This book has taken a long time to write and has been the product of the advice, inspiration and enthusiasm of many people, both directly and indirectly.

First and foremost I must thank my family for their patience and fortitude. I would also like to thank my students – past and present – for their interest and enthusiasm. I thank the numerous colleagues and friends who assisted with the preparation of draft chapters and who spent time discussing areas of expertise. A very special thank you to Noeleen McNamara for her input into Chapter 9. A huge thank you to Professors Ross Garnaut and Michael Crommelin for their vast knowledge and for taking the time to share it with me. And last, but not by any means least, thank you to the amazing team at Cambridge University Press for believing in me in the first place, and for their unstinting support, effort, and rigor in preparing the book for publication. I must particularly thank: Martina Edwards, Antonietta Anello, Jodie Fitzsimmons, and Ellie Gleeson.

ACRONYMS

ABARE   Australian Bureau of Agriculture and Resource Economics
AEMC   Australian Energy Market Commission
AEMO   Australian Energy Market Operator
AEMR   Annual Environmental Management Reports
AER   Annual Environmental Report
AER   Australian Energy Regulator
AIP   Aquifer Interference Policy
APPEA   Australian Petroleum Production Exploration Association
ARENA   Australian Renewable Energy Agency
BREE   Bureau of Resources and Energy Economics
BSAL   biophysical, strategic agricultural land
CBA   cost-benefit approach
CBM   coal bed methane (used internationally to describe CSG)
CCS   carbon capture and storage
CDM   clean development mechanisms
CEC   Code of Environmental Compliance
CEFC   Clean Energy Finance Corporation
CER   Clean Energy Regulator
CIC   critical industry cluster
CLCS   United Nations Committee on the Limits of the Continental Shelf
CMATS   *Treaty on Certain Maritime Arrangements in the Timor Sea*
COAG   Council of Australian Governments
CSG   coal seam gas
DEHP   Department of Environment and Heritage Protection
DMP   Department of Mines and Petroleum (WA)
DoE   Department of Energy (WA)
DoIR   Department of Industry and Resources (WA)
DTSSP   Declared Transmission System Service Provider
DWGMR   *Declared Wholesale Gas Market Rules*
EA   environmental authority
EC   *Energy Charter*
ECT   *Energy Charter Treaty*
EEA   *Environmental Effects Act 1978* (Vic)
EES   environmental effects statement
EEZ   exclusive economic zone
EIA   Environmental Impact Assessment

EIS environmental impact statement
EITC emissions-intensive trade-exposed
EM environmental management
EMPC *Environmental Management and Pollution Control Act 1994* (Tas)
EOR enhanced oil recovery
EPA *Environmental Planning Act 1986* (WA)
EPA *Environment Planning Act 1993* (SA)
EPAA *Environmental Planning and Assessment Act 1979* (NSW)
EPA *Environmental Protection Act 1994* (Qld)
EPA Environmental Protection Authority
EPBCA *Environment Protection and Biodiversity Conservation Act 1999* (Cth)
EPGR *Environment Protection (Greentape Reduction and other Legislation Amendment) Act 2012* (Qld)
EPHC Environment Protection and Heritage Council
ERA Economic Regulation Authority (WA)
ESD ecologically sustainable development
ESG2 Environmental Impact Assessment Guidelines for NSW
EURATOM European Atomic Energy Community
FLNG floating liquid natural gas
FM flexible mechanisms
FSMP fracture stimulation management plan
GGGSA *Greenhouse Gas Geological Sequestration Act 2008* (Vic)
GGSA *Greenhouse Gas Storage Act 2009* (Qld)
GHG greenhouse gas
GS Greater Sunrise oilfield
IEA International Energy Agency
IEC *International Energy Charter*
IESC Independent Expert Scientific Committee on CSG and Large Coal Mining Development
ILUA Indigenous Land Use Agreement
IMO International Maritime Organisation
IPCC Intergovernmental Panel on Climate Change
IUA International Unitisation Agreement
JI joint implementation
JORC Code Australasian Code for Exploration Results, Mineral Resources and Ore Resources
JPDA Joint Petroleum Development Area
LEP Local Environmental Plan
LGC Large-scale Generation Certificate
LNG liquid natural gas
LPG liquid petroleum gas

LREC   Large-scale Renewable Energy Certificate
LRET   Large Scale Renewable Energy Target
MA   *Mining Act 1971* (SA)
MA   *Mining Act 1978* (WA)
MA   *Mining Act 1992* (NSW)
MNES   matters of national environmental importance
MODU   Mobile Offshore Drilling Units
MOP   Mining Operations Plan
MRA   *Mineral Resources Act 1989* (Qld)
MRDA   *Mineral Resources Development Act 1995* (Tas)
MREMP   Mining, Rehabilitation and Environmental Management Process
MRRT   Minerals resource rent tax
MRSD   *Mineral Resources Sustainable Development Act 1990* (Vic)
MTA   *Mineral Titles Act* (NT)
MTPA   million tonne per annum
NCC   National Competition Council
NEL   National Electricity Law
NEM   National Electricity Market
NERL   National Energy Retail Law
NGC   National Gas Code
NGL   National Gas Laws
NGL   natural gas liquids
NGR   National Gas Rules
NNTT   National Native Title Tribunal
NOPSEMA   National Offshore Petroleum Safety Environment Management Agency
NOPTA   National Offshore Petroleum Titles Administrator
NPACSG   National Partnership Agreement on Coal Seam Gas
NTA   *Native Title Act 1993* (Cth)
NTPC   Native Title Protection Conditions
NWI   National Water Initiative
OCSG   Office of Coal Seam Gas (NSW)
OMA   *Offshore Minerals Act 1994* (Cth)
OPGGSA   *Offshore Petroleum and Greenhouse Gas Storage Act 2006* (Cth)
OPGGSA   *Offshore Petroleum and Greenhouse Gas Storage Act 2010* (Vic)
OSPAR   *Convention for the Protection of the Marine Environment of the North-East Atlantic*
PA   *Petroleum Act 1998* (Vic)
PDA   *Planning and Development Act 2005* (WA)
PEA   *Planning and Environment Act 1987* (Vic)
PEC   Partial Exemption Certificate
PEERA   Energy Charter Protocol on Energy Efficiency and Related Environmental Aspects

PEOA   *Protection of the Environment Operations Act 1997* (NSW)
PGEA   *Petroleum and Geothermal Energy Act 2000* (SA)
PGERA *Petroleum and Geothermal Energy Resources Act 1967* (WA)
PGPSA *Petroleum and Gas (Production and Safety) Act 2004* (Qld)
POA   *Petroleum (Onshore) Act 1991* (NSW)
PP   precautionary principle
PPSA   *Personal Property Securities Act 2009* (Cth)
PRRT   petroleum resource rent tax
PSC   profit sharing contract
PSLA   *Petroleum (Submerged Lands) Act 1967* (Cth)
PSLA   *Petroleum (Submerged Lands) Act 1992* (Tas)
QELRO Qualified Emission Limitation and Reduction Objective
REC   Renewable Energy Certificate
REEA   *Renewable Energy Electricity Act 2000* (Cth)
REF   Review of Environmental Factors
RET   Renewable Energy Target
RPP   Renewable Power Percentage
RSPT   Resource Super Profit Tax
SCER   Standing Council on Energy and Resources
SDPWOA *State Development and Public Works Organisation Act 1972* (NSW)
SEPP   State Environmental Planning Policy
SMP   safety management plan
SREC   Small-scale Renewable Energy Certificate
SRES   Small-scale Renewable Energy Scheme
SRET   Small-scale Renewable Energy Target
SRLUP Strategic Regional Land Use Plan
SSD   state significant development
SSI   state significant infrastructure
SSLA   *Sea and Submerged Lands Act 1973* (Cth)
STC   Small-scale Technology Certificates
STTM   Short Term Trading Market
TST   *Timor Sea Treaty 2003*
UNCED United Nations Conference on Environmental Development 1992
UNCLOS *United Nations Convention on the Law of the Sea*
UNEP   United Nations Environment Program
UNFCCC *United Nations Framework Convention on Climate Change*
WAL   Water Access Licence
WMA   *Water Management Act 2000* (NSW)
WMO   World Meteorological Organization

TABLE OF CASES

Adamson v Hayes (1973) 130 CLR 276 **15**
Anaconda Nickel Ltd v Tarmoola Australia Pty Ltd 165 (2000) 22 WAR 101 **70**
Bernstein v Skyviews & General Ltd [1978] QB 479 **9**
BHP Billiton Iron Ore Pty Ltd v The National Competition Council [2007] ATPR 42 **160**
BHP Petroleum Pty Ltd v Balfour (1987) 180 CLR 474 **431-3**
Bocardo SA v Star Energy UK Onshore Ltd [2010] 3 WLR 354 **6**
Bonsa v La Macchia (1970) 122 CLR 177 **103**
Brown v Western Australia (No 2) (2010) 268 ALR 149 **89**
Bursill Enterprises Pty Ltd v Berger Bros Trading Pty Ltd (1971) 124 CLR 73 **9**
Bury v Pope (1586) Cro Eliz 118; 78 ER 375 **5**
Cadia Holdings Pty Ltd v State of New South Wales (2010) 242 CLR 195 **18, 19-21**
Case of Mines [1568] 1 Plowd 310 **18-19**
Cherry Tree Wind Farm Pty Ltd v Mitchell Shire Council [2013] VCAT 521 **271-2**
Cherry Tree Wind Farm Pty Ltd v Mitchell Shire Council [2013] VCAT 1939 **272-3**
ChongHerr Investments Ltd v Titan Sandstone Pty Ltd (2007) Q ConvR 54-669 **81**
Commissioner for Railways v Valuer General [1974] 1 AC 382 **6**
Commissioner of Crown Lands v Page [1960] 2QB 274 **15**
Commissioner of State Revenue v Oz Minerals [2013] WASCA 239 **15**
Commonwealth v WMC Resources Ltd (1998) 194 CLR 1 **70-2, 73, 75**
Commonwealth v Yarmirr (2001) 208 CLR 1 **104-5, 305**
Connell v Santos NSW Pty Ltd [2014] NSWLEC 1 **210, 211-12**
Cougar Energy Ltd v Debbie Best, Chief Executive Under the EPA [2011] QPEC 150 **376**
Cudgen Rutile (No 2) Pty Ltd v Chalk PC [1975] AC 520 **408**
Doe d. Douglas v Lock (1835) 2 Ad & E 705 **10**
Duke of Sutherland v Heathcote [1892] 1 Ch 475 (CA) **69-70**
Edwards v Sims (1929) Ky 24 SW (2d) 619 **308**
Embrey v Owen (1851) 6 Ex 353 **24**
Epic Energy (WA) One Pty Ltd, Re v Commissioner of State Revenue (2011) 43 WAR 186 **32-4**
Finlay Stonemasonry Pty Ltd v JD & Sons Nominees Pty Ltd [2011] NTSC 37 **8**
Fullerton Cove Residents Action Group Inc v Dart Energy Ltd (No 2) [2013] NSWLEC 38 **225-6**
Gowan v Christie (1873) LR 2 Sc & Div 273 **74, 80**
Grant Pastoral Co Pty Ltd v Thorpes Ltd (1954) 54 SR (NSW) 129 **24**
Gray v The Minister for Planning [2006] NSWLEC 720 **395**
Grimaldi v Chameleon Mining NL (No 2) [2012] FCAFC 6 **416**
Gullen Range Wind Farm Pty Ltd v Minister for Planning [2010] NSWLEC 1102 **273**
Hancock Prospecting Pty Ltd v Wright Prospecting Pty Ltd [2012] WASCA 216 **16**
Hinkley v Star City Pty Ltd [2010] NSWSC 1389 **6, 17**
Hospital Products v United States Surgical Corporation (1986) 156 CLR 41 **416**
ICM Agriculture v The Commonwealth of Australia (2009) 240 CLR 140 **72, 75**
King v Minister for Planning [2010] NSWLEC 1102 **273**
Lord v Commissioners of Sydney (1859) 2 Legge 912 **24**
Mabo v Queensland [No 2] (1992) 175 CLR 1 **40, 305**
Meinhard v Salmon (1928) 164 NE 545 **417**

Metgasco Limited v Minister for Energy and Resources [2015] NSWSC 453 **387**
Millar v Wildish (1863) 2 W & W (E) 37 **19**
Minister for Resources, Re; Ex parte Cazaly Iron Pty Ltd (2007) WAR 403 **56, 76, 81**
Mitchell v Mosely [1914] 1 Ch 438 **7**
New South Wales v Commonwealth (1975) 135 CLR 337 **99-100, 103, 104, 106**
Newcrest Mining WA Ltd v The Commonwealth (1997) 190 CLR 513 **73-5, 81**
North Ganalanja Aboriginal Corporation v Queensland (1986) 185 CLR 595 **43-4**
NSW v Bardolph (1934) 52 CLR 455 **421**
O'Connor v Arrow (Daandine) Pty Ltd [2009] QSC 432 **198**
O'Keefe v Williams (1910) 11 CLR 171 **303**
Ownership of Offshore Mineral Rights, Reference re [1966] INSC 207; (1967) SCR 792; (1967) 65 DLR (2nd) 353 **99**
Pape v Federal Commissioner of Taxation (2009) 238 CLR 1 **19**
Parkesbourne-Mummel Landscape Guardians v Minister for Planning [2010] NSWLEC 1102 **273**
Pilbara Infrastructure Pty Ltd, The v Australian Competition Tribunal (2012) 246 CLR 379 **159**
Pilmer v The Duke Group Ltd (2001) 207 CLR 165 **416**
Pointe Gourde Quarrying & Transport Co Ltd v Sub-Intendent of Crown Lands [1947] AC 565 **8**
R v Earl of Northumberland [1568] 1 Plowd 310 (Case of Mines) **18-19**
R v Keyn (1876) 2 Ex D 63 **103, 305**
R v Wilson (1874) 12 SCR (L) (NSW) 258 **19**
Racal Communications Ltd, Re [1981] 1 AC 374 **432**
Red Hill Iron Ltd v API Management Pty Ltd [2012] WASC 323 **34, 416**
Rossmar Park Pastoral Co Pty Ltd v Coal Mines Australia Pty Ltd [2008] NSWSC 1385 **70**
Spencer v Commonwealth of Australia (2010) 241 CLR 118 **73**
Star Energy Onshore Ltd v Bocardo Ltd [2010] 1 Ch 100 **6-7, 16-17**
Star Energy Weald Basin Ltd v Bocardo SA [2011] AC 380 **7-8**
Stoneman v Lyons (1975) 133 CLR 550 **308**
Stow v Mineral Holdings (Australia) Pty Ltd (1977) 180 CLR 295 **70, 81**
Straits Exploration (Australia) Pty Ltd v Kokatha Uwankara Native Title Claimants [2012] SASCFC 121 **40**
Tate v Lyle Industries v Greater London Council [1983] 2 AC 509 **24**
TEC Desert Pty Ltd v Commissioner of State Revenue (2010) 241 CLR 576 **16**
Transport for London v Spirerose Ltd [2009] 1 WLR 1797 **8**
Ulan Coal Mines v Minister for Mineral Resources [2007] NSWSC 1299 **82-3**
Ulan Coal Mines v Minister for Mineral Resources [2008] NSWCA 174 **82**
United Dominions Corporation v Brian (1985) 157 CLR 1 **415**
Valuer-General v Perilya Broken Hill Ltd [2013] NSWCA 265 **81**
Wade v New South Wales Rutile Mining Co Pty Ltd (1969) 121 CLR 177 **15, 19, 80, 81-2**
Walker Superannuation Fund v Clough Property Fairmont Pty Ltd [2010] WASCA 232 **9**
Waters v Welsh Development Agency [2004] 1 WLR 1304 **8**
Western Australia v The Commonwealth (1995) 185 CLR 373 **42-3**
Western Australia v Ward (2002) 213 CLR 1 **16**
Westmoreland & Cambria Natural Gas Co v DeWitt 130 Pa. 235 (1889) **13**
Wik Peoples v Queensland (1996) 187 CLR 1 **303**
Williams, Re v De Biasi (Unreported, 18 August 1992) **70**
Williams v Commonwealth of Australia (2012) (1934) 248 CLR 156 **421**
Woolley v Attorney-General of Victoria (1877) 2 App Cas 163 **18, 19**
Wurridjal v The Commonwealth (2009) 237 CLR 309 **72-3**
Yarmirr v Northern Territory (2001) 208 CLR 1 **304**

TABLE OF STATUTES

Commonwealth

Aboriginal and Torres Strait Islander Heritage Protection Act 1984 45
Australian Energy Market Act 2004 138, 148
Australian Energy Market Commission Establishment Regulations 2005 144
Australian National Registry of Emissions Units Act 2011
    Pt 2 253
Australian Renewable Energy Agency Act 2011
    s 7 253
    s 8 253
Clean Energy Act 2011, Pt 5 253
Clean Energy Finance Corporation Act 2012
    s 3 254
    s 9 254
Clean Energy Regulator Act 2011 253
Coastal Waters (Northern Territory Powers) Act 1980 103
Coastal Waters (State Powers) Act 1980
    s 5 100-1
Coastal Waters (State Title) Act 1980 101
    s 4 101-2
Competition and Consumer Act 2010
    Pt IIIA 164
    Pt IIIAA 143
    s 86C 152
    s 86D 152
Constitution 106, 360, 361
    s 51(xxxi) 12, 70-5, 81, 90, 103, 306
    s 51(xxxviii) 100
    s 52 361
    s 61 19, 421
    s 64 421
    s 90 361
    s 96 73
Declared Wholesale Gas Market Rules 158
Environment Protection and Biodiversity Conservation Act 1999 45, 62, 178, 230-2, 265, 356, 361-2, 387-98
    Pt 3, Div 1 231
    Pt 3, Div 1, subdiv AA 45
    Pt 3, Div 1, subdiv F 125
    Pt 5 231, 393
    Pt 7, Div 1 231
    Pt 8, Div 3 231, 389
    s 3(1)(b) 230
    s 3A 393
    s 22 389
    s 24D 230, 390, 391
    s 24D(1) 231
    s 44 363
    s 45 363, 364
    s 46 363
    s 47 363
    s 136 396, 397-8
    s 136(1) 393
    s 136(2) 393
    s 136(2)(a) 393, 394
    s 178(1) 388
    s 523 389
    s 528 230, 231, 391
Environmental Protection (Impact of Proposals) Act 1974 388
    s 5 388
Financial Management and Accountability Act 1997
    s 32B 421
Gas Pipelines Access Law 144
Great Barrier Reef Marine Park Act 1975 101, 102
Minerals Resource Rent Tax Act 2012
    ss 1–10 436
Minerals Resource Rent Tax Repeal and Other Measures Act 2014 436
    Sch 1 436
National Electricity Law 144, 147
    Pt 6, Div 5 152
    s 7 261
National Electricity Rules 108, 144
National Energy Retail Law 144
National Gas Amendment (Price and Revenue Regulation of Gas Services) Rule 2012 170

National Gas Rules 144, 148, 152, 155, 158, 168, 170
   Pt 19 **158**
   r 42 **168**
   r 43 **168**
   r 45 **168**
   r 48 **169-70**
   r 118 **166**
   rr 118–119 **166-7**
   r 119 **166**
   r 369 **151**
National Greenhouse and Energy Reporting Act 2007 **253**
National Third Party Access Code for Natural Gas Pipeline Systems 144-7, 155, 168
   Sch 1, cl 1 **145**
   Sch A **146**
   s 3.1 **145**
   s 11 **145**
Native Title Act 1993 40-1, 45
   Pt 2, Div 3, Subdiv B **43**
   Div 3 **41**
   s 23 **43**
   s 24MD(6B) **56**
   s 26(1A) **42**
   s 26(2) **42-3**
   s 26(3)(b) **42**
   s 26A 40, 44
   s 28 **42**
   s 29 **43**
   s 31(1)(a) **44**
   s 31(1)(b) **44**
   s 32 **41**
   s 34EB(1)(b)(i) **265**
   s 35 **44**
   s 38 43, 44
   s 41 **44**
   s 42 **45**
   s 43 **42-3**
   s 43(1) **42**
   s 43(1)(b) **43**
   s 43(2)(1) **42**
   s 109 **43**
   s 139 **44**
   s 223 **40**
   s 235(8) **105**
   s 237 **41**
Navigation Act 1912 **118**
Navigation Act 2006, s 186A **116**
Northern Territory (Self-Government) Act 1978
   s 69(4) 10, 430
Offshore Constitutional Settlement 1979 97, 100, 106
Offshore Minerals Act 1994 **130-1**
   ss 29–32 **79**
   ss 29–34 **131**
   s 37 **131**
   ss 132–133 **79**
   s 137(1) **79**
   s 439A **69**
Offshore Petroleum Act 2006 **305**
Offshore Petroleum Agreement 1967 **97-8**
Offshore Petroleum and Greenhouse Gas Storage Act 2006 117-24, 126, 130, 294-5, 296, 305-7, 429
   Pt 1.3 79, 119
   Pts 2.2–2.9 **119**
   Pt 2.3 **79**
   Pt 2.9 **118**
   Pt 3.3 **306**
   Pt 3.4 **306**
   Pt 6.8, sch 3 **118**
   Pt 6.9 **119**
   Pt 6.10 **119**
   s 3 **294**
   s 5 97-8, 117-18
   ss 5(1)–5(2) **117-18**
   s 5(2)(a) **107**
   s 5(2)(b) **106**
   s 7 **117**
   s 8 **118**
   s 21(1) **306**
   s 56 **119**
   s 59(2) **119**
   s 80 **117**
   s 98 **120**
   s 110 **122**
   s 135 **120**
   s 141 80, 120
   s 142 **121**
   s 145 **121**
   s 148 **121**
   s 161 **122-3**
   s 165 **122**
   s 166 **122**
   s 168(2) **120**
   s 178 **122**
   s 198 **124**
   s 211 **124**
   s 215 **124**

s 217 124
s 226 124
s 227 124
s 297 295
s 304 295
s 319 306
s 357 306
s 357(1)(c)–(h) 306
s 358(3)(d) 309
s 358(3)(e) 309
s 358(3)(f) 309
s 358(3)(j) 309
ss 359–360 307
s 379(1) 310
s 379(1)(b) 310
s 380(1)(c) 310
s 380(1)(g) 310
s 380(1)(h) 310
s 383 306
s 458 306
ss 470–471 127
ss 616–23 118
ss 631–635 429
s 638 126
Sch 3 126
Offshore Petroleum and Greenhouse Gas Storage Amendment (Compliance Measures) Act 2013 126
Sch 2A 126
Offshore Petroleum and Greenhouse Gas Storage (Resource Management and Administration) Regulations 2011
Pt 5 126
Pt 6 295
Offshore Petroleum and Greenhouse Gas Storage (Safety) Regulations 2009 126
Offshore Petroleum (Royalty) Act 2006 429
s 5 429
s 6 429
ss 6–10 429
s 7 429
s 8 429
s 9 429
s 10 429
s 11 433
Personal Property Securities Act 2009 14, 209
s 8 15
s 10 14
Petroleum (Submerged Lands) Act 1967 71, 98, 117

s 42 431
Petroleum (Submerged Lands) Amendments Act 1982 126
Petroleum (Submerged Lands) (Royalty) Act 1967
s 5 431
s 8 431, 432
Petroleum (Timor Sea Treaty) Act 2003 127
Protection of Movable Cultural Heritage Act 1986 45
Racial Discrimination Act 1975 40
Renewable Energy (Electricity) Act 2000 240, 246, 249
s 3 251
s 6 250
s 17 254–5
s 39 250
Renewable Energy (Electricity) (Charge) Act 2000 249
Renewable Energy (Electricity) Regulations 2001 249
Renewable Energy (Electricity) (Small-Scale Technology Shortfall Charge) Act 2010 249
Sea Installations Act 1987 124–5
s 4 125
s 14 125
s 15 125
Seas and Submerged Lands Act 1973 99, 104, 106, 114, 304
Div 1A 107
Div 2A 107
s 6 103, 105
s 7 106
ss 10A–10C 109-10
s 11 103, 111
Water Act 2007, s 4 25
Work Health and Safety Act 2011 216

Australian Capital Territory

Environment Protection Act 1997 361
Heritage Act 2004 45
Heritage Objects Act 1991 45
National Gas (ACT) Act 2008 138, 148
Planning and Development Act 2007 361

New South Wales

Aboriginal Sacred Sites Act 1989 45
Aquifer Interference Policy 2012 221, 227-8
1.1 228

3.1 229
3.2 229
3.2.1 229
3.2.2 229
Pt 5 229
Coastal Powers (Coastal Waters) Act 1979 103
Coastal Protection Act 1979 228
Crown Lands Act 1884 430
  s 7 10
Crown Lands Consolidation Act 1913 430
Environmental Planning and Assessment Act 1979 202, 216, 223-6, 227, 263, 361, 366, 382, 383
  Pt 3A 264
  Pt 4 224, 226, 264
  Pt 4, Div 4.1 223, 224, 384
  Pt 5 264, 383
  Pt 5.1 223, 224, 384
  s 26 223, 382
  s 76A 224
  s 77a(2) 225
  s 78A(8)(a) 225
  s 89C 223, 384
  s 89C(3) 224
  s 89D 224
  s 89E 384
  s 89J(1)(g) 226, 228, 384
  s 111 225, 226, 383
  s 112 225, 383
  s 115G(1) 226
Environmental Planning and Assessment Regulations 2000 223, 264, 382
  Sch 1, Pt 1, reg 2(1)(c) 225
  Sch 1, reg 2(1)(e) 224
  Sch 2, reg 3(4A)(a) 384
  Sch 2, reg 3(4A)(b) 384
  Sch 2, reg 7 225
  Sch 3, reg 27 225
  reg 228 226, 383
Fisheries Management Act 1994 228
Fracture Stimulation Code 217-18
  1.2 45
  1.2(a) 217
  2.2(a) 218
  2.3(a) 218
  4.2(d) 217
  5.2(c) 218
  6.2 45, 217
  7.2(c) 218
Heritage Act 1977 45, 228

Heritage Conservation Act 1991 45
Mining Act 1992 4, 19-21, 60, 62, 70, 82, 83, 216, 225, 384
  Pt 11, Div 3 386
  s 11 17
  s 13 60
  ss 13–29 59
  s 22 61
  s 26 61
  s 27 61
  s 28 61-2
  s 29 62
  s 33 78
  s 45 77
  s 62(1) 83
  s 62(1)(c) 82
  s 62(6) 83
  s 73 86-7
  s 140 35
  s 141 64
  s 237 83
  s 239 385-6
  ss 263–267 38
  s 284 430
  s 291 428
  s 291A 428
  s 292 428
  Sch 1, Div 4 82
  Sch 1, Div 4, cl 21 82
  Sch 1, Div 4, cl 23A 83
  Sch 1, Div 4, cl 23B 82
Mining Regulation 2010
  reg 20 77
  regs 61A–65 428
National Gas (New South Wales) Act 2008 138, 148
National Parks and Wildlife Act 1974 228
National Parks and Wildlife Amendment (Aboriginal Ownership) Act 1996 45
New South Wales Gas Policy 2014 202-5, 207-12, 215-16, 221, 222
Petroleum (Offshore) Act 1982
  ss 20–39 60
Petroleum (Onshore) Act 1991 38, 202, 207-16, 226, 384
  s 3 12, 207
  s 6 5, 207
  s 6(1) 207
  s 6(2) 207
  s 7 207
  s 22 210, 211

s 22(1) 210
s 22(3A) 211
s 23 210
s 25 208
s 26 14, 209
s 28A 35
s 29 209
ss 29–32 60
s 33 209
s 38 209
s 41 209
s 45B(1) 210
s 45C 210
s 69A 14
s 69C 35
s 69C(1) 212
s 69D 14, 212
s 69D(1)(f) 215
s 69DB 213, 214
s 85(2) 429
s 107 38, 214
s 109 215
s 126B 209
s 136A 212
s 136A(1) 211
s 136A(3) 212
s 380 205
Petroleum (Onshore) Regulation 2007
   reg 23 429
Protection of the Environment Operations Act 1997 202, 226-7, 361
   s 79 227
   Sch 1 226
Rural Fires Act 1997 228
Snowy Hydro Corporatisation Act 1997 26
   s 26(1)(d) 27
State Environmental Planning Policy (Mining, Petroleum Production and Extractive Industries) 2007 217, 218, 221, 224, 382
   s 7(2) 224, 229
State Environmental Planning Policy (Mining, Petroleum Production and Extractive Industries) Amendment (Coal Seam Gas) 2014 221
   cl 4 220
   cl 9A(5) 220
State Environmental Planning Policy (State and Regional Development) 2011 223, 384
   Sch 1 223, 384

Sch 1, cl 5 223, 384
Sch 1, cl 6 224
Sch 2 223, 384
Sch 3 223, 384
Strategic Regional Land Use Policy 227
Water Management Act 2000 202, 226, 227-30, 384
   s 4 228
   ss 52–55 227
   ss 56–8 25
   s 65 228
   s 71M 26
   s 71T 26
   s 91 226, 384
   s 91F(2) 228
   s 115G(1) 226, 384
   s 601 228
Water Management (General) Regulation 2011
   reg 22 228
Water Rights Act 1896 25
   s 2 25
Well Integrity Code 217, 218
   4.1.1 218
   4.1.2 218
   4.1.3 218
Western Lands Act 1901
   Sch 4, cl 5 9
   Sch 4, cl 5(2) 9

Northern Territory

Coastal Waters (Northern Territory Title) Act 1980
   s 4(2) 104
Environmental Assessment Act 361
Mineral Royalty Act 1982
   Pt II 428
   s 4 427
Mineral Titles Act 2010
   s 9 4
Minerals (Acquisitions) Act
   s 3 10, 430
Mining Act 1980 71, 74
Mining Management Act 2001
   s 3(a) 83
National Gas (Northern Territory) Act 2008 138, 148
Planning Act 2009 361

Queensland

Aboriginal Cultural Heritage Act 2003 45
Coastal Powers (Coastal Waters) Act 1980 103
Energy and Water Ombudsman Act 2006 361
Environment Protection (Greentape Reduction) and Other Legislation Amendment Act 2012 361
Environmental Protection Act 1994 187, 361, 366, 374-5
   s 9 375, 376
   s 14 376
   s 16 376
   s 17 376
   ss 41–43 380
   s 125 375
   s 125(j) 375
   s 143 377-9
   s 319 377
   s 493A 376
   s 523 380
   s 539 380
Greenhouse Gas Storage Act 2009 295, 296, 301-3
   s 3 301
   s 27 12, 302
   s 27(1) 301
   s 27(2)(a) 302
   s 27(2)(b) 302
   s 28 303
   s 141(c)(i) 309
   s 158(c) 309
   s 161 303
   s 164 303
   s 181(2) 303
   s 338A 309
   s 363(1) 310
   s 363(1)(b) 310
   s 364(2) 310
Land Access Code 2010 196, 199-201, 213
Land Act 1923 196
Land Act 1994 196
Mineral Resources Act 1989 68-9, 187, 374
   s 6 4
   s 10 69
   s 43 17
   ss 126–136A 64
   s 129(1)(b) 69
   s 136E 191
   ss 137–178 64

   s 139 69
   s 140A 64
   s 179 78
   s 186 78
   s 194 78
   s 235 43, 84-5
   s 269 187
   s 276 84
   s 276(1A) 84
   s 308(1)(d) 84
   ss 320–325 428
   Sch 1, ss 1–3 64-5
   Sch 1, ss 5–11 65-6
   Sch 1, s 6 64
   Sch 1A 45
   Sch 2 57
Mineral Resources Regulation 2013
   Pt 3, Div 1 428
Mining and Other Legislation Amendment Act 2013 191
Mining on Private Land Act 1909
   s 6 10
   s 21 10
National Gas (Queensland) Act 2008 138, 148
Petroleum Act 1923
   s 10 189
   s 76K 189
Petroleum and Gas (Production and Safety) Act 2004 68, 187, 188-201, 374
   Pt 2 189
   Ch 2, Div 2 191
   Ch 3 68, 192, 193
   Ch 3, Div 1 68
   Div 2, ss 34–44 68
   Div 4 192
   s 5(2) 68
   s 8 45
   s 10 4, 12
   s 10(3) 4
   s 14 57
   s 24A 64, 199-200
   s 26 5, 189
   s 27(1) 196
   s 27(2)(b)(ii) 197
   s 35 190
   s 42 190
   s 48(1) 190
   s 72 191
   s 74 187
   s 89 191

s 108 34
s 109 34-5, 85-6, 189-90, 191
s 109(2) 191
s 110 189
s 111A 189
s 112 86
s 121 192
s 153 64
s 155(1) 429
s 185 192
s 185(2)(a) 192
s 295 192
s 299 189
s 305 193
s 305(2) 68
s 306 193
s 306(1) 190
s 313 193
ss 315–316 193
s 319 68
s 328 193
s 343 31
s 381 193
s 388 194
s 495 197, 198
s 495(2) 198
s 496 198
s 497(1)(c) 198
s 498 198
s 500 37, 197, 214
s 500A(a) 197
s 500A(e) 197
s 500A(e)(i) 197
s 500A(f) 197
s 500B 197
s 502 34, 198
s 503 214
s 503(2) 199
s 503(3) 198
s 506(2) 199
s 506(3) 199
s 532(1) 197
s 532(2) 199
s 532(4)(b) 199
s 534(2)(b)(i) 199
s 674 195
s 675 194
s 705B 194
s 804 34
Sch 2 38

Sch 2 Dictionary 197
Petroleum and Gas (Production and Safety) Regulations 2004 189
  reg 59 194
Petroleum (Submerged) Lands Act 1982 301
State Development and Public Works Organisation Act 1972 377
  s 27 377
Streamlining Act 2012 68
Sustainable Planning Act 2009 366
Torres Strait Islander Cultural Heritage Act 2003 45
Water Act 2000 192
  s 206 26
Water Supply (Safety and Reliability) Act 2008
  s 201A 180

South Australia

Aboriginal Heritage Act 1988 45
Australian Energy Market Commission Establishment Act 2004 144
Coastal Powers (Coastal Waters) Act 1979 103
Crown Lands Act 1888
  s 9 10
Environment Protection Act 1993 361
Gas Pipelines Access (South Australia) Act 1997 138, 144, 149, 152
  Sch 1 144, 145
  Sch 2 144, 145
Mining Act 1971
  Pt 3 428
  s 6 4
  s 16 10, 430
Mining Regulations 2011, pt 2 428
National Electricity (South Australia) Act 1996 144, 147
  s 7 261
National Electricity (South Australia) Regulations 2010 144
National Energy Retail Law (South Australia) Act 2011 144
National Energy Retail Regulations 2012 144
National Gas (South Australia) Act 2008 138, 144, 148
  Schedule: National Gas Law
    Pt 3, Div 1, s 23 162

National Gas Law 138, 144, 147-50, 152, 159-64, 167-9, 170-1
  Ch 8, Pt 7 152
  s 13 162-3
  s 13(2) 163
  s 14 163
  s 15 159, 160, 165
  s 15(a) 164
  ss 15–17 160
  s 21 161
  s 23 158
  s 27 161
  ss 27–28 148-50
  s 28 168
  s 28(2) 161
  s 31(1) 163
  s 44 167
  s 74 152-4
  s 91A 157
  s 98 161-2
  s 112(2) 165-6
  s 114 166
  s 116(2) 168-69
  s 132 168
  s 151 161
  s 151(1) 164
  s 191 166
  s 294 148
National Gas (South Australia) Regulations 2008 144
Petroleum and Geothermal Energy Act 2000 295
  s 4 5
  s 43 429
Water Resources Act 1997
  s 29(5) 25

Tasmania

Aboriginal Relics Act 1975 45
Coastal Powers (Coastal Waters) Act 1979 103
Crown Lands Act 1905
  s 27 10
Environmental Management and Pollution Control Act 1994 361, 366
Mineral Resources Development Act 1995
  Div 5 428
  s 3 4
  s 47(1) 77
  s 53(2) 77
  s 57 77
  s 59 77
  s 66(4) 77
  s 66(6) 77
  s 102 429
Mineral Resources Regulations 2006
  pt 3 428
  reg 7 429
National Gas (Tasmania) Act 2008 138, 148
Petroleum (Submerged Lands) Act 1982
  s 3 5
Water Management Act 1999
  s 60 25

Victoria

Aboriginal Heritage Act 2006 45
Charter of Human Rights and Responsibilities 2006 298
  s 20 298
Coastal Powers (Coastal Waters) Act 1980 103
Environment Effects Act 1978 361, 366
Environment Protection Act 1970 361
Greenhouse Gas Geological Sequestration Act 2008 297-301
  s 3 297
  s 6 310
  s 14 12, 297, 298
  s 14(1) 9, 297
  s 14(2) 298
  s 14(4) 9, 298
  s 16 300
  s 19(1) 299
  s 19(1)(b) 301, 303
  s 19(2) 299
  s 59 299
  s 59(b), 301, 303
  s 71 299
  s 71(c), 301, 303
  ss 89–91 310
  s 93 309
  s 182(c) 310
  s 182(f) 310
  s 182(g) 310
  s 187 309
  s 200 299
  s 201 299
  s 201(1)(a)–(g) 299
  s 203 299
  s 281 299
  s 406(1)(g) 310
  s 406(1)(h) 310
Heritage Act 1994 45

Irrigation Act 1886
  s 4 25
Land Act 1891
  s 12 10
Land Act 1958
  s 339(1) 297
Mineral Resources (Sustainable Development) Act 1990 38
  Pt 2, Div 1 78
  s 2A 83
  s 4 4, 12
  s 5A 37, 45
  s 9 5, 10, 430
  s 12 428
  s 14 80
  s 14(3) 78
  s 14C(1) 78
  s 40 36
  s 45(1)(a) 37
  s 45(3)(c) 37
  s 46 37
  s 70(3) 69
  s 70(4) 69
  ss 80–3 58, 59
  s 85 38-9
  Sch 4 4
Mineral Resources (Sustainable Development) (Minerals Industries) Regulations 2013
  ss 6–10 428
Mines Act 1891 (No 2)
  s 3 10
Mines (Amendment) Act 1983 430
National Gas (Victoria) Act 2008 138, 148
Offshore Petroleum and Greenhouse Gas Regulations 2011
  regs 157–161 307
  reg 158(2) 307
Offshore Petroleum and Greenhouse Gas Storage Act 2010 295, 296, 307
  Pt 3 307
  s 374(1) 309
  s 378 307
  s 383 307
  s 405(1) 310
  s 405(1)(b) 310
  s 406(1)(c) 310
Petroleum Act 1998 297, 299
  s 6 5
  s 18 37
  s 37 37
  s 46 37
  s 128 37
  s 136 45
  ss 149–150 429
Petroleum (Offshore) Act 1982
  s 53 87
Petroleum (Onshore) Act 1991
  s 41 87
Planning and Environment Act 1987 361
Victoria Planning Provisions
  s 78 270
  s 82 270
  s 91 270
Water Act 1989
  Div 5 26
  s 33F(2) 26
  s 33V(1) 26

Western Australia

Aboriginal Heritage Act 1972 45
Barrow Island Act 2003 296
Coastal Powers (Coastal Waters) Act 1979 103
Environmental Impact Assessment (Part IV Divisions 1 and 2) Administrative Procedures 2012
  s 7 369-70
Environmental Protection Act 1986 361, 366
  Pt IV 369, 371
  s 38(5) 367
Gas Pipelines Access (Western Australia) Act 1998
  Sch 1 144
  Sch 2 144
Mining Act 1904
  s 117 10
  s 273 15
Mining Act 1978 15-16, 55, 81, 368, 371
  Pt IV, Div 2A 76
  s 8 4
  s 46A 372
  s 48 36
  ss 56C–70 62
  s 58 55, 56
  s 60 55
  s 63AA 372
  s 66 36
  s 70D(7) 75
  s 70F 77
  s 70J 36, 77
  s 71 84

s 73(2) 62
s 76(1) 15
s 82 84
s 84 372-3
s 85 15, 88
s 85(3) 88
s 85(b) 15
s 85(d) 36
s 89 372
s 109 428
s 109A 428
s 114 16, 36
s 117 10
s 118 56
s 119 16
s 123 40
s 126 77
Mining Amendment Act 1993 76
Mining Regulations 1981
    Div 5 428
    reg 23G 77
    reg 112 77
Mining Rehabilitation Fund Act 2012
    s 9A 371
National Gas Access (WA) Act 2009 138, 148
National Third Party Access Code for Natural Gas Pipeline Systems
    s 1.9 146
Petroleum and Geothermal Energy Resources Act 1967 296
    s 5 12
    s 10 5
    s 16 36
    s 17 39
    ss 29–48 63
    s 29(3) 57
    s 38 63
    s 43D 63
    s 48C 36
    s 52 429
    s 62 36, 88-9
    s 106 36
    s 142 429
Petroleum and Geothermal Energy Resources (Environment) Regulations 2012 368
Petroleum Pipelines Act 1969 32
Petroleum Pipelines (Environment) Regulations 2012 368

Petroleum Submerged Lands (Environment) Regulations 2012 368
Planning and Development Act 2005 361
Stamp Act 1921 32-4
    Pt IIIBA 32
    s 76 32
    s 76(1) 33-4
    s 76AN 32
    s 76AP(2) 33
    s 76AP(2)(a) 33

Canada

Canadian Shipping Act (CSA 2001) 116
Northern Canada Vessel Traffic Services Regulations 116

United Kingdom

Mines (Working Facility and Supports) Act 1966 8

USA

Clean Air Act 283

California

Solar Rights Act 1978 29
Solar Shade Control Act 1978 29

Brazil

Constitution of Brazil
    art 20(XI) 13

Netherlands

Renewable Energy Act 243
Spatial Planning Act 2008 293

Australian National Agreements

Australian Energy Market Agreement 2004 143
    cl 2.1 147
Intergovernmental Agreement on the Environment (May 1992) 230
National Partnership Agreement on Coal Seam Gas and Large Coal Mining Development: Intergovernmental Agreement on Federal Financial Relations 217

Natural Gas Pipelines Access Inter-Government Agreement 1977 **144**
Offshore Constitutional Settlement 1975 **117-18**

International Agreements, Treaties and Conventions

Agreement between the Government of Australia and the Government of the Democratic Republic of Timor-Leste Relating to the Unitisation of the Sunrise and Troubadour Fields (International Unitisation Agreement) **128-9**
   art 9 **129**
Antarctic Treaty **112**
Convention on the Continental Shelf (1958) **110**
Convention for the Protection of the Marine Environment of the North East Atlantic 1992 **304**
Doha Amendment to the Kyoto Protocol **325**
   art 1, amendment F **326**
Energy Charter **329-30**
Energy Charter Protocol on Energy Efficiency and Related Environmental Aspects 1998 **330**
Energy Charter Treaty 1998 **329-33**
   Pt II, art 3 **333**
   Pt IV, art 7 **333**
   Pt IV, art 20 **333**
   Pt IV, art 23 **333**
   Pt IV, art 25 **333**
   art 1(2)(a) **333**
   art 1(2)(b) **333**
   art 3 **333**
   art 5 **333**
   art 8 **333**
   art 19 **331-3**
   art 19(1) **330**
European Energy Charter 1991 **329**
International Energy Charter 2015 **329, 333**
International Labour Organization Convention 169 1989
   art 15(2) **13**
International Unitisation Agreement 2007 **128-9**
   art 9 **129**
Kyoto Protocol 2005 **321, 324-9, 334**
   Annex I **325**
   art 3 **326**
   art 3.1 **326-7**
   art 3.3 **325**
   art 3.9 **325**
   art 5 **326, 327**
   art 6 **326**
   art 12 **326**
Petroleum Mining Code **128**
Protocol to the Convention on the Prevention of Marine Pollution by Dumping of Wastes and Other Matter 2006 **304**
Ramsar Convention **388, 390**
Timor Sea Treaty between the Government of East Timor and the Government of Australia 2003 **127-8**
   art 3 **128**
   art 4 **128**
   art 7(a) **128**
   art 9 **128-9**
Torres Strait Treaty 1985 **129-30**
Treaty on Certain Maritime Arrangements in the Timor Sea 2007 **129, 130**
United Nations Convention on the Law of the Sea **97, 99, 103-5, 106, 107, 111, 113-15, 116, 117, 129, 304**
   Preamble, art 136 **97, 105**
   Pt IV **105**
   Pt VI **111**
   Annex II, art 4 **111**
   Annex II, art 8 **111**
   art 4 **106**
   arts 4–5 **105**
   art 7 **105**
   art 17 **114**
   arts 17–19 **114-15**
   art 18 **114**
   art 33 **107**
   art 48 **119**
   art 55 **107, 108, 109**
   arts 55–58 **108-9**
   art 56 **108**
   art 57 **108, 130**
   art 58 **108**
   art 60(4) **118**
   art 60(5) **118**
   art 76 **110, 111**
   art 76(8) **111**
   art 77 **110, 130**
   art 78 **110**
   art 137(1) **114**

United Nations Framework Convention on Climate Change 1994 319, 320, 321, 328
   Preamble 322-4
   Annex I 325
   art 1[2] 318
   art 1[3] 318
   art 2 322-4
United Nations Rio Declaration on Environment and Development 1992 230, 394
World Heritage Convention, art 8 388

OWNERSHIP OF MINERALS AND NATURAL RESOURCES

1.1	Introduction	2
1.2	Ownership of the subsurface strata at common law	5
1.3	Public ownership for minerals and petroleum	10
1.4	The proprietary status of mining tenements	14
1.5	Royal minerals: Gold and silver	18
1.6	Ownership of renewable resources: Hydro-electricity, geothermal, solar, and wind	21
1.7	Division of land and resources: Overlapping tenures	31
1.8	Land access and compensation	34
1.9	Native title, cultural heritage, and mining rights	40
1.10	Review questions	46
1.11	Further reading	47

1.1 Introduction

1.1.1 The nature and scope of energy resources in Australia

Energy resources in Australia encompass a multitudinous range of different forms of renewable and non-renewable resources. These resources are largely utilised to generate energy for domestic and international customers. Australia has an abundance of non-energy mineral resources, which are utilised for other purposes, such as building and construction (iron ore for steel) or technology (copper for copper wiring). Western Australia has one of the world's largest economic reserves of iron ore. In 2011, the output for iron ore was 474 million tonnes and constituted 97 per cent of Australian production. The bulk of Western Australian iron ore was exported to China, which imported 70 per cent of production in 2010, followed by Japan with 19 per cent and South Korea with 10 per cent. Extraction and commercialisation of iron ore generates vast amounts of money for the government. In the financial year from 2011–2012, the Western Australian government received over A$3.9 billion in royalties from the iron ore mining industry.

The current energy market in Australia focuses extensively upon the commercialisation of fossil fuels, such as coal and gas. Coal is particularly abundant in the eastern states of New South Wales and Victoria. Both states have large reserves of black and brown coal, with the reserves constituting 10 per cent of the world's total resources. Australia also has more than one-third of the world's known uranium resources.[1] Australia's identified conventional gas resources have increased threefold over the past 20 years, with approximately 90 per cent of estimated recoverable reserves of conventional gas located off the west and north-west coast. Additionally, the commercial utilisation of Australia's resources of coal seam gas, located in the black coal deposits of Queensland and New South Wales, has expanded rapidly.[2]

In Western Australia, for example, the Gorgan natural gas project, which is located on Barrow Island off the Pilbara coast, is one of the largest natural gas projects in the world. By the time production of liquefied natural gas (LNG) commences at the end of 2015, it will represent one of Australia's most significant fossil fuel projects and one of the most effective for emission reduction, given its

---

[1] See Australian Government, Department of Agriculture, *Energy in Australia 2011* (2011), 3 http://www.daff.gov.au/ABARES/pages/publications/display.asp?url=http://143.188.17.20/anrdl/DAFFService/display.php?fid=pe_abares99001789_13f.xml. The report notes at p. 9 that Australia's energy consumption is primarily composed of fossil fuels (coal, oil, and gas), which represent 95 per cent of total energy consumption.

[2] Ibid 8.

incorporation of the latest carbon capture sequestration (CCS) technology. In contrast, transport fuels such as crude oil and liquefied petroleum are more limited in stock. This has meant that Australia has become increasingly dependant upon oil imports.

Renewable energy is essentially energy derived from natural processes that are replenished constantly. In its various forms, it derives directly or indirectly from the sun or from heat generated deep within the earth. Included within the definition of renewable energy is energy generated from solar, wind, biomass, geothermal, hydropower and ocean resources, and biofuels and hydrogen derived from renewable resources.[3]

Historically, renewable energy has been the only viable energy option. Wind and water were utilised to power ships and turn windmills and water wheels for mechanical needs. However, with the advent of the industrial revolution, expanding energy requirements led to the discovery and use of the first hydrocarbon-based fuel: coal. When the use of this and other fossil fuels became possible on a large scale, few renewable energy techniques were capable of competing.[4] More recently, however, with the increasing importance of climate change mitigation, energy security, the provision of inexpensive and uninterrupted energy supply to consumers, and the vast improvements in the performance and affordability of solar cells, wind turbines and biofuels, the large-scale commercialisation of renewable technologies have been reconsidered.[5]

Solar energy production in Australia is slowly gaining market presence. Additionally, wind, solar, geothermal, hydro, wave, tidal, and bio-energy are all readily available and are developing a stronger market presence, particularly given the impetus to commercialise energy resources with low greenhouse gas emissions.[6] To date, the difficulty with the renewable energy market in Australia has been the underdevelopment of technology to support the implementation of a strong and consistent production and this has impeded the capacity of the renewable sector to gain market share. This situation is rapidly changing as technology advances. It is predicted that by 2030 the energy mix in Australia is likely to incorporate a substantial range of different renewable energy resources given the climate change imperatives involved.[7]

---

3 Ibid 2.
4 See B Sorenson, 'A History of Renewable Energy Technology' (1991) (January/February) *Energy Policy* 8, 10–12. See also B Sorenson, *A History of Energy: Northern Europe from the Stone Age to the Present Day* (Routledge, 2012).
5 See D Kammen, 'The Rise of Renewable Energy' (2006) (September) *Scientific AM* 82, 85.
6 See the Australian Government, Department of Industry Geoscience Australia, Bureau of Resources and Energy Economics, *Australian Energy Resource Assessment Second Edition* (2014) http://www.ga.gov.au/webtemp/image_cache/GA21797.pdf.
7 Ibid 8–9.

In any discussion of the ownership framework that underpins minerals and energy resources in Australia, the primary focus will be upon corporeal fossil fuels residing within the subsurface strata. These resources have a tangible presence and are therefore amenable to control and ownership. They are also subject to statutory vesting provisions. Legislation in most states deals with mineral resources in a separate legislative framework to the regulatory framework that governs petroleum and hydrocarbon resources. The rationale for this bifurcation lies in the location in which the resources are found and also the fundamental difference in their corporeal characteristics. A hydrocarbon is an organic compound that consists entirely of hydrogen and carbon. The predominant use of hydrocarbons is as a combustible fuel source, although as a solid, hydrocarbons form asphalt or bitumen. Hydrocarbons may be located on and offshore. A mineral is an inorganic compound, usually abiogenic and with an ordered atomic structure.[8] There are over 4900 known mineral species; each is distinguished according to their chemical composition and crystal structure. Minerals tend to be located onshore, although there is a small sand and limestone offshore mineral industry in Brisbane and Western Australia.[9]

The Queensland framework provides a prime example of the separate regulation of minerals and hydrocarbons in Australia. The *Mineral Resources Act 1989* (Qld) defines minerals in s 6 to include a substance, normally occurring naturally as a part of the earth's crust, dissolved or suspended in water or within the earth's crust, or capable of being extracted from the earth's crust or water in the earth's crust. This includes clay, sand, coal seam gas, limestone, marble, peat, salt, oil shale, and rock mined in slabs for building purposes. Soil, sand, gravel, rock, living matter, steam or water is explicitly excluded from the definition of a mineral. Similar provisions exist in other states.[10]

By contrast, in the *Petroleum Gas (Production and Safety) Act 2004* (Qld), petroleum is defined in s 10 as a substance consisting of hydrocarbons (hydrocarbon can exist in a gaseous, liquid or solid state) which occurs naturally in the earth's crust, a substance that is extracted or produced as a by-product of hydrocarbon, or a fluid that is extracted from coal or oil shale and consists of hydrocarbons. Section 10(3) explicitly sets out, for the purposes of clarification, that petroleum does not include: alginate, coal, lignite, peat, oil shale, torbanite or water. Further, a substance will not cease to amount to petroleum merely because it is injected

---

8 The precise definition of a mineral is the subject of some debate. Minerals are mainly oxides and sulfides and, like most of the surface of the earth, organic in nature.
9 See the Australian Government, Geoscience Australia, *Offshore Minerals Fact Sheet* (2012) http://www.australianminesatlas.gov.au/education/fact_sheets/offshore_minerals.html.
10 *Mineral Resources (Sustainable Development) Act 1990* (Vic) s 4 and sch 4; *Mining Act 1992* (NSW) Dictionary; *Mining Act 1971* (SA) s 6; *Mineral Resources Development Act 1995* (Tas) s 3; *Mining Act 1978* (WA) s 8; *Mineral Titles Act 2010* (NT) s 9.

or reinjected into a natural underground reservoir. Similar provisions exist in other states.[11]

1.2 Ownership of the subsurface strata at common law

In all Australian states and territories, the ownership of subsurface non-renewable minerals has either been reserved or vested in the state pursuant to specific statutory vesting provisions or reservations on title provisions enacted under mineral and petroleum state legislation.[12]

The introduction of these provisions has significantly diminished the rights of the surface estate owner that existed at common law and which were essentially encapsulated within the maxim *cujus est solum, ejus est usque ad coelum et ad inferos*. Literally translated, the maxim states that the person who owns land owns it from the heavens above to the centre of the earth below.[13] This maxim is a fundamental component of the common law framework for land and mineral ownership. It presumes that ownership of the subsurface strata, which includes minerals and natural resources residing in that strata, belong to the surface estate owner.[14] The principle was recognised in English law in 1586 in the decision of *Bury v Pope* and was therefore a component of the common law inherited by Australia upon colonisation.[15]

In operation, the maxim prescribes to a surface estate owner an infinite stretch of ownership in the airspace above the land and in the subsurface strata below the land. Taken literally, this ownership assumption is unfeasible; hence, it has been interpreted as authorisation for the common law assertion of ownership over

---

11 *Petroleum (Onshore) Act 1991* (NSW) s 6; *Petroleum and Geothermal Energy Resources Act 1967* (WA) s 10; *Petroleum Act 1998* (Vic) s 6; *Petroleum (Submerged Lands) Act 1982* (Tas) s 3; *Petroleum and Geothermal Energy Act 2000* (SA) s 4.
12 See, eg, *Petroleum and Gas (Production and Safety) Act 2004* (Qld) s 26; *Petroleum (Onshore) Act 1991* (NSW) s 6; *Petroleum and Geothermal Energy Resources Act 1967* (WA) s 10; *Mineral Resources (Sustainable Development) Act 1990* (Vic) s 9.
13 See the discussion on the nature of the maxim by P Butt, *Land Law* (Thomson Reuters, 6th ed, 2010) [2.05]–[2.07]. See also J G Sprankling, 'Owning the Center of the Earth' (2008) 55 *University of California and Los Angeles Law Review* 979, 988–92. The author argues that the maxim is simply a shorthand approach confirming that a landowner owns the subsurface to the extent necessary to support normal and reasonable uses of the surface.
14 See A J Bradbrook, 'Relevance of the Cujus Est Solum Doctrine to the Surface Landowner's Claims to Natural Resources Located Above and Beneath the Land' (1987–1988) 11 *Adelaide Law Review* 462, 462–3.
15 *Bury v Pope* (1586) Cro Eliz 118; 78 ER 375. See also J R S Forbes and A G Lang, *Australian Mining and Petroleum Laws* (Butterworths, 2nd ed, 1987) ch 2.

the subsurface strata down to a reasonable level.[16] In this respect, the maxim has functioned as a general guide for common law principles rather than an exact measure.[17] In *Commissioner for Railways v Valuer General*, the Court noted that use of the Latin phrase, 'whether with reference to mineral rights, or trespass in the air space by projections, animals or wires, is imprecise and it is mainly serviceable as dispensing with analysis'.[18]

Similarly, the English Court of Appeal in *Star Energy Onshore Ltd v Bocardo Ltd* described the Latin 'brocard' as having relevance purely as 'an imperfect guide'.[19] In that case, Aikens LJ went on to conclude that the 'correct' common law position is that 'the registered freehold proprietor of the surface will also be the owner of strata beneath the surface including minerals unless there has been an express or implied alienation to another'.[20]

The facts of the *Bocardo* decision are interesting and provide a relevant outline of issues pertinent to sub-strata ownership. The landowner plaintiff sued Star Energy Onshore Ltd in trespass because it had been drilling for petroleum under the plaintiff's land. The well-head, which is the facility at the surface of an oil or gas well providing the structural and pressure-containing interface for drilling and production equipment, was located on neighbouring land. However, the drilling pipelines descended to a depth of 2800 feet and extended into the plaintiff's land. The company had obtained a licence to extract petroleum but the licence did not allow the company to lay pipelines on the neighbouring land and in not seeking the plaintiff's permission to do that, the company had breached the common law ownership rights of the plaintiff.

---

16 See Sprankling, above n 13, 1039 where the author concludes that 'productive human activity is only possible within the shallowest portion of the earth's crust' and that consequently, subsurface ownership should only extend down to a specified depth of 1000 feet. Cf J Howell, 'Subterranean Land Law: Rights Below the Surface of the Land' (2002) 53 *North Ireland Law Quarterly* 268, 270 where the author rejects the concept of ownership to a specific depth arguing that 'any intrusion in land which is not sanctioned by some counter-veiling property right will constitute a trespass and that, although the surface owner will not usually wish to or be able to utilise the ground below the surface, he has rights in the land which could be valuable'.

17 In *Commissioner for Railways v Valuer General* [1974] 1 AC 382 the Court concluded that the maxim was imprecise. In *Bocardo SA v Star Energy UK Onshore Ltd* [2010] 3 WLR 354, Lord Hope suggested that the latin maxim, whilst flawed, nevertheless retained some utility as a general guide to subsurface ownership under common law. See also P Butt, 'How Far Down Do You Own? The Final Word' (2010) 84 *Australian Law Journal* 746.

18 *Commissioner for Railways v Valuer General* [1974] 1 AC 382, 351.

19 *Star Energy Onshore Ltd v Bocardo Ltd* [2010] 1 Ch 100, [26] ('*Bocardo*'). See also *Hinkley v Star City Pty Ltd* [2010] NSWSC 1389, [226] (Ward J) who, in upholding *Bocardo*, notes that the paper title-holder of the surface estate is 'deemed' to have possession of the subsurface strata.

20 *Bocardo* [2010] 1 Ch 100, [59]. See also Butt, above n 17, 748.

The Court of Appeal concluded that a literal application of the maxim would lead to absurdities: if property rights continued down as far as the core of the earth, landowners would all have a 'lot of neighbours'.[21] Hence, in order to apply a sensible principle, their Lordships concluded that: 'the owner of the surface is the owner of the strata beneath it, including the minerals that are to be found there, unless there has been an alienation of them by a conveyance, at common law or by statute to someone else', and that this extended down as far as 'the point at which physical features such as pressure and temperature render the concept of the strata belonging to anybody so absurd as to be not worth arguing about'.[22]

On the facts, the entitlements of the licences included a right to use reasonable (ordinary and proper) means to extract the resource and this incorporated boring into the ground and laying down drilling pipelines. However, laying drilling pipelines in the subsurface strata of neighbouring property was beyond the scope of the licence and therefore went beyond what could be regarded as reasonable. The Court of Appeal concluded that Star Energy Onshore Ltd did commit a technical trespass, although on the facts the plaintiff had suffered no loss of enjoyment. Damages were assessed in the same manner as cases of compulsory land purchase and were therefore assessed strictly with the Court awarding only £1000.

The Supreme Court subsequently affirmed the decision of the Court of Appeal. Lord Hope agreed that the maxim, whilst not a literal tool, nevertheless retained some utility as a general guide for common law subsurface ownership and therefore remained 'good law'. His Lordship stated:

> The better view, as the Court of Appeal recognised, is to hold that the owner of the surface is the owner of the strata beneath it, including the minerals that are to be found there, unless there has been an alienation of it by a conveyance, at common law, or by statute to someone else. That was the view which the Court of Appeal took in *Mitchell v Mosely* [1914] 1 Ch 438. Much has happened since then, as the use of technology has penetrated deeper and deeper into the earth's surface. But I see no reason why its view should not still be regarded as good law. There must obviously be some stopping point, as one reaches the point at which physical features such as pressure and temperature render the concept of the strata belonging to anybody so absurd as to be not worth arguing about. But the wells that are at issue in this case, extending from about 800 feet to 2,800 feet below the surface, are far from being so deep as to reach the point of absurdity. Indeed the fact that the strata can be worked upon at those depths points to the opposite conclusion.

---

21 *Bocardo* [2010] 1 Ch 100, [60].
22 *Bocardo* [2010] 1 Ch 100, [13]–[14].

I would hold therefore that Bocardo's title extends down to the strata through which the three wells and their casing and tubing pass.[23]

The Supreme Court also upheld the Court of Appeal's conclusion regarding the calculation of damages. Lord Brown (with whom Lords Walker and Collins agreed) accepted Star Energy's submission that, in effect, the *Mines (Working Facility and Supports) Act 1966* provided for a compulsory purchase from Bocardo of a right of access. This meant that damages should be based on case law relating to compulsory land purchases. The core principle underpinning compulsory purchase valuations is that 'compensation for the compulsory acquisition of land cannot include an increase of value which is entirely due to the scheme underlying the acquisition'.[24] This effectively meant that the value is not what the grantee is gaining, but what the grantor is losing. As Bocardo had no right to the oil beneath its land, it had not actually lost or diminished any specific value. Lord Brown summarised the position:

> The correct analysis seems to me to be this: that by these provisions [i.e the 1934 and 1966 Acts] Parliament was at one and the same time extinguishing whatever pre-existing key value Bocardo's land may be thought to have had in the open market and creating a new world in which only the Crown and its licencees had any interest in accessing the oilfield and in which they had been empowered to do so (to turn the key if one wants to persist in the metaphor) compulsorily and thus on terms subject to the Pointe Gourde approach to compensation.[25]

Accordingly, Bocardo's appeal on quantum was dismissed and the Court of Appeal's determination that the compulsory purchase value of an access right through the substrata was £1000 was upheld.[26]

---

23 *Star Energy Weald Basin Ltd v Bocardo SA* [2011] AC 380, [26]–[28]. See also *Finlay Stonemasonry Pty Ltd v JD & Sons Nominees Pty Ltd* [2011] NTSC 37, [45] where Blokland J stated: 'Lord Hope takes a generous view of the legitimacy of the maxim for ownership ... below the surface, it is suggested this must yield to contrary intention, and to relevant rules of construction, including here, the purpose of the lease and the objectively determined intention of the parties. In my view the maxim must be applied with some caution ... must yield to the reasonable construction of the lease'. See also the general discussion by A Bradbrook, 'Relevance Of The *Cujuis Est Solum* Doctrine To The Surface Land Owners Claims To Natural Resources Located Above And Beneath The Land' (1988) 11 *Adelaide Law Review* 462.
24 See *Pointe Gourde Quarrying & Transport Co Ltd v Sub-Intendent of Crown Lands* [1947] AC 565, 572 (Lord McDermott). This was also discussed in *Waters v Welsh Development Agency* [2004] 1 WLR 1304, [40] (Lord Nicholls) and [124] (Lord Brown); and in *Transport for London v Spirerose Ltd* [2009] 1 WLR 1797, [19] (Lord Walker).
25 *Star Energy Weald Basin Ltd v Bocardo SA* [2011] AC 380, [90].
26 Lord Brown concluded that this calculation of damages was 'positively generous' in the circumstances: [2011] AC 380, [92].

Common law ownership of subsurface strata has been subject to a range of different qualifications. It has no application to surface estate grants that are subject to express height or depth limitations or to any express reservation contained in a Crown grant that concerns minerals.[27] It has also been substantially qualified by the introduction of a range of judicial and statutory modifications. The application of the maxim to airspace is severely limited because of its potential to interfere with air travel and satellite navigation.[28] Further, the introduction of a public ownership framework for subsurface minerals and petroleum has made the maxim virtually redundant in the sphere of mining and energy law in Australia and removed its core functionality.[29]

In a public ownership framework, the notion of land as a three-dimensional concept, with surface, subsurface and airspace domains, has facilitated the legal acceptance of what has been described as 'horizontal and vertical subdivisions'.[30] Surface land is vertically subdivided but subsurface strata may be horizontally divided so that it is possible for particular levels to be the subject of different mineral and petroleum ownership rights. Any common law rights of the landowner must therefore interact with the statutory entitlements of the Crown to minerals and petroleum and the statutory entitlements of resource title-holders to extract the minerals and petroleum. Interaction may also occur with the entitlements of third parties regarding pipeline access and carbon capture sequestration licences for the injection into subsurface reservoirs.[31]

---

27 Some Acts specifically incorporate this right. The *Western Lands Act 1901* (NSW) sch 4, cl 5 specifically sets out that the Minister may 'limit a grant to the surface of the land or to the surface and a state depth below the surface'. Clause 5(2) then sets out that land 'excluded by such a limitation is surrendered to the Crown'.
28 In *Bernstein v Skyviews & General Ltd* [1978] QB 479, 481 Griffiths J concluded that the rights of a surface owner to airspace should be restricted to 'such height as is necessary for the ordinary use and enjoyment of his land and the structures upon it'.
29 See T Hunter and M Weir, 'Property Rights and Coal Seam Gas Extraction: The Modern Property Law Conundrum' (2012) 2 *Property Law Review* 71, 77.
30 See *Walker Superannuation Fund v Clough Property Fairmont Pty Ltd* [2010] WASCA 232, [22] where Martin CJ quotes from Windeyer J in *Bursill Enterprises Pty Ltd v Berger Bros Trading Pty Ltd* (1971) 124 CLR 73 at 91:'Therefore, at common law he [the freeholder] could dispose of a part of his holding by horizontal subdivision, just as by vertical subdivision. There were objections to this in medieval times: see Challis's Real Property 3rd Ed (1911), p. 54. But by Coke's time these had disappeared. He said: 'A man may have an inheritance in an upper chamber though the lower buildings and soil be in another, and seeing it is an inheritance corporeal it shall pass by livery'.
31 See, eg, *Greenhouse Gas Geological Sequestration Act 2008* (Vic) s 14(1) which sets out that the 'The Crown owns all underground geological storage formations below the surface of any land in Victoria'; and in s 14(4) that 'The Crown is not liable to pay any compensation in respect of a loss' that this might cause.

1.3 Public ownership for minerals and petroleum

Towards the end of the nineteenth century the private ownership of minerals and petroleum was rejected in Australia in favour of state ownership. Commencing in New South Wales, all states and territories passed legislation reserving all minerals in land for future Crown grants.[32] This legislation operated prospectively, although some jurisdictions introduced retrospective vesting legislation.[33] Retrospective legislation, vesting minerals in the state, exists in South Australia, the Northern Territory, and Victoria.[34] The effect of the retrospective legislation is that the relevant minerals (with the exception of exempted minerals) are regarded as having always belonged to the Crown, rather than ownership being transferred to the Crown at the date when the legislation was introduced. In the states where prospective legislation has been introduced, some minerals continue to be owned privately, as a consequence of Crown grants issued in the nineteenth century.

The shift from private ownership of subsurface minerals and petroleum to public ownership reflects a shifting awareness of the open nature of natural resource interests. According to this perspective, benefits accruing from the exploitation and commercialisation of subsurface resources are best treated as belonging to the community as a whole rather than being treated as a fortuitous gift to the surface estate-holder that happens to own the land above them.[35] The core justification for implementing a public or state-based ownership framework was the perceived need to ensure that in a country of abundance, minerals and petroleum

---

[32] *Crown Lands Act 1884* (NSW) s 7; *Land Act 1891* (Vic) s 12; *Mines Act 1891 (No 2)* (Vic) s 3; *The Mining on Private Land Act 1909* (Qld) ss 6, 21; *Crown Lands Act 1888* (SA) s 9; *Mining Act 1904* (WA) s 117; *The Crown Lands Act 1905* (Tas) s 27. See also the discussion in Forbes and Lang, above n 15, 17–26. The effect of a reservation is that the Crown retains all rights to something specifically excluded by the terms of the grant: *Doe d Douglas v Lock* (1835) 2 Ad & E 705; 111 ER 271.

[33] See the discussion by S Christensen, P O'Connor, W Duncan and R Ashcroft, 'Early Australian Land Grants and Reservations: Any Lessons to the Sustainability Challenge to Land Ownership' (2008) 1 *James Cook University Law Review* 15 at 26 discussing the gradual changes to the regulatory framework that dealt with mineral reservations. See also N J Campbell Jr, 'Principles of Mineral Ownership in Civil Law and Common Law Systems' (1957) 3 *Tulane Law Review* 303.

[34] *Mining Act 1971* (SA) s 16; *Mineral Resources (Sustainable Development) Act 1990* (Vic) s 9; *Minerals (Acquisition) Act* (NT) s 3; *Northern Territory (Self-Government) Act 1978* (Cth) s 69(4).

[35] See especially A Cox, 'Land Access for Mineral Development in Australia' in R G Eggert (ed), *Mining and the Environment: International Perspectives on Public Policy* (Resources for the Future, Washington DC, 1994) 21. See also the discussion by P Babie, 'Sovereignty as Governance' (2013) 36 *University of New South Wales Law Journal* 1075, 1103.

are properly conserved and managed for the welfare and benefit of all citizens.[36] The ongoing justification for state ownership of minerals is, however, increasingly controversial. Statutory vesting facilitates the capacity of the state to issue resource titles for the commercial exploitation of energy resources. This generates enormous revenues for the title-holder and corresponding royalties for the state. Hence, whilst theoretically, the state reinjects these funds back into the community, the burgeoning profits made by mining proponents has stimulated concern that public resources are being exploited without proper consultation and community engagement and further, that this is occurring in a manner that may be ultimately deleterious to the longer term welfare of the community. This raises issues regarding the type of rights and responsibilities that should accompany public ownership of resources and the limitations that should be imposed upon the privileges granted to mining proponents.

State ownership of resources is derived from the regalian system which originated under Roman law. Pursuant to this framework, the *dominium directum* (dominion of the soil) vested in the sovereign whilst the *dominium utile* (the right to use and profit from the soil) remained separate. Within a regalian system, sovereign monarchs were entitled to assume ownership of subsurface minerals extracted from *dominium directum*.[37] The concept was subsequently integrated into the domanial system whereby the ownership of natural resources vests in the sovereign.[38] Hence, whilst ownership of minerals and petroleum are vested in the state, the landowner is left with nothing apart from a right to compensation which is only enforceable where supported by the constitutional framework. Under the domanial system, natural resources are treated in contradistinction to the land estate and ownership of the mineral and petroleum is statutorily vested in the state. The framework therefore depends upon the fragmentation of land and mineral ownership, despite their physical coalescence, through legislative intervention. Ownership of minerals contained in the subsoil is attributed to the state either as a juridical body or as the representative of the collective body.[39]

A public ownership regime is necessarily reliant upon the constitutional legitimacy of the vesting provisions as well as the implementation of a strong and

---

36 See the discussion by P Wieland, 'Going Beyond Panaceas: Escaping Mining Conflicts in Resource-Rich Countries Through Middle Ground Policies' (2013) 20 *New York University Law Journal* 199, 210.
37 See J K Boyce, 'From Natural Resources to Natural Assets' in J K Boyce and B G Shelley (eds), *Natural Assets: Democratizing Environmental Ownership* (Island Press, Washington DC, 2003) 7.
38 See the discussion by Y Omorogbe and P Oniemola, 'Property Rights in Oil and Gas under Domanial Regimes' in A McHarg et al (eds), *Property and the Law in Energy and Natural Resources* (Oxford University Press, 2010) 120.
39 See N J Campbell Jr, 'Principles of Mineral Ownership in the Civil Law and Common Law System' (1957) 31 *Tulane Law Review* 303.

effective concession framework for the granting of mining licences to third parties. The validity of the concession system will also depend upon the constitutional framework in which it is implemented. In Australia, all vesting provisions are derived from state and territory legislation. Significantly, this is not subject to the just terms provision within s 51(xxxi) of the *Australian Constitution* because no equivalent provision exists in state constitutions. As such, states and territories are constitutionally entitled to make laws regarding the land and resources existing within their jurisdictional domain and are not required to provide compensation to deprived landowners.[40]

Minerals vested in the state include all static and migratory minerals and, where expressly incorporated, coal seam gas.[41] Petroleum vested in the state includes all forms of hydrocarbons and, in most states, the definition is broad enough to incorporate unconventional gas, such as tight or shale gas.[42] These provisions effectively mean that commercially viable minerals and petroleum are now owned and therefore controlled by the states and territories.[43]

The vesting provisions affect statutory severance of the mineral ownership from the surface estate. Minerals coming within the scope of the vesting provisions are therefore presumed to have a separate ownership status to the land in which they reside and, in this sense, to be inherently transferable. There are two fundamental assumptions that underlie the statutory vesting of mineral and petroleum ownership in the state. First, that minerals are capable of being severed from the bundle of rights that comprise land ownership; and second, that the ownership rights acquired by the state only relate to the minerals and do not confer upon the state rights to the corporeal strata in which the minerals reside.[44] This statutory disaggregation represents a complete and fundamental rejection of the common law maxim and the core assumptions underlying the doctrine of accession which presumes that minerals existing within the strata belong to the strata.

---

[40] For a discussion of the non-application of the Commonwealth just terms provisions to state-based mining legislation see S Evans, 'When is an Acquisition of Property not an Acquisition of Property' (2001) 11 *Public Law Review* 183, 186.

[41] See, eg, *Mineral Resources (Sustainable Development) Act 1990* (Vic) s 4.

[42] See, eg, *Petroleum and Geothermal Energy Resources Act 1967* (WA) s 5; *Petroleum and Gas (Production and Safety) Act 2004* (Qld) s 10.

[43] See, eg, the definition of 'petroleum' in the *Petroleum (Onshore) Act 1991* (NSW) s 3 which means: (a) any naturally occurring hydrocarbon, whether in a gaseous, liquid or solid state; or (b) any naturally occurring mixture of hydrocarbons, whether in a gaseous, liquid or solid state; or (c) any naturally occurring mixture of one or more hydrocarbons, whether in a gaseous, liquid or solid state, and one or more of the following, that is to say, hydrogen sulphide, nitrogen, helium, carbon dioxide and water.

[44] But see the vesting of subsurface pore space for the storage of injected carbon pursuant to CCS projects in Victoria and Queensland: *Greenhouse Gas Storage Act 2009* (Qld) s 27; *Greenhouse Gas Geological Sequestration Act 2008* (Vic) s 14.

The public ownership framework is operational in many countries around the world. Most countries that uphold public ownership of minerals and resources have, however, explicitly implemented the system within their own constitutional framework. In Brazil, for example, the *Constitution of Brazil* explicitly vests lands traditionally occupied by the Indians in the federal government.[45] The removal of minerals and resources from the control of Indigenous people has been supported by the international law, provided governments maintain procedures pursuant to which they consult Indigenous communities before undertaking or permitting any programs for the exploration or exploitation of resources on Indigenous lands.[46]

The public ownership framework is, of course, not the only system pursuant to which minerals are regulated. In the United States, the core entitlements of the private surface estate owner remain. This means that individual owners continue to retain full control over minerals and are able to sever the minerals from the land and create a legally valid mineral interest.[47] Further, in many states, unitisation laws have been introduced for the exploitation of petroleum resources. Unitisation amounts to the joint and coordinated operation of a petroleum reservoir by all the different owners of the reservoir. Unitisation is a useful system in a framework where the private ownership of mineral interests has produced many small tracts of land that are subject to multiple ownership interests. Within a unitisation framework, the nature and scope of the mineral interest will depend entirely upon the rights and interests conferred within the deed and how those rights are subsequently interpreted and constructed.[48]

Under a public ownership system, the nature and scope of statutory mineral ownership is fully derived from the terms of the vesting provisions. Some statutes assume that the vesting provision merely transfers a preconceived and independent ownership interest in the minerals and/or resources to the Crown. Other statutes treat the vesting provisions conferring upon the Crown a new statute-based

---

45 See *Constitution of Brazil* art 20(XI).
46 The *International Labour Organization Convention* 169 art 15(2) states: 'In cases in which the State retains the ownership of minerals or sub-surface resources or rights to other resources pertaining to lands, governments shall establish or maintain procedures through which they shall consult with these peoples ... before undertaking or permitting any programmes for the exploration or exploitation of such resources pertaining to their lands.' See also the discussion by R Pereira and O Gough, 'Permanent Sovereignty over Natural Resources in the 21st Century' (2013) 14 *Melbourne Journal of International Law* 451, 476.
47 For a discussion of the nature of a common law mineral estate in the United States see *Westmoreland & Cambria Natural Gas Co v DeWitt* 130 Pa. 235, 18 Atl. 725 (1889) where the Court held that oil and gas that exist in the subsurface strata belong to the surface owner and will continue to do so until they are 'severed from the freehold'. The capacity of the surface estate owner to create a mineral estate is often referred to as the 'doctrine of severance.'
48 For an outline of unitisation laws see D Asmus and J L Weaver, 'Unitizing Oil and Gas Fields Around the World: A Comparative Analysis of National Laws and Private Contracts' (2006) 28 *Houston Journal of International Law* 125.

ownership right in the minerals and resources, whose internal characteristics are entirely dependant upon the interpretative dimensions of its statutory expression.[49]

1.4 The proprietary status of mining tenements

Ownership of minerals and petroleum gives the state government the power to issue resource titles/mining tenements to third party mining proponents.[50] The nature and scope of these titles will depend upon the specific statutory requirements regulating their issuance. Governments may only issue resource titles in circumstances where the particular legislative requirements governing approval are satisfied. These requirements are dealt with in subsequent chapters but include the submission of detailed work plans, rehabilitation plans, and environmental impact assessments. A resource title that confers rights to explore or produce minerals or petroleum may only be issued where the specific legislative requirements have been satisfied and the relevant authority is satisfied that a resource title should be issued.

Mining tenements are statutory creations. As such, they derive all of their rights and interests from the statutory provisions that support them. In most states, statutory mining tenements will confer upon the holder all ancillary rights necessary to support the activity for which the title is issued. Hence, an exploratory title will include rights to drill and construct appropriate exploratory equipment and structure. A production title on the other hand will include rights to access the land, rights to drill and extract the mineral, as well as rights to construct appropriate production operations.[51]

In most states, a statutory mining or petroleum title has either been statutorily defined or judicially interpreted to constitute personal rather than real property, although the exact nature and status of the interest and the rights conferred will depend upon the particular provisions of legislation.[52] The *Personal Property Securities Act 2009* (Cth) ('the *PPSA*') applies to all security interests in personal property in Australia. Personal property is defined under the *PPSA* to include any kind of

---

49 The different ways in which statutory mining rights may be expressed are explored by M Storey, 'Not of this Earth: The Extraterrestrial Nature of Statutory Property in the 21st Century' (2006) 25 *Australian Resources and Energy Law Journal* 51, 54.
50 See, eg, *Petroleum (Onshore) Act 1991* (NSW) ss 69A, 69D.
51 This is discussed in more detail in Chapter 2.
52 See, eg, s 26 of the *Petroleum (Onshore) Act 1991* (NSW) which sets out that all petroleum titles and any interest in petroleum titles amounts to personal property. Note that s 10 of the *PPSA* sets out that the Act will not apply to personal property if it is declared not to be personal property for the purposes of the *PPSA*.

property (including intangible property) other than land, fixtures, water rights, and certain statutory rights but significantly, it does not include resource titles.[53] It will, however, include the minerals and hydrocarbons that are extracted pursuant to a resource title as well as any income derived from dealing with those titles.

The nature and scope of a resource title issued under the *Mining Act 1978* (WA) was recently examined by the Western Australian Supreme Court in *Commissioner of State Revenue v Oz Minerals*.[54] In that case, the Court interpreted s 76(1) of the *Mining Act 1978* (WA), which sets out that an application for a mining lease that includes a portion of land already contained in a current mining tenement held by a person other than the applicant will be taken not to include the said portion of land. During the course of the judgment, Murphy JA adopted a broad approach to the interpretation of the proprietary status of the 'tenement' as referred to in s 76(1). His Honour held that the term was equivalent to the 'notion of property capable of being held in freehold'. His Honour stated:

> ... the word 'tenement' in the phrase 'tenement, right or interest' in the chapeau in par (c) of the definition of 'mining tenement' in s 76(1) of the Act, conveys the notion of any property capable of being held in freehold which is connected with land, including a tenure of a strata or seam of minerals, as well as any incorporeal hereditament involving the right to dig and carry away minerals.[55]

The proprietary status of a mining tenement in Western Australia has a significant history. Section 273 of the *Mining Act 1904* (WA) expressly defined all mining tenements to constitute 'chattel interests'. This phrase was subsequently judicially defined as a reference to personal property by the High Court in *Adamson v Hayes*.[56] These provisions reflect the tendency to characterise a mining lease as a 'sale of the mineral extracted rather than as a demise of the land from which they were taken'.[57]

There are no equivalent provisions in the subsequent *Mining Act 1978* (WA). Despite this, various provisions continue to suggest that the lease constitutes personalty. Section 85 of the *Mining Act 1978* (WA) describes the rights conferred by a mining lease as 'exclusive rights for mining purposes' in relation to the land over which the mining lease was granted. Section 85(b) also expressly confers a right upon the holder of a mining licence to take and remove all minerals lawfully mined from that land, subject to the Act and any conditions to which the mining lease is subject.

---

[53] *PPSA* (Cth) s 8.
[54] [2013] WASCA 239.
[55] [2013] WASCA 239, [28].
[56] (1973) 130 CLR 276, 312.
[57] *Wade v New South Wales Rutile Mining Co Pty Ltd* (1969) 121 CLR 177, 192–3.

The conferral of such exclusive rights has been judicially interpreted to be directed at the prevention of others carrying out mining activities on land which is the subject of the mining lease rather than evidence of an intention to create an interest in real property.[58] Further, s 114 allows the holder of a mining tenement which expires, is surrendered, or forfeited to remove mining plant whether that plant is affixed to the land or not. Section 119 allows mining tenements to be sold or disposed of and to be the subject of legal and equitable interests provided the dispositions are effected by a signed written instrument. In *TEC Desert Pty Ltd v Commissioner of State Revenue* the High Court held that the exercise of equitable jurisdiction with respect to mining tenements issued under the *Mining Act 1978* (WA) 'is not necessarily indicative of the character of those tenements as interests in realty rather than as personalty'.[59]

In *Hancock Prospecting Pty Ltd v Wright Prospecting Pty Ltd* the Court concluded that if a mining lease under the *Mining Act 1978* did not give rise to an estate or interest in land, 'it must be the case, by parity of reasoning, that an exploration licence granted under the *Mining Act 1978* (an interest which permits the holder to conduct exploratory activities prior to full production and extraction) also fails to confer an estate or interest in land'.[60] McLure P held that the exploration licences constitute property:

> ... but are not land or an interest in land and are not choses in possession. They are, in my view, legal choses in action.[61]

Despite the alignment of resource titles with institutional common law estates and/or interests, the statutory foundation of the title means that the interest cannot always be regulated by common law principles that have developed to respond to the character and form of the common law. For example, it has been held that exploration and extraction rights are not affected by the doctrine of non-derogation from grant. In *Star Energy Onshore Ltd v Bocardo Ltd*, one of the issues considered by the Court of Appeal was whether the doctrine of non-derogation from grant applied to support a mining licence (as opposed to a lessee or recipient of a proprietary interest in land). Non-derogation of grant is a fundamental common law principle which sets out that a grantor of land cannot refuse the grantee something that is obviously necessary for the reasonable use of the land. Hence, if a grantor grants or leases land to be used for a specific purpose and then does something to prevent the land from being

---

58 *Western Australia v Ward* (2002) 213 CLR 1, [308].
59 (2010) 241 CLR 576, [36].
60 [2012] WASCA 216, [72]–[73] (McLure P).
61 [2012] WASCA 216, [73] (McLure P).

used for that purpose, the common law will not allow the grantor to derogate from the grant.

As discussed above, the facts of the case involved the conferral of exploration and extraction rights in relation to an oil field under a statutory licence. The issue was whether the licensee could enforce such rights against a private third party who owned the land. The Court determined that the third party's ownership of land extended below the surface of the soil to include the oil field and therefore supported an action in trespass. The issue was whether the principle of non-derogation from grant provided a defence to the trespass action. The Court held that where the licence was not granted by the owner seeking to bring the claim in trespass, but rather was a statutory grant, the principle of non-derogation did not arise and could not provide a defence to trespass. Lord Hope stated at [32]:

> The principle of non-derogation from grant would prevent the appellant from doing anything that would hamper the respondents' use of the strata for the purpose that both parties contemplated at the time of the grant. But the right to search and bore for and get the petroleum was obtained by the respondents under licence from the Crown. I do not think that there is any common law principle that the respondents can invoke in that situation to regulate their position in relation to a landowner who was not a party to that arrangement.[62]

All statutory resource titles are issued on the assumption that the mineral or petroleum to which they relate has been severed from the underlying strata in which it resides at the point of statutory vesting. Ownership of the resource therefore continues to reside with the state until the point of extraction, at which point a transfer of ownership will occur; however, any such transfer will be subject to the right of the state government to receive royalties.[63] Once the ownership of the produced mineral is transferred from the Crown to the resource title-holder, state ownership in the resource is extinguished. In some states, this transfer of ownership is dealt with explicitly within the legislation.[64]

---

62 See also *Hinkley v Star City Pty Ltd* [2010] NSWSC 1389, [240].
63 The scope and enforceability of royalties in this sector is discussed in further detail in Chapter 10.
64 See, eg, *Mineral Resources Act 1989* (Qld) s 43 which sets out that all minerals lawfully taken pursuant to a prospecting permit will cease to be the property of the Crown and become the property of the holder subject to the right of the Crown to the rights to royalty payments. See also *Mining Act 1992* (NSW) s 11 which provides that upon lawful severance of minerals from the land, such minerals become the property of the miner.

1.5 Royal minerals: Gold and silver

Gold and silver are treated by the common law as royal minerals. This means that the Crown retains the prerogative right of ownership in these minerals although the right does not extend to base minerals. The existence of this right is derived from the decision of the *Case of Mines* in 1568.[65] In that case, the judges held by a majority, that 'all ores or mines of copper ... containing or bearing gold or silver, belong to the King'.[66] The judges went on to hold that this prerogative right was an incident which was inseparable from the Crown and therefore could not be granted or severed from it by the use of express words.

The justification given for the assertion of this prerogative was the 'excellence of the monarch's person which draws to it things of an excellent nature', the need for the Crown to obtain sufficient funds to finance such vital areas as defence and the royal right to control coinage.[67] The Court also held that the assertion of the prerogative over gold and silver avoided the undue concentration of financial power in the King's subjects. The Court concluded that the prerogative has now become a settled part of the common law of England.[68]

The position was well summarised by Sir James Colvile in *Woolley v Attorney-General of Victoria*:

> Now whatever may be the reasons assigned in the case in Plowden for the rule thereby established, and whether they approve themselves or not to modern minds, it is perfectly clear that ever since that decision it has been settled law in England that the prerogative right of the Crown to gold and silver found in mines will not pass under a grant of land from the Crown, unless by apt and precise words the intention of the Crown be expressed that it shall pass.[69]

---

65 (1568) 1 Plowden 310 [75 ER 472]. The right was classified in Hale's scheme of the prerogatives as *Census Regalis* ('of the King's Revenue'): D E C Yale (ed), *Sir Matthew Hale's The Prerogatives of the King*, vol 92 (Selden Society, Fellow of Christ's College, Cambridge, 1976) xvii–xix, See also W Blackstone, *Commentaries on the Laws of England* (Legal Classics, England 1765), bk 1, c 8, 284–5.
66 (1568) 1 Plowden 310, 336 [75 ER 472, 511].
67 (1568) 1 Plowden 310, 315 [75 ER 472, 480].
68 See Blackstone, above n 65, 266–7. See also J Chitty, *A Treatise on the Law of the Prerogatives of the Crown*, (J Butterworth and Son, 1820) 145. This is discussed by the Australian High Court in *Cadia Holdings Pty Ltd v State of New South Wales* (2010) 242 CLR 195, [13]–[17] (French CJ).
69 *Woolley v Attorney-General of Victoria* (1877) 2 App Cas 163, 166 ('*Woolley*').

One of the fundamental incidents of the Crown prerogative is the right to enter the land and the 'liberty to dig and carry away the ores thereof, and with other such incidents thereto as are necessary to be used for the getting of the ore.'[70] As such, the prerogative right of the Crown to gold and silver has, traditionally, included incidental rights of access and extraction.

The application of the royal prerogative in mines of gold and silver to the Australian colonies was first confirmed in New South Wales in *R v Wilson* and then in Victoria in *Millar v Wildish*.[71] Subsequently, in *Wade v New South Wales Rutile Mining Co Pty Ltd*, in outlining the history of mining law in New South Wales, Windeyer J held that the decision in *Woolley* established beyond doubt that:

> ... gold in the Australian colonies belonged always to the Crown, whether it was in Crown land or in lands alienated by the Crown ... [that] [n]o express reservation was necessary to preserve the Crown's rights ... [and] that [t]hey depended upon prerogative rights recognized by the common law.[72]

Today, the royal prerogative powers and rights enjoyed by the Crown represent an integral component of the executive powers held by the Commonwealth and state governments.[73] The Australian High Court has affirmed the validity of the prerogative right to royal minerals in *Cadia Holdings Pty Ltd v State of New South Wales*.[74]

> **Facts:** The decision in *Cadia Holdings* dealt with royalties that were payable to New South Wales by Cadia Holdings Pty Ltd ('Cadia') with respect to copper that had been mined. Cadia and Newcrest Operations Ltd ('Newcrest') owned the land but it was subject to the 'reservations and conditions' contained in the Crown grants. The Crown grants had been issued between 1852 and 1881. Only one of the grants reserved 'all gold and mines of gold' to the Crown and there was no express reservation of copper to the Crown in any of the grants.
>
> Cadia were the recipients of four resource titles, issued pursuant to the *Mining Act 1992* (NSW). Cadia recovered both iron ore and copper from two of the mines. The copper was often combined with gold which was not able to be mined separately. The weight of the

---

70 (1568) 1 Plowden 310, 336.
71 (1874) 12 SCR (L) (NSW) 258, 269–71 (Martin CJ), 280 (Hargrave J), 281 (Faucett J); (1863) 2 W & W (E) 37, 43 (Molesworth J).
72 (1969) 121 CLR 177, 186.
73 For further discussion on this see s 61 of the *Constitution*. See also *Pape v Federal Commissioner of Taxation* (2009) 238 CLR 1, 60–4 [126]–[133] (French CJ), 83–92 [214]–[245] (Gummow, Crennan and Bell JJ).
74 *Cadia Holdings Pty Ltd v State of New South Wales* (2010) 242 CLR 195 ('*Cadia Holdings*').

extracted copper vastly exceeded the weight of the extracted gold; however, the value of the gold significantly exceeded the value of the recovered copper.

The issue for the High Court was whether the copper which was mined could be characterised as a privately owned mineral – which belonged to Cadia and Newcrest as owners of the freehold – having been conveyed to the grantee at the time of the original freehold grants. If this was the case, the state was required to repay seven-eighths of the royalties to Cadia and Newcrest. Royalties are payable pursuant to the *Mining Act 1992* (NSW) by holders of mining leases upon recovery of publicly owned minerals. Where the minerals are privately owned, the lessee will be liable to pay royalties as if they are publicly owned but the minister has to repay seven-eighths of the royalties to the owner of the minerals. On the other hand, if the copper mined qualified as a royal mineral, which had not passed over with the Crown grant, the state was entitled to retain all the royalties which were paid for the copper.

Held: In a separate judgement, French CJ indicated that a determination of the royalty amount payable in NSW depended upon a detailed evaluation of events which had occurred 'more than three centuries ago' in Tudor England because consideration had to be given to the nature and scope of the royal prerogative and its reception into colonial NSW.

French CJ considered the following issues: whether the royal prerogative to gold and silver included other metals, such as copper mixed with gold and silver; whether statutory changes in NSW impacted upon the reception and enforceability of the royal prerogative in NSW; whether the royal prerogative formed a part of the executive powers of the Commonwealth or the state governments and whether subsequently enacted NSW mining legislation impacted upon the capacity of the Crown to make a grant of the minerals pursuant to a freehold estate.

At first instance, the Supreme Court of New South Wales held that Cadia and Newcrest could recover seven-eighths of the royalties because, on the facts, copper came within the definition of a mineral under the *Mining Act 1992* (NSW) and could therefore be treated as privately owned. The State appealed successfully to the Court of Appeal. The majority held that the royal prerogative had not been modified and continued to apply to an indivisible ore mineral, and this mineral included copper. This meant that the intermingled copper could not, per se, be treated as a mineral capable of being privately owned and that it did come within the scope of the royal prerogative so there was no obligation to repay seven-eighths of the royalties.

On appeal to the High Court, French CJ reviewed the ongoing validity of the royal prerogative to royal minerals. His Honour held that at the time of colonisation, NSW inherited a modified royal prerogative. The rights of the Crown to intermingled copper had, however, been conveyed pursuant to Crown grants which had been issued in the mid-nineteenth century. This meant that on the facts, copper did constitute a privately owned mineral and seven-eighths of the royalty did need to be repaid.

> His Honour further held that the broad prohibition against mining by any person without relevant authority, as contained in the *Mining Act 1992* (NSW), precluded the historical right of entry that accompanied the royal prerogative but did not abolish it altogether. The right of entry was found to be a 'logical incident' of the prerogative but not a 'necessary condition of its existence' and hence, the inability of the Crown to enter land without authority should not be regarded as denying the 'existence of the prerogative'.[75]

Following the decision in *Cadia Holdings* the status of the royal prerogative in Australia may now be summarised as follows:

- The Crown retains ownership of gold and silver by way of royal prerogative which endures despite the statutory preclusion of a right of entry
- A grant of freehold in land will not convey the gold and silver that may be contained within that land to the grantee unless the rights to these royal resources are expressly conferred
- A grant of freehold in land will convey the copper, tin, lead or other base metals that may be contained within that land to the grantee unless the ownership of such metals has been expressly reserved to the state or territory in accordance with specific legislative provisions
- The royal prerogative over gold and silver may only be modified or abrogated expressly or where implication is clear and, in the circumstances, necessary.

1.6 Ownership of renewable resources: Hydro-electricity, geothermal, solar, and wind

Non-renewable, fossil fuel energy sources, such as coal and gas, have a significant advantage over wind and solar energy because their physical characteristics make them more susceptible to private ownership and control. It is not feasible to impose ownership rights upon emancipated resources such as wind and solar, which lack any tangible or defined presence. By contrast, ownership rights in water, which may constitute a source of renewable energy through its usage in

---

75 *Cadia Holdings Pty Ltd v New South Wales* (2010) 242 CLR 195, [22].

hydro-electricity or via its geothermal capacities, is crucial to the functionality of renewable industries dependent upon this resource.[76]

1.6.1 Hydro-electrical power

Hydro-electrical power constitutes the production of electricity through the use of the gravitational force of falling or flowing water. The moving water rotates a turbine shaft and this movement is converted to electricity with an electrical generator. Hydropower is the most advanced and mature renewable energy technology, although no new large-scale hydro-electric projects are proposed for Australia and it is unlikely that this form of renewable energy will overtake conventional electricity generation.[77] Despite this, in 2012, hydro-electricity constituted 17 per cent of the world's total electricity production. Australia has over 100 hydro-electricity power stations, with the majority being located in Tasmania and New South Wales.[78] The hydro-electricity project in the Snowy Mountains is the largest scheme in Australia with sixteen major dams, seven power stations, a pumping station and 145 km of tunnels and aqueducts. The scheme collects and stores water and diverts it via mountain tunnels and power stations. The water is subsequently released into the Murray and Murrumbidgee Rivers for irrigation. The scheme accounts for around half of Australia's total hydro-electricity generation capacity and provides base load and peak load power to the eastern mainland grid of Australia.[79]

Hydro-electricity has, however, been criticised because of its broad-ranging environmental impacts including landscape destruction, contamination of rivers and fish stock with mercury and this – combined with limited potential for the creation of new hydro projects in Australia – means that it is unlikely to constitute a primary source of renewable energy in the future.[80]

1.6.2 Geothermal energy

Geothermal energy is essentially derived from the heat underneath the earth. The upper 3 m of the earth's surface maintains a virtually constant temperature of

---

76 For an interesting discussion on the privatisation of renewable resources see M T Stoeven and M F Quaas, 'Privatizing Renewable Resources: Who Gains, Who Loses?' (Economics Working Paper No 2012-02, Christian-Albrechts-Universität zu Keil, 3 February 2012).

77 See D Harries, 'Hydro-Electricity in Australia: Past, Present and Future' (2011) March/April Eco-Generation http://ecogeneration.com.au/news/hydroelectricity_in_australia_past_present_and_future/055974/.

78 See especially the discussion at Australian Government, Geoscience Australia, *Hydro Energy*. <http://www.ga.gov.au/scientific-topics/energy/resources/other-renewable-energy-resources/hydro-energy>.

79 Ibid.

80 See the discussion by D Rosenberg, R A Bodaly and P J Usher, 'Environmental and Social Impacts of Large-Scale Hydro-Electric Development: Who Is Listening?' (1995) 5 *Global Environmental Change* 127, 128–9.

between 10° and 16 °C (50° and 60 °F). Geothermal resources can vary markedly from bare heat found within shallow ground surfaces, to boiling subsurface water and/or hot dry rock capable of heating water, located 5 km into the subsurface strata. Geothermal resources can be hot water, hot dry rock, hot carbon dioxide, or any hot geologic resource. Fluids and water are most commonly found near active tectonic plate boundaries where volcanic activity has occurred. This is particularly true in areas such as New Zealand and Iceland. Hydrothermal systems can also form in sedimentary rocks above areas of hot basement rocks and this is the type of system that is generally located within Australia. Much further down into the strata, molten rock known as 'magma' can generate extremely high temperatures. Utilising the heat from subsurface rocks involves injecting cold water down one well, circulating it through hot fractured rock, and drawing off the heated water from another well.[81]

A general lack of traditional hydrothermal resources and the relative infancy of geothermal technology in Australia means that this form of power generation is not well developed. Part of the reason for this stems from the fact that existing drilling technology limits economic development of geothermal resources to a maximum depth of about 5 km and temperatures at this depth vary according to endogenous geologic factors.[82] There are, however, a number of pilot projects; for example, a geothermal power plant has been periodically in operation in Birdsville, Queensland, since 1992. The plant uses a bore that produces water from the Great Artesian Basin at 98°C to generate about 80 kW net and this supplies about 30 per cent of the plant output. The remainder is fuelled by diesel and liquefied petroleum gas.[83]

1.6.3 Ownership of water in Australia

Unlike many other fossil fuel resources, water is fluid. This means that at common law ownership rights incorporate not only rights to the drops constituting the body of water but, more fundamentally, rights to the actual flow of water, measured according to its quality, velocity, and level.[84] Water, unlike any other resource, is amenable to variations generated by third party users. Water is in constant motion and this makes it difficult for a holder to retain exclusive control.

---

[81] See the discussion by C F Austin, 'Technical Overview of Geo-Thermal Resources' (1977) 13 *Land and Water Law Review* 9, 9 where the author sets out that geo-thermal resources represent the natural heat of the earth's crust, derived from the distribution of temperatures and thermal energy in the subsurface strata.

[82] See Geoscience Australia, above n 78.

[83] See Australian Government, National Water Commission, *Groundwater Essentials* (2012), 42 http://www.nwc.gov.au/__data/assets/pdf_file/0020/21827/Groundwater_essentials.pdf.

[84] See the discussion by A Scott and G Coustalin, 'The Evolution of Water Rights' (1995) 35 *Natural Resources Journal* 821, 824–5.

Under common law, rights to water were generally associated with pre-existing land ownership. Hence, ownership rights to water in a stream could be upheld where a person retained ownership of the banks of the stream. This 'riparian' form of water ownership was inextricably connected with land ownership. Hence, if the physical dimensions of the land were such that it incorporated banks of a stream which came into direct contact with running water, the 'riparian' owner would automatically acquire 'super-added' use and access entitlements.[85] The riparian water entitlement was therefore connected to the ownership spectrum of riparian land owners *ex jure naturae;* this meant that the rights could not be transferred separately from that land, nor could they be lost by abandonment or acquired by prescription.[86] Common law also recognised a usufructuary-based right to water in circumstances where the holder has legal access to a water resource. This type of water entitlement could be protected against interference by the tortious action of nuisance.[87]

In Australia today, the 'fugitive' status of water as a vital and diminishing resource means that statutory regulation of its usage is increasingly imperative.[88] In contemporary Australia, private common law water rights have been completely abolished and replaced by statutory entitlements.[89] Land-based riparian rights have been replaced with water permits and usufructuary water entitlements replaced by appropriative rights, such as water licences. The statutory regulation of water entitlements commenced with the statutory vesting of all riparian entitlements to water in the Crown.

---

85 See *Grant Pastoral Co Pty Ltd v Thorpes Ltd* (1954) 54 SR (NSW) 129, 145 where Kinsella J stated: 'at common law riparian rights are not and never were the only rights of riparian owners in relation to riparian lands, but were of a special class super-added to the ordinary rights incident to the possession of the land'. The English common law approach to riparian rights is a derivation of the Roman civil law where a river, which was regarded as common property, conferred free usage rights upon all without restriction. See generally J Getzler, *A History of Water Rights at Common Law* (Oxford University Press, 2004) 65–75.

86 See D Fisher, 'Rights of Property in Water: Confusion or Clarity' (2004) 21 *Environmental Planning Law Journal* 200, 215. See also *Tate v Lyle Industries v Greater London Council* [1983] 2 AC 509, 516 (Lord Templeman); and *Lord v Commissioners of Sydney* (1859) 2 Legge 912, 927.

87 See the discussion by D Clark and I A Renard, 'The Riparian Doctrine and Australian Legislation' (1970) 7 *Melbourne University Law Review* 475, 476–9.

88 See the discussion by J Getzler, *A History of Water Rights at Common Law* (Oxford University Press, 2004) ch 4 where the author discusses the complications of owning water and examines the Roman principles which set out that water – like the air, the sea, and wild animals – is owned by everyone and therefore a corpus of water may only be owned individually where a person has taken the resource into his possession, or appropriated it, or captured it from the source. As outlined by Chancellor Kent, vital natural resources are 'bestowed by Providence for the common benefit of man': quoted by Parke B in *Embrey v Owen* (1851) 6 Ex 353, 372; [155 ER 579, 587] quoting from J Kent, *Commentaries on American Law* (Little Brown, 1828) vol 3, lecture 51, 354.

89 See Fisher, above n 82, 215 who stated: 'The fundamental principle is clear beyond doubt. The right to the flow, use and control of water has been given to a public sector agency'.

By the end of the nineteenth century in Australia, each state jurisdiction had introduced a statutory framework for the control and management of water usage.[90] These frameworks were superimposed upon the existing common law, their primary objective being to vest ownership and control of all private water entitlements in the Crown and, in doing so, create a more equalised and accessible public water management regime.[91] Most of the early Acts incorporated vesting provisions that explicitly transferred usufructuary rights in flowing water to the Crown. For example, the *Water Rights Act 1896* (NSW) set out that the 'right to the use and flow and to the control of the water in all rivers and lakes which flow through or past or are situate within the land of two or more occupiers ... shall ... vest in the Crown'.[92] These provisions divested riparian owners of their common law entitlements and replaced them with qualified, statutory rights.[93]

The statutory regulation of water resources was imperative in Australia because the unfixed nature of running water made misuse and waste difficult to monitor.[94] This is a particular concern for water utilised to generate energy. Unlike common law entitlements, statutory water entitlements do not depend upon the holder proving ownership of adjacent land.[95] Successful applicants acquire a conditional entitlement conferring a right to access and use specified rivers and/or lakes for a prescribed period of time in accordance with defined terms and conditions.[96]

---

90 See L Godden, 'Water Law Reform in Australia and South Africa: Sustainability, Efficiency and Social Justice' (2005) 17 *Journal of Environmental Law* 181, 189. See also T Garry, 'Water Markets and Water Rights in the United States: Lessons from Australia' (2007) 4 *Macquarie Journal of International and Comparative Environmental Law* 23, 35.

91 See Godden, above n 90, 187 where the author notes that 'Australia's resource management is distinguished by the early vesting of natural resources in the Crown'. See also the early conclusions in the New South Wales Royal Commission on the Conservation of Water 1887 where the importance of introducing a more equalised distribution of water was specifically highlighted: New South Wales, *Royal Commission on the Conservation of Water* (1887).

92 See also the *Irrigation Act 1886* (Vic) s 4 which effectively deemed surface water to be vested in the Crown until a valid, contrary right could be proven. An important aspect of this legislation was, as the parliamentary debates suggest, the abolition of common law riparian entitlements: Victoria, *Parliamentary Debates*, Legislative Assembly, 24 June 1886, 441–2.

93 See, eg, the *Water Rights Act 1896* (NSW) s 2 which originally conferred specific rights upon landowners to utilise flowing or groundwater only for domestic purposes which meant watering stock or gardens not exceeding 5 acres. For similar provisions in Victoria see the *Irrigation Act 1886* (Vic) s 4.

94 See especially R Glennon, 'Water Scarcity, Marketing, and Privatization' (2005) 83 *Texas Law Review* 1873, 1890 where the author notes that wasteful water usage is already a significant cause of the world's water scarcity.

95 Statutory property has been described as existing where 'the title itself (as opposed to the authority to grant it) exists by virtue of the statute, absent satisfaction of the relevant statutory requirements, there is no title': see M Storey, 'Not of this Earth: The Extraterrestrial Nature of Statutory Property' (2006) 25 *Australian Resources and Energy Law Journal* 51, 53.

96 See, eg, *Water Management Act 2000* (NSW) ss 56–58 which sets out a range of water licensing categories and accords different priority to different categories, without specifically

Usually the entitlement is a permissory licence and generally it is accompanied by a specific water allocation grant.[97]

The administrative agencies conferring water licences upon successful applicants retain the capacity to cancel or vary entitlements at any time, without compensation to the holder. The transferability of statutory water licences is also an important aspect of most regulatory schemes.[98] Transfer provisions were introduced with the aim of encouraging a more active trade in environmental entitlements.[99] For a transfer to be effective it must be conducted in accordance with the terms and conditions of the specific legislative provisions. Common requirements include ministerial approval, the recording of obligations on title and the introduction of specific provisions, qualifying the range of appropriate transferees. For example, in s 33V(1) of the *Water Act 1989* (Vic), the holder of a water share may transfer the whole or a part of the water allocation available to that person to either the owner or occupier of land specified in the water-use licence, the owner or occupier of land in another state or territory of the Commonwealth, a representative of the Crown, or the environment minister.

Most hydro-electric projects operate pursuant to an issued water licence. For example, the Snowy River hydro-electric scheme was issued with the Snowy River Water licence in 2002 pursuant to the *Snowy Hydro Corporatisation Act 1997* (NSW). The licence defines the rights and obligations of the corporation, allowing for the collection, diversion, storage and release of water by and from the works of the Snowy Scheme for the 75-year duration of the licence. The licence defines the rules for

---

defining the proprietary status of these categories. See also *Water Act 2007* (Cth) which applies specifically to the Murray-Darling Basin where water access rights are broadly defined by s 4 to be rights which may *include* water access rights, water allocation rights and riparian rights, but which also are not specifically defined. The exception to this is South Australia and Tasmania. In the *Water Resources Act 1997* (SA) s 29(5) a water licence, including the allocation entitlement, is specifically described as 'personal property' vested in the licensee and may be transferred in accordance with the provisions of the Act or general provisions concerning the transfer of personal property. An identical provision exists in the *Water Management Act 1999* (Tas) s 60.

97 See, eg, *Water Act 2000* (Qld) s 206 which describes the access entitlement as a 'water licence' and the allocation right as a 'water allocation grant'. The water allocation grant is to be distinguished from the licence because it describes the volumetric amount of water that can be taken. In Victoria, the *Water Act 1989* (Vic) s 33F(2) describes a statutory allocation of water as a 'water share' which authorises the taking of water under the water allocation for the share during the water season for which the water allocation is allocated or, with the approval of the Minister, in a subsequent season.

98 See, eg, *Water Act 1989* (Vic) div 5 which permits the transferral of water shares subject to its recording on the register and the approval of the Minister. See also *Water Management Act 2000* (NSW) s 71M (access licences) and s 71T (rights under access licences).

99 The evolution of environmental markets is discussed by M Woolston, 'Registration of Water titles: Key Issues in Developing Systems to Underpin Market Development' in J Bennett (ed), *The Evolution of Markets for Water: Theory and Practice in Australia* (Edward Elgar, United Kingdom, 2005) 78.

releases into the Murray and Murrumbidgee Rivers and imposes a range of different environmental flow release obligations upon the project operators. The water licence is amenable to variation. For example, in 2011, the NSW Office of Water along with both the Victorian and Commonwealth Governments realised that there was a need to amend parts of the water licence as a consequence of the extreme drought and above average rainfall. In accordance with s 26(1)(d) of the *Snowy Hydro Corporatisation Act 1997* (NSW), the licence was varied by the Water Administration Ministerial Corporation (now the NSW Office of Water). These amendments established a 'drought account' which could be utilised where inflows reach critically low levels and to allow for greater usage of water savings where environmental flows are needed.[100]

Water licensing is also relevant to geothermal energy; however, unlike hydroelectricity, geothermal water resides in the subsurface in underground aquifers. An aquifer is an underground geological formation that transmits and contains appreciable quantities of groundwater. Geothermal technology has the potential to affect groundwater by connecting previously unconnected aquifers. Geothermal power generation plants may also use water for cooling, which increases demand and pressure on existing water resources. As such, any project seeking to commercialise the geothermal usage of hot water must obtain a groundwater extraction licence. The issuance of such water licences are subject to strict approvals. Between 2004 and 2006, all Australian governments signed the National Water Initiative ('NWI'), which called for a 'whole of water cycle' approach to water management and which sought to improve groundwater management. The NWI has committed all governments: to improve knowledge of groundwater–surface water connectivity by managing connected systems as one integrated resource; to return all currently overallocated or overused systems to sustainable levels of extraction; to improve understanding of sustainable extraction rates and regimes; and to promote sustainability by developing a better understanding of the relationship between groundwater resources and groundwater-dependent ecosystems.[101]

---

100 For a full discussion of the 2011 amendments to the Snowy Water Licence see the NSW Government, Department of Primary Industries, Office of Water, *Proposed Variations to the Snowy Water Licence: Revised requirements for release of water accumulated under dry inflow sequence during drought years* (June 2011) http://www.water.nsw.gov.au/Water-licensing/Corporate-licences/Snowy-Hydro-Limited/Snowy-Hydro.
101 To progress the groundwater reforms agreed under the NWI, the Australian Government established the $82 million National Groundwater Action Plan. Under the Plan, the National Water Commission is investing in projects to improve knowledge, understanding, planning and management of Australia's groundwater resources at all levels. For further discussion of this see National Water Commission, Australian Government, *Groundwater Essentials* (March 2012) http://www.nwc.gov.au/__data/assets/pdf_file/0020/21827/Groundwater_essentials.pdf.

1.6.4 Access entitlements for wind and solar

Renewable sources of energy like wind and solar are not amenable to traditional corporeal ownership principles because they lack definable, excludable characteristics. The intangible nature of sunlight and wind preclude them from being touched or handled and this means they cannot be owned in the same way as minerals. The natural, kinetic process of generating this type of renewable energy necessarily operates beyond the parameters of private control.

The core rights to construct and develop renewable wind and solar projects centre on access entitlements. The wind and the sun are not resources which may, in themselves, be controlled; however, the right to access solar light and the right to access land for the construction of a wind farm is vital. In relation to solar, the sun is rarely directly overhead and this means that sunlight that reaches a solar device will invariably have to pass through one or more neighbouring properties.[102] Vegetation may shade or block the sun and, for this reason, is a significant problem with energy generation. Similarly, with wind power, a wind generator may be rendered ineffective if a building development or vegetation restrict the natural flow of wind to the generator.

Access entitlements for solar rights or wind power are not protected under common law because the right to light and the right to wind access are characterised as negative rights, and therefore new rights that restrict the way in which the land 'burdened' by the easement may be used may not be created.[103] Further, ownership of a right to wind power or solar access is vague and indistinct and difficult to propertise. Despite this, the Victorian Law Reform Commission has argued that solar access easements should be recognised under common law because:

> ... the potential gains in facilitating the development of solar energy collection given the current concern in relation to fossil fuel use, both as to stocks and effects, more than justify facilitating the means by which solar energy easements may be acquired. It is thought that any difficulties that might arise in relation to property development raise issues that relate to the removal of these easements rather than their creation.[104]

---

102 See A J Bradbrook, 'Australian and American Perspectives on the Protection of Solar and Wind Access' (1988) 28 *Natural Resources Journal* 229, 232.
103 This is the general view; however, note that no Australian case has directly indicated a prohibition on the creation of new negative easements. See the discussion in the Tasmanian Law Commission, *Law of Easements in Tasmania*, Final Report (2010) 43 noting that the distinction between negative and positive easements under common law was 'troublesome' for new easements that would confer access entitlements to solar and wind power.
104 See Law Reform Commission of Victoria, *Easements and Covenants*, Report No 41 (1992) 27–8.

A similar argument may be made in terms of wind access.

Regulation also has the ability to promote more efficient and effective solar access in a number of ways. First, it can promote the energy efficiency and the maximisation of solar access in the construction of dwellings. Second, statutory solar access easements may be created and enforced, which confer upon the holder ownership of a right of access over the burdened land.

Private solar and wind access agreements may be privately entered into between landowners and project proponents; however, enforceability is limited as such agreements are voluntary and do not, by force of agreement, generate a property entitlement.[105]

Statutory interests may be created; however, to date, apart from limited local government actions, no specific statutory solar or wind power easements have been recognised.[106] This position may be directly contrasted with the United States. One of the world leaders in this area has been the state of California. The *Solar Rights Act* was enacted in 1978 in California. This Act sought to balance the needs of individual solar energy system owners with other property owners by developing solar access rights. The Act implemented a range of objectives:

- It prohibited covenants, conditions and restrictions which would unreasonably restrict use or installation of solar energy systems
- It established the legal right to a solar easement, which protects access to sunlight across adjacent properties. It also describes the minimum requirements needed to create a solar easement
- The Act covers active solar devices and passive solar design strategies
- The Act discourages local governments from adopting an ordinance that would unreasonably restrict the use of solar energy systems. It also requires local governments to use a non-discretionary permitting process for solar energy systems
- The Act requires certain subdivisions to provide for future passive and natural heating and cooling opportunities to the extent feasible
- The Act allows regions to impose, by ordinance, solar easements in particular subdivision developments as a condition of tentative map approval.

See also the *Solar Shade Control Act 1978* (Cal) which precludes a person from owning or controlling property that allowed a tree or shrub to cast a shadow covering more than 10 per cent of a collection absorption area for a solar collector on the property of another.

---

[105] See A B Klass, 'Property Rights on the New Frontier: Climate Change, Natural Resource Development and Renewable Energy' (2011) 38 *Ecology Law Quarterly* 63, 97.
[106] See the discussion by Bradbrook, above n 102, 232–3.

1.6.5 Market progression

The ownership principles that underlie fossil fuel energy resources have promoted a flourishing market. Ownership rights give the state full control to implement a licensing framework for the exploration and commercialisation of fossil fuel resources. Ownership rights promote certainty and help investors mitigate hurdles, such as land access and compensation. In light of this, changes to the property framework are sometimes necessary in order to ensure that property facilitates rather than hinders energy innovation. Burgeoning demand for energy combined with concerns about the environmental impact of fossil fuel consumption and climate change have generated strong interest in new energy technologies, particularly renewable forms of energy. Efficient and fair property rules capable of accommodating the unique characteristics of emerging energy technologies play an important part in promoting energy sustainability.[107]

The progression of renewable energy projects has been significantly hindered by regulation and landowner objections. For example, when a wind farm is constructed on private lands, the state has little power to interfere with the right of a landowner to refuse access. This means that the wind energy company must negotiate privately with the landowner and cannot simply allocate development rights in the absence of landowner consent. These negotiations can be arbitrary and expensive. The difficulties experienced by the renewable sector reflect what has been described as the 'energy paradox', whereby, despite the fact that investment in efficient renewable energies will return fuel savings that significantly outweigh the initial investment cost over the lifetime of the purchase, investment in such energies is often rejected by businesses and consumers.[108] This is a product of consumer uncertainty and cognitive barriers but can also be exacerbated by a lack of cooperation between property owners. For example, opposition by neighbours is a particularly acute impediment to the progression of solar and wind energy because of entrenched fears and the objection many private landowners have to any restrictions.[109]

To date, the energy framework in Australia has been largely predicated on the use of fossil fuels. All of the significant infrastructure and development has been connected with the construction of coalmines, gasfields, railroads, pipelines, and ships for the purposes of extracting and transporting fossil fuels. This is unlikely to

---

[107] See the discussion by T A Rule, 'Property Rights and Modern Energy' (2013) 20 *George Mason Law Review* 803, 813–16. See also M Pappas, 'Energy vs Property' (2014) 41 *Florida State University Law Review* 435, 440–4.

[108] See A J Krupnick et al, *Towards a New National Energy Policy: Assessing the Options – Executive Summary* (Resources for the Future and National Energy Policy Institute, Washington DC, 2010) 9.

[109] See the discussion by U Outka, 'The Renewable Energy Footprint' (2011) 30 *Stanford Environmental Law Journal* 241, 253.

change in the foreseeable future, which means that renewable energy resources, such as wind and solar, are more likely to supplement rather than supplant conventional energy sources. As technology advances and climate change imperatives make investment and industry progression in renewables more viable this position may alter.

The renewable energy sector in Australia has always relied significantly upon strong regulatory incentives and market mechanisms in order to provide investor stimulus.[110] For example, providing farmers with significant government subsidies for the approval of wind farm development or implementing subsidies for consumers installing solar devices provide crucial support for industries that do not have the certainty of established property entitlements and which are amenable to market distortions.[111] The market incentives specifically introduced to promote the renewable sector in Australia are outlined and discussed in detail in Chapter 7.

1.7 Division of land and resources: Overlapping tenures

The different forms of ownership recognised in the subsurface region can generate competition between overlapping entitlements. The rights of the surface estate owner to subsurface strata may coalesce with a range of different resource titles that include rights to extract coal or gas, rights to inject carbon in subsurface reservoirs, and rights to lay pipelines for the transportation of gas. In most cases, the separate resource and injection entitlements are reconciled by statutory provisions that establish regimes for prioritising rights and regulating interference. For example, s 343 of the *Petroleum and Gas (Production and Safety) Act 2004* (Qld) sets out that the Minister cannot make a call for tenders for a petroleum lease over land already included within a coal or oil shale mining lease.

---

110 See, eg, the REL in China which promotes the development and utilisation of renewable energy. See further Hao Zhang, 'China's Low Carbon Strategy: The Role of Renewable Energy Law in Advancing Renewable Energy' (2011) 2 *Renewable Energy Law and Policy Review* 133, 139.

111 See the discussion by K Weismantle, 'Building a Better Solar Energy Framework' (2014) 26 *St Thomas Law Review* 221, 225 where the author notes that solar easements allow owners of solar energy systems to secure access to sunlight from neighbouring parties whose property could potentially restrict such access. These rights are, however, likely to generate compensation issues where their enforcement reduces the value of the land. See also P Babie, 'How Property Law Shapes Our Landscape' (2012) 38 *Monash University Law Review* 1; and A Bradbrook, 'Future Directions in Solar Access Protection' (1989) 19 *Environmental Law* 167.

The common law doctrine of fixtures is also relevant in this context. It has a particular application in circumstances where pipelines are being laid in the land. In this situation, the holder of a pipeline licence has a right to lay the pipelines, although the pipelines themselves, once affixed, may be treated as a component of the land. In this way, the doctrine of fixtures deems an object that is affixed to the subsurface strata of the land to lose its independent property identity and henceforth become a part of the ownership rights of the surface estate owner.[112] In this context, however, it is important to determine what 'land' the attached fixture relates to.

This issue was explored by the Supreme Court of Western Australia in *Re Epic Energy (WA) One Pty Ltd v Commissioner of State Revenue*.[113] In this case, the Western Australian Court of Appeal dismissed an appeal by the taxpayer and held that 'land' for the purposes of the landholder provisions of the former *Stamp Act 1921* (WA) included a taxpayer's beneficial entitlement to land and beneficial entitlement to pipelines fixed to the land.

Facts: Epic Energy Pty Ltd acquired the shares in the taxpayer companies which were holders of pipeline licences for certain natural gas pipelines in the Pilbara. In their statements to the Commissioner under s 76AN of the former *Stamp Act 1921* (WA), the taxpayers stated that the estimated unencumbered value of all 'land' in Western Australia to which they were entitled as at 2 June 2004 was nil, on the basis that the pipelines did not constitute land within the meaning of Pt IIIBA of the *Stamp Act 1921* (WA).

In 2007, the Commissioner assessed stamp duty on the transfer of shares on the basis that landholdings held by the companies included the pipeline that was the subject of the licence. Section 76 of the *Stamp Act 1921* (WA) defined land to include any estate or interest in land and anything fixed to the land including anything that was, or purported to be, the subject of ownership separate from the ownership of the land.

The pipeline licences were owned pursuant to the *Petroleum Pipelines Act 1969* (WA) and pipeline was defined as not only the pipe or system of pipes, but a range of other associated infrastructure related to the system of pipes. The pipeline was partly buried under the surface of the land through which it was laid and partly attached to the surface of the land. It crossed several roads, railway lines, and major rivers.

A variety of registered easements had been granted to the previous owners of the pipeline over parcels of Crown land to which the pipeline licences applied. Other land on which

---

[112] For a detailed overview of the common law doctrine of fixtures and its relationship to common law principles of accession see T W Merrill, 'Accession and Original Ownership' (2009) 1 *Journal of Legal Analysis* 459.
[113] (2011) 43 WAR 186.

parts of the pipeline were located was not the subject of any easement. For this area of land, access agreements with pastoralists holding Crown leases had been granted in order to facilitate the construction, operation, and maintenance of the pipelines.

The taxpayers objected to the assessments and applied to the state Administrative Tribunal of Western Australia for review of the Commissioner's objection decisions. Before the Tribunal, the parties agreed that the preliminary issue that needed to be determined was whether or not each of the taxpayers could be defined as a landholder for the purposes of assessment.

The Commissioner argued that the easements constituted an interest in land and that the pipelines were included in that land interest because they were affixed to it. In areas where the pipeline was fixed to land the subject of a pipeline easement and also the subject of a pastoral lease, the pipeline was regarded as affixed to the easement.

Epic Energy One Pty Ltd rejected this argument. It contended that s 76(1) of the *Stamp Act 1921* (WA) did not operate to include in the definition of land something fixed to the land, unless the rights of the relevant corporation in the thing which was fixed to the land were derived or conferred from the estate or interest which they held in the land. Where the estate or interest in land conferred no interest in the pipelines affixed to the land, because the rights amounted to easements or statutory licences, the pipelines could not be included within the definition of land and therefore retained their independent proprietary status as a chattel.

Held: McLure J concluded that s 76AP(2)(a) of the *Stamp Act* specifies that 'land' to which a corporation is '(beneficially) entitled' includes 'land' situated in Western Australia of which the corporation is 'a co-owner of the freehold or of a lesser estate in the land'. The provisions in s 76AP(2)(a) and the definition of 'land' in s 76(1) make it clear that Pt IIIBA applies to all of the multiplicity of estates and interests that may be created in 'land' (as defined) at common law, in equity or by statute. The definition of land set out in s 76(1) draws a distinction between actual or purported 'ownership' of a thing fixed to land, on the one hand, and the 'ownership of the land', on the other. The actual or purported 'ownership' of the thing relates to the particular thing as distinct from the physical land to which it is fixed. Hence, the 'ownership of the land' relates to the estates and interests that have been created in the physical land.

A thing fixed to land will not be, and will not purport to be, the subject of ownership 'separate from the ownership of the land', in accordance with s 76(1) unless the source or derivation of the actual or purported ownership is separate from all estates and interests created in the land in question. Ownership of a thing fixed to land will not be 'separate from the ownership of the land' if the actual or purported ownership of the thing is conferred by or arises from an estate or interest in the land to which the thing is fixed.

McLure J therefore held that an easement was sufficient to attract the application of s 76AP(2). The easements existed expressly for the purpose of installing and maintaining the pipelines and the land, which was the subject of the easements, was utilised expressly

> for the purpose of installing and maintaining the pipelines. The Court found, however, that a pipeline licence did not constitute an interest in land because it was a statutory entitlement and therefore did not come within the definition of land in s 76(1).[114]

1.8 Land access and compensation

As statutory creatures, resource titles are regulated by the provisions of the legislation from which they are issued. Legislation in each state confers specific rights upon resource title-holders to explore for or commercially produce minerals and petroleum. Where the title does not confer a right of access to the land, access must be privately negotiated. The provisions regulating the scope and range of rights of access and compensation vary according to the nature of the resource title issued and the jurisdiction in which the title is issued.

In Queensland, s 108 of the *Petroleum and Gas (Production and Safety) Act 2004* (Qld) allows the authorised activities of a licence-holder to be carried out despite the rights of an owner or occupier of land on which they are exercised. This is qualified by the terms of s 804 which sets out that an authorised activity must not 'unreasonably interfere' with anyone else carrying out a lawful activity. Section 502 then confers upon the holder of a petroleum authority the right to 'cross the land if it is reasonably necessary to allow the holder to enter the area of the authority and to carry out activities on the land that are reasonably necessary to allow the crossing of the land.'

The specific entitlements conferred under exploration, production and storage licences are set out in s 109:

Petroleum and Gas (Production and Safety) Act 2004 (Qld)

109 Exploration, production and storage activities

(1) The lease holder may carry out the following activities in the area of the lease –
   (a) exploring for petroleum;
   (b) subject to section 152 –
      (i) testing for petroleum production; and
      (ii) evaluating the feasibility of petroleum production; and

---

114 See also *Red Hill Iron Ltd v API Management Pty Ltd* [2012] WASC 323, [127] noting that important statutory definitions should not be construed in isolation of their operative provisions.

      (iii) testing natural underground reservoirs for storage of petroleum or a prescribed storage gas;
- **(c)** petroleum production;
- **(d)** evaluating, developing and using natural underground reservoirs for petroleum storage or to store prescribed storage gases, including, for example, to store petroleum or prescribed storage gases for others;
- **(e)** plugging and abandoning, or otherwise remediating, a bore or well the lease holder reasonably believes is a legacy borehole and rehabilitating the surrounding area in compliance with the requirements prescribed under a regulation.

(2) However, the holder must not carry out any of the following –
- **(a)** extraction or production of a gasification or retorting product from coal or oil shale by a chemical or thermal process;
- **(b)** exploration for coal or oil shale to carry out extraction or production mentioned in paragraph (a);
- **(c)** GHG stream storage.

(3) The rights under subsection (1) may be exercised only by or for the holder.

(4) The right to store petroleum or prescribed storage gases for others is subject to part 6.

In New South Wales, s 28A of the *Petroleum (Onshore) Act 1991* (NSW) sets out that every petroleum title confers on its holder the right to carry on such operations as necessary. Section 69C sets out that a holder of a prospecting title must conduct operations in accordance with an access arrangement.

Petroleum (Onshore) Act 1991 (NSW)

69C Prospecting to be carried out in accordance with access arrangement

(1) The holder of a prospecting title must not carry out prospecting operations on any land except in accordance with an access arrangement or arrangements applying to the land:
- **(a)** agreed (in writing) between the holder of the prospecting title and each landholder of the land, or
- **(b)** determined by an arbitrator in accordance with this Part.

(2) Separate access arrangements may (but need not) be agreed or determined with different landholders of the same area of land, for different areas of the same landholding or with respect to the different matters to which access arrangements relate.

(3) Separate access arrangements may be made to preserve the confidentiality of provisions of the arrangements, to deal with persons becoming landholders at different times or for any other reason.

Similar provisions exist in the *Mining Act 1992* (NSW) where s 140 requires prospecting titles, defined as 'exploration titles', to be carried out in accordance with an access arrangement. Further, s 69DB of the Petroleum (Onshore) Amendment Bill 2013 (NSW) sought to introduce an access code to regulate all access activities over private land and impose a range of best practice activities to ensure that private landowners and their surface industries and livelihoods are not unduly disrupted by the exercise of authorised access activities.

In Western Australia, s 85(d) of the *Mining Act 1978* (WA) sets out that the holder of a mining lease is entitled to do 'all acts and things that are necessary to effectually carry out mining operations in or under the land.' A prospecting, exploration or retention licence gives the holder the right 'to enter and re-enter the land which is the subject of the licence with such agents, employees, vehicles, machinery and equipment as may be necessary or expedient for the purpose of further exploring for minerals in, on or under the land'. See also ss 48, 66 and 70J. Section 114 of the the *Mining Act 1978* (WA) gives the holder of a mineral tenement the right to enter land for the purpose of conducting remedial work after the expiry, surrender, or forfeiture of the mining tenement.

Section 16 of the *Petroleum and Geothermal Energy Resources Act 1967* (WA) sets out that a drilling reservation, access authority, or special prospecting authority lessee or licensee may not access land without first obtaining the written consent of the owners. The Act authorises the holders of retention leases and production licences to carry on such operations and execute such works in the licence area as are necessary for those purposes.[115] Section 106 allows the holder of a petroleum drilling reservation, petroleum lessee, petroleum licensee or holder of a petroleum special prospecting authority to apply for access authority to carry out petroleum activities within an area not covered by the grant.

Section 106 states:

Petroleum and Geothermal Energy Resources Act 1967 (WA)

106 Access authorities

(1) A petroleum permittee, holder of a petroleum drilling reservation, petroleum lessee, petroleum licensee or holder of a petroleum special prospecting authority may make an application to the Minister for the grant of a petroleum access authority to enable him to carry on, in an area being part of the State that is not part of the permit area, drilling reservation, lease area or licence area or area of the blocks specified in the special prospecting authority, petroleum exploration operations or operations related to the recovery of petroleum in or from the permit area, drilling reservation, lease area or licence area or area of the blocks so specified.

In Victoria, s 40 of the *Mineral Resources (Sustainable Development) Act 1990* (Vic) sets out that holders of exploration licences, mining, prospecting and retention licences may access land in accordance with the work program outlined in the authorised work plan. The only qualifications to this are that holders of native title

---

115 See ss 48C and 62 *Petroleum and Geothermal Energy Resources Act 1967* (WA).

rights are entitled to procedural rights.[116] Further, a licence-holder has no right to exercise work under a licence within 100 metres of a pre-existing dwelling house without consent, although the Minister may authorise access for work carried out in consultation with council and community groups.[117]

Similar provisions exist in the *Petroleum Act 1998* (Vic) which sets out that holders of exploration, retention and production licences are entitled to do anything within the licence area necessary or incidental to the purpose of the licence.[118] Prior to any petroleum operation commencing on private land, consent of the owner must be obtained or the parties must have entered into a compensation agreement. This is set out in s 128:

Petroleum Act 1998 (Vic)

128 Consent of, or compensation agreement with, owner etc. needed before operation on private land starts

(1) A person must not carry out any petroleum operation on private land unless –
  (a) it has obtained the consent of the owners and occupiers of the land to the operation; or
  (b) it has entered into a compensation agreement with the owners and occupiers of the land in relation to the operation; or
  (c) the Tribunal has determined the amount of compensation that is payable to the owners and occupiers of the land under this Act in relation to the operation.

Statutory compensation provisions for landholders affected by mining exist in each state. Broadly, the compensation which is payable concerns damage or injury caused to land or land-based activities, disturbance that might arise as a result of tenement-holders exercising access to the land, or disturbance flowing from the construction and usage of operational facilities for mineral exploration or production. Significantly, no compensation is payable for the extraction and removal of subsurface minerals because they do not belong to the landholder. Generally, compensation agreements will be entered into between landholders and mining proponents to outline the scope of compensation payable, which may be individually negotiated as well as the manner in which financial payments are to be made. Compensation agreements are not mandatory in all states but where they exist they do help to reduce land access conflicts.

In Queensland, s 500 of the *Petroleum and Gas (Production and Safety) Act 2004* sets out that a compensation and conduct agreement must be entered into whenever an advanced activity is planned. An advanced activity is defined in

---

116 See *Mineral Resources (Sustainable Development) Act 1990* (Vic) s 5A.
117 Ibid ss 45(1)(a), 45(3)(c), 46.
118 See *Petroleum Act 1998* (Vic) ss 18, 37, 46.

sch 2 to constitute activities such as levelling of drill pads, vegetation clearing, geophysical surveying, and access track construction.

In New South Wales, the *Petroleum (Onshore) Act 1991* sets out that the holder of a petroleum title must compensate every person injuriously affected by reason of any operations conducted pursuant to this act. Section 107 states:

Petroleum (Onshore) Act 1991 (NSW)

107 Compensation

(1) The holder of a petroleum title, or a person to whom an easement or right of way has been granted under this Act, is liable to compensate every person having any estate or interest in any land injuriously affected, or likely to be so affected, by reason of any operations conducted or other action taken in pursuance of this Act or the regulations or the title, easement or right of way concerned.

(1A) A native title holder within the meaning of the Commonwealth Native Title Act is to be treated as having an estate or interest in land for the purposes of subsection (1).

(2) The holder of a petroleum title is liable to compensate any other holder of a petroleum title whose operations under the title are detrimentally affected, or likely to be so affected, by the grant under this Act of an easement or right of way through, on or over the land comprised in the title held by that other holder or by the use of any such easement or right of way.

(3) Compensation is not payable under this Act by the holder of a petroleum title, or a person to whom an easement or right of way has been granted under this Act, where the operations of the holder or person do not affect, and are not likely to affect, any portion of the surface of any land.

(4) Any compensation agreed on or determined under Subdivision M or P of Division 3 of Part 2 of the Commonwealth Native Title Act for essentially the same act as an act in respect of which compensation is payable under this Part must be taken into account in the assessment of compensation for the act under this Part.

Similar provisions exist in ss 263–7 of the *Mining Act 1992* (NSW).

In Victoria, the *Mineral Resources (Sustainable Development) Act 1990* makes it clear that compensation is payable by a licensee to the owner or occupier of private land affected by loss or damage which is a natural and reasonable result of the approval of the work plan. Section 85 states as follows:

Mineral Resources (Sustainable Development) Act 1990 (Vic)

85 What compensation is payable for

(1) Compensation is payable by the licensee to the owner or occupier of private land that is land affected for any loss or damage that has been or will be sustained as a direct, natural

and reasonable consequence of the approval of the work plan or the doing of work under the licence including –
- **(a)** deprivation of possession of the whole or any part of the surface of the land; and
- **(b)** damage to the surface of the land; and
- **(c)** damage to any improvements on the land; and
- **(d)** severance of the land from other land of the owner or occupier; and
- **(e)** loss of amenity, including recreation and conservation values; and
- **(f)** loss of opportunity to make any planned improvement on the land; and
- **(g)** any decrease in the market value of the owner or occupier's interest in the land; and
- **(h)** loss of opportunity to use tailings disposed of with the consent of the Minister under section 14(2).

In Western Australia, s 17 of the *Petroleum and Geothermal Energy Resources Act 1967* sets out that the owner of private land may agree with a licencee regarding the amount of compensation to be paid for occupying the land. Compensation relates to deprivation of possession and damage to the surface of the land. Section 17 states:

Petroleum and Geothermal Energy Resources Act 1967 (WA)

17 Compensation to owners and occupiers of private land

- **(1)** A permittee, holder of a drilling reservation, lessee or licensee may agree with the owner and occupier respectively of any private land comprised in the permit, drilling reservation, lease or licence as to the amount of compensation to be paid for the right to occupy the land.
- **(2)** Subject to subsections (3) and (5), the compensation to be made to the owner and occupier shall be compensation for being deprived of the possession of the surface or any part of the surface of the private land, and for damage to the surface of the whole or any part thereof, and to any improvements thereon, which may arise from the carrying on of operations thereon or thereunder, and for the severance of such land from other land of the owner or occupier, and for rights-of-way and for all consequential damages.
- **(3)** In assessing the amount of compensation no allowance shall be made to the owner or occupier for any gold, minerals, petroleum, geothermal energy resources or geothermal energy known or supposed to be on or under the land.
- **(4)** If within such time as may be prescribed the parties are unable to agree upon the amount of compensation to be paid, either party may apply to the Magistrates Court at the place nearest to where the land is situated to fix the amount of compensation.
- **(5)** In determining the amount of compensation, the Court shall take into consideration the amount of any compensation which the owner and occupier or either of them have or has already received in respect of the damage for which compensation is being assessed, and shall deduct the amount already so received from the amount which they would otherwise be entitled to for such damage.

Under s 18, compensation may also be paid to any private land owner adjoining or in the vicinity of the licensed area which is injured or depreciated in value by operations carried out.

The *Mining Act 1978* (WA) contains similar provisions, with compensation being payable pursuant to s 123 for any loss or damage suffered or likely to be suffered as a result of the mining, although the provision specifically sets out that no compensation is payable purely for the issuance of a mining permit or for the value of the mineral. The above extracts are representative of the scope and nature of the access and compensation provisions that exist in each state and territory of Australia.

1.9 Native title, cultural heritage, and mining rights

The issuance of mining licences for exploration, retention, and extraction purposes may also intersect with native title and cultural heritage rights. Since the decision in *Mabo v Queensland [No 2]*[119] and the subsequent enactment of the *Native Title Act 1993* (Cth) and the various state acts, Indigenous communities who are able to establish that their traditions and custom retain a continuing connection with the land, which has not been extinguished by the grant of a common law or statutory estate, may apply to the Native Title Tribunal for recognition of their native title interests. Native title rights include rights to live on the area; access the area for traditional purposes such as camping or conducting ceremonies; visit and protect important places and sites; hunt, fish and gather food or traditional resources, such as water, wood and ochre; and teach law and custom on country.[120] Once upheld, these interests confer upon the holder important statutory rights. Mining licenses issued over land that is subject to native title must take into account the requirements of the relevant native title legislation.

The *Native Title Act 1993* (Cth) (the '*NTA*') sets up a scheme whereby the Commonwealth and the states may legislatively validate acts attributable to them which extinguish native title notwithstanding the provisions of the *Racial Discrimination Act 1975* (Cth). Acts occurring after 1 January 1994, which extinguish native title, are referred to as 'future acts'. The *NTA* prescribes and limits the circumstances in which future acts may validly affect native title.

Prior to the issuance of a mining licence, the relevant minister must take account of s 26A of the *NTA* that confers upon native title-holders and claimants a range of procedural rights. These rights seek to strike a balance between the rights and interests of Indigenous Australians to the land and the public interest and economic advantages associated with the exploitation of mineral and resource deposits that reside beneath that land.[121]

---

[119] (1992) 1 CLR 175.
[120] See *NTA* (Cth) s 223 for the statutory definition of native title.
[121] See *Straits Exploration (Australia) Pty Ltd v Kokatha Uwankara Native Title Claimants* [2012] SASCFC 121, [11] (Kourakis CJ).

The *NTA* sets out that the issuance of a mining or petroleum title over land which is subject to native title rights must be classified as a 'right to mine'.[122] All future acts under the *NTA* must outline any potential impacts on native title rights and interests. The process set out by the *NTA* will depend upon the nature of the resource title issued, the character of the mining or extraction process, and the amount of land affected.

Where the land which is subject to the resource title is fully subject to native title rights, the applicant will be entitled to determine the preferred *NTA* process. The applicant may choose from the following:

- an expedited process
- the right to negotiate
- entering into a private Indigenous land use agreement
- entering into a state Indigenous land use agreement
- adopting a blend of the above procedures.

In Queensland, where applications are made via the MyMinesOnline lodgement interface, the mapping system will indicate what land is subject to native title and will trigger a native title process.[123] Compliance with the native title process provisions is not necessary on land where native title is taken to have been extinguished in accordance with the doctrine of extinguishment.

The expedited procedure is a mechanism to address native rights and interests in a manner that is more expeditious than the full right-to-negotiate process and occurs where the state anticipates the activities will not have a significant impact on native title rights and interests.[124]

In Queensland, the state may conduct an expedited procedure where it proposes to attach the Native Title Protection Conditions ('NTPCs') to the permit upon grant.[125] The NTPCs are conditions placed on exploration permits for minerals and coal – and some mineral development licences – granted under the expedited procedure.

The NTPCs identify:

- the native title parties which are engaged by the expedited procedure
- what the native title parties must do before and during any exploration activity
- what happens when the native title parties do not meet specified time frames.

---

122 *NTA* (Cth) div 3, s 26D.
123 MyMinesOnline was established by the Queensland Government and is available at the business and industry portal: https://www.business.qld.gov.au/industry/mining/mining-online-services/myminesonline.
124 See *NTA* (Cth) ss 32, 237.
125 The NTPC is a strict condition which generally specifies strict time frames pursuant to which actions, such as notices, meetings, and inspections, must be taken by the parties. In many cases, a failure by the native title parties to respond, or act, within the time frame will result in the explorer being able to proceed with the exploration activities without any further consultation.

The NTPC will generally require the holder of an exploration licence to give notice to the native title parties prior to any exploration activities commencing. This will necessarily include all relevant maps and details of the mining project. Native title parties have 20 business days to respond to the notice and are entitled to request a meeting with the holder of the exploration licence and to require the holder to conduct a field inspection. Low impact activities which the parties agree upon – for example, geological and survey work that does not involve land clearing – will not be subject to any field inspection. Native title parties are entitled to engage an anthropologist or archaeologist; however, there is no requirement that the explorer pay fees for this unless this is specifically agreed to. The explorer must, however, pay annual administration fees to native title groups within the permit area and will be required to pay inspection fees. The NTPC should specify strict time frames in which actions should be taken by the parties; for example, notices, meetings, and inspections. Failure to respond within this time frame will enable the explorer to proceed without further consultation.

A right to negotiate is conferred upon registered native title bodies, corporations, and registered native title claimants. It specifically applies to the renewal of an existing mining lease or licence or the creation of a new right to mine.[126] Section 43 of the NTA contemplates the enactment of alternative state legislative regimes for the validation of future acts which would 'have effect' where the Commonwealth Minister is satisfied that the alternative regime meets the criteria prescribed by s 43(2)(1).

In *Western Australia v The Commonwealth ('Native Title Act Case')*, Mason CJ, Brennan, Deane, Toohey, Gaudron and McHugh JJ described the operation of s 43 in these terms:

> Section 43(1) raises the same problem as that raised by sub-s (3)(b) of s 26. It provides, inter alia:
>
> If: (a) a law of a State or Territory makes alternative provisions to those contained in this Subdivision in relation to acts covered by this Subdivision that are attributable to the State or Territory; and
>
> (b) the Commonwealth Minister determines in writing that the alternative provisions comply with subsection (2);
>
> the alternative provisions have effect.
>
> Sub-section (2) prescribes a list of requirements with which the "alternative provisions" must, in the opinion of the Commonwealth Minister, comply if the Subdivision B procedure and the s 28 conditions are not to apply.

---

126 *NTA* (Cth) s 26(1A).

Section 43 states the Parliament's intention that acts covered by s 26(2) should not be governed exclusively by Commonwealth law when the Minister is of the opinion that State laws and executive action taken under them satisfy the requirements which the Parliament specifies. When the Minister makes a determination under s 43(1)(b), the Commonwealth law (Subdivision B) is withdrawn pro tanto and the State law is left with a corresponding field of effective operation. The Minister is not empowered to engage but to exercise a power to disengage the operation of s 109. Section 43 of the Native Title Act is valid.[127]

Section 23 of the *NTA* sets out that permissible future acts are valid 'subject to subdiv B'. Subdivision B of pt 2, div 3 (ss 26–44) of the *NTA* applies where a State proposes a permissible future act falling within the scope of s 26(2) of the *NTA*, such as granting rights to mine or explore for minerals on shore. The requirements of subdiv B included a determination made under s 38(1) of the *NTA* by the National Native Title Tribunal.

The right-to-negotiate process requires the relevant minister to first notify the registered native title bodies, corporations, and registered native title claimants.[128] In *North Ganalanja Aboriginal Corporation v Queensland*[129] ('*Ganalanja*') the High Court considered the nature of the 'right to negotiate' which a claimant enjoyed as an incident of the registration of his or her claim. The majority observed:

> Thus, once an application for determination is accepted, the Act maintains the status quo as between the registered native title claimant on the one hand and the Government and those having proprietary interests or seeking rights to mine on the other, unless the parties negotiate and agree on the resolution of their respective claims or a competent authority makes a binding decision. It is erroneous to regard the registered native title claimant's right to negotiate as a windfall accretion to the bundle of those rights for which the claimant seeks recognition by the application. If the claim is well founded, the claimant would be entitled to protection of the claimed native title against those powers and interests which are claimed or sought by persons with whom negotiations might take place under the Act. Equally, it is erroneous to regard the acceptance of an application for determination of native title as a stripping away of a power otherwise possessed by Government to confer mining rights and the other rights to which Sub-div B applies. If the claim of native title is well founded, the power was not available to be exercised to defeat without compensation the claimant's native title. The Act simply preserves the status quo pending determination of an accepted application claiming native title in land subject to the procedures referred to. The mere acceptance

---

[127] *Western Australia v The Commonwealth* (1995) 185 CLR 373, 473.
[128] *NTA* (Cth) s 29.
[129] (1986) 185 CLR 595.

of an application for determination of native title does not otherwise affect rights, powers or interests.[130]

The native title parties will then, together with the relevant minister and the mining proponent, become parties to a negotiation process. That process requires that the native title parties be given an opportunity to make submissions and that they negotiate in good faith for the purpose of obtaining the agreement of the native title parties to the issuance of the licence or lease.[131] If no agreement has been reached after six months, the parties may apply to the arbitral body (the National Native Title Tribunal) for a determination of whether the licence or lease may be issued.[132] The arbitral body will make a determination as soon as practicable after holding an inquiry.[133] The arbitral body can determine whether the act can be done, whether it cannot be done, or whether it can be done subject to conditions.[134] Any conditions that may be imposed by the arbitral body will be enforced as contractual terms between the negotiation parties.[135]

Generally, for right to mine processes, the state and the National Native Title Tribunal ('the NNTT') engage in the following ways: ·

- The NNTT manages the process for objections to the expedited procedure
- The NNTT undertakes mediation at the request of any negotiating party
- The NNTT registers Indigenous land use agreements ('ILUA')
- The NNTT conducts overlap analysis of claims for proposed tenements
- The NNTT manages the process and makes determinations
- The NNTT produces native title information and resources.

It is possible for exploration licences and rights to prospect or fossick to be excluded from the right to negotiate under s 26A of the *NTA*. For this to occur, the Commonwealth Minister must be satisfied that alternative procedural rights are available and that the act is unlikely to have a significant impact on the land or waters concerned. The procedural rights which must be made available in such circumstances are: rights to be notified of the act; a right to be heard by an independent body; and consultation about how the impact of the act upon the exercise of native title rights may be minimised, including the protection of traditionally significant sites, access, and the conducting of exploration processes.

---

130 (1986) 185 CLR 595, 616.
131 *NTA* (Cth) s 31(1)(a), (b).
132 *NTA* (Cth) s 35.
133 *NTA* (Cth) s 139.
134 *NTA* (Cth) s 38.
135 *NTA* (Cth) s 41.

The Commonwealth Minister may overrule a determination of the NNTT. An appeal is allowed against the determination of the NNTT to the Federal Court on errors of law alone.[136]

The rights conferred under the *NTA* are protected in every state by relevant mining legislation which precludes the provisions from operating inconsistently with the *NTA*.[137] Resource title-holders should also be cognisant of the need to protect cultural heritage rights of Indigenous groups. Australian governments have a range of laws to protect Indigenous heritage, including: the *Environmental Protection Biodiversity Conservation Act 1999* (Cth); the *Aboriginal and Torres Strait Islander Heritage Protection Act 1984* (Cth) and the *Protection of Movable Cultural Heritage Act 1986* (Cth).

Under the *Environmental Protection Biodiversity Conservation Act 1999* (Cth) (*EPBCA*), there are penalties for anyone who takes an action – such as issuing a mineral or petroleum licence – that has or will have a significant impact on the national heritage values of a place.[138] The Indigenous Advisory Committee provides advice to the Commonwealth Minister on the operation of the *EPBCA*, taking into account the knowledge of the Indigenous holders of the land regarding the nature of the land, conservation, and use of biodiversity.

State and territory governments also have broad responsibilities for recognising and protecting Australia's Indigenous heritage, including archaeological sites. Most state and territory acts protect various areas or objects, while enabling mining proponents to apply for a permit or certificate to allow them to proceed with activities that might affect Indigenous heritage. The relevant legislation in each state is: *Heritage Act 2004* (ACT); *Heritage Objects Act 1991* (ACT); *Heritage Act 1977* (NSW); *National Parks and Wildlife Amendment (Aboriginal Ownership) Act 1996* (NSW); *Aboriginal Sacred Sites Act 1989* (NSW); *Heritage Conservation Act 1991* (NSW); *Aboriginal Cultural Heritage Act 2003* (Qld); *Torres Strait Islander Cultural Heritage Act 2003* (Qld); *Aboriginal Heritage Act 1988* (SA); *Aboriginal Relics Act 1975* (Tas); *Aboriginal Heritage Act 2006* (Vic); *Heritage Act 1994* (Vic); *Aboriginal Heritage Act 1972* (WA).

The *Aboriginal and Torres Strait Islander Heritage Protection Act 1984* (Cth) enables the Australian Government to respond to requests to protect important Indigenous areas and objects that are under threat if it appears that state or territory laws have not provided effective protection. Major reforms to this Act have been

---

136 *NTA* (Cth) s 42.
137 See *Mineral Resources (Sustainable Development) Act 1990* (Vic) s 5A; *Petroleum Act 1998* (Vic) s 136; *Petroleum and Gas (Production and Safety) Act 2004* (Qld) s 8; *Mineral Resources Act 1989* (Qld) sch 1A.
138 *Environment Protection Biodiversity Conservation Act 1999* (Cth) pt 3, div 1, subdiv AA.

proposed with the aim of improving the protection of the traditional heritage of Indigenous Australians.[139]

The Commonwealth Government can also make special orders, known as 'declarations', to protect traditional areas and objects of particular significance to Aboriginals in accordance with Aboriginal tradition from threats of injury or desecration. However, the government cannot make a declaration unless an Indigenous person (or a person representing an Indigenous person) has requested it. The power to make declarations is meant to be used as a last resort, once all of the relevant processes of the state or territory have been exhausted.

1.10 REVIEW QUESTIONS

1. Discuss the scope and ongoing relevance of the *Cujus est solum ejus est usque ad coelum et ad inferos* maxim in Australian land law.

2. Explain the corporeal difference between a hydrocarbon and a mineral and consider why they are generally regulated by different regulatory frameworks.

3. Did the Court of Appeal in *Star Energy Onshore Ltd v Bocardo Ltd* award Bocardo any damages for the trespass by Star Energy Onshore in drilling wells in the subsurface of land he owned? Explain the rationale for this decision.

4. Are the states and territories constitutionally entitled to introduce vesting provisions transferring the ownership of minerals and resources to themselves?

5. Consider the following issue:

    Santos holds a production lease over land in NSW entitling it to commercially produce gas. The area covered by the lease does not include the adjacent land which is home to many endangered species. The drilling and extraction process conducted by Santos results in water resources that feed the endangered birdlife on the adjacent property being diminished as the subsurface aquifer underlying the area covered by the production lease feeds into the water resource for the adjacent property. Can the local environment group obtain an injunction to prevent the extraction process from continuing?

6. What are 'royal minerals' and do they still belong to the Crown where they are 'intermingled' with copper or other minerals?

---

139 See Australian Government, Department of Environment, Water, Heritage and the Arts, *Indigenous Heritage Law Reform Paper: Possible Reforms to the Legislative Arrangements for Protecting Traditional Areas and Objects* (August, 2009) <http://www.environment.gov.au/system/files/consultations/2ba035da-5005-4e25-bc67-f80418645d9e/files/discussion-paper.pdf>.

7. Consider the following issue:

   Infigen Energy own a wind farm in South Australia on land that belongs to three different local farmers. The area upon which it is located is subject to strong inland winds. In late 2013, a group of wildlife activists seeking to have the wind farm closed down complain about the devastating loss of birdlife that the wind farm is causing. The activists are supported by FossilFuel Future, a group seeking to promote the continued use of brown coal in the area. FossilFuel Future obtain planning permission to construct a large building that will significantly affect the wind tunnel and reduce the energy output of the wind farm. Can Infigen Energy seek an injunction to prevent the construction of the building given the level of interference it is likely to cause?

8. Will the holder of an exploration licence automatically be entitled to access private land belonging to a third party in order to carry out authorised exploration activities?

9. What is the nature and scope of compensation payable for the exercise of rights authorised pursuant to issued resource titles in Victoria?

10. In what circumstances will a right to negotiate with Indigenous title-holders arise and what must mining proponents do in order to comply with this right?

1.11 FURTHER READING

D Asmus and J L Weaver, 'Unitizing Oil and Gas Fields Around the World: A Comparative Analysis of National Laws and Private Contracts' (2006) 28 *Houston Journal of International Law* 125.

C F Austin, 'Technical Overview of Geo-Thermal Resources' (1977) 13 *Land and Water Law Review* 9.

Australian Government, Department of Agriculture, *Energy in Australia 2011* (2011) http://www.daff.gov.au/ABARES/pages/publications/display.aspx?url=http://1.

Australian Government, Department of Environment, Water, Heritage and the Arts, *Indigenous Heritage Law Reform Paper: Possible Reforms to the Legislative Arrangements for Protecting Traditional Areas and Objects* (August 2009) http://www.environment.gov.au/system/files/consultations/2ba035da-5005-4e25-bc67-f80418645d9e/files/discussion-paper.pdf.

Australian Government, Department of Industry Geoscience Australia, Bureau of Resources and Energy Economics, *Australian Energy Resource Assessment Second Edition* (2014) http://www.ga.gov.au/webtemp/image_cache/GA21797.pdf.

Australian Government, Geoscience Australia, *Hydro Energy*. <http://www.ga.gov.au/scientific-topics/energy/resources/other-renewable-energy-resources/hydro-energy>.

Australian Government, Geoscience Australia, *Offshore Minerals Fact Sheet* (2012) http://www.australianminesatlas.gov.au/education/fact_sheets/offshore_minerals.html.

Australian Government, National Water Commission, *Groundwater Essentials* (2012), 42 http://www.nwc.gov.au/__data/assets/pdf_file/0020/21827/Groundwater_essentials.pdf.

P Babie, 'How Property Law Shapes Our Landscape' (2012) 38 *Monash University Law Review* 1.

P Babie, 'Sovereignty as Governance' (2013) 36 *University of New South Wales Law Journal* 1075.

W Blackstone, *Commentaries on the Laws of England* (Legal Classics, England 1765).

J K Boyce, 'From Natural Resources to Natural Assets', in J K Boyce and B G Shelley (eds) *Natural Assets: Democratizing Environmental Ownership* (Island Press, Washington DC, 2003).

A Bradbrook, 'Australian and American Perspectives on the Protection of Solar and Wind Access' (1988) 28 *Natural Resources Journal* 229.

A Bradbrook, 'Relevance of the *Cujus Est Solum* Doctrine to the Surface Landowner's Claims to Natural Resources Located above and beneath the Land' (1988) 11 *Adelaide Law Review* 462.

A Bradbrook, 'Future Directions in Solar Access Protection' (1989) 19 *Environmental Law* 167.

P Butt, 'How Far Down Do You Own? The Final Word' (2010) 84 *Australian Law Journal* 746.

P Butt, *Land Law* (Thomson Reuters, 6th ed, 2010).

N J Campbell Jr, 'Principles of Mineral Ownership in the Civil Law and Common Law System' (1957) 31 *Tulane Law Review* 303.

Challis's Real Property 3rd Ed (1911).

J Chitty, *A Treatise on the Law of the Prerogatives of the Crown*, (J Butterworth and Son, 1820).

S Christensen, P O'Connor, W Duncan and R Ashcroft, 'Early Australian Land Grants and Reservations: Any Lessons to the Sustainability Challenge to Land Ownership' (2008) 1 *James Cook University Law Review* 15.

S V Ciriacy-Wantrup, 'Common Property as a Concept in Natural Resources Policy' (1975) 15 *Natural Resources Journal* 713.

D Clark and I A Renard, 'The Riparian Doctrine and Australian Legislation' (1970) 7 *Melbourne University Law Review* 475.

A Cox, 'Land Access for Mineral Development in Australia' in R G Eggert (ed), *Mining and the Environment: International Perspectives on Public Policy* (Resources for the Future, Washington DC, 1994).

H Demsetz, 'Towards a Theory of Property Rights II: The Competition between Private and Collective Ownership' (2002) 31 *Journal of Legal Studies* 653.

R C Ellickson, 'Property in Land' (1993) 102 *Yale Law Journal* 1315.

S Evans, 'When is an Acquisition of Property not an Acquisition of Property' (2001) 11 *Public Law Review* 183.

D Fisher, 'Rights of Property in Water: Confusion or Clarity' (2004) 21 *Environmental Planning Law Journal* 200.

J R S Forbes and A G Lang, *Australian Mining and Petroleum Laws* (Butterworths, 2nd ed, 1987).

T Garry, 'Water Markets and Water Rights in the United States: Lessons from Australia' (2007) 4 *Macquarie Journal of International and Comparative Environmental Law* 23.

J Getzler, *A History of Water Rights at Common Law*, (Oxford University Press, 2004).

R Glennon, 'Water Scarcity, Marketing, and Privatization' (2005) 83 *Texas Law Review* 1873.

L Godden, 'Water Law Reform in Australia and South Africa: Sustainability, Efficiency and Social Justice' (2005) 17 *Journal of Environmental Law* 181.

D Harries, 'Hydro-Electricity in Australia: Past, Present and Future' (2011) March/April *Eco-Generation* http://ecogeneration.com.au/news/hydroelectricity_in_australia_past_present_and_future/055974/.

J Howell, 'Subterranean Land Law: Rights Below the Surface of the Land' (2002) 53 *North Ireland Law Quarterly* 268.

T Hunter and M Weir, 'Property Rights and Coal Seam Gas Extraction: The Modern Property Law Conundrum' (2012) 2 *Property Law Review* 71.

D Kammen, 'The Rise of Renewable Energy' (2006) *Scientific AM* 82.

J Kent, *Commentaries on American Law* (Little Brown, 1828).

A B Klass, 'Property Rights on the New Frontier: Climate Change, Natural Resource Development and Renewable Energy' (2011) 38 *Ecology Law Quarterly* 63.

A J Krupnick et al, *Towards a New National Energy Policy: Assessing the Options – Executive Summary* (Resources for the Future and National Energy Policy Institute, Washington DC, 2010).

J Landau, 'Who Owns the Air – The Emission Offset Concept and its Implications' (1979) 9 *Environmental Law* 575.

J G Latos, 'The Problem with Wilderness' (2008) 32 *Harvard Environmental Law Review* 503.

Law Reform Commission of Victoria, *Easements and Covenants*, Report No 41 (1992).

T W Merrill, 'Accession and Original Ownership' (2009) 1 *Journal of Legal Analysis* 459.

New South Wales, *Royal Commission on the Conservation of Water* (1887).

NSW Government, Department of Primary Industries, Office of Water, *Proposed Variations to the Snowy Water Licence: Revised requirements for release of water accumulated under dry inflow sequence during drought years* (June 2011) http://www.water.nsw.gov.au/Water-licensing/Corporate-licences/Snowy-Hydro-Limited/Snowy-Hydro.

Y Omorogbe and P Oniemola, 'Property Rights in Oil and Gas under Domanial Regimes' in A McHarg et al (eds), *Property and the Law in Energy and Natural Resources* (Oxford University Press, 2010).

U Outka, 'The Renewable Energy Footprint' (2011) 30 *Stanford Environmental Law Journal* 241.

M Pappas, 'Energy vs Property' (2014) 41 *Florida State University Law Review* 435.

R Pereira and O Gough, 'Permanent Sovereignty over Natural Resources in the 21st Century' (2013) 14 *Melbourne Journal of International Law* 451.

D Pulver, 'This is Your Land, This is my Land: Allowing Third Party Standing to Address Environmental Harms on the Federal Public Lands' (2012) 39 *Ecology Law Quarterly* 507.

C Rose, 'The Comedy of the Commons: Custom, Commerce and Inherently Public Property' (1986) 53 *University of Chicago Law Review* 711.

C Rose, 'Re-thinking Environmental Controls: Management Strategies for Common Resources' (1991) *Duke Law Journal* 1.

D Rosenberg, R A Bodaly and P J Usher, 'Environmental and Social Impacts of Large-Scale Hydro-Electric Development: Who is Listening?' (1995) 5 *Global Environmental Change* 127.

T A Rule, 'Property Rights and Modern Energy' (2013) 20 *George Mason Law Review* 803.

A Scott and G Coustalin, 'The Evolution of Water Rights' (1995) 35 *Natural Resources Journal* 821.

B Sorenson, 'A History of Renewable Energy' (1991) *Energy Policy* 8.

B Sorenson, *A History of Energy: Northern Europe from the Stone Age to the Present Day* (Routledge, 2012).

J G Sprankling 'Owning the Center of the Earth' (2008) 55 *University of California and Los Angeles Law Review* 979.

M T Stoeven and M F Quaas, 'Privatizing Renewable Resources: Who Gains, Who Loses?' (Economics Working Paper No 2012–02, Christian-Albrechts-Universität zu Keil, 3 February 2012).

M Storey, 'Not of this Earth: The Extraterrestrial Nature of Statutory Property in the 21st Century' (2006) 25 *Australian Resources and Energy Law Journal* 51.

Tasmanian Law Commission, *Law of Easements in Tasmania*, Final Report (2010).

K Weismantle, 'Building a Better Solar Energy Framework' (2014) 26 *St Thomas Law Review* 221.

P Wieland, 'Going Beyond Panaceas: Escaping Mining Conflicts in Resource-Rich Countries Through Middle Ground Policies' (2013) *New York University Law Journal* 199.

M Woolston, 'Registration of Water Titles: Key Issues in Developing Systems to Underpin Market Development' in J Bennett (ed), *The Evolution of Markets for Water: Theory and Practice in Australia* (Edward Elgar, United Kingdom, 2005).

D E C Yale (ed), *Sir Matthew Hale's The Prerogatives of the King: Vol 92* (Selden Society, Fellow of Christ's College, Cambridge, 1976).

Hao Zhang, 'China's Low Carbon Strategy: The Role of Renewable Energy Law in Advancing Renewable Energy' (2011) 2 *Renewable Energy Law and Policy Review* 133.

2

RESOURCE TITLES: PERMITS, LICENCES, AND LEASES

2.1	Introduction	52
2.2	Mining approval process: Exploration, assessment, and extraction phases	52
2.3	Exploration licences and permits	54
2.4	Retention licences and assessment leases	75
2.5	Mining and production leases	80
2.6	Review questions	90
2.7	Further reading	92

2.1 Introduction

This chapter examines the regulatory framework underpinning the issuance of resource titles in Australia. No operational activities may be conducted in Australia for mining and resource projects in the absence of a resource title. Ownership of land does not in itself give any rights to the owner to explore for or extract resources from the land. This is because, as we know from the discussion in Chapter 1, ownership of the resource resides with the state. A resource title constitutes the authorisation from the state, as owner of the resource, for a mining proponent to conduct exploratory, extraction, or production activities within the authorised area. Hence the state, as owner of the resource, gives permission to the mining proponent to explore for, extract, and produce the resource. This authorisation occurs through the interface of a resource title. The exact status of the title will depend upon the statutory terms and conditions that inform it. The focus of this chapter is upon the differences between different forms of resource (mining) titles, in terms of character, scope, and entitlement. The nature and character of a resource title reflects the particular stage that the resource project has reached. Hence, exploration titles will vary markedly in terms of duration, scope, and range to full-scale commercial production titles. The issuance of resource titles in every state and territory is based upon a structured regulatory framework. The process involves relevant applications, the submission of work plans, rehabilitation plans, environmental reviews, and any other permission or statutory obligation relevant to the particular project given its size, location, and focus. The environmental review relevant to the approval of resource titles is considered in more detail in Chapter 9. The primary focus of this chapter is a legal explication of the different forms of resource titles and their scope and characteristics.

2.2 Mining approval process: Exploration, assessment, and extraction phases

The type of resource title that a mining proponent applies for will depend completely upon the project's stage of development. Exploration and assessment entitlements reflect the preliminary investigative stages of a mining project. These entitlements confer rights upon a title-holder to explore for potential minerals and resources and, if and when they are discovered, to apply for rights allowing the proponents to assess and retain their interests in the areas covered by the licence whilst determining the suitability and capacity of commercialising those resources.

By contrast, extraction and production rights are applied for once exploration and assessment is completed and the decision to commercialise the resource has been made. As such, these entitlements are conferred upon mining proponents who already hold exploration or assessment licences, have determined that commercial resources exist, and have made the considerable decision to move into the production phase of a mining project. An applicant for a production licence will have thought carefully about the strategic and financial implications of proceeding to the production phase. Producing and commercialising a resource generally involves extensive capital costs because the operational infrastructure necessary to extract or produce large quantities of the resource must be constructed. It is important to bear in mind that exploration does not necessarily lead to production.

Exploration can be a high-risk activity. It involves studying the geological history of the authorised area. Fieldwork, magnetic surveys, gravity surveys, and seismic surveys are all common. Most international oil and gas companies have large portfolios of exploration interests and each of those interests have their own geological and fiscal characteristics and differing probabilities regarding the scope and nature of the resource. The management of these exploration assets is a major responsibility. If a resource deposit is found, it has to be economically viable for production. Viability will depend on a variety of factors, such as size of the find, the quality of the resource and its reservoir, access to the deposit, distance from core facilities relevant to the construction and operation of a working mine site, and the ability to find a buyer for the product. Viability will also depend upon whether the political and fiscal conditions favour the commercial success of exploration ventures, the distance to potential markets, and the availability or otherwise of a skilled workforce.[1] In this regard, the technical, political, economic, social, and environmental aspects of the region are all highly relevant factors to the question of whether and how exploration assets should be developed.

The actual process of exploration can involve a range of techniques and is generally conducted in different stages, which can include:

- **Reconnaissance**: Under this stage, a geologist may assess: rock outcrops to map the geology, vehicle access to a property, measurements to be taken and recorded, samples from rock outcrops, soils or streams for chemical analysis, and air surveys may be conducted.
- **Follow-up Investigations**: The reconnaissance stage may result in the identification of areas, which are in need of further investigation. This may involve taking additional surveys and samples, conducting geophysics surveys using electronic instruments, and undertaking detailed airborne surveys.

---

1 See the discussion of the risks associated with petroleum exploration by F Jahn, M Cook and M Graham, *Hydrocarbon Exploration and Production* (Elsevier, 2008) 2 where the author notes that the importance of understanding and responding to the social and political conditions of the country in which the mine is located and that investment in exploration is generally made many years prior to there being any opportunity of producing the oil.

- **Detailed Investigations**: If an area of a potential mineral resource were discovered, the next stage of exploration would usually involve drilling. Drilling is expensive, so the number of holes drilled to test an area of interest is kept to a minimum. Explorers normally use truck-mounted drill rigs. At this stage, trenches or test pits may also be dug to take a bulk sample. Bulk sampling is generally only required as a component of feasibility investigations for a mining proposal.

The advancement of exploration techniques has significantly improved the nature and process of the exploration phase as resources may now be targeted more accurately and with much greater success. The appraisal or assessment phase will generally follow the exploration phase. Once a resource has been discovered, considerable appraisal will still be required to ascertain its potential. This necessarily involves considerable risk assumption. The mining proponent may decide to proceed with the project, with the aim of generating income over a short-term period. Alternatively, an appraisal program may be conducted with the aim of optimising development and longer-term profitability. Finally, the proponent may decide to either sell the discovery – and many companies specialise purely in exploration projects with no intention of ever investing capital and labour in producing and commercialising the resource – or simply abandon the project and do nothing. Generally, at this stage, a feasibility study will be conducted so that a field development plan may be generated. The field development plan allows a resource proponent to determine a range of factors including development objectives, operating and maintenance principles, engineering facilities, project planning, and project economics.

The production phase of a mining or resource project marks a strong shift in focus because production of a resource generates cash flow. The production phase of a mining or energy project will therefore commence with the emergence of the first commercial quantities of the extracted and produced resource. The resource titles issued to support each phase of the mining project reflect the objectives of the activity and the different phases.

2.3 Exploration licences and permits

2.3.1 Approval process

A mining proponent who wants to investigate and determine the availability of a resource within a particular location will need to apply for an exploration licence in order to carry out exploratory activities. The approval process for an exploration licence will generally involve the submission of a clear statement of claim as well as the deposit of a bond. To illustrate the usual processes involved in making an application for an exploration licence, the framework for applying for an onshore exploration licence in Western Australia is outlined.

In Western Australia, pursuant to the *Mining Act 1978* (WA), an exploration licence for an onshore mining project may be lodged physically at the relevant office or electronically using the Department of Mines and Petroleum's *Mineral Titles Online* website.[2] A graticular boundary (or block) system applies to determine the size of an exploration licences. The surface of the earth is divided by predetermined lines of latitude and longitude into regular units of land. The lines are known as 'graticules' and the units of land created are known as 'graticular sections'. The dimension of a single basic graticular section amounts to one minute of latitude by one minute of longitude. A part of a graticular section is counted as a full graticular section and both a full and a part graticular section are known as a 'block'. Each block has a unique reference number which is comprised of three elements: a plan name, a primary number, and a particular section. In Western Australia, there are 22 1:1 000 000 plans which cover the entirety of the state. The primary number is determined by dividing each 1:1 000 000 plan into five minute by five minute areas. Each of the areas are numbered from 1 to 3456 to give each block the primary number. The graticular section is determined by dividing each five-by-five minute area into 25 one minute by one minute areas and these areas are identified by the letters *a–z* (but the letter *i* is omitted). Hence, for example, the Hammersley range block in Western Australia is identified as follows: 1:1000;2300;h.[3]

The minimum size of an exploration licence is one block, and the maximum size is 70 blocks, except in areas not designated as mineralised areas, where the maximum size is 200 blocks. Unlike other resource titles, an exploration licence is not marked out.

A security of $5000 must be lodged within 28 days of lodging the application.[4] Further, a statement must be lodged with an application for an exploration licence which specifies the method of exploration, the details of the proposed work program, an estimate of the proposed expenditure on the licence, and the technical and financial resources of the applicant.[5]

Where the application relates to land which is the subject of a pastoral lease, a copy of the application must be sent by registered post to the pastoral lessee. If the application relates to private land, copies of the application must be served

---

2 See Government of Western Australia, Department of Mines and Petroleum *Mineral Titles Online* (10 August 2010) http://www.dmp.wa.gov.au/3968.aspx.
3 See Government of Western Australia, Department of Mines and Petroleum, *Exploration Licence Graticular Boundary System* (2013) 2 http://www.dmp.wa.gov.au/documents/132289_Graticular_Boundary_System.pdf.
4 *Mining Act 1978* (WA) s 60.
5 Ibid s 58.

on the relevant municipal council, the owner and occupier of the land, and every registered mortgagee.[6] Applications must also be sent to any registered native title claimants or holders.[7] Objections may be lodged within 35 days of the application.[8] Any disputes that may arise regarding objections lodged against exploration licence applications – or between competing applicants – are determined by the Mining Registrar in the Warden's Court. The Mining Registrar in the Warden's Court performs both judicial and administrative functions. The primary administrative function of the Warden is to either grant or recommend the issuance of resource titles. The judicial function involves hearing any objections against an application and making a determination. Most objections are lodged by either existing application holders or, alternatively, other competing applicants. A Warden has the power, for example, to reject an application for an exploration licence if he feels that the public interest or right would be prejudicially affected.

The nature and scope of objections to an exploration licence was reviewed by the Western Australian Court of Appeal in *Re Minister for Resources; Ex parte Cazaly Iron Pty Ltd*.[9] In that case, the Court upheld a decision by the relevant Minister to determine an application for an exploration licence by Cazaly Iron Pty Ltd on the basis that the previous applicant, RRJV, had allowed their exploration application to inadvertently expire and that this type of unintentional oversight should not lead to the unexpected loss of an exploration licence by a bona fide miner. During the course of his judgement, Buss JA made the following comments:

> The Minister clearly had in mind the factors relevant to his decision, namely the circumstances in which RRJV failed to renew the Expired Licence and the fact that RRJV had signalled its genuine intention to extend the Expired Licence, the need to avoid circumstances where a minor oversight might have disproportionate consequences, the need to be even-handed or fair to the competing parties and the interest in promoting investment in the resources industry in Western Australia. The Minister did not take into account any irrelevant considerations and he did not misdirect himself on the law.[10]

In most situations, priorities between exploration licences are determined according to the date of the application. Once issued, an exploration title confers upon the title-holder priority for a subsequent application for a full production or mining lease. Expenditure on exploration licences will depend upon the number of blocks

---

6 Ibid s 118.
7 *Native Title Act 1993* (Cth) s 24MD(6B).
8 *Mining Act 1978* (WA) s 58.
9 (2007) WAR 403.
10 (2007) WAR 403, [27].

which are granted. Hence, for the first three years of the term of the exploration licence: $1000 per block must be spent, with a minimum of $10000 where the licence is only subject to one block, $15000 where the licence is subject to two to five blocks, and $20000 where six or more blocks are involved.[11]

2.3.2 The character of an exploration licence/permit

An exploration title is a preliminary, investigative right, authorising the holder to carry out exploratory activities to determine whether minerals exist in the licence area and, if so, the feasibility of extracting those minerals for commercial production. Exploration is defined variously by different states. For example, the *Petroleum and Gas (Production and Safety) Act 2004* (Qld) defines 'exploration' in s 14 as 'carrying out an activity for the purpose of finding petroleum.'

The definition in the *Mineral Resources Act 1989* (Qld), sch 2 is more extensive as 'explore' is defined to mean:

> take action to determine the existence, quality and quantity of minerals on, in or under land, or in the waters or sea above land by prospecting, using instruments, equipment and techniques appropriate to determine the existence of any mineral, extracting and removing from the land for sampling and testing an amount of material, mineral or other substance in each case reasonably necessary to determine its mineral bearing capacity or its properties as an indication of mineralization.

Similar definitions exist in other states.[12]

The exploratory title is, in many ways, unique to the resource industry. As outlined by Professor Crommelin:

> The importance of exploration sets mining apart from most other industries. It is this phase of operations that gives mining its traditional flamboyance and mystique: the risks are high but so are the stakes.[13]

The exploration licence confers upon the title-holder an exclusive right to explore for the minerals specified in the licence. In most cases an exploration licence will be granted for between two and five years and it has the potential to encompass

---

11 *Exploration Licence Graticular Boundary System*, above n 3, 3.
12 See, eg, *Petroleum and Geothermal Energy Resources Act 1978* (WA) s 29(3) where 'explore for' in relation to geothermal energy resources or petroleum includes to conduct any geophysical survey, the data from which are intended for use in the search for petroleum or geothermal energy resources.
13 M Crommelin 'Mineral Exploration in Australia and Western Canada' (1974) 9 *University of British Columbia Law Review* 38, 41.

large areas of land. The scope of works covered by the licence will not include any commercial extraction or production activities. The purpose of the exploration licence is to confer upon the title-holder rights consistent with conducting preliminary, investigative activities. This may involve some drilling or resource extraction, but that is purely for the purposes of sampling and analysis rather than commercialisation.

Crucially, the exploration activities that are carried out on the land will not always generate positive results. Even when they do, this will not automatically mean that the discovered resources should be subject to commercial production. Exploration licences authorise the holder to conduct activities that are consistent with exploration activities. Economically recoverable resources may not be located. Even if they are, commercial production may be unviable in the circumstances due, for example, to investor uncertainty, social licensing concerns, difficulties generating capital or, as is the case with coal seam gas mining, the imposition of a government moratorium.

Most exploration licences will include detailed conditions aimed at protecting the environment and the holder will generally need to lodge a substantial security deposit to encourage compliance with the licence conditions and to promote rehabilitation and remediation of any areas which have been disturbed during the exploration process. For example, in Victoria the *Mineral Resources Sustainable Development Act 1990* requires rehabilitation bonds for onshore mining projects to be lodged prior to the granting of a work authority or exploration licence and this bond is payable whether the land is private or Crown land.[14] The bond is paid to secure performance of the obligations set out within the rehabilitation plan. The rehabilitation plan must take into account any special characteristics of the land, the surrounding environment, the need to stabilise the land, the desirability or otherwise of returning agricultural land to a state that is as close as is reasonably possible to its state before a mining licence was granted, and any potential long-term degradation of the environment. Progressive rehabilitation may be imposed as a component of the rehabilitation plan where this is deemed appropriate given the character of the land and the ongoing potential for environmental degradation.

The amount of the rehabilitation bond, which is payable as a bank guarantee, is to be fully determined by the Minister. Consultation with private landowners and councils is required if the land includes private land. The Minister may require the licensee to enter into a further bond during the operation of the licence if the Minister considers that the current bond is insufficient. A rehabilitation liability assessment may also be undertaken if there is some doubt regarding the adequacy of the rehabilitation bond.

---

14 *Mineral Resources (Sustainable Development) Act 1990* (Vic) ss 80–83.

Non-compliance with this requirement can result in the Minister serving a notice prohibiting the licensee from doing any further work. A penalty of $100 000 applies if a corporation fails to comply with such a notice.

The licensee must rehabilitate the land in the course of doing work and must, as far as practicable, complete the rehabilitation before the cessation of the exploration licence. When satisfied that either the land has been rehabilitated or that rehabilitation is likely to be successful, the rehabilitation bond is returned to the licensee. However, the Minister has the right to require that, as a condition of returning the bond, a further rehabilitation bond be entered into which relates to any part of the land that requires further rehabilitation. If the land is not rehabilitated as needed, the Minister may carry out the necessary rehabilitation and seek to recover expenses for such rehabilitation as a debt due to the Crown.

Liability for contamination of land can arise even though a mining licence has ceased and the rehabilitation bond has been returned to the licensee. A claim for compensation for loss or damage in relation to Crown land may be made up to three years after the loss or damage was sustained, or the licence expires, whichever is earlier.[15]

An exploration licence will not automatically entitle the holder to enter any of the lands in the area covered by the licence in the absence of the consent of the landowner. The issue of consent is important. In all states and territories a right to access the land is not granted as a component of an exploration licence. This reflects the inherent character of the public ownership framework that underpins minerals and resources in Australia because, as discussed in Chapter 1, the surface estate owner retains control of the land whilst the resource is vested in the state.

2.3.3 Relevant legislative provisions

Each state has different statutory provisions articulating the application requirements for exploration licences, the requirements for granting an exploration licence, and the rights and duties that will accompany the issuance of an exploration licence. A brief overview of some of the relevant legislative provisions for New South Wales, Western Australia, and Queensland is set out below.

2.3.3.1 New South Wales

In New South Wales, the relevant provisions for exploration licences are ss 13–29 of the *Mining Act 1992* (NSW) which outline the entitlement to apply for an

---

15 *Mineral Resources (Sustainable Development) Act 1990* (Vic) ss 80–83. See also Australian Government, Department of Industry, Tourism and Resources, *Mine Rehabilitation: Leading Practice Sustainable Development Program for the Mining Industry* (October 2006) http://www.industry.gov.au/resource/Documents/LPSDP/LPSDP-MineRehabilitationHandbook.pdf.

exploration licence, the application process, and the powers of the Minister to issue exploration titles and impose conditions. Some of these provisions are extracted below. Equivalent provisions also exist in the *Petroleum (Onshore) Act 1991* ss 29–32 and the *Petroleum (Offshore) Act 1982* (NSW) ss 20–39. A reference to 'prospect' in the *Mining Act 1992* (NSW) has a similar meaning to 'explore' in that it refers to the carrying out of works on or the removal of samples from land for the purpose of testing the mineral bearing qualities of the land.

Mining Act 1992 (NSW)

13 Application for exploration licence

(1) Any person may apply for an exploration licence.

(2) To avoid doubt, the owner of privately owned minerals may apply for an exploration (mineral owner) licence or any other exploration licence with respect to those minerals.

Note. The owner of privately owned minerals may choose to apply for an ordinary exploration licence with respect to those minerals, rather than an exploration (mineral owner) licence. In relation to exploration (mineral owner) licences see section 24(4).

(3) An application that relates to land in a mineral allocation area may not be made, except with the Minister's consent, in relation to any group of minerals that includes an allocated mineral.

(4) An application for an exploration licence must:
  (a) specify the group or groups of minerals in respect of which the application is made, and
  (b) be lodged with the Secretary, and
  (c) be accompanied by the required information and the application fee prescribed by the regulations, and
  (d) if the application is for an exploration (mineral owner) licence with respect to privately owned minerals that have more than one owner, be made by all the owners.

(5) The required information is as follows:
  (a) a description, prepared in the approved manner, of the proposed exploration area,
  (b) particulars of the financial resources and relevant technical advice available to the applicant,
  (c) particulars of the program of work proposed to be carried out by the applicant in the proposed exploration area,
  (d) particulars of the estimated amount of money that the applicant proposes to spend on prospecting in that area,
  (e) if the application is for an exploration (mineral owner) licence, evidence that the minerals to which the application relates are owned by the applicant,
  (f) any other information that is prescribed by the regulations.

(6) If there is more than one applicant for the licence, a reference in subsection (5) to the applicant is a reference to each applicant.

Division 3 Granting of exploration licences

22 Power of decision-maker in relation to applications

(1) After considering an application for an exploration licence, the decision-maker:
   (a) may grant to the applicant an exploration licence over all or part of the land over which a licence was sought, or
   (b) may refuse the application.
(2) Without limiting the generality of subsection (1) or any other provision of this Act, an application may be refused on any one or more of the following grounds:
   (a) that the decision-maker is satisfied that the applicant (or, in the case of an applicant that is a corporation, a director of the corporation) has contravened this Act or the regulations (whether or not the person has been prosecuted or convicted of any offence arising from the contravention) or has been convicted of any other offence relating to mining or minerals,
   (b) that the decision-maker is satisfied that the applicant provided false or misleading information in or in connection with the application.
(3) The decision-maker may grant a single exploration licence in respect of 2 or more applications or 2 or more exploration licences in respect of a single application.

26 Conditions of exploration licence

(1) An exploration licence is subject to such conditions (if any) as the decision-maker imposes when the licence is granted, or at any other time under a power conferred by this Act.
(2) Without limiting the generality of subsection (1), the conditions of an exploration licence may include any of the following:
   (a) a condition requiring the holder of the licence to pay royalty to the Crown on any minerals recovered under the licence (but only if it is not an exploration (mineral owner) licence),
   (b) a condition with respect to cores and samples obtained in the course of drilling.
(3) Part 14 applies: (a) to royalty payable under a condition referred to in subsection (2) (a) in the same way as it applies to royalty payable on a mineral recovered under a mining lease, and (b) to the person by whom royalty is payable as if the person were the holder of a mining licence.

27 Term of exploration licence

An exploration licence:

(a) takes effect on the date on which it is granted or on such later date, or on the occurrence of such later event, as the decision-maker may determine, and
(b) ceases to have effect on the expiration of:
   (i) 2 years after the date on which it took effect, in the case of an exploration (mineral owner) licence, or
   (ii) such period (not exceeding 5 years) as the decision-maker determines, in the case of any other exploration licence.

28 Form of exploration licence

An exploration licence is to be in the approved form and is to include the following particulars:

(a) a description of the land over which it is granted,
(b) a list of the group or groups of minerals in respect of which it is granted,

(c) the conditions to which it is subject,
(d) the period for which it is to have effect.

Division 4 Rights and duties under an exploration licence

29 Rights under exploration licence

(1) The holder of an exploration licence may, in accordance with the conditions of the licence, prospect on the land specified in the licence for the group or groups of minerals so specified.
(2) If an application for an assessment lease, mining lease or mineral claim made by the holder of an exploration licence is not finally dealt with before the date on which the licence would otherwise cease to have effect, the licence continues to have effect, in relation only to the land to which the application relates, until the application is finally dealt with.
(3) Subsection (2) does not operate to extend an exploration licence for more than 2 years, or such further period as the Minister may approve in a particular case, after the date on which it would otherwise expire.

The NSW Government has now introduced new regulations expediting the time frame for processing applications under the *Mining Act 1992* (NSW). As of 2013, new time frames will apply to applications for exploration licences, assessment leases, and mining leases. For applications lodged from 1 July 2013, the following target time frames are applicable:

- for all mineral applications and renewals: 45 business days
- for all coal applications: 95 business days
- for all coal renewals: 55 business days.

Exceptions will apply in circumstances beyond the control of the New South Wales Department of Resources and Energy, where processing must be deferred to take account of other legally mandated requirements which include:

- *Environment Protection and Biodiversity Conservation Act 1999* (Cth) ('the *EPBCA*') referral
- native title right to negotiate process
- development consent that is required prior to *Mining Act 1992* approval
- compliance issues of direct relevance to the application or mineral title
- any significant unresolved issue with a third party of direct relevance to the application or mineral title
- for mining leases only: a significant improvement determination, an agricultural land determination and an outstanding survey.

2.3.3.2 Western Australia

In Western Australia, the relevant provisions exist in the *Mining Act 1978* ss 56C–70. Unlike New South Wales, s 73(2) of the *Mining Act 1978* (WA) confers an express right upon the holder of a mining lease, and any agent or employee whilst the lease is in force, to enter the authorised area and do anything 'so authorised and required'. Unlike other states, this precludes any need for the holder of the licence to obtain the express consent of the landowner. This statutory right of entry only applies to 'mining leases' and a mining lease is defined to exclude exploration licences. Similar provisions exist in the *Petroleum and Geothermal Energy Resources Act 1967* (WA), ss 29–48 which separately regulates offshore petroleum exploration. Sections 38 and 43D, which outline the specific statutory rights that accompany offshore petroleum exploration permits and offshore drilling reservations, are extracted below:

Petroleum and Geothermal Energy Resources Act 1967 (WA)

38 Rights conferred by permit

(1) A petroleum exploration permit, while it remains in force, authorises the permittee, subject to this Act and in accordance with the conditions to which the permit is subject, to explore for petroleum, and to carry on such operations and execute such works as are necessary for that purpose, in the permit area.

(2) A geothermal exploration permit, while it remains in force, authorises the permittee, subject to this Act and in accordance with the conditions to which the permit is subject –
   (a) to explore for geothermal energy resources in the permit area; and
   (b) to recover geothermal energy in the permit area for the purpose of establishing the nature and probable extent of a discovery of geothermal energy resources; and
   (c) to carry on such operations and execute such works in the permit area as are necessary for those purposes.

43D Rights conferred by drilling reservation

(1) A petroleum drilling reservation, while it remains in force, authorises the holder of the drilling reservation, subject to this Act and in accordance with the conditions to which the drilling reservation is subject, to drill for petroleum, and to carry on such operations and execute such works as are necessary for that purpose, in the drilling reservation area.

(2) A geothermal drilling reservation, while it remains in force, authorises the holder of the drilling reservation, subject to this Act and in accordance with the conditions to which the drilling reservation is subject –
   (a) to drill for geothermal energy resources in the drilling reservation area; and
   (b) to recover geothermal energy in the drilling reservation area for the purpose of establishing the nature and probable extent of a discovery of geothermal energy resources; and
   (c) to carry on such operations and execute such works in the drilling reservation area as are necessary for those purposes.

2.3.3.3 Queensland

In Queensland, the relevant provisions for exploration licences are contained in the *Mineral Resources Act 1989* (Qld) ss126–136A. Further conditions regarding the grant and exercise of exploration permits are contained in ss 137–178.

Schedule 1, s 6 requires holders of exploration licences to give affected landowners an entry notice, outlining the details, duration, and activities proposed under the licence. The entry notice must include a copy of the exploration tenement as well as a copy of the Queensland Land Access Code. The Land Access Code was established in 2010 pursuant to s 24A of the *Petroleum and Gas (Production and Safety) Act 2004*. The Code imposes mandatory conditions concerning the conduct of authorised mining activities on private land and imposes best practice guidelines for communication between title-holders and private landholders. Advanced activities require the parties to enter into a conduct and compensation agreement.[16] In this respect, the approach in Queensland has been replicated in New South Wales, where the legislation has now been amended to mandate access agreements prior to the commencement of any CSG activity.[17]

The relevant provisions in the *Mineral Resources Act 1989* (Qld) are extracted below:

Mineral Resources Act 1989 (Qld)

Schedule 1: Access and compensation provisions for exploration permits and mineral development licences

1 Meaning of exploration tenement
   An exploration tenement is any exploration permit or mineral development licence.
2 What is a preliminary activity
   (1) A preliminary activity, for a provision about an exploration tenement, means an authorised activity for the tenement that will have no impact, or only a minor impact, on the business or land use activities of any owner or occupier of the land on which the activity is to be carried out.
   (2) However, the following are not preliminary activities –
      (a) an authorised activity carried out on land that –
         (i) is less than 100ha; and
         (ii) is being used for intensive farming or broadacre agriculture;
      (b) an authorised activity carried out within 600m of a school or an occupied residence;
      (c) an authorised activity that affects the lawful carrying out of an organic or bio-organic farming system.

---

16 *Petroleum and Gas (Production and Safety) Act 2004* (Qld) s 153 and *Minerals Resources Act 1989* (Qld) s 140A.
17 *Mining Act 1992* (NSW) s 141.

3 What is an advanced activity

An advanced activity, for a provision about an exploration tenement, means an authorised activity for the tenement other than a preliminary activity for the tenement.

....

5 Entry notice requirement for particular authorised activities

(1) A person must not –
   (a) enter private land in an exploration tenement's area to carry out a preliminary activity for the tenement; or
   (b) enter private land in an exploration tenement's area to carry out an advanced activity for the tenement if either of the following applies for the entry –
      (i) the deferral agreement exemption;
      (ii) the Land Court application exemption; or
   (c) enter public land in an exploration tenement's area to carry out any authorised activity for the tenement;

   unless the exploration tenement's holder has given each owner and occupier of the land a written notice of the entry that complies with section 6 (an entry notice).

(2) The entry notice must be given –
   (a) generally – at least 10 business days before the entry; or
   (b) if, by a signed endorsement on the notice, the relevant owner or occupier has agreed to a shorter period – the shorter period.

(3) The holder must give the chief executive a copy of the entry notice immediately after the notice is given and before entry is made under the exploration tenement.

(4) A contravention of subsection (3) does not affect the validity of the notice or the entry.

(5) This section is subject to section 7.

(6) In this section –

   deferral agreement exemption, for an entry, means that the conduct and compensation agreement requirement does not apply for the entry because of section 11(c)(i).

   give, for an entry notice, includes publishing it in a way approved under section 9.

Land Court application exemption, for an entry, means that the conduct and compensation agreement requirement does not apply for the entry because of section 11(c)(ii).

6 Required contents of entry notice

(1) An entry notice must state the following –
   (a) the land proposed to be entered;
   (b) the period during which the land will be entered (the entry period);
   (c) the activities proposed to be carried out on the land;
   (d) when and where the activities are proposed to be carried out;
   (e) contact details for –
      (i) the relevant exploration tenement holder; or
      (ii) another person the holder has authorised to discuss the matters stated in the notice.

(2) Also, the first entry notice from the exploration tenement holder to a particular owner or occupier must be accompanied by or include a copy of –
    (a) the exploration tenement; and
    (b) the land access code; and
    (c) if the exploration tenement is for a small scale mining activity – the small scale mining code; and
    (d) any code of practice made under this Act applying to authorised activities for the exploration tenement; and
    (e) any relevant environmental authority for the exploration tenement.
(3) The entry period cannot be longer than –
    (a) generally – 6 months; or
    (b) if the relevant owner or occupier agrees in writing to a longer period – the longer period.
(4) Subject to subsections (1) to (3), an entry notice may state an entry period that is different to the entry period stated in another entry notice given by the exploration tenement holder to another owner or occupier of the land.

7 Exemptions from entry notice requirement

(1) The requirement under section 5(1) to give an entry notice does not apply for an entry to land to carry out an authorised activity if any of the following apply –
    (a) the exploration tenement holder owns the land;
    (b) the holder has the right other than under this Act to enter the land to carry out the activity;
    (c) if –
        (i) there is a conduct and compensation agreement relating to the land; and
        (ii) each eligible claimant for the land is a party to the agreement; and
        (iii) the agreement includes a waiver of entry notice;
    (d) the entry is to preserve life or property or because of an emergency that exists or may exist;
    (e) the relevant owner or occupier has, by signed writing, given a waiver of entry notice.
(2) A waiver of entry notice mentioned in subsection (1) must comply with section 8(1).

8 Provisions for waiver of entry notice

(1) A waiver of entry notice mentioned in section 7 must –
    (a) if it does not form part of a conduct and compensation agreement, be written and signed; and
    (b) state the following –
        (i) that the relevant owner or occupier has been told they are not required to agree to the waiver of entry notice;
        (ii) the authorised activities proposed to be carried out on the land;
        (iii) the period during which the land will be entered;
        (iv) when and where the activities are proposed to be carried out.

     (2) The relevant owner or occupier cannot withdraw the waiver of entry notice during the period.
     (3) The waiver of entry notice ceases to have effect at the end of the period.
9 Giving entry notice by publication
  (1) The chief executive may approve an exploration tenement holder giving an entry notice for the tenement by publishing it in a stated way.
  (2) The publication may relate to more than 1 entry notice.
  (3) The chief executive may give the approval only if –
     (a) for a relevant owner or occupier who is an individual, it is impracticable to give the owner or occupier the notice personally; and
     (b) the publication will happen at least 20 business days before the entry.
10 Conduct and compensation agreement requirement for particular advanced activities
  (1) A person must not enter private land in an exploration tenement's area to carry out an advanced activity for the tenement (the relevant activity) unless each eligible claimant for the land is a party to an appropriate conduct and compensation agreement.
  Maximum penalty – 500 penalty units.

  Note– If a corporation commits an offence against this provision, an executive officer of the corporation may be taken, under section 412B, to have also committed the offence.

  (2) The requirement under subsection (1) is the conduct and compensation agreement requirement.
  (3) In this section –
  appropriate conduct and compensation agreement, for an eligible claimant, means a conduct and compensation agreement about the holder's compensation liability to the eligible claimant of at least to the extent the liability relates to the relevant activity and its effects.
11 Exemptions from conduct and compensation agreement requirement
  The conduct and compensation agreement requirement does not apply for an entry to land to carry out an advanced activity if any of the following apply –
  (a) the exploration tenement holder owns the land;
  (b) the holder has the right other than under this Act to enter the land to carry out the activity;
  (c) each eligible claimant for the land is –
     (i) a party to an agreement, complying with section 12, that a conduct and compensation agreement can be entered into after the entry (a deferral agreement); or
     (ii) an applicant or respondent to a Land Court application under section 22 relating to the land;
  (d) the entry is to preserve life or property or because of an emergency that exists or may exist.

These provisions distinguish between preliminary and advanced activities and impose entry notice requirements for both preliminary and advanced activities but only impose an obligation to enter into a conduct and compensation agreement for advanced activities. The conduct and compensation agreement is an important element underpinning the Queensland regulatory framework. The legislation requires both parties to enter into negotiations regarding the way in which access is to be exercised, and this must include times, locations, and levels of disturbance. It also requires parties to enter into negotiations regarding the nature and scope of compensation payable for disturbances to surface estate activities. Similar provisions exist in the *Petroleum and Gas (Production and Safety) Act 2004* (Qld) ('the *PGPSA*').[18]

Chapter 3 of the *PGPSA* is significant because it articulates how overlapping exploration titles for petroleum and unconventional gas are to be prioritised. Division 1 of ch 3 contains specific provisions outlining the requirements for obtaining a petroleum lease over land in an area where a coal or shale exploration tenement has already been issued. In particular, the Minister has discretion, pursuant to s 319 of the *PGPSA*, to decide whether to grant the petroleum lease application or whether to give preference to coal or oil shale development.

In making this determination, the assessment criteria set out in s 305(2) of the *PGPSA* must be taken into account. This criteria mandates careful evaluation of the proposed timing and rate of petroleum production and of the technical and fiscal viability associated with the implementation of a coordinated development of coal or oil shale reserves. Other factors relevant to the Ministerial discretion in this context include the value, public interest, and functionality of the proposed petroleum project.

The objectives of ch 3 of the *PGPSA* are to ensure that petroleum exploration (and production) does not compromise the ability to mine coal seams economically in the future, and further, to ensure if the issuance of a petroleum title is commercially and technically feasible, the grant of petroleum leases that may affect coal or oil shale mining, or proposed coal or oil shale mining, optimises the commercial use of coal, oil shale, and petroleum resources in a safe and efficient way.

Amendments to the *Mineral Resources Act 1989* (Qld) and the *PGPSA* have been introduced which expedite the approval process for exploration permits.[19] The approval process used to take approximately 22 months. Under the new *Streamlining Act 2012* (Qld), a series of key reforms were introduced. One of the key reforms was the implementation of an online platform for the application of mining licences: MyMinesOnline. Once all requirements have been finalised, an exploration permit which is subject to native title may be determined in less than

---

18 *PGPSA* (Qld) div 2, ss 34–44.
19 *Streamlining Act 2012* (Qld) has an application to both the *Mineral Resources Act 1989* (Qld) and the *PGPSA*.

12 months and an application for an exploration permit which is not subject to native title may be determined in six months or less. Other reforms include initial terms under the *Mineral Resources Act 1989* (Qld) being reduced from 10 years to five years. Section 139 of *Mineral Resources Act 1989* (Qld) now reduces by 40 per cent land covered by an exploration permit by the end of the first three years after the permit is granted, and by a further 50 per cent by the end of the first five years. Further, s 129(1)(b) of the *Mineral Resources Act 1989* (Qld) expressly authorises the holder of an exploration permit to enter permit land for the purpose of doing all things necessary to apply for a mineral development licence or a mining lease and to comply with associated requirements for environmental assessments.

2.3.4 Proprietary status of the exploration licence

The issue of whether an exploration right actually creates a legal estate or interest in the land depends entirely upon the legislative provision in issue. In some jurisdictions the character of the interest is expressly defined; however, for the most part, the proprietary character of an exploratory licence has not been legislatively defined.[20] As such, a determination of this issue involves an assessment of the different judicial approaches that have been employed and, in this regard, some resort to the common law may be made. Whilst it is important to emphasise that the common law acts in a supplementary capacity in this context, it is clear from the cases that common law property principles can be relevant. Exploration titles confer rights, which have a similar quality to institutional property rights in the sense that they are enforceable *in rem* against the community at large.[21]

Many cases have characterised the exploration licence as a *profit à prendre*: holding that the right to enter land and remove minerals for the purpose of exploration and investigation is akin to the common law incorporeal hereditament. In the early decision of *Duke of Sutherland v Heathcote*,[22] Lindley LJ held that, 'A right to work mines is something more than a mere licence: it is a *profit à prendre*,

---

20 For legislation setting out that mining tenements do not constitute property, see *Offshore Minerals Act 1994* (Cth) s 439A where exploration licences and the rights which attach to them are expressly set out to not constitute personal property. See also the *Mineral Resources Act 1989* (Qld) s 10 which sets out that the grant of a mining tenement does not create an estate or interest in land. For statutory provisions setting out that mineral tenements do constitute property see *Mineral Resources (Sustainable Development) Act 1990* (Vic) s 70(3) (a registered exploration or mining licence confers a proprietary interest in the land). Section 70(4) sets out that the proprietary character of this interest is for 'the purpose of assisting the licensee to exercise the rights and discharge the obligations under the licence'.
21 See the discussion by M Crommelin, 'Economic Analysis of Property' in D J Galligan (ed), *Essays in Legal Theory* (Melbourne University Press, 1984) 78.
22 [1892] 1 Ch 475 (CA).

an incorporeal hereditament lying in grant'.[23] This approach has been approved in Australia. In *Rossmar Park Pastoral Co Pty Ltd v Coal Mines Australia Pty* Ltd, Rothman J held that an exploration licence issued pursuant to the *Mining Act 1992* (NSW) was more than a mere licence and was 'analogous to a *profit à prendre*' and therefore amounted to an interest in land.[24] His Honour noted that the effect of this was a recalibration of the rights of the surface estate owner, concluding that:

> Modern legislatures regularly enact laws that take away or modify common law rights. This particular statute is concerned with the facilitation of mining and/or exploration rights and the balancing of that economic imperative and the rights of persons granted such licences, on the one hand, with the holders of land otherwise entitled to the enjoyment of that land, on the other.[25]

Exploration licences have also been analogised as proprietary in circumstances where the ancillary rights attached to them have proprietary characteristics. For example, in *Anaconda Nickel Ltd v Tarmoola Australia Pty Ltd*, Ipp J concluded that the transferable nature of an exploration licence gave it a 'proprietorial character'.[26] In *Commonwealth v WMC Resources Ltd*, the High Court concluded that 'an exploration right could be classified as a statutory property right'.[27] However, as a statutory property right, its continued existence is completely dependent upon legislative support and any institutional rights, such as a right to transfer the licence, may only be exercised in accordance with the relevant provisions regulating this process.

2.3.5 Legal status of offshore exploration permits: s 51(xxxi) *Constitution*

The High Court of Australia has concluded[28] that the legal status of an offshore exploration permit amounts to a bare statutory entitlement, rather than a property interest, because of the absence of radical title in the offshore area.[29] This means that any subsequent excision of an offshore exploration title, which occurs as the result of a provision within a Commonwealth Act, will not constitute an acquisition of property for the purposes of s 51(xxxi) of the *Constitution*.

---

23 *Duke of Sutherland v Heathcote* [1892] 1Ch 475 (CA), 483. This was approved in *Re Williams v De Biasi* (Unreported, Federal Court of Australia, Drummond J, 18 August 1992).
24 *Rossmar Park Pastoral Co Pty Ltd v Coal Mines Australia Pty Ltd* [2008] NSWSC 1385, [26]. See also *Stow v Mineral Holdings (Australia) Pty Ltd* (1977) 180 CLR 295, 410 (Aickin J).
25 *Rossmar Park Pastoral Co Pty Ltd v Coal Mines Australia Pty Ltd* [2008] NSWSC 1385, [31].
26 (2000) 22 WAR 101 at [103]. See also *Commonwealth v WMC Resources Ltd* (1998) 194 CLR 1 ('*WMC Resources*'), [14] (Brennan CJ).
27 (1998) 194 CLR 1, [14] (Brennan CJ).
28 *Commonwealth v WMC Resources Ltd* (1998) 194 CLR 1.
29 For a full discussion on the laws relevant to the offshore area see Chapter 3.

> **Facts:** This was the conclusion of the High Court in *Commonwealth v WMC Resources Ltd*.[30] This is an interesting case, focusing upon whether offshore exploration permits may be characterised as proprietary interests and therefore capable of expropriation pursuant to s 51(xxxi) of the *Constitution*. At issue for the Court was whether, because of the absence of Crown radical title in the offshore region, a statutory exploration permit attracted the same Commonwealth constitutional protection as an onshore exploration title. The facts of the case involved the excision in 1990 by the Commonwealth Government of blocks from an offshore exploration permit which was held by WMC Resources Ltd and issued under the *Petroleum (Submerged Lands) Act 1967* (Cth). The excision was carried out in order to give effect to a treaty between Australia and Indonesia in an offshore area known as the 'Timor Gap'. The Commonwealth argued in the High Court that the excision did not constitute an acquisition within s 51(xxxi) of the Constitution.
>
> **Held:** The High Court held (Brennan CJ, Gaudron, McHugh and Gummow JJ; Toohey and Kirby JJ in dissent) that the excision did not amount to an acquisition of property from WMC Resources Ltd and that no compensation was payable. Brennan CJ, Gaudron and Gummow JJ distinguished the offshore exploration permit granted under the *Petroleum (Submerged Lands) Act 1967* (Cth) '*PSL Act*' from an onshore exploration lease issued under the *Mining Act 1980* (NT). Brennan CJ held that the offshore exploration permit amounted to a pure statutory entitlement that was inherently susceptible to change, because statutory rights are defeasible and variable.[31] This type of interest did not attract the application of s 51(xxxi) of the *Constitution*. Gaudron and Gummow JJ agreed with Brennan CJ. Toohey and Kirby JJ strongly dissented. Their Honours concluded that the absence of radical title should not preclude the capacity to make laws creating proprietary estates.[32]

Brennan CJ argued that the offshore exploration permit, being located on the continental shelf beyond the outer limits of the territorial sea, did not generate an interest that was capable of attracting s 51(xxxi) because it did not create a proprietary title as the Crown does not hold radical title in this area. This position may be contrasted with onshore petroleum titles because the issuance of a statutory grant over land impacts upon the underlying radical title held by the Crown. His Honour stated:

> It is erroneous to regard the PSL Act as the offshore equivalent of those provisions, which, in Australia, authorise the Crown to alienate interests in

---

30 (1998) 194 CLR 1.
31 (1998) 194 CLR 1, [16].
32 (1998) 194 CLR 1 [53]–[56] (Toohey J); [237]–[241] (Kirby J).

the waste lands of the Crown (provisions which I shall call "Land Acts"). If it were the equivalent of Land Acts, it would be arguable that the extinguishing of a permittee's proprietary rights relieves the Commonwealth of a reciprocal burden on its title to land within the permit area and thus constitutes an acquisition of property. Land Acts assume the existence of the Crown's radical title to land lying above the low water mark, a title which is sufficient to support the alienation of interests in that land and to found the Crown's full beneficial title to that land when there are no other interests or when other interests have been extinguished or are exhausted ... [T]he extinguishing of an interest in land above the low water mark necessarily results in the enhancement of the title which was subject to the interest extinguished. The position in relation to interests in or over the continental shelf is quite different.[33]

Gummow J came to a similar conclusion holding that the petroleum exploration permit

suffered from the "congenital infirmity" that its scope and incidents were subject to the PSL Act in the form it might from time to time thereafter assume. Any proprietary rights which were created generated in respect of the Permit were liable to defeasance.[34]

An excision or variation of this type of entitlement would not, by reason of its very nature, amount to an acquisition of property pursuant to s 51(xxxi).[35]

The susceptibility of 'pure' statutory entitlements to subsequent modification or extinguishment and their capacity to attract s 51(xxxi) was further reviewed by Crennan J in *Wurridjal v The Commonwealth* (2009) 237 CLR 309 at [363]–[364] who stated:

It can be significant that rights, which are diminished by subsequent legislation, are statutory entitlements. Where a right, which has no existence apart from statute, is one that, of its nature, is susceptible to modification, legislation which effects a modification of that right is not necessarily legislation with respect to an acquisition of property within the meaning of s 51(xxxi). It does not follow, however, that all rights which owe their

---

33 *Commonwealth v WMC Resources Ltd* (1998) 194 CLR 1, [20].
34 *Commonwealth v WMC Resources Ltd* (1998) 194 CLR 1, [203].
35 See also *ICM Agriculture v The Commonwealth of Australia* (2009) 240 CLR 140, [152] where Hayne, Kiefel and Bell JJ concluded that 'the statutes by which the mining tenements were created carved those interests out of the radical title of the Commonwealth to the land. The mining tenements were a species of property in the land and in the minerals which, when the rights under the mining tenements came to an end, enlarged the Commonwealth's radical title to the land'.

existence to statute are ones which, of their nature, are susceptible to modification as the contingency of subsequent legislative modification or extinguishment does not automatically remove a statutory right from the scope of s 51(xxxi).

Whether the statutory modification of a pre-existing entitlement constitutes an acquisition of property and therefore generates an entitlement to compensation under s 51(xxxi) depends entirely upon the nature of the statutory entitlement. The express terms of the statute may indicate that the entitlement is susceptible to statutory variation. Alternatively, the statutory provision itself may impose limitations upon the scope of the entitlement resulting in a situation where what initially appeared to be a modification must be classified as a pre-existent and inherent limitation. It is also possible that the statutory entitlement is a constituent of an overall scheme of entitlements that may require modification or reform over time.

These issues only have a direct application to offshore exploration permits because onshore exploration permits are characterised as land-based entitlements with equivalent property characteristics. It is also significant to bear in mind that onshore exploration permits are generally issued pursuant to state rather than Commonwealth legislation (see the provisions outlined above) and therefore do not attract the operation of s 51(xxxi), which does have an application to Commonwealth Acts which apply, in the context of the *WMC Resources* desicion, to exploration titles issued in Commonwealth waters.

Legislative provisions seeking to modify or extinguish exploration titles issued from state legislation will only attract compensation where compensation provisions are specifically included in that state act. Just compensation for an acquisition of property flowing from a state act is not available under s 51(xxxi). It was argued in *Spencer v Commonwealth of Australia*[36] that a state law which is directly derived from a Commonwealth Act, such as a law providing for grants to be made to a state under s 96 of the *Constitution*, or for agreements under which such grants could be made, might be characterised by reference to informal arrangements between the Commonwealth and the state as a law with respect to the acquisition of property. This issue was not, however, determined by the High Court on the facts.

The decision in *WMC Resources*[37] can be directly contrasted with the earlier decision of the High Court in *Newcrest Mining WA Ltd v The Commonwealth* ('*Newcrest*'), which dealt with an onshore mining lease rather than an offshore exploration title.[38]

---

36 (2010) 241 CLR 118, [32].
37 (1998) 194 CLR 1.
38 (1997) 190 CLR 513.

> **Facts:** On the facts of *Newcrest*, the Federal Government extended the boundaries of Kakadu National Park. This resulted in land which had been included in 25 mining leases issued under the *Mining Act 1980* (NT) and held by Newcrest Mining (WA) Ltd being amended so that mining operations could no longer be conducted in Kakadu National Park. No compensation was given to Newcrest for this act by the Commonwealth and one of the issues was whether the impact upon the mining lease by the actions of the Federal Government attracted the application of s 51(xxxi) of the *Constitution*.
>
> **Held:** Unlike the decision in *WMC Resources*, the High Court held by a majority that the extension of the boundaries of Kakadu National Park amounted to an acquisition of property and therefore an infringement of the guarantee in s 51(xxxi) of the *Constitution*.
>
> Gummow J concluded that the mineral lease did confer a title upon the title-holder. During the course of his judgement, Gummow J referred to the earlier decision of Lord Cairns in *Gowan v Christie*[39] where his Lordship concluded that what we call a 'mineral lease' is really 'when properly considered, a sale out and out, of a portion of land'. Whilst a mining lease is not identical to a common law title, it nevertheless confers dual rights of using land that belongs to another and appropriating minerals at the point of severance. In this sense, Gummow J held that the mining lease was not entirely statutory and that it did have some link to institutional common law leases because mining leases were an emanation of the Crown's proprietorship of land. Hence, whilst statute provided the means of creating the mining lease, the ultimate source of the ownership was the radical title of the Crown over land in the Northern Territory. Toohey, Gaudron and Kirby JJ agreed in substance with the conclusions of Gummow J in this regard.

Gummow J went on to hold that the Commonwealth derived 'identifiable and measurable' advantage from the change in boundaries to the Kakadu National Park as it released the Commonwealth from the burden of the right of Newcrest to occupy the land and extract minerals from it. This attracted the application of s 51(xxxi) of the *Constitution*. Kirby J and Brennan CJ agreed. Brennan CJ stated:

> Under its mining leases, Newcrest had the right exercisable against the Commonwealth as reversioner to mine for, extract and take away minerals from the leased land during the term of the lease. When that land was included in Kakadu National Park, Newcrest's rights to carry on operations for the recovery of minerals were extinguished.[40]

---

39 (73) LR 2 Sc & Div 273, 284.
40 *Newcrest Mining WA Ltd v The Commonwealth* (1997) 190 CLR 513, 1350–1.

McHugh J dissented, holding that the denial of a right to mine did not constitute an acquisition because, whilst the proclamations impinged on Newcrest's rights to exploit the minerals, there was no gain by the Commonwealth. Hence as a matter of substance and form 'the Commonwealth obtained nothing which it did not already have'.[41]

The conclusions of the majority of the High Court in *Newcrest* were subsequently upheld in *WMC Resources Ltd*[42] where Brennan CJ concluded at [17] that

> the law which sterilised Newcrest's right under its mining lease to carry on "operations for the recovery of minerals" on land vested in the Commonwealth was, in my opinion, a law for the acquisition of property because it extinguished the liability of the Commonwealth to have those minerals extracted from its land and thereby enhanced the property of the Commonwealth.

Subsequently, in *ICM Agriculture Pty Ltd v Commonwealth*[43], Hayne, Kiefel and Bell JJ held at [152] that the property which Newcrest had was

> more than a statutory privilege under a licensing system. The statutes by which the mining tenements were created carved those interests out of the radical title of the Commonwealth to the land. The mining tenements were a species of property in the land and in the minerals which, when the rights under the mining tenements came to an end, enlarged the Commonwealth's radical title to the land.

2.4 Retention licences and assessment leases

In most states and territories it is possible for mining proponents to apply for a retention or assessment title. This title is generally relevant once an exploration licence or permit is issued and it is clear that commercial production of minerals in a particular region is viable.[44] The purpose of the retention or assessment title

---

41 *Newcrest Mining WA Ltd v The Commonwealth* (1997) 190 CLR 513, 1375. For a more detailed discussion of a regulatory taking constituting an acquisition of property pursuant to s 51(xxxi) see P O'Connor, 'The Changing Paradigm of Property and the Framing of Regulation as a Taking' (2010) 36(2) *Monash University Law Review* 50; S Brennan, 'Section 51(xxxi) and the Acquisition of Property Under Commonwealth-State Arrangements: The Relevance to Native Title Extinguishment on Just Terms' (2011) 15(2) *University of New South Wales Law Review* 74.
42 (1998) 194 CLR 1.
43 (2009) 240 CLR 140.
44 *Mining Act 1978* (WA) s 70D(7).

is to allow a mining proponent to retain rights to mineral resources which are not, at the point of applying for the title, economically viable to mine but which may become so in the future.

The fundamental character of the retention licence was described by the Western Australian Government during the second reading of pt IV, div 2A of the *Mining Act 1978* (WA) as an

> ... intermediate form of tenure between the exploration licence and the mining lease. Its primary purpose will be to provide secure tenure, for a limited time, to enable an explorer to hold an identified mineral resource which is not a commercially viable proposition in the short term but for which there is a reasonable prospect for development in the longer term. From time to time deposits are identified for which no further exploration or mining is warranted in the short term. The identified resource may be sub-economic or cannot be mined for some other reason. In these circumstances the current mining tenements are inadequate. The exploration licence is for the exploration of the ground for mineral resources; it is not a holding title, and the mining lease is inappropriate and too expensive. A less expensive title is needed with a work program determined by the Minister after taking into account economic, technological and policy factors. A retention licence would clearly indicate to all parties that a resource had been identified and that different commitments to and expenditure on the working of the ground applied.

The utility of the retention licence lies in the fact that it:

- provides the holder with improved security of title for mineral deposits which cannot be mined for the time being;
- facilitates the implementation of an appropriate work and/or research program designed to achieve economic development of the identified resource;
- highlights areas of identified mineralisation which may be developed at some future stage; and
- reduces the administration costs and simplifies administrative procedures associated with such mineral deposits.[45]

In Western Australia, a retention licence may also be issued pursuant to the *Mining Act 1978* (WA) pt IV, div 2A. The holder of a prospecting licence, an exploration licence, or a mining lease may apply for a retention licence which authorises further development and exploration of an area, including the right to extract a specified

---

45 See the second reading speech for the *Mining Amendment Act 1993* (WA) by C J Barnett: Western Australia, *Parliamentary Debates*, Legislative Assembly, 12 August 1993, 2439 (C J Barnett). See also Buss JA in *Re Minister for Resources; Ex parte Cazaly Iron Pty Ltd* (2007) 34 WAR 403, [174] (Wheeler and Pullin JJA agreeing).

amount of minerals from a designated area.[46] In deciding whether to issue a retention licence to the holder of a prospecting or exploration licence, the Minister must be satisfied that mineral deposits have been located within the area and that the development of those minerals is not currently feasible.[47]

In Tasmania, the *Mineral Resources Development Act 1995* (Tas) authorises the issuance of a retention licence to the holder of an exploration licence or a mining lease where the holder is able to establish justifiable reasons for not proceeding to mine in the area.[48] A retention licence may be issued for no more than five years over an area of land between 10 and 50 km^2.[49] During the currency of the licence, the licence-holder may evaluate the potential of the mine. The Minister is entitled to revoke a retention licence should the licence-holder breach a condition of the licence although the licensee has a right of appeal to the Mining Tribunal against a decision.[50]

In New South Wales, an applicant may apply for a similar title, known in that state as an 'assessment lease'. Like the retention licence, the assessment lease is essentially a holding or 'interim' resource title, conferring upon the title-holder the right to make ongoing assessment of the commerciality of a particular area where mineral deposits have been located and mining is not commercially viable in the short term; however, there is a reasonable prospect that it will become so in the longer term.

The assessment lease entitles the title-holder to continue prospecting activities authorised under an initial exploration licence and to recover minerals in the course of assessing the viability of commercial mining. In essence, an assessment lease is designed to allow the developer to retain a title over a potential project area, without necessarily having to commit to further exploration as would normally be the case with an exploration licence. However, an assessment lease is not intended to allow a holder to restrict resources indefinitely and it is expected that direct expenditure on the area will occur during the currency of the lease. An assessment lease may not exceed five years.[51]

In making an application for an assessment lease, an applicant should include their environmental performance record during the currency of the exploration licence.[52] It must also be established that exploration of the area has been completed to a level of confidence, which allows the resource in issue to be

---

46 *Mining Act 1978* (WA) s 70J; *Mining Regulations 1981* (WA) reg 23G.
47 *Mining Act 1978* (WA) ss 70F, 126. The security amount for a retention licence in Western Australia is $5000. See *Mining Regulations 1981* (WA) reg 112.
48 *Mineral Resources Development Act 1995* (Tas) ss 47(1), 53(2).
49 Ibid ss 57, 59.
50 *Mineral Resources Development Act 1995* (Tas) ss 66(4), (6).
51 *Mining Act 1992* (NSW) s 45.
52 *Mining Regulation 2010* (NSW) reg 20.

classified as either indicated or measured under the Australian Code for Reporting of Exploration Results, Mineral Resources, and Ore Resources (2012) ('JORC Code').[53] Further, a basic mining plan should be drafted, taking into account a range of geological, geographical, and cultural issues relevant to resource recovery, including a feasibility study outlining core costs and project economics and a work program.[54]

In Queensland, a mineral development licence may be issued where a mineral deposit is located pursuant to an exploration permit; however, further investigation of the deposit is required. As with the retention licence and assessment lease, the mineral development licence is essentially a holding title, conferring the right to conduct further and more detailed examination of minerals which have been discovered pursuant to an exploration right to determine their commercial viability. Only the holder of an exploration licence can apply for a mineral development licence and it will generally only be granted where the Minister is satisfied that more detailed investigation is necessary.[55]

A mineral development licence – like a retention licence or an assessment lease – is generally issued for a term of five years but may, at the direction of the Minister, be transformed into a full mining licence and will be prioritised over other claims.[56]

In Victoria, a retention licence may be issued pursuant to pt 2, div 1 of the *Mineral Resources Sustainable Development Act 1990*. A retention licence may be issued under this Act for the purpose of mining a mineral resource in the future. The holder of a retention licence is entitled to retain rights to a mineral resource in the land covered by the licence, which is not presently economically viable to mine but may become so in the future, or for the purpose of sustaining mining operations and exploring or carrying out other work to establish the economic viability of mining a mineral resource.[57] A retention licence issued may be granted for 10 years and may be renewed for a similar period twice.[58]

Retention titles may also be issued in the offshore region. As with the onshore retention title, offshore retention titles confer holding entitlements upon the mining

---

53 The JORC Code is a professional code of practice that sets minimum standards for public reporting of minerals, exploration results, mineral resources, and ore reserves. Classification will occur in accordance with the level of confidence in geological knowledge, technical assessment, and economic relevance. The JORC Classification Code must be satisfied prior to the issuance of any retention title.
54 *Mining Act 1992* (NSW) s 33.
55 *Mineral Resources Act 1989* (Qld) ss 179, 186.
56 Ibid s 194.
57 *Mineral Resources (Sustainable Development) Act 1990* (Vic) s 14C(1).
58 Ibid s 14(3).

proponent within the area of the proponent's exploration licence for petroleum discoveries which are not, at that point, commercially viable.[59] A joint Federal-State authority regulates the offshore area for every state and the Northern Territory.[60] The authority includes both the state and federal ministers who will determine whether an offshore licence may be issued. The relevant provisions are extracted below:

Offshore Minerals Act 1994 (Cth)

132 Retention licences

This Part provides for the grant of retention licences over blocks in an offshore area.

Note: A retention licence is designed to allow an exploration licence holder to retain rights over an area if:

- the holder has identified and evaluated a significant mineral deposit in the exploration licence area; and
- mining the deposit is not commercially viable in the short term for some reason (for example, the political situation, the prevailing situation in the commodity market for particular minerals, the need to arrange finance or build up capital reserves, the need to develop new technologies or the impending development of new technologies); and
- there is a reasonable prospect of development of the deposit in the longer term.

133 Activities authorised by a retention licence

(1) Subject to subsections (2) and (3), a retention licence holder may:
  (a) explore for minerals in the licence area; and
  (b) recover minerals in the licence area.
(2) A retention licence does not authorise the recovery of minerals as part of a commercial mining operation.
(3) If the licence is expressed to restrict the kind of minerals covered by the licence, the holder is not permitted to explore for, or to recover, minerals not covered by the licence.
(4) A restriction on the kind of minerals covered by the licence may be inclusive (for example, only minerals A, B and C) or exclusive (for example, all minerals except A, B and C).
(5) For the purposes of subsection (3), the holder does not recover an excluded mineral if, in the course of exploring for, or recovering, another mineral, the holder recovers some excluded mineral.

---

59 See *Offshore Minerals Act 1994* (Cth), ss 132, 137(1). See also the *Offshore Petroleum and Greenhouse Gas Storage Act 2006* (Cth) pt 2.3 which provides for the issuance of a petroleum retention lease.

60 *Offshore Minerals Act 1994* (Cth) ss 29–32. See also the *Offshore Petroleum and Greenhouse Gas Storage Act 2006* (Cth) pt 1.3. NOPSEMA (National Offshore Petroleum Safety Environment Management Agency) is a Commonwealth agency with jurisdiction over the environmental management of all offshore petroleum resources.

2.5 Mining and production leases

A mining or production lease confers statutory rights upon the holder to access the minerals in an area and to extract and produce them for commercial purposes.[61] Once the minerals have been extracted, ownership in those minerals is generally transferred from the state to the mining proponent. A mining or production lease is, as with all resource titles, a statutory creation and therefore derives all entitlements from legislative provisions. In this sense, whilst the title is generally described as a lease, it is not identical to an institutional common law lease. In particular, a mining licence does not confer upon the mining proponent an automatic entitlement to exclusive possession of the land in which the minerals reside. However, the mining lease represents the highest form of resource title that can be issued under the legislative framework and is therefore the most durable form of title, conferring upon the holder the most extensive rights to extract and produce the resource.

The nature and purpose of exploration, retention, and production titles, and the role they play in authorising the private development of state owned minerals and petroleum, is a crucial policy issue. Judicial processes and statutory interpretation must be fully cognisant of this impact. Resource titles straddle the public and private law divide and, in so doing, generate both property and administrative concerns. One of the biggest challenges for the future lies in the attempt to integrate the disparate economic and management components associated with resource exploitation.[62]

2.5.1 Statutory character of a mining lease

The legal character of the mining lease was explored by the High Court in *Wade v New South Wales Rutile Mining Co Pty Ltd*[63] where Windeyer J concluded that, 'A mining lease ... is really a sale by the Crown of minerals reserved to the Crown to be taken by the lessee at a price payable over a period of years as royalties'. Earlier, in *Gowan v Christie*[64], Lord Cairns described a 'mineral lease' in the following way:

---

[61] The only state to refer to a mining or production title as a 'licence' rather than a 'lease' is Victoria: *Mineral Resources (Sustainable Development) Act 1990* (Vic) s 14.

[62] See M Crommelin, 'Mining and Petroleum Titles' (1988) 62 *Australian Law Journal* 863, 867. H Williams, 'Comments on Oil and Gas Jurisprudence in Canada and the United States' (1965) 4 *Alberta Law Review* 189, 192–3 where the author suggests that 'excessive conceptualism can have unfortunate consequences' when evaluating mineral tenements and that common law principles should be applied 'selectively rather than blindly' to take account of the fact that the mining lease is fundamentally a 'commercial instrument'.

[63] (1969) 121 CLR 177, 192 (Windeyer J).

[64] (1873) LR 2 Sc & Div 273 at 284 (Lord Cairns).

> ... liberty given to a particular individual, for a specific length of time, to go into and under the land, and to get certain things there if he can find them, and to take them away, just as if he had bought so much of the soil.[65]

The statutory origin of the mining lease means that, despite connecting to the radical title of the Crown for the purposes of an acquisition of property under s 51(xxxi) of the *Constitution*,[66] the title does not actually generate an estate in land and, whilst ownership in the minerals is conferred pursuant to the lease, this entitlement is subject to a range of statutory obligations. In *ChongHerr Investments Ltd v Titan Sandstone Pty Ltd*[67], Keane JA explained this:

> Importantly, the grant of a mining lease does not create an estate or interest in the land. The effect of the mining lease is to vest ownership of any minerals and property brought onto the land in the lessee. The lessee is obliged to pay compensation to the owner of the land and to pay rental, royalties, rates and security deposits. If the lessee fails to comply with these conditions the Minister may cancel the lease. A lessee must not sublease the mining lease without the written consent of the Minister.

This means that the mining lease does not attract common law principles and therefore is not, as Leeming JA stated in *Valuer-General v Perilya Broken Hill Ltd*,[68] 'a lease in any conventional sense'.

Historically, mining statutes have primarily focused upon the specific issues relevant to the development of the mine and given scant attention to the impact of mining rights on broader land uses and the surrounding communities.[69] Modern statutes have developed differently, and the courts have ensured that the interpretation of the provisions take into account not only the economic imperatives of mining, but the social imperatives of protecting private land rights.

One of the core manifestations of this lies in the reiteration of the fundamental principle of statutory construction that it should not be assumed, in the absence of clear and express provisions, that parliament has derogated from the common law rights held by citizens. As outlined by Barwick CJ in *Wade v New South Wales Rutile Mining Co Pty Ltd*:

---

65 See also *Newcrest Mining (WA) Ltd v Commonwealth* (1997) 190 CLR 513, 616 (Gummow J).
66 See Part 2.3.5 above.
67 (2007) Q ConvR 54–669 at [42] (Keane JA).
68 [2013] NSWCA 265, [27].
69 See especially the discussion by the High Court in *Stow v Mineral Holdings (Aust) Pty Ltd* (1977) 180 CLR 295 where the Court examines the legislative history of mining legislation and its focus upon the economic imperatives of mining development. See also *Re Minister for Resources; ex parte Cazaly Iron Pty Ltd* (2007) 34 WAR 403, [70] (Buss JA) who stated: 'The primary object of the [*Mining Act 1978* (WA)] is to encourage and promote the prospecting and exploration for, and mining of, mineral deposits in the State'.

> ... the fundamental principle that if Parliament intends to derogate from the common law right of the citizen it should make its law in the respect plain is pertinent to the question whether any such implication should be sought to be made. The courts are not entitled, and ought not, to eke out a derogation of such private rights by implications not rendered necessary by the words used by Parliament but merely considered to be consistent with the policy which the courts conclude or suppose Parliament to have intended to implement.[70]

The importance of balancing these issues was carefully examined by the New South Wales Court of Appeal in *Ulan Coal Mines v Minister for Mineral Resources* where the reciprocal entitlements of a private landholder and a mining proponent were considered.[71] The issue in this case was whether or not the Minister has the capacity to issue a mining lease over land coming within categories not authorised under the *Mining Act 1992* (NSW) when a landholder objection to such an issuance has not been received within the prescribed time frame.[72]

Section 62(1)(c) of the *Mining Act 1992* (NSW) confers a right on a landholder not to have a mining lease granted over the surface of any land on which is situated an improvement (as defined by that Act), except with his or her written consent.

Division 4 of sch 1 of the *Mining Act 1992* (NSW) provides a mechanism for a landholder, whose land is the subject of an application for a mining lease, to make a claim to the Minister that something on the land is a valuable work or structure. This may include, for example, important agricultural infrastructure or valuable historical buildings. The mechanism is triggered by the service on the landholder of a notice by the mining lease applicant.[73] The notice is required to state that claims with respect to valuable works or structures on the land must be made to the Minister within 28 days after service of the notice. In the event a claim is made and the applicant for the mining lease objects to the claim, the Director-General of the Department is to refer the objection to a warden for inquiry and report.[74] Anything identified in the claim is taken to be a valuable work or structure unless, as the result of the warden's inquiry, it is declared not to be.

On the facts of the *Ulan Coal Mines* case, the landholder was required to lodge written notice of objection to the issuance of a mining title over lands upon which there were structural improvements; however, this had not occurred within the prescribed time frame. In determining the validity of the lease, the Court examined the

---

70 (1969) 121 CLR 177, 181.
71 [2007] NSWSC 1299 ('*Ulan Coal Mines*').
72 [2008] NSWCA 174.
73 *Mining Act 1992* (NSW) cl 21.
74 Ibid cl 23B.

character of statutory mining leases and considered the objectives of the *Mining Act 1992* (NSW). It was held that the legislation seeks to balance two competing public interest policies: mining enterprises and the protection of private property.[75]

Hodgson, Bell and Tobias JJA concluded that the legislation did not permit the issuance of a mining lease over lands upon which structural improvements existed and the fact that the landholder did not lodge a written objection to this within the prescribed time frame did not mean that the lease was valid. Their Honours noted that a mining title is a statutory creation and is therefore bound by the provisions of the legislation. In interpreting the effect of s 62(1) of the *Mining Act 1992* (NSW), their Honours found that it was reasonable to expect that the applicant for a mining lease would ascertain what is on the land and disclose the existence of any works or structures in the lease application. Their Honours further found that in a case in which the applicant does not have the landholder's written consent to the grant of a lease over the surface of land on which are works or structures which the applicant contends are not substantial and valuable improvements (and the landholder has not made a claim under cl 23A), it is open to the applicant to invite the Minister to refer the question to the warden pursuant to s 62(6) of the *Mining Act 1992* (NSW)

2.5.2 General terms and conditions of a mining lease

In all states and territories other than the ACT, a mining lease may be issued subject to terms and conditions and may also be subject to further approvals being obtained. For example, in New South Wales, s 237 of the *Mining Act 1992* (NSW) makes it clear that a mining lease may only be granted where account of the need to conserve and protect flora, fauna and features of Aboriginal architectural, archaeological, historical or geological interest have been given. In Victoria, s 2A of the *Mineral Resources (Sustainable Development) Act 1990* (Vic) requires regard to be given, in the issuance of a mining licence, to the principles of sustainable development. In the Northern Territory, s 3(a) of the *Mining Management Act 2001* (NT) requires the issuance of mining leases to be carried out in accordance with best practice environmental standards. This effectively means that all leases must be cognisant of environmental and health risks and should adopt appropriate review mechanisms to ensure that these standards are incorporated. The environmental review of mining leases is examined in detail in Chapter 9.

---

75 See *Ulan Coal Mines Ltd v Minister for Mineral Resources* [2007] NSWSC 1299, [46g].

In Western Australia, s 71 of the *Mining Act 1978* (WA) sets out that the Minister may grant a mining lease on such terms and conditions as the Minister considers reasonable. Section 82 states that every mining lease must contain core statutory covenants and conditions including an obligation to pay rents and royalties, an obligation to use the land for the purposes specified in the mining licence, an obligation to comply with prescribed expenditure conditions, and an obligation not to use disturbing equipment unless this has been approved in a work program.

In Queensland, s 276 of the *Mineral Resources Act 1989* (Qld) sets out that a mining lease may be granted provided particular conditions are complied with. These include, inter alia, that the title-holder must use the area for the purposes set out in the mining lease; the holder must carry out improvement restoration for the mining lease; prior to the termination of the mining lease all buildings and structures created for the purpose of mining must be removed; all reports, returns, documents and statements must be furnished; and all rental and royalty payments must be made. Further, s 276(1A) sets out that the Minister may determine a condition of a mining lease if the Minister considers the condition to be in the public interest. Where conditions in a lease are breached, the right to surrender or repudiate the lease is regulated by legislation. For example, a breach of an environmental condition will give the relevant Minister the discretion to determine the appropriate sanction which, in some rare instances, may include termination.[76]

2.5.3 General entitlements of a mining lease

As outlined in Part 2.5.2, the entitlements conferred upon the holder of a mining lease are all completely derived from statute. Generally, given that the object of a mining lease is for the purpose of extracting and commercially producing minerals, the statutory rights that accompany the issuance of a mining licence will complement this. Hence, in Queensland, s 235 of the *Mineral Resources Act 1989* (Qld) makes it clear that the holder of a mining lease has a right of access and may enter the area of the mining lease and do all things permitted and required under the lease or in accordance with the statutory provisions of the act. Section 235 is extracted below:

Mineral Resources Act 1989 (Qld)

235 General entitlements of holder of mining lease

(1) Subject to section 236 and chapter 8, part 8, division 1, during the currency of a mining lease, the holder of the mining lease and any person who acts as agent or employee of

---

[76] See *Mineral Resources Act 1989* (Qld) s 308(1)(d) which authorises the Minister to cancel a mining lease for a failure to comply with a condition.

the holder (or who delivers goods or substances or provides services to the holder) for a purpose or right for which the mining lease is granted –
- (a) may enter and be –
  - (i) within the area of the mining lease; and
  - (ii) upon the surface area comprised in the mining lease;
  for any purpose for which the mining lease is granted or for any purpose permitted or required under the lease or by this Act;
- (b) may do all such things as are permitted or required under the lease or by this Act.
(2) During the currency of the mining lease, the rights of the holder relate, and are taken to have always related, to the whole of the land and surface area mentioned in subsection (1).
(3) Where any Act provides that water may be diverted or appropriated only under authority granted under that Act, the holder of a mining lease shall not divert or appropriate water unless the holder holds that authority.

A similar provision exists in the *Petroleum and Gas (Production and Safety) Act 2004* (Qld) s 109 which entitles a petroleum leaseholder to carry out exploration, production and storage activities consistent with the activities authorised within the lease. Section 109 is extracted below:

Petroleum and Gas (Production and Safety) Act 2004 (Qld)

109 Exploration, production and storage activities
(1) The lease holder may carry out the following activities in the area of the lease –
- (a) exploring for petroleum;
- (b) subject to section 152 –
  - (i) testing for petroleum production; and
  - (ii) evaluating the feasibility of petroleum production; and
  - (iii) testing natural underground reservoirs for storage of petroleum or a prescribed storage gas;
- (c) petroleum production;
- (d) evaluating, developing and using natural underground reservoirs for petroleum storage or to store prescribed storage gases, including, for example, to store petroleum or prescribed storage gases for others.
- (e) plugging and abandoning, or otherwise remediating, a bore or well the lease holder reasonably believes is a legacy borehole and rehabilitating the surrounding area in compliance with the requirements prescribed under a regulation.
(2) However, the holder must not carry out any of the following –
- (a) extraction or production of a gasification or retorting product from coal or oil shale by a chemical or thermal process;
- (b) exploration for coal or oil shale to carry out extraction or production mentioned in paragraph (a);
- (c) GHG stream storage.

(3) The rights under subsection (1) may be exercised only by or for the holder.
(4) The right to store petroleum or prescribed storage gases for others is subject to part 6.

The *Petroleum and Gas (Production and Safety) Act 2004* (Qld) also confers upon leaseholders any incidental rights reasonably necessary or incidental to the authorised activity of the lease. This is set out in s 112, which is extracted below:

Petroleum and Gas (Production and Safety) Act 2004 (Qld)

112 Incidental activities

(1) The lease holder may carry out an activity (an incidental activity) in the area of the lease if carrying out the activity is reasonably necessary for, or incidental to –
   (a) another authorised activity for the lease; or
   (b) an authorised activity for another petroleum lease or an authority to prospect.

   Examples of incidental activities–
   1   constructing or operating plant or works, including, for example, communication systems, compressors, power lines, pumping stations, reservoirs, roads, evaporation or storage ponds and tanks
   2   constructing or using temporary structures or structures of an industrial or technical nature, including, for example, mobile and temporary camps
   3   removing vegetation for, or for the safety of, exploration or testing under section 152(1)

   Note – See also part 10, section 239, chapter 5 and section 20(2).
(2) However, constructing or using a structure, other than a temporary structure, for office or residential accommodation is not an incidental activity.

Similar authorising provisions exist in other states. In New South Wales, s 73 of the *Mining Act 1992* (NSW) sets out that the holder of a mining lease may carry out any mining purpose consistent with the grant including a right to access the area subject to the mining licence, primary treatment operations and other authorised activities. Section 73 is extracted below:

Mining Act 1992 (NSW)

73 Rights under mining lease

(1) The holder of a mining lease granted in respect of a mineral or minerals may, in accordance with the conditions of the lease:
   (a) prospect on the land specified in the lease for, and mine on that land, the mineral or minerals so specified, and
   (b) carry out on that land such primary treatment operations (such as crushing, sizing, grading, washing and leaching) as are necessary to separate the mineral or minerals from the material from which they are recovered, and
   (c) carry out on that land any mining purpose.

(1A) The holder of a mining lease granted in respect of a mining purpose or mining purposes only may, in accordance with the conditions of the lease, carry out the mining purpose or mining purposes specified in the lease.

(2) While a mining lease is in force, the holder of the lease and any person acting as agent or employee of the holder, or delivering goods or providing services to the holder, for the purpose of a requirement of or an activity authorised by the lease may:

(a) for that purpose enter and be on the mining area, and

(b) do anything so authorised or required.

(9) In this section:

"**mining area**" includes, in relation to a lease that does not include the surface of land, any part of the surface of land on which the holder of the lease is authorised, in accordance with section 81, to carry out activities.

Similar provisions exist in s 53 of the *Petroleum (Offshore) Act 1982* (Vic) which confers rights to explore for and recover petroleum in the licence area as well as s 41 of the *Petroleum (Onshore) Act 1991* (Vic) which confers exclusive rights to conduct petroleum mining operations on the land included in the lease together with the right to construct and maintain works and equipment necessary for mining operations. Both sections are extracted below:

Petroleum (Offshore) Act 1982 (Vic)

53 Rights conferred by licence

A licence, while it remains in force, authorises the licensee, subject to this Act and the regulations and in accordance with the conditions to which the licence is subject:

(a) to recover petroleum in the licence area and to recover petroleum from the licence area in another area to which the licensee has lawful access for that purpose,

(b) to explore for petroleum in the licence area, and

(c) to carry on such operations and execute such works in the licence area as are necessary for those purposes.

Petroleum (Onshore) Act 1991 (Vic)

41 Rights of holders of production leases

The holder of a production lease has the exclusive right to conduct petroleum mining operations in and on the land included in the lease together with the right to construct and maintain on the land such works, buildings, plant, waterways, roads, pipelines, dams, reservoirs, tanks, pumping stations, tramways, railways, telephone lines, electric power lines and other structures and equipment as are necessary for the full enjoyment of the lease or to fulfill the lessee's obligations under it.

In Western Australia, s 85 of the *Mining Act 1978* (WA) confers rights to work and mine the land which is the subject of the mining lease, to take and remove minerals, and to do all acts necessary to effectually carry out mining operations in or under the land. Section 85(3) makes it clear that these entitlements are exclusive statutory rights conferred specifically for mining purposes over the land, which is the subject of the mining lease. Section 85 is extracted below:

Mining Act 1978 (WA)

85 Rights of holder of mining lease

(1) Subject to this Act and to any conditions to which the mining lease is subject, a mining lease authorises the lessee thereof and his agents and employees on his behalf to –
   (a) work and mine the land in respect of which the lease was granted for any minerals; and
   (b) take and remove from the land any minerals and dispose of them; and
   (c) take and divert subject to the *Rights in Water and Irrigation Act 1914*, or any Act amending or replacing the relevant provisions of that Act, water from any natural spring, lake, pool or stream situate in or flowing through such land or from any excavation previously made and used for mining purposes, and subject to that Act to sink a well or bore on such land and take water therefrom and to use the water so taken for his domestic purposes and for any purpose in connection with mining for minerals on the land; and
   (d) do all acts and things that are necessary to effectually carry out mining operations in, on or under the land.
(2) Subject to this Act and to any conditions to which the mining lease is subject, the lessee of a mining lease –
   (a) is entitled to use, occupy, and enjoy the land in respect of which the mining lease was granted for mining purposes; and
   (b) owns all minerals lawfully mined from the land under the mining lease.
(3) The rights conferred by this section are exclusive rights for mining purposes in relation to the land in respect of which the mining lease was granted.

Similar provisions exist in the *Petroleum and Geothermal Energy Resources Act 1967* (WA) for offshore petroleum licences. Section 62 confers rights to explore for and recover petroleum from the licence area and to carry on such operations and execute such works necessary for those purposes. Section 62 is extracted below:

Petroleum and Geothermal Energy Resources Act 1967 (WA)

62 Rights conferred by licence

(1) A petroleum production licence, while it remains in force, authorises the licensee, subject to this Act and in accordance with the conditions to which the licence is subject –

- (a) to recover petroleum in the licence area and to recover petroleum from the licence area in another area to which he has lawful access for that purpose; and
- (b) to explore for petroleum in the licence area; and
- (c) to carry on such operations and execute such works in the licence area as are necessary for those purposes.

(2) A geothermal production licence, while it remains in force, authorises the licensee, subject to this Act and in accordance with the conditions to which the licence is subject –

- (a) to recover geothermal energy in the licence area and to recover geothermal energy from the licence area in another area to which the licensee has lawful access for that purpose; and
- (b) to explore for geothermal energy resources in the licence area; and
- (c) to carry on such operations and execute such works in the licence area as are necessary for those purposes.

The title-holder may only exercise the statutory rights which have been expressly conferred by the mining lease provisions. In this sense, the rights are exclusive because other persons not holding a mining lease are unable to exercise the same rights. However, the conferral of these statutory rights does not mean that they are exclusive of all rights. As outlined by Bennett J in *Brown v Western Australia (No 2)*: 'A mining lease, of its nature, grants a right to exclude other miners from exercising mining rights but does not necessarily entail a right to exclude all others'.[77]

Professor Crommelin has carefully summarised the core method of ascertaining the legal character of the resource title:

> In essence, the task of characterization is no more than one of statutory interpretation. The difficulty, however, lies in the fact that legislatures do not usually provide much assistance in the performance of this task. Seldom is there any express statement upon the legal character of exploration and production titles. Instead, answers must be derived by inference, a familiar but nevertheless speculative aspect of legal analysis. Context is important to characterization, as it is to other instances of statutory interpretation. Dealings with titles give rise to the largest number of characterization matters, mainly because of deficiencies in the statutory provisions relating to dealings. These provisions typically impose restraints upon dealings, such as a requirement of Ministerial consent. They also establish registration systems for titles and interests therein. However, those who deal with mineral and

---

[77] (2010) 268 ALR 149, [181].

petroleum titles need to know whether those titles are proprietary in nature, and sometimes also the type of property. These matters will determine the range of dealings, which may occur with titles, and the legal effect of those dealings.[78]

An application for a resource title amounts to an application to create a new interest rather than a determination of any pre-existing legal entitlement. The characterisation of the legal status of this new interest is important for a number of reasons. First, it is important to determine whether the resource title constitutes an interest in land or whether it amounts to personal property and whether the relevant legislation sets this out. Second, if the title is a property interest in land, consideration should be given to the question of whether the title is to be classified by reference to common law estates or whether it should be treated as a new and unique statutory interest, whose entitlements are fully derived from statute. Third, where the statutory entitlement is subject to the vicissitudes of statutory change, a modification or extinguishment made by the Commonwealth Government may attract the just terms protection (s 51(xxxi)) of the *Constitution*. This will depend upon the nature of the lease in issue and whether, in accordance with just terms jurisprudence, the resource title may be characterised as inherently variable. Finally, if the resource title is recognised as purely statutory in nature, consideration needs to be given to the substantive difference between on and offshore exploration titles. The absence of radical title in the offshore zone effectively precludes any institutional characterisation of the resource title as a common law estate.

2.6 REVIEW QUESTIONS

1. Discuss the range of activities companies carry out during the exploration phase of a mining project and consider the type of entitlements that a resource title should authorise to support these activities?
2. How are the physical dimensions of an exploration title determined?
3. Once an exploration title has expired, does the holder have any entitlement to seek renewal?
4. There are a number of clear and distinct stages which differentiate the development of a mining project. Which stage do you think is the most crucial in terms of (i) investment and (ii) viability?

---

78 M Crommelin, 'The Legal Character of Resource Titles' (1998) 17 *Australian Mining and Petroleum Law Journal* 57, 58.

5. Does a mining proponent require a rehabilitation plan in order to acquire an exploration licence? Explain the purpose of a rehabilitation plan and its relevance to the resource title framework.
6. How does a production title differ from an exploration title? Explain in terms of (i) legal status and (ii) practical purpose.
7. GDG Sadi holds an exploration title which entitles the company to explore for gas in Central Queensland. At the time when the title was issued, the Minister imposed a condition stating that during the term of the title, the holder 'must do everything necessary to protect the environment, comply with best practice standards and adhere to specific requirements set out in an attached environmental management plan'. The environmental management plan specifically required annual monitoring of the exploration activities to reduce the risk of environmental degradation. In the second year of a five-year exploration licence, GDG Sadi failed to properly monitor the chemical levels in an adjoining water resource following exploratory drilling and the removal of chemically contaminated water. The heightened chemical contaminants resulted in adjoining farming properties experiencing significant agricultural loss. What rights, if any, does the affected farm owner have to terminate the exploration licence in such circumstances?
8. Why would a change to an offshore exploration licence be unlikely to attract compensation under the *Constitution* in accordance with s 51(xxxi)? Discuss the conclusions and rationales in *Commonwealth v WMC Resources Ltd* (1998) 194 CLR 1.
9. What is the purpose of an assessment or 'retention' licence and in what circumstances would a mining proponent apply for such a licence?
10. Is a production licence the equivalent of a fee simple in land? If, for example, a mining proponent owns land outright, why would it be necessary for the holder to apply, in addition to that ownership, for a production licence?
11. Consider the following situation:

> BHP Petroleum (Bass Strait) Pty Ltd and Esso Australia Resources Pty Ltd each own one-half of the petroleum leases in Bass Strait. These leases have been acquired for the purpose of recovering and producing petroleum and gas from the area covered by the leases. BHP and Esso hold the leases jointly. Pursuant to an operating agreement entered into between BHP and Esso, Esso has constructed gas processing facilities and a crude oil stabilisation plant at Longford in Victoria and has also established processing and storage facilities at Long Island Point in Victoria. Gas liquids and stabilised crude oil are transferred by a pipeline from the Longford facilities to the Long Island Point facilities. BHP wishes to terminate its arrangement with Esso and requests a transfer of the joint petroleum leases in the Bass Strait from Esso. Esso refused to negotiate any such transfer and argued that a petroleum lease confers exclusive possession upon both lessees for the duration of time that they exist. Esso also argues that the operational facilities that have been constructed in the lease area are included within the scope of the petroleum lease. BHP seek your advice.

2.7 FURTHER READING

Australian Government, Department of Industry, Tourism and Resources, *Mine Rehabilitation: Leading Practice Sustainable Development Program for the Mining Industry* (October 2006) http://www.industry.gov.au/resource/Documents/LPSDP/LPSDP-MineRehabilitationHandbook.pdf.

A J Bradbrook, 'The Contents of the New Geothermal Legislation' (1987) 5 *Journal of Energy and Natural Resources Law* 81.

S Brennan, 'Section 51(xxxi) and the Acquisition of Property Under Commonwealth-State Arrangements: The Relevance to Native Title Extinguishment on Just Terms' (2011) 15(2) *University of New South Wales Law Review* 74.

M Crommelin 'Mineral Exploration in Australia and Western Canada' (1974) 9 *University of British Columbia Law Review* 38.

M Crommelin, 'Economic Analysis of Property' in D J Galligan (ed), *Essays in Legal Theory* (Melbourne University Press, 1984).

M Crommelin, Mining and Petroleum Titles (1988) 62 *Australian Law Journal* 863.

M Crommelin, 'The Legal Character of Resource Titles' (1998) 17 *Australian Mining and Petroleum Law Journal* 57.

L Fraser, 'Property Rights in Environmental Management: The Nature of Resource Consents in the Resource Management Act of 1991' (2008) 12 *New Zealand Journal of Environmental Law* 145.

A Gardner, 'Dealings with Mining Titles Under the Mining Act 1978 (WA): Part 1 – Requirements of Form, Consent and Registration' (2005) 24 *Australian Resources and Energy Law Journal* 342.

A Gardner, 'Dealings with Mining Titles Under the Mining Act 1978 (WA): Part 2 – The Effect of Registration and Caveats' (2006) 25 *Australian Resources and Energy Law Journal* 41.

T Glover, 'Defects in Marking out Applications for Mining Titles: A Comparison of the Requisite Levels of Compliance under the General Law and Mineral Resources Act 1989 (Qld)' (2000) 19 *Australian Mining and Petroleum Law Journal* 230.

Government of Western Australia, Department of Mines and Petroleum, *Mineral Titles Online* (10 August 2010) http://www.dmp.wa.gov.au/3968.aspx.

Government of Western Australia, Department of Mines and Petroleum, *Exploration Licence Graticular Boundary System* (2013) http://www.dmp.wa.gov.au/documents/132289_Graticular_Boundary_System.pdf.

F Jahn, M Cook and M Graham, *Hydrocarbon Exploration and Production* (Elsevier, 2008).

A B Klass, 'Climate Change and the Convergence of Environmental and Energy Law' (2013) 24 *Fordham Environmental Law Review* 180.

P O'Connor, 'The Changing Paradigm of Property and the Framing of Regulation as a Taking' (2010) 36(2) *Monash University Law Review* 50.

A J Wildermuth, 'The Next Step: The Integration of Energy Law and Environmental Law' (2011) 31 *Utah Environmental Law Review* 369.

H Williams, 'Comments on Oil and Gas Jurisprudence in Canada and the United States' (1965) 4 *Alberta Law Review* 189.

AUSTRALIAN OFFSHORE PETROLEUM AND MINERALS REGULATION

3.1	Introduction	95
3.2	Constitutional arrangements for offshore regulation	97
3.3	Review questions	131
3.4	Further reading	132

3.1 Introduction

The earth's oceans are expansive and cover over 70 per cent of the planet. This makes the strong commercial interest in offshore drilling for petroleum and gas and, increasingly, offshore mining for mineral resources, unsurprising. The offshore regulation for petroleum, gas mining, and carbon capture storage has, however, a fundamentally different jurisdictional and policy focus to onshore regulation given its location. Offshore drilling primarily concentrates upon the extraction and production of petroleum and gas resources through advanced drilling techniques that take account of the underwater environment. Offshore drilling involves drilling a wellbore through the seabed in order to explore for and extract the petroleum that resides within rock formations that exist beneath the seabed. Drilling activities commonly occur on the continental shelf, although projects may also be developed over inland seas. Offshore drilling operations may be conducted from drilling rigs, floating platforms, or deepwater mobile offshore drilling units (MODU) which have the capacity to operate in water depths of up to 3000 metres.[1]

Offshore drilling allows mining proponents to access potentially vast reserves of petroleum. There are strong political and economic imperatives associated with petroleum production within modern society. Technological developments have enhanced the capacity of mining proponents to recover oil at greater depths. For example, in 1975, Shell for the first time made a huge discovery: an estimated 100 million-barrel reserve in 305 m of water in an area in the Gulf of the Mississippi Canyon that reached just beyond the edge of the continental shelf. This momentous find came to be referred to as the 'flex trend' which was the modern definition of 'deepwater'.[2]

Offshore drilling in Australia has significant presence in the Bass Strait region. Esso/BHP drilled the first offshore well in the Bass Strait region off Victoria's Gippsland coast in 1965 which was known as the Barracouta gas field. Subsequently in 1967, the Kingfish offshore oilfield was discovered and this oilfield remains the largest ever discovered in Australia. Today, there are 23 offshore platforms and installations in the Bass Strait, which feed a network of 600 km of underwater pipelines.[3] As a component of a new $4.5 billion project, Esso Australia is drilling

---

1 For an overview of offshore drilling processes see the discussion by National Commission on the BP Horizon Oil Spill and Offshore Drilling, 'Chapter 2: The History of Offshore Oil and Gas' in *Deepwater: The Gulf Oil Disaster and the Future of Offshore Drilling* (2011) http://www.gpo.gov/fdsys/pkg/GPO-OILCOMMISSION/pdf/GPO-OILCOMMISSION.pdf.
2 Ibid.
3 See ExxonMobil Australia, *About us: Bass Strait* http://www.exxonmobil.com.au/Australia-English/PA/about_what_gipps_bs.aspx.

five new wells at the Kipper Tuna Turrum field which holds an estimated 1 trillion cubic feet of natural gas and 110 million barrels of oil and gas.[4]

It is crucial that the social and economic gains associated with offshore drilling are offset by appropriate legal mechanisms that minimise harm to the marine environment. Offshore drilling has strong environmental implications given the fragile ecologies that exist within the marine environment and the potential devastation that may occur through oil spills. The potential environmental hazards associated with conventional offshore drilling include underwater explosions, and contamination from drilling projects with toxic chemicals and byproduct waste. For example, the Deep Water Horizon oil spill which occurred on 20 April 2010 in the Gulf of Mexico resulted in 11 deaths and is considered to be the largest accidental marine oil spill in the history of the petroleum industry. The month-long spill caused extensive damage to marine and wildlife habitats and was found to have been the result of a methane bubble that escaped from a well into the drill column and leaked through a faulty cement well before igniting a 'blowout'.[5] The subsequent national commission report found that deepwater 'energy exploration and production, particularly at the frontiers of experience, involve risks for which neither industry nor government has been adequately prepared, but for which they can and must be prepared in the future'.[6] Ensuring that mining companies are adequately prepared has necessitated strong regulatory oversight of energy exploration, leasing and production in the offshore region in order to ensure that technical expertise and political autonomy are appropriately balanced with issues relevant to environmental protection. The National Commission Report on the Deepwater Horizon Oil Spill found that it is particularly important to ensure that the oil and gas industry implement their own, unilateral processes to dramatically increase safety and self-policing mechanisms.[7]

In addition to offshore petroleum and gas drilling, offshore mining for mineral resources is developing increasing momentum.[8] A vast array of different minerals

---

[4] See the discussion at: http://www.exxonmobil.com.au/Australia-English/PA/news_releases_20141216.aspx.

[5] National Commission on the BP Horizon Oil Spill and Offshore Drilling, *Deepwater: The Gulf Oil Disaster and the Future of Offshore Drilling* (2011) 103 http://www.gpo.gov/fdsys/pkg/GPO-OILCOMMISSION/pdf/GPO-OILCOMMISSION.pdf. BP has awarded over US$4 billion in claims from those who suffered economically but is also subject to thousands of individual lawsuits. See the discussion by J E Darling, 'Addendum: The Story of the BP Oil Spill (2012) 3 *Elon. Law Rev.* 254, 257–9.

[6] National Commission on the BP Horizon Oil Spill and Offshore Drilling, above n 5, 7.

[7] National Commission on the BP Horizon Oil Spill and Offshore Drilling, above n 5, 10. For a further discussion of this see K M Murchison, 'Beyond Compensation for Offshore Drilling Accidents: Lowering Risks, Improving Response' (2012) 30 *Mississippi College Law Review* 277, 282–9.

[8] See especially the company Neptune Minerals, which holds a number of exploration licences in New Zealand, Vanuatu, and the Federal States of Micronesia for seafloor minerals and resource development and has conducted several exploration programs for mineral resources

may reside in the seabed and it is possible that such minerals could be extracted and mined in the future.[9] The offshore zone was only realised as a potentially fertile region for mineral exploration during the 1960s when plate tectonic research 'posited that ocean basins are not merely passive collections of minerals eroded from land, but rather active sources of volcanic mineralization'.[10] Since this period, offshore mining has rapidly evolved. Some of the most promising offshore mineral reserves exist within the high seas region, and these areas are regulated according to the international framework established by the *United Nations Convention on the Law of the Sea* ('*UNCLOS*'). *UNCLOS* was signed in 1982 and came into effect in 1994 and treats the international high seas waters as the 'common heritage of mankind'.[11] The constitutional and jurisdictional framework for offshore drilling and mining is discussed further below.

3.2 Constitutional arrangements for offshore regulation

The offshore laws for petroleum and gas mining and greenhouse gas storage are regulated by complementary state and federal legislation. This reciprocal regulatory regime was first established by the *Offshore Petroleum Agreement 1967* (Cth), which set out that the Commonwealth and the states would each introduce complementary (mirror) legislation with the aim of establishing a regime for offshore petroleum exploration and for the sharing of royalties. The *Offshore Petroleum Agreement* also required the Commonwealth, the states, and the territories to agree on a Common Mining Code. The *Offshore Petroleum Agreement of 1967* was subsequently encapsulated in the later *Offshore Constitutional Settlement 1979* (Cth), which is now enshrined in s 5 of the *Offshore Petroleum and Greenhouse Gas Storage Act 2006* (Cth) ('the *OPGGSA*'). Section 5 of the *OPGGSA* is extracted below:

---

off the coast of Japan, Palau and Italy. See Neptune Minerals, *Seafloor Minerals Exploration and Resource Development* http://www.neptuneminerals.com/.

9 The likelihood of copper and rare earth minerals residing in the seabed makes the development prospects attractive given the global demand for these resources. For example, global copper demand has risen over 35 per cent in the last 10 years. It rose from US$1843.85 per metric tonne in January 2009 to US$8053.74 in January 2013. The reason for this appears to be its utility given that it is used, inter alia, for electrical cabling, piping and values. See Index Mundi, *Commodity Prices* http://www.indexmundi.com/commodities/?conunodity=copper&months=300.

10 See P A Rona, 'Resources of the Sea Floor' (2003) 299 *Science* 673, 673.

11 *United Nations Convention on the Law of the Sea*, opened for signature 10 December 1982, 1833 UNTS 397 (entered into force 16 November 1994), Preamble, para 7, art 136.

Offshore Petroleum and Greenhouse Gas Storage Act 2006 (Cth)

5 Commonwealth-State agreement (the Offshore Constitutional Settlement)

(1) This section explains the agreement known as the Offshore Constitutional Settlement, to the extent to which that agreement relates to exploring for, and exploiting, petroleum.
(2) The Commonwealth, the States and the Northern Territory have agreed that:
   (a) Commonwealth offshore petroleum legislation should be limited to the area that is outside the coastal waters of the States and the Northern Territory; and
   (b) for this purpose, the outer limits of State and Northern Territory coastal waters should start 3 nautical miles from the baseline of the territorial sea; and
   (c) the States and the Northern Territory should share, in the manner provided by this Act in the administration of the Commonwealth offshore petroleum legislation; and
   (d) State and Northern Territory offshore petroleum legislation should apply to State and Northern Territory coastal waters; and
   (e) the Commonwealth, the States and the Northern Territory should try to maintain, as far as practicable, common principles, rules and practices in regulating and controlling the exploration for, and exploitation of, offshore petroleum beyond the baseline of Australia's territorial sea.

The implementation of the *Offshore Petroleum Agreement 1967* (Cth) has its genesis in the international conventions that preceded it and the importance of ensuring that Australia, as a signatory, complied with its international responsibilities regarding the regulation of pollution from seabed activities. It was also crucial that, within a federal framework, the regulatory regime be codified under one single regime rather than a variety of state regimes.[12]

The first Act to establish the jurisdictional framework was the *Petroleum (Submerged Lands) Act 1967* (Cth) which conferred jurisdiction upon the Commonwealth over natural resources existing beyond the three-mile limit of the territorial sea and which extended out to the 12-mile limit. The states and the Northern Territory all had jurisdiction within the three-mile limit and were required to share in the administration of petroleum resources. All parties were required to develop common principles, rules, and practices for the regulation of offshore resources within their three-nautical mile territorial limits. The *Petroleum (Submerged Lands) Act 1967* (Cth) defined the area within the first three miles of the territorial sea as the 'coastal area' and the area beyond this as the 'adjacent area'.

---

12 See the discussion in M White, 'Australia's Offshore Legal Jurisdiction: History, Development and Current Situation' (2011) 25(1) *Australia and New Zealand Maritime Law Journal* 1. See also R Cullen, *Federalism in Action: The Australian and Canadian Offshore Disputes* (Federation Press, 1990) chs 3 and 4; M White and N Gaskell, 'Australia's Offshore Constitutional Law: Time for Revision' (2011) 85 *Australian Law Journal* 504.

Prior to the passing of *UNCLOS* in 1982, and in order to give effect to the conventions that were already operative, Australia passed the *Seas and Submerged Lands Act 1973* (Cth). This Act claimed Commonwealth sovereignty from the low-water mark and challenged the traditional assumption of the states that they had sovereignty in the three-mile territorial area.

Subsequently, in the High Court case of *New South Wales v Commonwealth*[13] the Commonwealth claims for sovereignty and jurisdiction in the offshore area from the low-water mark were upheld. In reaching this decision, the majority concluded that the fundamental features of a federation lie in the power to protect and control areas of the marginal sea, seabed, and the continental shelf. Barwick CJ made the following comments at [53]–[56]:

> A consequence of creation of the Commonwealth under the Constitution and the grant of the power with respect to external affairs was, in my opinion, to vest in the Commonwealth any proprietary rights and legislative power which the colonies might have had in or in relation to the territorial sea, seabed and airspace and continental shelf and incline. Proprietary rights and legislative powers in these matters of international concern would then coalesce and unite in the nation. That, in my opinion, was the intendment of the Constitution. It is far easier to conclude that the Act of the Imperial Parliament setting up the federal Constitution intended to vest such matters of international consequence in the new Commonwealth, withdrawing them from the former colonies, than it was to decide that when an American State, already an independent nation in possession of international rights, entered the Union, these rights became vested in the United States. Yet that is received doctrine in the United States expressed in decisions which have recently been affirmed... The Supreme Court's reasons were applicable to the circumstances of the States originally entering the Union. These were then independent nation states. Yet without so clear an indication as the grant of the power with respect to external affairs, those States did not retain any rights or legislative power over the territorial sea, subsoil, etc. Later entrants to the Union, coming in on an "equal footing", were in the same situation. This result conforms, in my opinion, to an essential feature of a federation, namely, that it is the nation and not the integers of the federation which must have the power to protect and control as a national function the area of the marginal seas, the seabed and airspace and the continental shelf and incline. This has been decided by the Supreme Courts of the United States and of Canada: see above citations and Reference re Ownership of Offshore Mineral Rights [1966] INSC 207; (1967) SCR 792; (1967) 65 DLR (2d) 353. I am satisfied with the reasons given by those Courts for their conclusions. The Canadian Supreme Court reached its conclusion after a close

---

13 (1975) 135 CLR 337 ('*Seas and Submerged Lands Act Case*').

examination of the case law. I do not disagree with anything that is said in the Supreme Court's judgment about that law, though for my part I have found it unnecessary to deal with it in these my reasons. However, the Supreme Court's conclusion depends in no small degree upon the fact of Canada's independent nationhood and its recognition as such by the nations of the world. Appropriately, it is concluded that such international rights and obligations as derive from the Convention on the Territorial Sea devolve on Canada and not on any province of the federation. I can find no reason to differentiate in relevant respects the circumstances of this federation from those of the other great federations, except to say that the result of the cases to which I have referred more obviously flows in the case of our Constitution. It is my opinion, therefore, that upon the enactment of the Constitution, any rights or powers which the former colonies might have had in the territorial sea, seabed and airspace or in the continental shelf and incline became vested in the Commonwealth. The emergence of Australia as an independent nation state confirmed this situation. Therefore, however the matter be viewed, in my opinion, the Act is valid in enacting the sovereignty and sovereign rights in respect of the territorial sea and contiguous zone and of the continental shelf for which the respective Conventions provide. That sovereignty and those sovereign rights are exercisable in and in respect of the territorial sea and the continental shelf. The Act, in my opinion, validly vests that sovereignty and these sovereign rights in the Crown in right of the Commonwealth but any Act or law operating within Australia to implement either of those Conventions or the powers they give must be itself a valid law of the Commonwealth. But if there is such a law, it may operate on matters and things which otherwise could not be the subject of a law of the Parliament.

The Commonwealth subsequently decided, however, to enter into negotiations to return this jurisdiction to the states as they ultimately felt that, despite the superior claim of the Commonwealth, the states were in a better position to regulate local issues existing close to their shores.

This eventually resulted in the implementation of the *Offshore Constitutional Settlement 1979* (Cth), which was given effect in a number of acts passed by the states pursuant to the s 51(xxxviii) constitutional power, which entitled the states to request and consent to Commonwealth legislation.

In response to this, the Commonwealth passed two substantial acts. The first was the *Coastal Waters (State Powers) Act 1980* (Cth). Section 5 of that Act gave the states legislative powers over particular coastal and adjacent waters as if they were within the jurisdiction of the state. The states were given specific power over ports, harbours, shipping facilities, installations, dredging, and powers relating to fisheries beyond the coastal waters as may be agreed with the Commonwealth. Section 5 is extracted below:

Coastal Waters (State Powers) Act 1980 (Cth)

5 Legislative powers of States

The legislative powers exercisable from time to time under the constitution of each State extend to the making of:

(a) all such laws of the State as could be made by virtue of those powers if the coastal waters of the State, as extending from time to time, were within the limits of the State, including laws applying in or in relation to the sea-bed and subsoil beneath, and the airspace above, the coastal waters of the State;

(b) laws of the State having effect in or in relation to waters within the adjacent area in respect of the State but beyond the outer limits of the coastal waters of the State including laws applying in or in relation to the sea-bed and subsoil beneath, and the airspace above, the first-mentioned waters, being laws with respect to:
 (i) subterranean mining from land within the limits of the State; or
 (ii) ports, harbours and other shipping facilities, including installations, and dredging and other works, relating thereto, and other coastal works; and

(c) laws of the State with respect to fisheries in Australian waters beyond the outer limits of the coastal waters of the State, being laws applying to or in relation to those fisheries only to the extent to which those fisheries are, under an arrangement to which the Commonwealth and the State are parties, to be managed in accordance with the laws of the State.

The second Act passed by the Commonwealth was the *Coastal Waters (State Title) Act 1980* (Cth). Pursuant to s 4, this Act conferred on the states the same rights and title to the seabed beneath the coastal waters and the water column and airspace above it, as if those waters, water column and airspace were within the jurisdictional limits of the states. This provision was, however, subject to any pre-existing title in another person as well as the right of the Commonwealth to use the area for communications, safety of navigation, quarantine or defence, or to authorise the construction and use of undersea pipelines and the provisions of the *Great Barrier Reef Marine Park Act 1975* (Cth). This is an interesting provision as it seeks to vest in the state rights and titles to the sea and seabed, not pursuant to any transfer of right, but rather, on the basis of an assumed state of affairs, that being, that sea and sea-bed were beneath state waters. Section 4 is extracted below:

Coastal Waters (State Title) Act 1980 (Cth)

4 Vesting of title in States

(1) By force of this Act, but subject to this Act, there are vested in each State, upon the date of commencement of this Act, the same right and title to the property in the sea-bed beneath the coastal waters of the State, as extending on that date, and the same rights in respect of the space (including space occupied by water) above that sea-bed, as would belong to

the State if that sea-bed were the sea-bed beneath waters of the sea within the limits of the State.

(2) The rights and title vested in a State under subsection (1) are vested subject to:
   (a) any right or title to the property in the sea-bed beneath the coastal waters of the State of any other person (including the Commonwealth) subsisting immediately before the date of commencement of this Act, other than any such right or title of the Commonwealth that may have subsisted by reason only of the sovereignty referred to in the *Seas and Submerged Lands Act 1973*;
   (b) a right of the Commonwealth, or an authority of the Commonwealth authorized by the Commonwealth or by a law of the Commonwealth, to use the sea-bed and space referred to in subsection (1) for purposes in relation to communications, the safety of navigation, quarantine or defence, and to place, construct and maintain equipment and structures for the purposes of such use; and
   (c) a right of the Commonwealth to authorize the construction and use of pipelines for the transport across the sea-bed referred to in subsection (1) of petroleum (including petroleum in gaseous form), recovered, in accordance with a law of the Commonwealth, from any area of the sea-bed beyond the coastal waters of the State.

(3) The rights and title vested by subsection (1) are vested subject to the operation of the *Great Barrier Reef Marine Park Act 1975* and accordingly are so qualified that nothing contained in, or done under, that Act shall be taken to constitute an infringement of, or derogation from, any such right or title.

(4) Where, after the commencement of this Act, a change takes place in the baseline from which the breadth of the territorial sea of Australia is measured:
   (a) if, by reason of the change, the coastal waters of a State extend to an area to which they did not previously extend – subsections (1) and (2) have effect in relation to that area as if the references in those subsections to the date of commencement of this Act were references to the date on which the change occurs; or
   (b) if, by reason of the change, the coastal waters of a State cease to extend to an area to which they previously extended – neither the State, nor any person claiming through the State, continues to have, by virtue of the operation of this Act, any right or title in relation to that area.

(5) It is the intention of the Parliament that, subject to subsections (2) and (3), any right or title vested in a State by this section may be disposed of or otherwise dealt with in accordance with the laws of the State.

(6) In this section the *Great Barrier Reef Marine Park Act* 1975 means that Act as amended from time to time, in its application to any area that is, at the date of commencement of this Act, part of, or capable of being prescribed as part of, the Great Barrier Reef Region as defined in section 3 of that Act.

The states have each subsequently passed mirror legislation to that of the Commonwealth.[14] The Northern Territory, not being a state, is covered by separate Commonwealth legislation.[15]

A similar jurisdictional framework was subsequently confirmed by *UNCLOS* in 1982. As a coastal state, Australia retains jurisdiction over its onshore territories as well as its territorial waters. *UNCLOS* does not alter the constitutional agreements and subsequent legislation that has been implemented. The right to regulate within waters beyond the territorial sea is, however, subject to the international maritime law as well as the laws of the sea.

3.2.1 United Nations Convention on the Law of the Sea

The right of the Commonwealth to exercise sovereignty of power over the territorial sea was described by Windeyer J in *Bonsa v La Macchia*, as 'residing in the imperial Crown' and, when Australia became a nation, this right passed automatically to the Commonwealth.[16] This formal position was statutorily endorsed in 1973 when the Commonwealth asserted jurisdiction over the territorial sea and the continental shelf pursuant to s 6 of the *Seas and Submerged Lands Act 1973* (Cth) ('the *SSLA*').[17]

The validity of the *SSLA* was confirmed by the High Court in *New South Wales v The Commonwealth* where it was held that the Act was valid and that the jurisdiction of the states terminated at the low-water mark.[18] The High Court held that the external affairs power, under s 51(xxix) of the *Constitution*, established federal jurisdiction over the territorial sea and the continental shelf.[19]

---

14 *Coastal Powers (Coastal Waters) Act 1979* (NSW); *Coastal Powers (Coastal Waters) Act 1980* (Qld); *Coastal Powers (Coastal Waters) Act 1979* (SA); *Coastal Powers (Coastal Waters) Act 1979* (Tas); *Coastal Powers (Coastal Waters) Act 1980* (Vic); *Coastal Powers (Coastal Waters) Act 1979* (WA).
15 See *Coastal Waters (Northern Territory Powers) Act 1980* (Cth).
16 (1970) 122 CLR 177, 223. Windeyer J relied upon the earlier English decision of *Regina v Keyn* (1876) 2 Ex D 63.
17 Section 6 of the *Seas and Submerged Lands Act 1973* (Cth) declares that 'the sovereignty in respect of the territorial sea, and in respect of the airspace over it and in respect of its bed and subsoil, is vesting in an exercisable by the Crown in right of the Commonwealth'. The Governor-General can declare the limits of the territorial sea, as long as the declaration is 'not inconsistent with the Territorial Sea Convention'. See also s 11 which states that 'the sovereign rights of Australia as a coastal State in respect of the continental shelf of Australia, for the purpose of exploring it and exploiting its natural resources, are vested in and exercisable by the Crown in right of the Commonwealth'.
18 (1975) 135 CLR 337, 368.
19 (1975) 135 CLR 337, 364.

Significantly, however, whilst the *SSLA* deals with the regulatory power of the Crown over the territorial sea and seabed, it does not confer actual ownership upon the Crown in this area. The Crown retains sovereignty of power (jurisdiction) over the territorial sea and within the zones established pursuant to *UNCLOS*[20], but not sovereignty of title (ownership) because such ownership is inconsistent with the recognition of the right of innocent passage under international law.[21] The ownership rights that the Crown holds in the territorial sea are akin to the concept of radical title which does not amount to a full beneficial ownership. The scope and nature of the rights and interests held by the Crown within the territorial sea were explored in some detail in *Commonwealth v Yarmirr* where the High Court made it clear that:

> It is unnecessary to decide what was the right and title that was vested in the Territory. If it is appropriate to speak of that right and title in the language of the real property lawyer, the right and title thus vested in the Territory was no more than a radical title; it was not full ownership of the sea-bed or space above it. (We need not and do not decide whether it is appropriate to adopt such terms as radical title in this context.) There are several reasons why the right and title that was vested does not amount to full ownership. First, the right and title was vested by an Act of the Parliament which was itself an exercise of the sovereignty which had been asserted by the *Seas and Submerged Lands Act* and earlier by Acts of the Imperial and later the federal executive. It would be inconsistent with the public rights to fish and to navigate that were recognised as qualifying those sovereign rights, for purposes of municipal law, to treat the right and title vested as absolute and unqualified ownership. Further, it would be inconsistent with the international obligations which Australia had undertaken in the Convention on the Territorial Sea and the Contiguous Zone to afford innocent passage to ships of all States through the territorial sea to vest absolute and unqualified ownership in the area in the Territory. Secondly, as the *NT Title Act* makes plain, the right and title which was vested in the Territory was identified as the same right and title the Territory had over the sea-bed beneath waters of the sea within the limits of the Territory. It was not submitted that the right and title to areas of the latter kind was any greater than radical title to land. It is unnecessary to stay to consider whether it is less than a radical title. If the title thus vested is not larger than a radical title, that title is not inconsistent with the continued existence of native title rights and interests. Finally, s 4(2) of the *NT Title Act* expressly made the right and title vested in the Territory subject to "any right

---

20 *United Nations Convention on the Law of the Sea,* opened for signature 10 December 1982, 1833 UNTS 397 (entered into force 16 November 1994).
21 See especially *Commonwealth v Yarmirr* (2001) 208 CLR 1, [57]–[59]. See also *Seas and Submerged Lands Act Case* (1975) 135 CLR 337, 493.

or title to the property in the sea-bed ... of any other person ... subsisting immediately before the date of commencement" of the *NT Title Act*. Any native title rights and interests in relation to the sea-bed that then existed were, therefore, expressly preserved.[22]

3.2.2 Territorial waters

Australia acquired sovereignty over the territorial sea pursuant to s 6 of the *Seas and Submerged Lands Act 1973* (Cth). Sovereignty rights are also encapsulated within the specifically designated zones established in the territorial sea pursuant to *UNCLOS*.[23] The baseline for determining the commencement of the internal or territorial waters of Australia is articulated within arts 4 and 5 of *UNCLOS*, both of which are extracted below:

Article 4 Outer limit of the territorial sea

The outer limit of the territorial sea is the line every point of which is at a distance from the nearest point of the baseline equal to the breadth of the territorial sea.

Article 5 Normal baseline

Except where otherwise provided in this Convention, the normal baseline for measuring the breadth of the territorial sea is the low-water line along the coast as marked on large-scale charts officially recognized by the coastal State.

The normal baseline is the low-water line along the coast.[24] However, because much of the northern coastline of Australia is dotted with atolls and reef islands, in localities where the coastline is deeply indented and cut into, or if there is a fringe of islands along the coast, the method of straight baselines joining appropriate points may be employed when drawing the baseline from which the breadth of the territorial sea is measured.[25]

In particular, pt IV of *UNCLOS* provides for straight baselines to be drawn in archipelagic waters. Archipelagic waters are relevant where states are surrounded by many small islands. For such states, the baselines may be drawn around the

---

22 (2001) 208 CLR 1, [70]–[72] (Gleeson CJ, Gaudron, Gummow and Hayne JJ).
23 *United Nations Convention on the Law of the Sea,* opened for signature 10 December 1982, 1833 UNTS 397 (entered into force 16 November 1994), Preamble, art 136. The extension of Australia's territorial limit from 12 to 200 nautical miles, as provided by the entry into force in Australia on 16 November 1994 of the *United Nations Convention on the Law of the Sea* constitutes a permissible future act for the purposes of s 235(8) of the *Native Title Act 1993* (Cth).
24 *UNCLOS* art 5.
25 *UNCLOS* art 7.

outer edge of the islands as a group rather than individual, separate islands. The baselines may run from the outermost points of the islands and across the intervening waters.

The internal waters of Australia represent the waters on the landward side of the baselines. As mentioned above, Australia has proclaimed its sovereignty over these waters under the *SSLA* and in accordance with *UNCLOS*. Most of the internal waters to the landward side of the baselines lie within the jurisdiction of the states and the Northern Territory.

The territorial sea now runs from the baselines out for 12 nautical miles. The outer limits of the territorial sea were extended to 12 nautical miles from 3 nautical miles in 1990, pursuant to s 7 of the *SSLA*.[26]

3.2.3 State jurisdiction in the territorial sea

As outlined above, the states only retain jurisdiction within a three-mile limit of the territorial sea. Beyond this, the Commonwealth has full jurisdiction. This jurisdictional framework was agreed upon under the *Offshore Constitutional Settlement 1979* (Cth) and was confirmed by the High Court in the *Seas and Submerged Lands Act Case*. The High Court articulated the position as follows:

> ... when a nation state has dependent territories, international law concedes to the nation state the same dominion over an area of the high seas which washes the shores of its territories as it does in relation to waters which wash the territorial margins of the homeland.[27]

The Court went on to note that nation states only retain dominion over sea defined as 'territorial sea', and that 'the very existence of a territorial sea depends on international agreement, established in earlier times by custom or practice amongst nations or a significant number and range of them, but now most definitively by international convention'.[28]

The High Court stated that the appropriate repository of international rights and obligations was the Crown in right of the Commonwealth and this was supported by the *Constitution*. As discussed above, the Commonwealth subsequently re-conferred jurisdiction to the states and territory to a three-mile territorial sea limit.

This framework is now encapsulated in s 5(2)(b) of the *Offshore Petroleum and Greenhouse Gas Storage Act 2006* (Cth) ('the *OPGGSA*') which is extracted

---

26 The right to extend the territorial sea up to 12 nautical miles is recognised in pt 2, art 4 of *UNCLOS*.
27 (1975) 135 CLR 337, [21].
28 *Seas and Submerged Lands Act Case* (1975) 135 CLR 337, [25].

above. Pursuant to the *OPGGSA*, Commonwealth offshore petroleum legislation now applies within the waters beyond the territorial limits of the states and the Northern Territory and also within any extended zones set up pursuant to the terms of *UNCLOS*.[29]

3.2.4 Contiguous zone, exclusive economic zone, and the continental shelf

In the waters beyond the territorial sea, the Commonwealth only retains sovereignty within newly declared areas of jurisdiction that have been set up pursuant to the provisions of *UNCLOS*. The new zones of Commonwealth jurisdiction established by *UNCLOS* within the waters beyond the state and territorial limits are: the contiguous zone, the exclusive economic zone, and the continental shelf.[30]

The contiguous zone is recognised under both div 2A of the *SSLA* and art 33 of *UNCLOS*. Within this zone, a coastal state may prevent and punish any infringement of its customs, fiscal, immigration, or sanitary laws and regulations. The contiguous zone must not extend beyond 24 nautical miles from the baselines from which the breadth of the territorial sea is measured.

Article 33 of *UNCLOS* which defines the contiguous zone and sets out the limits of Commonwealth jurisdiction is extracted below:

Article 33 Contiguous zone

1. In a zone contiguous to its territorial sea, described as the contiguous zone, the coastal State may exercise the control necessary to:
    (a) prevent infringement of its customs, fiscal, immigration or sanitary laws and regulations within its territory or territorial sea;
    (b) punish infringement of the above laws and regulations committed within its territory or territorial sea.
2. The contiguous zone may not extend beyond 24 nautical miles from the baselines from which the breadth of the territorial sea is measured.

The exclusive economic zone (EEZ) is recognised under div 1A of the *SSLA* and art 55 of *UNCLOS*. The EEZ is an area both beyond and adjacent to the territorial sea. The outer limit of the exclusive economic zone cannot exceed 200 nautical miles from the baseline from which the breadth of the territorial sea is measured. Within the

---

29 See *OPGGSA* (Cth) s 5(2)(a).
30 For a discussion on these zones see I Brownlie, *Principles of Public International Law*, (Oxford University Press, 7th ed, 2008) chs 9 and 10; and Evans, 'The Law of the Sea' in M Evans (ed) *International Law*, (Oxford University Press, 2003) ch 20. See also M White, *Australia Offshore Areas* (Federation Press, 2009).

EEZ, Australia has acquired sovereign rights for the purpose of exploring, exploiting, conserving and managing all natural resources of the waters superjacent to the seabed and of the seabed and its subsoil together with other activities, such as the production of energy from water, currents, and wind. Jurisdiction also extends to the establishment and use of artificial islands, installations and structures, marine scientific research, the protection and preservation of the marine environment, and other rights and duties. The jurisdiction conferred upon the Commonwealth within the EEZ is therefore particularly important for offshore petroleum exploration and production.

A specific legal regime has been set up for the EEZ that confers upon a coastal state rights for the purpose of exploring, exploiting, conserving, and managing the natural resources and this is outlined in art 55. Rights, jurisdictions, and duties of the Commonwealth are outlined in art 56 and they include not only exploiting natural resources but also protecting and preserving the marine environment. The breadth of the EEZ is set out in art 57 and art 58 outlines the entitlements of all states (as confirmed in art 87) and which include the navigation, overflight, and the laying of pipelines. Articles 55–58 of *UNCLOS* are extracted below:

Article 55 Specific legal regime of the exclusive economic zone

The exclusive economic zone is an area beyond and adjacent to the territorial sea, subject to the specific legal regime established in this Part, under which the rights and jurisdiction of the coastal State and the rights and freedoms of other States are governed by the relevant provisions of this Convention.

Article 56 Rights, jurisdiction and duties of the coastal State in the exclusive economic zone

1. In the exclusive economic zone, the coastal State has:
    (a) sovereign rights for the purpose of exploring and exploiting, conserving and managing the natural resources, whether living or non-living, of the waters superjacent to the seabed and of the seabed and its subsoil, and with regard to other activities for the economic exploitation and exploration of the zone, such as the production of energy from the water, currents and winds;
    (b) jurisdiction as provided for in the relevant provisions of this Convention with regard to:
        (i) the establishment and use of artificial islands, installations and structures;
        (ii) marine scientific research;
        (iii) the protection and preservation of the marine environment;
    (c) other rights and duties provided for in this Convention.
2. In exercising its rights and performing its duties under this Convention in the exclusive economic zone, the coastal State shall have due regard to the rights and duties of other States and shall act in a manner compatible with the provisions of this Convention.

3. The rights set out in this article with respect to the seabed and subsoil shall be exercised in accordance with Part VI.

Article 57 Breadth of the exclusive economic zone

The exclusive economic zone shall not extend beyond 200 nautical miles from the baselines from which the breadth of the territorial sea is measured.

Article 58 Rights and duties of other States in the exclusive economic zone

1. In the exclusive economic zone, all States, whether coastal or land-locked, enjoy, subject to the relevant provisions of this Convention, the freedoms referred to in article 87 of navigation and overflight and of the laying of submarine cables and pipelines, and other internationally lawful uses of the sea related to these freedoms, such as those associated with the operation of ships, aircraft and submarine cables and pipelines, and compatible with the other provisions of this Convention.
2. Articles 88 to 115 and other pertinent rules of international law apply to the exclusive economic zone in so far as they are not incompatible with this Part.
3. In exercising their rights and performing their duties under this Convention in the exclusive economic zone, States shall have due regard to the rights and duties of the coastal State and shall comply with the laws and regulations adopted by the coastal State in accordance with the provisions of this Convention and other rules of international law in so far as they are not incompatible with this Part.

Jurisdiction is also conferred in the EEZ for the establishment and use of artificial islands, installations and structures, marine scientific research, and the protection and preservation of the marine environment. In exercising its rights and performing its duties in the exclusive economic zone, the coastal state must have due regard to the rights and duties of other states.

The resources in the EEZ are valuable to Australia and art 55 of *UNCLOS* allows the Commonwealth to set up a legal framework for the issuance of petroleum and gas permits in this area. Section 10A of the *SSLA* declares that the rights and jurisdiction in the EEZ are vested and exercisable by the Crown in right of the Commonwealth. Sections 10B and 10C set out the jurisdictional limits of the EEC. Sections 10A, 10B, and 10C are extracted below:

Seas and Submerged Lands Act 1973 (Cth)

10A Sovereign rights in respect of exclusive economic zone

It is declared and enacted that the rights and jurisdiction of Australia in its exclusive economic zone are vested in and exercisable by the Crown in right of the Commonwealth.

10B Limits of exclusive economic zone

The Governor-General may, from time to time, by Proclamation declare, not inconsistently with:

(a) Article 55 or 57 of the Convention; or

(b) any relevant international agreement to which Australia is a party;

the limits of the whole or of any part of the exclusive economic zone of Australia.

10C Charts of limits of exclusive economic zone

(1) The Minister may cause to be prepared such charts as he or she thinks fit showing any matter relating to the limits of the exclusive economic zone of Australia

(2) The mere production of a copy of a paper purporting to be certified by the Minister to be a true copy of such a chart is prima facie evidence of any matter shown on the chart relating to the limits of the exclusive economic zone of Australia.

The continental shelf comprises the seabed and subsoil of the submarine areas that extend beyond the territorial sea throughout the natural prolongation of its land territory to the outer edge of the continental margin, or to a distance of 200 nautical miles from the baselines from which the breadth of the territorial sea is measured.[31] The outer continental shelf was established as a maritime zone pursuant to the 1958 *Continental Shelf Convention*.[32] The entire continental shelf region is largely coextensive with the exclusive economic zone within 200 nautical miles from the territorial sea baselines although there are certain areas between Australia and Indonesia and Australia and Papua New Guinea where they are not coextensive.

Australia has jurisdiction over the continental shelf for the purposes of exploring and exploiting the mineral and other non-living resources of the seabed and subsoil, together with sedentary organisms. Within the continental shelf area, Australia also has jurisdiction to conduct marine scientific research as well as other rights and responsibilities. Article 76 of *UNCLOS* confers upon the coastal state the exclusive right to explore and exploit the natural resources of the continental shelf without occupation or proclamation. The natural resources in this context consist of mineral and other non-living resources of the seabed as well as sedentary living organisms.[33] Significantly, the rights that the coastal state has over the continental shelf do not affect the legal status of the superjacent waters or the air space above those waters. Further, the exercise of those rights must not infringe or result in any unjustifiable interference with navigation and other rights and freedoms of other states as provided for in *UNCLOS*.[34] There is a significant overlap between the EEZ and the continental shelf as the continental shelf lies under much of the Australian EEZ.

---

31 See *UNCLOS* art 76.
32 *Convention on the Continental Shelf*, opened for signature 29 April 1958, 499 UNTS 311 (entered into force 10 June 1964).
33 See *UNCLOS* art 77.
34 See *UNCLOS* art 78.

To support a claim for delineation of the outer limit of the extended continental shelf beyond 200 nautical miles, as measured from the territorial sea baseline, Australia submitted details and supporting scientific information to the United Nations Commission on the Limits of the Continental Shelf ('the CLCS'). The Commission was established pursuant to the provisions of *UNCLOS* in pt VI as well as the *SSLA* s 11.[35] A coastal state must establish the outer limits of its continental shelf, where it extends beyond 200 nautical miles, on the basis of the recommendations of the CLCS. Where, however, the area of extended continental shelf is within 200 nautical miles of another country, Australia may not make a claim with the CLCS. Further, the claim of a coastal state to an extension of the continental shelf cannot extend beyond 350 miles from the baselines.[36]

The primary purpose of the CLCS is to facilitate the implementation of the *UNCLOS* provisions regarding the establishment of the outer limits of the continental shelf, beyond 200 nautical miles from the baselines, from which the breadth of the territorial sea is to be measured. The CLCS must make recommendations to coastal states on matters related to the establishment of those limits; its recommendations and actions must not prejudice matters relating to the delimitation of boundaries between states with opposite or adjacent coasts. Annex II, art 4 notes that within 10 years of entry into force of *UNCLOS*, the states which intend to 'establish the outer limits of their continental shelf beyond 200 miles' are to submit to the CLCS scientific and technical data to support their proposed establishment of such limits. The CLCS will consider such data and make recommendations to the submitting state with respect to the outer limits. Where a submitting state is in disagreement with the CLCS, the state must make a revised or new submission to the CLCS.[37] The final sentence in art 76(8) makes it clear that the limits of the shelf established by the coastal state, on the basis of recommendations made by the CLCS, shall be 'final and binding'.

On 15 November 2004, Australia lodged its submission with the CLCS to extend the outer limits of its continental shelf. On 9 April 2008, the CLCS met at the United Nations in New York and adopted recommendations confirming Australia's entitlement to an extended continental shelf of some 2.56 million km^2 and encompassing

---

35 The Commission on the Limits of the Continental Shelf was established pursuant to *UNCLOS* in accordance with art 76(8). Pursuant to that Article, '... the Commission shall make recommendations to coastal States on matters relating to the establishment of the outer limits of their continental shelf'. The limits of the shelf established by these recommendations are to be final and binding. Annex II of *UNCLOS* sets out that the Commission is to be composed of an elected group of 21 technical specialists. See the discussion by T L McDorman, 'The Role of the Commission on the Limits of the Continental Shelf: A Technical Body in a Political World' (2002) 17(3) *International Journal of Marine and Coastal Law* 301.
36 *UNCLOS* art 76.
37 *UNCLOS* annex II, art 8. See also T L McDorman, 'The Entry into Force of the 1982 LOS Convention and the Article 76 Outer Continental Shelf Regime' (1995) 10 *International Journal of Marine and Coastal Law* 165.

nine distinct regions. The submission was large and complex, covering a total of 3.37 million km². [38] The region adjacent to the Australian Antarctic Territory was not considered because of the special legal and political status of Antarctica under the *Antarctic Treaty*. [39] Seven of the nine regions submitted by Australia were confirmed, with only slight amendment. A further two regions were adopted with marginal reductions in the areas sought. The confirmation of this extension has produced a secure regime within which to manage the resources and environment of virtually all parts of the continental margins, associated plateaus and ridges, and adjacent ocean basins. It has also generated significant commercial opportunities for petroleum production.

One of those offshore regions captured in the expansion of Australia's continental shelf is the Wallaby Plateau, a basin situated approximately 500 km west of Carnarvon, Western Australia. This basin is situated within the westward Southern Carnarvon Basin. Initial exploratory data has indicated the presence of significant petroleum deposits in this region and this represents an important opportunity for commercial development. The Wallaby Plateau is regarded as the deepwater frontier of the Carnarvon Basin. In 2008, three out of six deepwater wells from this region revealed significant gas deposits. [40]

In addition to the Wallaby Plateau, two other significant regions have been acquired under Australia's expanded outer continental shelf. Lord Howe Rise, a region between Lord Howe and Norfolk Islands 800 km east of Brisbane have both been acquired. These regions are also predicted to have a strong presence of oil and hydrocarbons within the seabed although there are difficulties with exploiting these resources given the fact that reports suggest that hydrocarbons are likely to be located within deeper sections, thereby increasing the risk and cost associated with exploration and production. [41]

The petroleum and mineral resource rich areas gained through the expansion of Australia's outer continental shelf may be further exploited in time, when deep-sea technology and investment is sufficiently advanced to provide for effective well construction, particularly in deeper areas where the resources are more likely to exist.

---

38 See the submission documents at United Nations, Oceans and Laws of the Sea, *Commission on the Limits of the Continental Shelf (CLCS): Selected documents of the Commission* http://www.un.org/depts/los/clcs_new/commission_documents.htm#Statements by the Chairman of the Commission.

39 *The Antarctic Treaty*, opened for signature 1 December 1959, 402 UNTS 71 (entered into force 23 June 1961).

40 G Briggs, 'Australia's Extended Continental Shelf and Northwest Offshore Energy Prospects' (Strategic Analysis Paper, Future Directions International, 2010) http://www.futuredirections.org.au/files/1271732207-Australias%20Expanded%20Continental%20Shelf%20and%20Northwest%20Offshore%20Energy%20Prospects.pdf.

41 See D Jongsma, Department of Minerals and Energy, Bureau of Mineral Resources, Geology and Geophysics, *A Review of Marine Geophysical Investigations Over the Lord Howe Rise and Norfolk Ridge* (1976) 33 http://www.ga.gov.au/corporate_data/13432/Rec1976_012.pdf

Offshore exploration is encouraged via the release of offshore acreages. The 2009 acreage release was comprised of 31 areas, plus two special areas, each spanning five offshore sedimentary basins in Commonwealth waters. Once assessed under a 'work program bidding system', mining proponents may be awarded an initial exploration term of six years.

Significantly, Australia is required to pay a percentage of the value or volume of production in any area of extended continental shelf to the International Seabed Authority which was set up to regulate deep seabed mining and seeks to ensure that the marine environment is protected against harmful effects which may arise from mining exploration and production.[42] Any member state that wishes to explore for minerals from the international deep seabed is only entitled to do so in accordance with a contract with the International Seabed authority. The activity must be carried out for the benefit of mankind as a whole because the authority is responsible for regulating to prevent marine pollution and damage to the marine environment as well as encouraging scientific research in the deep seabed.

Contractual payments to the International Seabed Authority must be made by the holder of the resource title and must commence the sixth year after petroleum production commences. The initial rate is 1 per cent of the value or volume of production at the site. The rate increases by 1 per cent for each subsequent year until the twelfth year and then remains at 7 per cent.[43]

3.2.5 The high seas

The high seas represent the waters beyond the maritime zones of the EEZ. There is no precise definition of the high seas in *UNCLOS* and in these waters there is no coastal state ownership or jurisdiction. The absence of Crown ownership in the high seas is a direct consequence of the fact that this area of water is not subject to any domestic ownership principles because it is treated as reserved for the 'common heritage of mankind'.[44] The common heritage principle has its source in the Roman *res communes* principle which, in this context, prevents

---

[42] As at 15 January 2015 there are 167 member states of the International Seabed Authority, including Australia. See International Seabed Authority, *Scientific Activities and Promotion* http://www.isa.org.jm/scientific-activities.

[43] See 'Australia's Offshore Jurisdiction: Explanation of Terminology in Relation to Petroleum Exploration and Development', at 6. Available at: http://www.industry.gov.au/resource/Documents/upstream-petroleum/Australia±%C3%A5s_Offshore_Jurisdiction.pdf

[44] See generally E Guntrip, 'The Common Heritage of Mankind: An Adequate Regime for Managing the Deep Seabed?' (2003) 4 *Melbourne Journal of International Law* 376, 382. See also B Larschan and B C Brennan, 'The Common Heritage of Mankind Principle in International Law' (1983) 21 *Columbia Journal of Transnational Law* 305.

extra-jurisdictional natural resources in the high seas from being owned or apportioned otherwise than in accordance with rules that promote the common interest of all nations.[45]

The common heritage principle has been codified in *UNCLOS* art 137(1) which provides that 'No State shall claim or exercise sovereignty or sovereign rights over any part of the Area or its resources, nor shall any State or natural or juridical person appropriate any part thereof'. The *SSLA* does not make any claim to rights in the high seas. This means that within these waters the Commonwealth will only have specific power over Australian ships and Australian nationals.

UNCLOS provides for freedom of navigation in international waters and this principle is recognised under international law. It is also consistent with the international principle of innocent passage, which entitles the ships of all nation states to pass through the territorial sea without the consent of the coastal state in accordance with ordinary navigation or as deemed necessary by the force of circumstance.[46] *UNCLOS* also adheres to the prerogative limitations that existed under municipal law, which confer public rights of fishing in the sea and tidal rivers[47] and rights of navigation.[48] The relevant *UNCLOS* provisions for innocent passage are arts 17–19 and are extracted below:

---

45 Larschan and Brennan, above n 44, 306. There are four established elements to the common heritage principle: first, no nation may apportion the area to which the doctrine applies; second, all countries must share in the management of the area; third, all countries are entitled to share in the profits that may be derived from the natural resources in the area; and fourth, the area must be preserved for peaceful purposes. This summary was set out by D Goedhuis, 'Some Recent Trends in the Interpretation and the Implementation of the Rules of International Space Law' (1981) 19 *Columbia Journal of Transnational Law* 213, 219. See also M Harry, 'The Deep Seabed: The Common Heritage of Mankind or Arena for Unilateral Exploitation?' (1992) 40 *Naval Law Review* 207, 226.

46 The international right of innocent passage was discussed in *R v Keyn* (1876) 2 Ex D 63, 70 (Sir R Philimore). See also *Foreman v Free Fishers and Dredgers of Whitstable* (1869) LR 4 HL 266. This right is also set out in *UNCLOS* arts 17 and 18. See also the discussion in I Brownlie, *Principles of Public International Law* (Oxford University Press, 5th ed, 1998) 191–5.

47 The public right to fish refers to a right to fish marine life on the seabed and free-swimming fish and includes ancillary rights of access and limited control. For a discussion on the nature of public fishing rights see *Attorney-General for British Columbia v Attorney-General for Canada* [1914] AC 153 at 170–1 where the Judicial Committee of the Privy Council stated that 'the right of the public to fish in the sea has been well established in English law for many centuries and does not depend upon the assertion or maintenance in the Crown of any title in the subjacent land'. See also *Harper v Minister for Sea Fisheries* (1989) 168 CLR 314, 330.

48 The public right to navigate refers to a right to pass and repass over the water and includes rights to anchor, moor, and ground where necessary in accordance with ordinary navigation or as deemed necessary by the force of circumstance. See the discussion by Beaumont and von Doussa JJ in *Commonwealth v Yarmirr* (1999) 10 FCR 171, [213]–[218].

Article 17 Right of innocent passage

Subject to this Convention, ships of all States, whether coastal or land-locked, enjoy the right of innocent passage through the territorial sea.

Article 18 Meaning of passage

1. Passage means navigation through the territorial sea for the purpose of:
    (a) traversing that sea without entering internal waters or calling at a roadstead or port facility outside internal waters; or
    (b) proceeding to or from internal waters or a call at such roadstead or port facility.
2. Passage shall be continuous and expeditious. However, passage includes stopping and anchoring, but only in so far as the same are incidental to ordinary navigation or are rendered necessary by force majeure or distress or for the purpose of rendering assistance to persons, ships or aircraft in danger or distress.

Article 19 Meaning of innocent passage

1. Passage is innocent so long as it is not prejudicial to the peace, good order or security of the coastal State. Such passage shall take place in conformity with this Convention and with other rules of international law.
2. Passage of a foreign ship shall be considered to be prejudicial to the peace, good order or security of the coastal State if in the territorial sea it engages in any of the following activities:
    (a) any threat or use of force against the sovereignty, territorial integrity or political independence of the coastal State, or in any other manner in violation of the principles of international law embodied in the Charter of the United Nations;
    (b) any exercise or practice with weapons of any kind;
    (c) any act aimed at collecting information to the prejudice of the defence or security of the coastal State;
    (d) any act of propaganda aimed at affecting the defence or security of the coastal State;
    (e) the launching, landing or taking on board of any aircraft;
    (f) the launching, landing or taking on board of any military device;
    (g) the loading or unloading of any commodity, currency or person contrary to the customs, fiscal, immigration or sanitary laws and regulations of the coastal State;
    (h) any act of wilful and serious pollution contrary to this Convention;
    (i) any fishing activities;
    (j) the carrying out of research or survey activities;
    (k) any act aimed at interfering with any systems of communication or any other facilities or installations of the coastal State;
    (l) any other activity not having a direct bearing on passage.

These freedoms enable holders of resource titles to conduct a range of activities connected with their authorised activities including navigating, conducting marine scientific research, laying submarine cables and pipelines, and fishing, subject to conservation and management measures imposed by regional fisheries management. The siting of installations to drill for hydrocarbons or mine seabed minerals on the extended continental shelf must observe the established shipping routes in the area, the location of submarine cables and pipelines, and the existence of equipment related to marine scientific research on the seabed. Provided drilling and mining obstructions are properly marked, oil, gas and mineral development in the continental shelf that crosses the main shipping channels will have minimal impact on transiting vessels. This can, however, be subject to increasing obligations to protect the marine environment.

In 2006, Australia introduced legislation requiring the compulsory use of a pilot by certain foreign vessels transiting through the Torres Strait: a strait commonly used for international navigation.[49] Compulsory pilotage was proposed as an associated protective measure arising from the Strait's designation through the International Maritime Organization (IMO) as a particularly sensitive sea area. Whilst this legislation interfered with the right of innocent passage, it was argued that the use of pilots familiar with the specific hazards of the Strait was an important requirement for the protection of an environmentally sensitive area. The proposal by Australia was rejected by the IMO as interfering with the right of innocent passage. Despite this, in Canada, the *Northern Canada Vessel Traffic Services Zone Regulations* require certain foreign and Canadian vessels to register with and report to the Canadian Coast Guard when entering and travelling through Canadian-claimed Arctic waters.[50] These regulations require a vessel to gain permission to enter Canada's claimed EEZ and territorial sea and were justified as necessary protection for the marine environment.

The European Commission has challenged *UNCLOS* arguing that the legal system relating to oceans and seas as set up under *UNCLOS* needs to be developed to better respond to new challenges. The *UNCLOS* regime for EEZ and international straits makes it harder for coastal states to exercise jurisdiction over transiting ships, despite the fact that any pollution incident that occurs within these zones presents an imminent risk for them. This, in turn, makes it difficult to comply with the general obligations (themselves set up by *UNCLOS*) of coastal states to protect their marine environment against pollution.[51]

---

49 See *Navigation Act 2006* (Cth) s 186A.
50 The *Northern Canada Vessel Traffice Service Zone Regulations*, SOR/2010-127 are authorised by the *Canadian Shipping Act* (CSA 2001).
51 See the discussion by the Commission of the European Communities Green Paper, *Towards a Future Marine Policy for the Union: A European Vision for the Oceans and Seas* (COM 2006) 275, adopted by the European Commission on 7 June 2006.

All proponents of petroleum production will also need to factor into investment calculations the possibility of disputes arising between Australia and other states with multiple interests in the water column above the exploration sites. Additionally, investors need to be aware of the annual payments payable for production in the extended continental shelf within the high sea after five years. As outlined above, these payments must be made to the International Seabed Authority (ISA) and, in accordance with the ISA provisions, are to be distributed to countries which have ratified *UNCLOS*, taking into account the interests and needs of developing states.

There are also important logistical and security challenges involved in establishing and protecting exploitation activities in the extended continental shelf and the high seas. Offshore installations located in remote outer continental shelf areas are more susceptible to terrorist attack, illegal exploitation, and severe weather events attributable to climate change. Risk management of such projects can be extensive, particularly when costs of surveillance and enforcement are taken into account.

3.2.6 *Offshore Petroleum and Greenhouse Gas Storage Act 2006* (Cth)

The *Offshore Petroleum and Greenhouse Gas Storage Act 2006* (Cth) ('the OPGG-SA') is now the primary Commonwealth Act regulating the offshore jurisdiction for petroleum mining and greenhouse gas storage.[52] This Act is the controlling Act, having replaced the framework set out in the earlier *Petroleum (Submerged Lands) Act 1967* (Cth). Each of the state Acts are now structured to comply with the framework of the *OPGGSA*.[53]

Section 5 of the *OPGGSA* implements the *Offshore Constitutional Settlement (1979)* (Cth), whereby the Commonwealth and states undertook negotiations which resulted in the states acquiring responsibility for areas – known as 'coastal waters' – up to three nautical miles from the territorial sea baseline. In accordance with this section, the *OPGGSA*, sets out that the states and Northern Territory must share in the administration of the Commonwealth offshore petroleum legislation.[54]

Unlike the *Petroleum (Submerged Lands) Act 1967* (Cth), the *OPGGSA* introduces the use of two new terms to delineate between state territorial waters and Commonwealth territorial waters. The references are 'coastal waters' and 'offshore area'.[55] 'Coastal waters' refers to the area which extends beyond the internal waters of the state and out to the three-mile limit and includes waters between the

---

52 The *OPGGSA* was given assent on 29 March 2006 and replaced its predecessor, the *Petroleum (Submerged Lands) Act 1967* (Cth).
53 See *OPGGSA* (Cth) s 80.
54 See *OPGGSA* (Cth) s 5.
55 The definition is contained in *OPGGSA* (Cth) s 7.

outer limits of a state and the baselines.[56] 'Offshore areas' refer to waters seaward from the outer limit of the coastal waters to the outer limit of the EEZ or the outer continental shelf.[57]

The *OPGGSA* is the primary Act regulating the thriving offshore oil and gas framework for Australia. Petroleum is defined in s 5 of the *OPGGSA* as 'any naturally occurring hydrocarbons, whether in a gaseous, liquid or solid state'. The breadth of this definition means that the *OPGGSA* has an application to both oil and gas. It also covers petroleum in liquid, gaseous, or solid state and therefore covers offshore gas and gas pipelines as well as liquid petroleum.[58]

State and territory courts have jurisdiction under the *OPGGSA* in matters arising within their offshore geographical limits.[59] The *OPGGSA* has an application to persons, equipment, rigs, and pipelines but it does not apply to ships.[60]

The *OPGGSA* regulates the steps involved in exploration for and exploitation of offshore petroleum including the grant and administration of exploration permits, retention leases, production licences, infrastructure licences, and pipeline licences. The legislation also provides for the issuance of petroleum scientific investigation consents that authorise the holder to carry on exploration operations in the course of a scientific investigation.[61]

With the aim of reducing accidents and collisions involving offshore rigs, the *OPGGSA* sets out that all petroleum wells and installation equipment must be surrounded by a 500 m safety zone and that a designated authority will regulate which vessels are entitled to enter this zone.[62] Petroleum safety zones and greenhouse gas safety zones may be established and unauthorised vessels that attempt to enter these zones may be detained.[63]

The *OPGGSA* also sets up a licensing framework for the regulation and issuance of greenhouse gas licences whereby liquefied greenhouse gases may be injected into authorised offshore greenhouse gas storage formations.[64] The *OPGGSA* has an application to all individuals and companies conducting petroleum activities within coastal waters and offshore areas and it seeks to maintain good oilfield practice and to ensure that operations are compatible with optimum long-term recovery of

---

56 See *OPGGSA* (Cth) s 7.
57 See *OPGGSA* (Cth) ss 7, 8. Section 8 of the *OPGGSA* (Cth) specifically defines the actual offshore areas for the states and territories.
58 See *OPGGSA* (Cth) s 7.
59 See *OPGGSA* (Cth) pt 6.8, sch 3.
60 The major Act regulating shipping in Australia is the *Navigation Act 1912* (Cth).
61 See *OPGGSA* (Cth) pt 2.9
62 This is reinforced in *UNCLOS* art 48 which sets out that a coastal state may establish safety zones to ensure the safety of navigation and artificial islands, installations and structures. The safety zones must not exceed 500 m: see art 60(4), (5).
63 See *OPGGSA* (Cth) ss 616–23.
64 The framework for greenhouse gas injection and licensing is examined in detail in Chapter 7.

petroleum. Good oilfield practice reflects the underlying principles of conservation, energy efficiency, and optimum recovery. The underlying objective is that the production of crude oil or raw gas should be sustained at a rate that is feasible without adversely affecting the petroleum reservoir or the marine environment.

3.2.7 Overview of the *OGGSA* title framework

In pts 2.2–2.9, the *OPGGSA* establishes a range of resource permits with a specific application to the offshore areas.[65] All titles are issued by the joint authority which is entitled to grant licences subject to such terms and conditions as it deems appropriate. A joint authority is established for each offshore area of a state (other than Tasmania) and the authority is comprised of the state and Commonwealth minister. This collaborative arrangement is known as the 'Commonwealth (relevant state) offshore petroleum joint authority'.[66] The *OPGGSA* sets up a separate Commonwealth Tasmania offshore petroleum joint authority, a Commonwealth Northern Territory offshore petroleum joint authority, and a Greater Sunrise offshore petroleum joint authority which has authority over the Sunrise and Troubadour fields in association with Timor-Leste. The offshore petroleum joint authority has the power to make decisions pursuant to the provisions of the *OPGGSA* and, in the event that the state and Commonwealth minister are unable to agree, the decision of the Commonwealth minister will prevail.[67]

The National Offshore Petroleum Titles Administrator (NOPTA), set up pursuant to pt 6.10 of the *OPPGSA*, is responsible for the issuance and management of resource titles. The National Offshore Petroleum Safety and Environment Management Authority (NOPSEMA), set up pursuant to pt 6.9 of the *OPPGSA*, is responsible for environment and integrity issues including safety issues, environment plans, and well operation management plans. All applications for offshore resource titles are received by NOPTA on behalf of the joint authorities. NOPTA is responsible for assessing applications and providing recommendations to the joint authorities, taking into account consultation with relevant stakeholders that may include petroleum companies and other offshore agencies and environmental groups. The recommendations made by NOPTA must address title administration and resource management issues and, once received, Commonwealth and state departments are then responsible for assessing the NOPTA recommendations and incorporating any relevant policy analysis.[68]

---

[65] This section overviews petroleum exploration permits, petroleum retention leases, petroleum production licences, infrastructure licences and pipeline licences. Greenhouse gas licences are examined in more detail in Chapter 7.
[66] See *OPGGSA* (Cth) pt 1.3, s 56.
[67] See *OPGGSA* (Cth) s 59(2).
[68] See the Australian Government, Department of Resources, Energy and Tourism, *Guidelines for Offshore Petroleum Joint-Authority Decision Making Procedures* (1 January 2012) http://www.nopta.gov.au/_documents/guidelines/JA-Decision_guidelines.pdf.

3.2.7.1 Petroleum exploration permit

The petroleum exploration permit confers a right to explore for and recover, on an appraisal, basis petroleum in the permit area and to carry on such operations and execute such works in the permit area that are necessary to support these rights.[69] In the event that petroleum is discovered, the holder of a permit may nominate a block or blocks for declaration as a 'location' and may then apply for a retention lease or a production licence in respect of that location. Only the holder of a petroleum exploration permit is entitled to apply for a retention lease or a production licence.[70] Section 141 of the *OPGGSA* is extracted below:

Offshore Petroleum and Greenhouse Gas Storage Act 2006 (Cth)

141 Application for petroleum retention lease by the holder of a petroleum exploration permit

(1) If a petroleum exploration permit is in force over a block that constitutes, or the blocks that constitute, a location, the permittee may, within the application period, apply to the Titles Administrator for the grant by the Joint Authority of a petroleum retention lease over that block or over one or more of those blocks.

(2) An application under this section must be accompanied by details of:
  (a) the applicant's proposals for work and expenditure in relation to the area comprised in the block or blocks specified in the application; and
  (b) the current commercial viability of the recovery of petroleum from that area; and
  (c) the possible future commercial viability of the recovery of petroleum from that area.

(3) The *application period* for an application under this section is:
  (a) the period of 2 years after the day (the *declaration day*) on which the block that constitutes the location concerned was, or the blocks that constitute the location concerned were, declared to be a location; or
  (b) such longer period, not more than 4 years after the declaration day, as the Titles Administrator allows.

(4) The Titles Administrator may allow a longer period under paragraph (3)(b) only on written application made by the permittee within the period of 2 years mentioned in paragraph (3)(a).

3.2.7.2 Petroleum retention lease

A petroleum retention lease confers upon the holder a right to explore for and recover petroleum on an appraisal basis and to carry out operations in the lease area for those purposes.[71] A petroleum exploration permit will cease to be in force

---

69 See *OPGGSA* (Cth) s 98.
70 See *OPGGSA* (Cth) ss 141, 168(2).
71 See *OPGGSA* (Cth) s 135.

as soon as a petroleum retention lease becomes effective.[72] A retention lease may also be granted over areas covered within a production licence in which petroleum recovery has ceased because recovery at that stage is less than commercial.[73]

A retention lease provides security of title for petroleum resources that are not currently viable but which have genuine development potential; that is, are likely to become commercially viable within 15 years. In this context, 'commercially viable' means that petroleum could be developed given existing knowledge in the field, having regard to prevailing market conditions, and using proven technology available within the industry, such that the commercial rates of return for the recovery of the petroleum either meet or exceed the minimum return acceptable for the type of project involved by a reasonable petroleum developer.[74] Any retention leases application must include details of the possible future commercial viability of the recovery of petroleum from the area.[75] Once a petroleum retention lease is issued, a petroleum exploration permit will cease to be in effect.[76]

Offshore Petroleum and Greenhouse Gas Storage Act 2006 (Cth)

142 Grant of petroleum retention lease – offer document

If:

(a) an application for a petroleum retention lease has been made under section 141; and
(b) the Joint Authority is satisfied that:
 (i) the area comprised in the block, or any one or more of the blocks, specified in the application contains petroleum; and
 (ii) the recovery of petroleum from that area is not, at the time of the application, commercially viable but is likely to become commercially viable within 15 years after that time;

the Joint Authority must give the applicant a written notice (called an *offer document*) telling the applicant that the Joint Authority is prepared to grant the applicant a petroleum retention lease over the block or blocks as to which the Joint Authority is satisfied as mentioned in paragraph (b).

3.2.7.3 Petroleum production licence

A petroleum production licence confers upon a holder a right to explore for and produce petroleum resources and to carry on operations and execute works that support those rights.[77] A production licencee can be directed to maintain, increase,

---

72 See *OPGGSA* (Cth) s 145.
73 See *OPGGSA* (Cth) s 148.
74 See Australian Government, Department of Resources, Energy and Tourism *Offshore Petroleum Guideline for Grant and Administration of a Retention Lease* (May 2012) 5 http://www.nopta.gov.au/_documents/guidelines/offshorepetroleumguidelinegrantadministrationretentionlease.pdf.
75 See *OPGGSA* (Cth) s 141.
76 See *OPGGSA* (Cth) s 145.
77 See *OPGGSA* (Cth) s 161.

or reduce the rate of recovery. A petroleum production licence, where issued after July 1998, exists indefinitely, although in practice the licence will continue for the life of the field.[78] A petroleum production licence must, however, be utilised and may be terminated where there have been no operations for the recovery of petroleum or geothermal energy for a continuous period of five years.[79]

The *OPGGSA* framework also provides for competitive cash bidding with respect to a location that has ceased to be subject to either an exploration permit, a retention lease, or a production licence. A Joint Authority may invite applications for a cash-bid petroleum production licence in respect of blocks within the area of the permit, lease, or licence.[80]

Section 161 of the *OPGGSA* setting out the statutory rights conferred by a petroleum production title is extracted below:

Offshore Petroleum and Greenhouse Gas Storage Act 2006 (Cth)

161 Rights conferred by petroleum production licence

(1) A petroleum production licence authorises the licensee, in accordance with the conditions (if any) to which the licence is subject:
  (a) to recover petroleum in the licence area; and
  (b) to recover petroleum from the licence area in another area to which the licensee has lawful access for that purpose; and
  (c) to explore for petroleum in the licence area; and
  (d) to carry on such operations, and execute such works, in the licence area as are necessary for those purposes.

(2) Express references in this Act to the injection or storage of a substance do not imply that subsection (1) does not operate so as to authorise the licensee:
  (a) to carry on operations to inject a substance into the seabed or subsoil of an offshore area; or
  (b) to carry on operations to store (whether on a permanent basis or otherwise) a substance in the seabed or subsoil of an offshore area.

(3) The regulations may provide that a petroleum production licence authorises the licensee, in accordance with the conditions (if any) to which the licence is subject:
  (a) to explore in the licence area for a potential greenhouse gas storage formation; and
  (b) to explore in the licence area for a potential greenhouse gas injection site; and
  (c) to carry on such operations, and execute such works, in the licence area as are necessary for those purposes.

(4) The regulations may provide that, if:
  (a) petroleum is recovered in the licence area of a petroleum production licence (the *first licence*); and

---

78 See *OPGGSA* (Cth) s 165.
79 See *OPGGSA* (Cth) s 166.
80 See *OPGGSA* (Cth) ss 110, 178.

(b) operations for the recovery or processing of the petroleum are carried on using a facility located in the licence area of another petroleum production licence (the *second licence*); and

(c) a prescribed substance (which may be a hydrocarbon) is recovered as an incidental consequence of the recovery of the petroleum;

the second licence authorises the licensee of the second licence, in accordance with the conditions (if any) to which the second licence is subject:

(d) to inject the substance into the seabed or subsoil of the licence area of the second licence; and

(e) to store (whether on a permanent basis or otherwise) the substance in the seabed or subsoil of the licence area of the second licence; and

(f) to carry on such operations, and execute such works, in the licence area of the second licence as are necessary for those purposes.

(5) Subsections (3) and (4) do not limit subsection (1).

(6) The rights conferred on the licensee by or under subsection (1), (3) or (4) are subject to this Act and the regulations.

3.2.7.4 Infrastructure licence

The *OPGGSA* also provides for the issuance of infrastructure licences and pipeline licences. Infrastructure licences authorise the licensee to construct and operate an infrastructure facility within the licence area, which relates to petroleum or a greenhouse gas substance. An application for an infrastructure licence would ordinarily accompany an application for a production licence given the significant capital infrastructure required to commercial produce offshore petroleum.

The National Offshore Petroleum Titles Administrator guidelines indicate three primary situations where an infrastructure licence is required:

1. Where an infrastructure facility is being used to produce petroleum and is not located within the production licence area from which the petroleum is being produced. This generally occurs where the infrastructure facility is located in a depleted reserve.

2. Where an infrastructure facility is being used to produce petroleum and is located within the production licence area but the ownership of the infrastructure facility differs from that of the production licence; and

3. Where a hydrocarbon processing facility is undertaking conversion activities; for example, for LNG production, and is located either inside or outside the production licence area.[81]

---

81 See Australian Government, Department of Resources, Energy and Tourism, *Guidelines for Offshore Petroleum Joint-Authority Decision-Making Procedures* (1 January 2012) http://www.nopta.gov.au/_documents/guidelines/JA-Decision_guidelines.pdf.

3.2.7.5 Pipeline licence

The jurisdiction of the Commonwealth and states over the sea and the seabeds means that the *OPGGSA* and its mirror state Acts may also regulate the issuance of pipeline licences for the construction of seabed pipelines utilised for the transportation of oil and gas. As with all *OPGGSA* titles, the application and issuance of the licence is regulated by the relevant joint authority pursuant to recommendations by the National Offshore Petroleum Titles Administrator. Where, however, a pipeline crosses more than one jurisdiction, an application must be submitted to the relevant joint authority within each jurisdiction in order to gain separate approvals from each relevant state or territory.

A pipeline licence is required where an applicant wishes to construct, reconstruct, alter, and operate a pipeline in an offshore area for the purposes of conveying either petroleum or greenhouse gas substances.[82] It is also required in order to construct and operate pumping stations, tank stations, and valve stations associated with the pipeline and to carry on other operations or works as necessary to the pipeline.[83] In making an application for a pipeline licence, the applicant needs to outline the pipeline design as well as the route suggestion.[84]

A pipeline licence will confer upon the holder a right to construct and operate the pipeline in accordance with the route and specifications in the licence. A pipeline licence remains in force indefinitely, although the relevant Joint Authority may terminate the licence in circumstances where no construction work has been carried out under the licence or the pipeline has not been used for a continuous period of five years.[85] Significantly, in constructing the pipeline, if the route is not exactly as specified in the licence, the licensee must seek a variation before the pipeline can be completed.[86]

3.2.8 Sea installations

The *OPGGSA* does not regulate the installation and management of sea installations or flotation devices. Rather, the *Sea Installations Act 1987* (Cth) establishes a system of permits for the due regulation and administration of offshore sea installations that do not constitute 'petroleum installations' and therefore do not come under the *OPGGSA* framework.

The *Sea Installations Act 1987* (Cth) applies to man-made structures, floating or anchored. A 'sea installation' is defined as any man-made structure that can be used

---

[82] See *OPGGSA* (Cth) s 198.
[83] See Australian Government, Department of Resources, Energy and Tourism, *Offshore Petroleum Guideline for Grant and Administration of a Pipeline Licence* (January 2012) 5 http://www.nopta.gov.au/legislation/guidelines.html.
[84] See *OPGGSA* (Cth) s 217.
[85] See *OPGGSA* (Cth) ss 211, 215.
[86] See *OPGGSA* (Cth) ss 226, 227.

for an environmental activity, either when in physical contact with the seabed or whilst floating.[87] The Act makes it an offence to install, use, or work on an installation in an offshore area otherwise than in accordance with a permit.[88] Further, the Commonwealth may take action in accordance with the *Environmental Protection Biodiversity Conservation Act 1999* (Cth) where the sea installation constitutes a threat to the safety of persons or is likely to have an 'adverse effect' on safety.[89]

3.2.9 Offshore petroleum safety: NOPSEMA

The regulatory framework for Australia's offshore petroleum underwent significant reform following two catastrophic oil spills: the Montara and the Deep Water Horizon oil spills in the northern waters of Australia. In 2009, there was an uncontrolled release of oil and gas from the Montara Wellhead Platform in the Timor Sea that continued for 10 weeks. This spill has been described as the worst incident of its kind in Australia's offshore petroleum history.[90] Prior to this, in 2008, there was an explosion at the Varanus Island gas pipeline.

The oil spill generated many issues that required further investigation. This task was given to the Montara Commission of Inquiry, which was established in 2009 and given the powers of a royal commission in order to ensure that the investigation was rigorous and comprehensive.[91] The Commission concluded that the operator of the Montara Wellhead Platform, the Australian subsidiary of Thai-owned PTTEP, did not follow good oilfield practices and that the shortcomings in protocol were 'widespread and systematic' and led directly to the blowout. It also found that the relevant Northern Territory regulator, the Northern Territory Department of Resources, was not a sufficiently diligent regulator, as the drilling program which had been put forward did not reflect best oilfield practice.[92]

The final government response to the June 2010 Report of the Montara Commission Inquiry ('Montara Report') issued a range of recommendations. The two primary recommendations, now implemented, were the establishment of the National Offshore Petroleum Safety and Environmental Management Authority

---

87 See *Sea Installations Act 1987* (Cth) s 4.
88 See *Sea Installations Act 1987* (Cth) ss 14, 15.
89 See *Environment Protection and Biodiversity Conservation Act 1999* (Cth), div 1, sub-div F.
90 See S Marsden, 'Regulatory Reform of Australia's Offshore Oil and Gas Sector After the Montara Commission of Inquiry: What About Transboundary Environmental Impact Assessment?' (2013) 15 *Flinders Law Journal* 21. See also T Hunter, 'The Montara Oil Spill and the Marine Oil Spill Contingency Plan: Disaster Response or Just a Disaster?' (2010) 24 *Australian and New Zealand Maritime Law Journal* 46.
91 Marsden, above 90, 43.
92 See Commonwealth, Report of the Montara Commission of Inquiry, *National Report (2010)* http://www.industry.gov.au/resource/Documents/upstream-petroleum/approvals/Montara-Report.pdf.

(NOPSEMA), which came into effect at the start of 2012, and the creation of a new national titles administrator, the National Offshore Petroleum Titles Administrator (NOPTA). NOPSEMA is now the national regulatory body in Commonwealth waters for safety, well integrity, environmental regulation, and day-to-day operations of petroleum activities. NOPTA is now the national regulatory body responsible for managing the issuance of petroleum leases and licences.

The mandate of NOPSEMA is based upon provisions within the *OPGGSA*. Section 638 of *OPGGSA* sets out the laws that NOPSEMA and its occupational health and safety (OHS) inspectors are required to administer and enforce in Commonwealth waters. The obligations of NOPSEMA are set out in sch 3 of the *OPGGSA*, the *Offshore Petroleum and Greenhouse Gas Storage (Safety) Regulations 2009* (Cth), and pt 5 of the *Offshore Petroleum and Greenhouse Gas Storage (Resource Management and Administration) Regulations 2011*.[93]

Schedule 3 of the *OPGGSA* establishes an OHS regime for petroleum and greenhouse gas activities at facilities located within Commonwealth waters. The primary features of the regime are:

1. **Duties of care**: Certain persons involved in offshore activities at facilities are required to 'take all reasonably practicable steps' to protect the health and safety of the facility workforce and other affected persons.
2. **Consultation provisions**: The regime mandates effective consultation between each facility operator, relevant employers, and the workforce regarding OHS.
3. **Powers of inspectors**: Inspectors are granted powers to enter offshore facilities, conduct inspections, interview people, seize evidence, and take action to ensure compliance by parties holding OHS legal responsibilities.

Recent changes to the *Offshore Petroleum and Greenhouse Gas Storage Amendment (Compliance Measures) Act 2013* (Cth) go further in implementing the recommendations of the final government response to the Montara Report. These changes seek to strengthen the operating practices of the offshore petroleum industry and provide greater enforcement powers to NOPSEMA. In particular, the creation of a new Schedule 2A, which confers monitoring powers upon NOPSEMA inspectors regarding the obligations of petroleum lease and licence holders pursuant to environmental management laws. These provisions give a NOPSEMA inspector a range of

---

93 Amendments to the *OPGGSA* (Cth) effective from 1 January 2012 require states and the Northern Territory to confer well integrity functions to NOPSEMA in order to continue to be subject to the regulation of OHS in their coastal waters under the *Petroleum (Submerged Lands) Amendments Act 1982* (Cth). As at 1 January 2013, only Victoria had conferred well integrity functions in designated coastal waters on NOPSEMA and, at that point, was the only state or territory to have validly conferred OHS functions.

powers including the capacity to issue 'do not disturb notices', as well as the power to monitor an area without obtaining a warrant where it is deemed necessary.

The other regulatory body set up pursuant to the recommendations of the Montara Report is NOPTA. The key functions of NOPTA, as briefly outlined above, are to provide information, assessments, analysis, reports, advice, and recommendations to members of the Joint Authorities and the responsible Commonwealth Minister. NOPTA collects and manages the release of data, titles administration, approval, and registration of transfers and dealings. Significantly, NOPTA is responsible for keeping the register of petroleum and greenhouse gas storage titles which includes both the issuance and registration of new titles as well as the cessation of pre-existing ones.[94]

3.2.10 Joint petroleum area and Greater Sunrise: Australia and Timor-Leste

The Joint Petroleum Development Area (JPDA) and the Greater Sunrise fields are areas of the Timor Sea in which Australia and Timor-Leste work together to facilitate development of resources.[95] The Timor Sea is rich in oil and gas resources. In particular, the Timor Gap is an area of sea and continental shelf located in the Timor Sea between Australia and Timor-Leste which has an abundancy of petroleum resources and has great economic significance to both countries.[96]

The JPDA was formally established under the *Timor Sea Treaty 2003* ('the *TST*'). The *TST* provides for petroleum exploration and development within the JPDA. The aim of the *TST* was for Australia and Timor-Leste to enter into an interim agreement regarding the scope of the maritime boundaries with the aim of reducing boundary disputes. The *TST* was intended to endure for 30 years or until a permanent maritime boundary could be established. The *TST* is crucial for petroleum exploration and development because mining proponents can rely upon it to determine the jurisdictional requirements of particular regions. The *TST* is also given effect in the *Petroleum (Timor Sea Treaty) Act 2003* (Cth).

In addition to articulating the maritime boundaries, the *TST* sets out that upstream taxation revenue from petroleum production in the JPDA is to be divided

---

[94] See *OPGGSA* (Cth) ss 470–71.
[95] Prior to declaring its independence in 1975, Timor-Leste was known as East Timor. The name change occurred in 2002 with the Declaration of Independence.
[96] See especially J Nevins, 'Contesting the Boundaries of International Justice: State Countermapping and Offshore Resource Struggles between East Timor and Australia' (2004) 80(1) *Economic Geography* 1; A Symon, 'Economic Issues No 2: Timor Sea Natural Gas Development: Still in Embryo' (2) *Economic Issues* (August 2001) http://www.adelaide.edu.au/saces/publications/papers/issues/EconIssuesTwo.pdf.

between Timor-Leste and Australia on a 90:10 basis.[97] Further, the Joint Commission is established pursuant to the *TST*.[98] The core function of the Joint Commission is to establish and oversee policy and regulation for petroleum activities in the JPDA; it is comprised of one Australian representative and two Timorese representatives.

Australia regulates the fiscal regime for petroleum entities in the JPDA, pipeline construction, marine protection, surveillance, quarantine and customs, the regulations under the *Petroleum Mining Code*, set up in accordance with art 7(a) of the *TST* to govern the exploration, development and exploitation of petroleum in the JPDA as well as the exportation of petroleum from the JPDA and may also apply relevant Australian taxation laws.[99] The day-to-day regulation of the JPDA is, however, managed by the Designated Authority, which is currently the Timor-Leste offshore regulator: the Autoridade Nacional do Petróleo (ANP). The ANP regulates operations in the JPDA on behalf of both countries.

Currently, there are two operational fields in the JPDA. The first field is 'Bayu-Undan', which is estimated to contain 3.4 trillion cubic feet (TCF) of liquid natural gas and 110 000 barrels of condensate. This field is operated pursuant to a joint venture, with ConocoPhillips (58 per cent) together with Eni (11 per cent), Santos (11 per cent), INPEX Timor Sea Ltd (11 per cent), and Tokyo Timor Sea Resources Pty (9 per cent).

The second field is 'Kitan Oilfield', which commenced production in October 2011. It is estimated that up to 34.6 million barrels of oil will be recoverable from the Kitan Oilfield. The Kitan Oilfield is also operated pursuant to a joint venture with Eni (40 per cent) and joint venture partners INPEX Timor Sea Ltd (35 per cent) and Talisman Resources Pty Ltd (25 per cent).[100]

The Greater Sunrise (GS) gas and condensate fields are important as they overlap a combined area of exclusive Australian jurisdiction (79.9 per cent) and the JPDA (20.1 per cent). These fields are significant as it is estimated they contain up to 5.1 TCF of liquefied natural gas and 226 million barrels of condensate.

The problem of overlapping is dealt with expressly in art 9 of the *TST*, where Timor-Leste and Australia agreed that any deposit that extends beyond the boundary of the JPDA is not to be developed as a single entity in terms of management and development, thereby precluding either Timor-Leste or

---

97 *Timor Sea Treaty between the Government of East Timor and the Government of Australia*, opened for signature 20 May 2002, ATS 13 (entered into force 2 April 2003), art 4.
98 Ibid art 3.
99 The *Petroleum Mining Code* was established to complement the *TST*. It is available at: http://www.resourcegovernance.org/sites/default/files/Petroleum%20Mining%20Code%20 for%20Joint%20Petroleum%20Development%20Area.pdf.
100 See the outline of the Kitan Oilfield: Offshore Technology.com, *Kitan Oilfield, Australia* http://www.offshore-technology.com/projects/kitanoilfield/

Australia from exploiting the resources unilaterally. Article 9 is given specific effect by the International Unitisation Agreement ('the IUA'), which was entered into in 2007.[101]

In accordance with the IUA, the GS must be developed pursuant to a commercial joint venture. The joint venture may only proceed once a field development plan has been submitted and approved. The relevant Australian authority regulates 79.9 per cent of the fields within the JPDA, whilst the ANP regulates 20.1 per cent of the fields on behalf of both Australia and Timor-Leste.

All proposed field development plans need to be approved by both regulators which effectively precludes either country from unilaterally developing the field. Day-to-day arrangements concerning the administration of GS, dispute resolution, and strategic direction are conducted by the Sunrise Commission which is set up under art 9 of the IUA to facilitate the provisions of the IUA and this specifically includes reviewing and making recommendations to regulators regarding exploration and exploitation of petroleum resources.

The commercial joint venture for GS is led by Woodside Petroleum Limited (operator and 33.44 per cent shareholder) ('Woodside'), and includes Royal Dutch Shell (26.56 per cent), ConocoPhillips (3 per cent), and Osaka Gas (10 per cent).

In 2010, Woodside announced that it would prefer to develop GS by utilising a floating LNG ('FLNG') processing plant. There are only two plants of this nature that are operational (one owned by Woodside and the other by Royal Dutch Shell): the Shell-owned and operated FLNG processing plant in the Browse Basin off the North West coast of Western Australia is scheduled to commence production in 2016.

A further important agreement for GS is the *Treaty on Certain Maritime Arrangements in the Timor Sea* ('the *CMATS*') which was entered into in 2007 between Australia and Timor-Leste.[102] *CMATS* should be distinguished from the *Torres Strait Treaty*, entered into in 1978 and ratified in 1985. The *Torres Strait Treaty* was negotiated to be consistent with *UNCLOS*. It defines the border between Australia and Papua New Guinea and provides a framework for the management of the common border area, defining the maritime boundaries between Papua New Guinea and Australia, and protecting the ways of life of traditional inhabitants in the Torres Strait Protected Zone (TSPZ). There are two main boundaries described by the *Torres Strait Treaty*:

- **The Seabed Jurisdiction Line:** Australia has rights to all things on or below the seabed south of this line and Papua New Guinea has the same rights north of the line.

---

101 *Agreement between the Government of Australia and the Government of the Democractic Republic of Timor-Leste Relating to the Unitisation of the Sunrise and Troubadour Fields*, opened for signature 6 March 2003, ATS 11 (entered into force 23 February 2007) ('IUA').

102 *Treaty on Certain Maritime Arrangements in the Timor Sea,* opened for signature 12 January 2006, ATS 12 (entered into force 23 February 2007).

- **The Fisheries Jurisdiction Line:** Australia has rights over swimming fish south of this line and Papua New Guinea has the same rights north of the line. The two countries have agreed under the Treaty to share these rights.

By contrast, *CMATS* sets out that all upstream petroleum revenues from the GS must be shared equally between the two countries. Prior to the implementation of this treaty, East Timor would only have received about 18 per cent of the revenue from the field.

Further, *CMATS* does not resolve boundary disputes but merely suspends discussion on them and, in fact, defers determination of the boundaries until the exhaustion of the disputed resources.[103] Hence, *CMATS* suspends the right by both Australia and Timor-Leste to claim sovereign rights, discuss maritime boundaries, or engage in any legal process in relation to maritime boundaries or territorial jurisdiction for 50 years, which is the duration that *CMATS* is in effect.[104]

CMATS is effectively a response to the long-running dispute between Australia and Timor-Leste regarding natural resources within the Timor Gap. The foundation of this dispute lies in the fact that both Australia and East Timor are coastal states and are therefore entitled to claim sovereign rights in accordance with arts 57 and 77 of *UNCLOS*. Both countries have claimed the full extent of these sovereign rights under these articles. Timor-Leste claimed a 200 nautical mile exclusive economic zone and continental shelf rights to the greatest possible distance.

Similarly, Australia claimed a 200 nautical mile EEZ as well as continental shelf rights up to the Timor Trough, which is a deep trench in the seabed 250–300 nautical miles off the Australian coast. The sovereign entitlements which were claimed by Australia and Timor-Leste overlap significantly as they lie less than 400 nautical miles apart. In fact, at their nearest point, the distance is only 130 nautical miles.[105] Where countries are located within 400 nautical miles of each other, the international practice is to establish a maritime boundary which is equidistant between the two. However, Australia was reluctant to follow this practice as it would result in a loss of the GS. The implementation of *CMATS* has helped to determine entitlements associated with this boundary overlap but it has not resolved the dispute.

3.2.11 *Offshore Minerals Act 1994* (Cth)

Minerals which do not come within the definition of 'petroleum' and therefore are not covered by the *OPGGSA* are regulated and administered pursuant to

---

103 See the discussion by C Scheiner, 'Drilling East Timor: Australia's Oil Grab in the Timor Sea' (2006) 27(1) *MultiNational Monitor* 30, 31.
104 See especially M J Smith, 'Australian Claims to the Timor Sea's Petroleum Resources: Clever, Cunning or Criminal?' (2011) 37 *Monash University Law Review* 42, 58.
105 See P Cleary, *Shakedown: Australia's Grab for Timor Oil* (Allen and Unwin, 2007) 4–8.

the *Offshore Minerals Act 1994* (Cth) ('the *OMA*') which has an application to all individual or corporations involved in offshore mining.[106] In Australia, the offshore minerals industry is small, with only two offshore dredging operations in production. These operations produce: (i) sand for construction purposes in Moreton Bay near Brisbane; and (ii) lime sand for cement manufacture south of Fremantle.

The jurisdictional framework for the *OMA* is similar to that under *OPGGSA*. Hence, the *OMA* provides the statutory framework for the exploration and production of minerals over the continental shelf three nautical miles beyond the territorial baseline of the states and territories.

Applications for a mineral exploration licence must be made to the Designated Authority, usually the state or territory minister responsible for mining. As with the *OPGGSA*, under the *OMA*, Joint Authority consists of the responsible state or territory minister and the responsible Commonwealth minister. The Joint Authority, which acts through different 'designated authorities' for those offshore territory areas, is responsible for major decisions relating to *OMA* licensing and the resource titles available under the *OMA* include exploration licences, retention licences and mining licences.[107]

3.3 REVIEW QUESTIONS

1. What is the primary offshore resource that mining proponents have sought to explore for and commercialise to date?
2. Explain the constitutional arrangement for the regulation of the territorial seas.
3. How is the constitutional arrangement articulated within the *OPGGSA*?
4. *UNCLOS* articulates a framework for extended Commonwealth jurisdiction within the high seas. Explain the jurisdictional basis for the EEZ and how this is set out within *UNCLOS*.
5. How broad is the definition of 'petroleum' in the *OPGGSA*?
6. Explain the difference between an offshore petroleum exploration permit under the *OPGGSA* and an offshore geological storage formation injection licence under the *OPGGSA*.
7. Consider the following problem:
    > Esso Petroleum Pty Ltd hold a petroleum production lease in Bass Strait. They seek to acquire a greenhouse gas injection licence in order to develop a carbon capture project which will involve the transportation of carbon dioxide from

---

106 See *OMA* (Cth) s 37.
107 See *OMA* (Cth) ss 29–34.

the operational site for petroleum production to geological storage formations existing under the seabed. The liquidised carbon dioxide will be transported via underground pipelines. The storage formation for the carbon dioxide is located in the area covered by the production lease. Esso Petroleum Pty Ltd are unsure whether the production lease they already hold automatically entitles them to utilise the geological storage area as this area is immediately adjacent to their drilling operations. They seek your advice. In your advice, please refer to the relevant provisions of the *OPGGSA*.

8. What is the purpose of the Commission on the Limits of the Continental Shelf which was set up by *UNCLOS* and are its decisions binding or may they be appealed?

9. What factors does an applicant for an offshore production licence need to take into account in determining whether to commercialise the production of offshore petroleum?

10. Consider the following problem:

    In 2010 ExxonMobil, who hold an exploration licence for an area off the Western Australian coast, make a significant petroleum discovery utilising new, state-of-the-art deep sea exploratory equipment. This is an important find because the trend in Australia's production of liquid petroleum (crude oil and condensate) has been steadily going downwards from a peak in 1999. This has produced an increasing gap between Australia's petroleum production and its consumption of petroleum products. ExxonMobil need to determine whether to commercialise the discovery. But ExxonMobil knows that Australia is generally perceived to offer low prospectivity for oil, with relatively low discovery rates and small average field sizes. They are unsure in the circumstances whether to apply to NOPTA and the relevant joint authority for a retention/assessment title. If you were required to prepare a risk analysis for ExxonMobil, what factors would you take into account in making this decision?

3.4 FURTHER READING

Australian Government, Department of Resources, Energy and Tourism, *Guidelines for Offshore Petroleum Joint Authority Decision-Making Procedures* (1 January 2012) http://www.nopta.gov.au/_documents/guidelines/JA-Decision_guidelines.pdf.

Australian Government, Department of Resources, Energy and Tourism, *Offshore Petroleum Guideline for Grant and Administration of a Pipeline Licence* (January 2012) http://www.nopta.gov.au/legislation/guidelines.html.

Australian Government, Department of Resources, Energy and Tourism, *Offshore Petroleum Guideline for Grant and Administration of a Retention Lease* (May 2012) http://www.nopta.gov.au/_documents/guidelines/offshorepetroleumguidelinegrantadministrationretentionlease.pdf.

Australian Government, Department of Industry and Science, *Resources: Joint Petroleum Development Area and Greater Sunrise* http://www.industry.gov.au/resource/UpstreamPetroleum/Pages/JointPetroleumDevelopmentAreaandGreaterSunrise.aspx.

D Bothwick, *Report of the Montara Commission of Inquiry* (Commonwealth of Australia, 2010).

P Brazil, 'The Establishment of a National Offshore Petroleum Safety Authority' (2005) 24 *Australian and Resources Energy Law Journal* 90.

G Briggs, 'Australia's Extended Continental Shelf and Northwest Offshore Energy Prospects' (Strategic Analysis Paper, Future Directions International, 2010) http://www.futuredirections.org.au/files/1271732207-Australias%20Expanded%20Continental%20Shelf%20and%20Northwest%20Offshore%20Energy%20Prospects.pdf.

I Brownlie, *Principles of Public International Law* (Oxford University Press, 5th ed, 1998).

I Brownlie, *Principles of Public International Law* (Oxford University Press, 7th ed, 2008).

P Cleary, *Shakedown: Australia's Grab for Timor Oil* (Allen and Unwin, 2007).Commonwealth of Australia, Report of the Montara Commission of Inquiry, *National Report* (2010) http://www.industry.gov.au/resource/Documents/upstream-petroleum/approvals/Montara-Report.pdf.

R Cullen, *Federalism in Action: The Australian and Canadian Offshore Disputes* (Federation Press, 1990).

J E Darling, 'Addendum: The Story of the BP Oil Spill' (2012) 3 *Elon. Law Rev.* 254.

M Evans, *The Law of the Sea* in M Evans (ed) *International Law* (Oxford University Press, 2003) ch 20.

ExxonMobil Australia, *About us: Bass Strait* http://www.exxonmobil.com.au/Australia-English/PA/about_what_gipps_bs.aspx.

D Goedhuis, 'Some Recent Trends in the Interpretation and the Implementation of the Rules of International Space Law' (1981) 19 *Columbia Journal of Transnational Law* 213.

J Griffiths, 'Rewriting the *Petroleum (Submerged Lands) Act* 1967 (Cth)' (2005) 24 *Australian Resources and Energy Law Journal* 372.

E Guntrip, 'The Common Heritage of Mankind: An Adequate Regime for Managing the Deep Seabed?' (2003) 4 *Melbourne Journal of International Law* 376.

M Harry, 'The Deep Seabed: The Common Heritage of Mankind or Arena for Unilateral Exploitation?' (1992) 40 *Naval Law Review* 207.

C D Hunt, 'The Offshore Petroleum Regimes of Canada and Australia: Some Comparative Observations' (1988) 7 *AMPLA Bulletin* 103.

T Hunter, 'The Montara Oil Spill and the Marine Oil Spill Contingency Plan: Disaster Response or Just a Disaster?' (2010) 24 *Australian and New Zealand Maritime Law Journal* 46.

T Hunter, 'Australian Offshore Petroleum Regulation After the Varanus Island Explosion and the Montara Blowout – Drowning in a Sea of Federalism' (2011) 25 *Australian and New Zealand Maritime Law Journal* 69.

T Hunter, 'Offshore Petroleum Facility Incidents Post-Varanus Island, Montara and Macondo: Have We Really Addressed the Root Cause' (2014) 38 *William and Mary Environmental Law and Policy Review* 585.

Index Mundi, *Commodity Prices* http://www.indexmundi.com/commodities/?conunodity=copper&months=300.

D Jongsma, Department of Minerals and Energy, Bureau of Mineral Resources, Geology and Geophysics, *A Review of Marine Geophysical Investigations Over the Lord Howe Rise and Norfolk Ridge* (1976) 33 http://www.ga.gov.au/corporate_data/13432/Rec1976_012.pdf.

M Kashubsky, 'Can an Act of Piracy be Committed Against an Offshore Petroleum Installation' (2012) 26 *Australian and New Zealand Maritime Law Journal* 163.

G Kevin, 'The Development of Outer Continental Shelf Energy Resources' (1984) 11 *Pepperdine Law Review* 9.

B Larschan and B C Brennan, 'The Common Heritage of Mankind Principle in International Law' (1983) 21 *Columbia Journal of Transnational Law* 305.

S Marsden, 'Regulatory Reform of Australia's Offshore Oil and Gas Sector After the Montara Commission of Inquiry: What About Transboundary Environmental Impact Assessment?' (2013) 15 *Flinders Law Journal* 21.

T L McDorman, 'The Entry into Force of the 1982 LOS Convention and the Article 76 Outer Continental Shelf Regime' (1995) 10 *International Journal of Marine and Coastal Law* 165.

K M Murchison, 'Beyond Compensation for Offshore Drilling Accidents: Lowering Risks, Improving Response' (2012) 30 *Mississippi College Law Review* 277.

National Commission on the BP Horizon Oil Spill and Offshore Drilling, 'Chapter 2: The History of Offshore Oil and Gas' in *Deepwater: The Gulf Oil Disaster and the Future of Offshore Drilling* (2011). http://www.gpo.gov/fdsys/pkg/GPO-OILCOMMISSION/pdf/GPO-OILCOMMISSION.pdf.

National Commission on the BP Horizon Oil Spill and Offshore Drilling, *Deepwater: The Gulf Oil Disaster and the Future of Offshore Drilling* (2011) 103 http://www.gpo.gov/fdsys/pkg/GPO-OILCOMMISSION/pdf/GPO-OILCOMMISSION.pdf.

Neptune Minerals, *Seafloor Minerals Exploration and Resource Development* http://www.neptuneminerals.com/.

J Nevins, 'Contesting the Boundaries of International Justice: State Countermapping and Offshore Resource Struggles between East Timor and Australia' (2004) 80(1) *Economic Geography* 1.

Offshore Technology.com, *Kitan Oilfield, Australia* http://www.offshore-technology.com/projects/kitanoilfield/.

P A Rona, 'Resources of the Sea Floor' (2003) 299 *Science* 673.

C Scheiner, 'Drilling East Timor: Australia's Oil Grab in the Timor Sea' (2006) 27(1) *MultiNational Monitor* 30.

M J Smith, 'Australian Claims to the Timor Sea's Petroleum Resources: Clever, Cunning or Criminal?' (2011) 37 *Monash University Law Review* 42.

A Symon, 'Economic Issues No 2: Timor Sea Natural Gas Development: Still in Embryo' (2) *Economic Issues* (August 2001) http://www.adelaide.edu.au/saces/publications/papers/issues/EconIssuesTwo.pdf.

United Nations, Oceans and Laws of the Sea, *Commission on the Limits of the Continental Shelf (CLS): Selected documents of the Commission* http://www.un.org/depts/los/clcs_new/commission_documents.htm#Statements by the Chairman of the Commission.

A Warwyk, 'Recent Changes to the Commonwealth Offshore Petroleum Legislation: Strengthening Environmental Liability, Compliance and Enforcement Provisions' (2013) 27 *Australian and New Zealand Maritime Law Journal* 49.

M White, *Australia Offshore Areas* (Federation Press, 2009).

M White, 'Australia's Offshore Legal Jurisdiction: History, Development and Current Situation' (2011) 25(1) *Australia and New Zealand Maritime Law Journal* 19.

M White and N Gaskell, 'Australia's Offshore Constitutional Law: Time for Revision' (2011) 85 *Australian Law Journal* 504.

P Yuan, 'China's Offshore Petroleum Resources Law' (1985) 1 *Journal of Law and Environment* 105.

NATURAL GAS REGULATION

4.1	Introduction	137
4.2	What is natural gas?	139
4.3	The Australian gas market	140
4.4	The regulatory framework for natural gas	143
4.5	Review questions	171
4.6	Further reading	172

4.1 Introduction

This chapter overviews the regulation of natural gas in Australia. It examines the nature of natural gas as a fossil fuel and hydrocarbon and seeks to outline the scope and capacity of the domestic natural gas and liquid natural gas market in Australia, including some of the current challenges faced by this market in light of the growing international LNG market. The particular focus of the chapter, however, is upon the natural gas transportation industry in Australia. Transportation issues are crucial for natural gas as they ensure that natural gas is able to travel from the well-head through gathering systems and midstream processing facilities and into the storage or transportation pipelines ready for delivery to the consumer. Regulating the management and access arrangements associated with the transportation of gas in gas pipelines across Australia through a nationally coordinated legislation helps to promote regulatory certainty and clarity for the industry.

Following the findings of the 1990 Commonwealth National Gas Strategy and the 1991 Industry Commission Report on Energy Generation and Distribution, the Australian gas industry underwent major restructuring. Both reports indicated a strong need for reform in order to increase competition and efficiency and to lower energy prices to both industry and domestic users. To date, the changes which have been made involve the breakup of a number of government-owned single entity organisations into competing corporatised bodies. Third party access legislation enabling independent operators to negotiate their entry into transmission pipelines on commercial terms has also been implemented. The private sector dominates the supply of gas from the upstream sector – that is, exploration for and development of gas reserves. The marketing of gas – and this includes transmission, distribution, and retailing – involves both public and private sector enterprises. The operations associated with the transmission of gas via high-pressure pipelines and its distribution via low-pressure reticulation systems is subject to monopolisation. The owner of the facilities acquires a natural monopoly which, absent government intervention, allows the owner to charge above competitive prices, often at the disadvantage of consumers and energy efficiency. This problem is compounded where the owner of the pipeline is vertically integrated, meaning their ownership and control extends to both upstream and downstream markets.[1] At present, Origin and AGL and, increasingly, EnergyAustralia retain this vertical integration power in the gas industry.[2]

---

1 Western Australia, *Parliamentary Debates*, Legislative Assembly, 18 June 2008, 2 (John Kobelke).
2 See COAG Energy Council, *Gas Transmission Pipeline Capacity Trading* (15 May 2013) 9 http://www.scer.gov.au/workstreams/energy-market-reform/gas-market-development/gtpct/. This report indicates that large vertically integrated retailers, such as AGL, Origin, and to a lesser extent

This chapter examines the scope and operation of the National Gas Laws ('the NGL'). The NGL is enacted as a law of South Australia and is set out in the *National Gas (South Australia) Act 2008* (SA). The NGL is a cooperative framework and mirror legislation has been enacted in all states and territories.[3] The NGL regulates Australian gas pipelines. It seeks to reform gas access regulation and replaces the *Gas Pipelines Access (South Australia) Act 1997* (SA), which was commonly referred to as the 'Gas Code'.

The NGL provides for the law to be supported by the National Gas Rules ('the NGR'). The NGR may be amended by the Australian Energy Market Commission ('the AEMC'). The primary purpose of the NGL is to regulate access to natural pipeline services and elements of broader natural gas markets. Australia's pipeline infrastructure is deemed an 'essential facility' in that the owner of that facility holds the key to the efficient operation of upstream and downstream markets.[4] The reforms implemented within the NGL were a component of the broader National Competition Policy Review that sought to implement a more competitive structure for, inter alia, the transmission and reticulation of natural gas through natural gas pipelines. In particular, it sought to introduce regulatory reforms that ensured third party access to essential facilities, such as gas pipelines, in order to promote competition and, in so doing, generate economic efficiency and consumer benefit.[5]

This chapter briefly reviews the operational dimensions of the *National Gas Code* that preceded the implementation of the NGL in 2008. The subsequent discussion of the NGL includes a detailed outline of the powers and functions of the regulatory bodies which operate under that law, the duties and functions of those bodies, the criterion by which it is determined that a pipeline is regulated and classified by the NGL, and an outline of the access information that must be disclosed according to the classification of the pipeline. The chapter then examines the enforcement powers of the regulators and the means by which access determination disputes are resolved. Relevant extracts from the NGL are also included to highlight the scope and detail of the regulatory framework.

    EnergyAustralia, control access to downstream domestic market channels including their own power stations.

3 See *National Gas (Queensland) Act 2008* (Qld); *National Gas (New South Wales) Act 2008* (NSW); *National Gas (ACT) Act 2008* (ACT); *National Gas (Victoria) Act 2008* (Vic); *National Gas (Tasmania) Act 2008* (Tas); *National Gas (Northern Territory) Act 2008* (NT); *Australian Energy Market Act 2004* (Cth); *National Gas Access (WA) Act 2009* (WA).

4 This is outlined by the Independent Committee of Inquiry into Competition Policy in Australia, Commonwealth Government, *National Competition Policy* (1993) 239 ('the Hilmer Report').

5 See W Pengilley 'Hilmer and Essential Facilities' (1994) 17 *University of New South Wales Law Journal* 1, 19 discussing how denial of access to an essential facility could, under Australian law, constitute a misuse of power.

4.2 What is natural gas?

'Natural gas' is the term used to describe naturally occurring methane gas that can be sourced from a range of naturally occurring geological formations in the earth.[6] Natural gas, crude oil, and coal are all referred to as 'hydrocarbons'. Hydrocarbons are comprised of hydrogen, carbon, and other impurities. The fewer carbon molecules that are contained within hydrocarbon, the more likely the hydrocarbon will be found in the gaseous phase. Crude oils contain longer chains of carbon molecules and are therefore heavier than gas and more likely to be found in a liquid state.

Methane is the primary component of natural gas and generally comprises 70–90 per cent of the total volume produced. Where the gas contains more than 95 per cent methane content, it will produce very little liquids. By contrast, anything containing less than 95 per cent methane and more than 5 per cent denser, heavier hydrocarbon molecules, such as ethane, propane and butane, will generally produce liquids during production.

Methane is the most common component of natural gas and is the main component consumed by power stations and industrial and residential users. It is methane that is converted to liquid natural gas – known as 'LNG' – and transported via pipelines and ships to consumers. LNG is the liquid product that is produced by cooling methane to -161.5°C. The liquid product is more readily transportable to overseas markets. Once it arrives, the liquid product is then heated back to standard temperature and pressure and converted to gaseous methane.

Liquefied petroleum gas – known as 'LPG' – is propane and butane that is stored, transported, and marketed in pressurised containers. Propane and butane gases liquefy at -43°C. Natural gas liquids – known as 'NGL' – include components that remain gaseous in both the reservoir as well as the above surface. This includes ethane, propane, and butane, as well as components that liquefy on the surface, such as condensates and natural gasoline. Condensates are low-density liquid mixtures of heavier hydrocarbons such as pentanes. Heavier, non-methane products have significant value given their capacity to generate heat and energy and a significant gas project will generally produce both methane and non-methane products for commercial markets.

Natural gas may also contain non-hydrocarbon impurities including carbon dioxide, hydrogen sulphide, hydrogen, nitrogen, helium, and argon. These

---

6 See the discussion by V Chandra, *Fundamentals of Natural Gas: An International Perspective* (Pennwell Book Publishing, Tulsa, 2006) 4. For further discussion on the nature and composition of natural gas see H P Pressler, 'Legal Problems Involved in Cycling Gas in Gas Fields' (1945) 24 *Texas Law Review* 19, 20–1; J C Jacobs, 'Unit Operation of Oil and Gas Fields' (1948) 57 *Yale Law Journal* 1207, 1210.

impurities must be separated from the natural gas prior to its commercialisation. In particular, carbon dioxide is a greenhouse gas and causes significant pollution and contributes to global warming and hydrogen sulphide can corrode pipelines.

Natural gas is the cleanest burning fossil fuel.[7] Burning methane releases carbon dioxide and water and few other impurities. Within a carbon economy, where emissions are costly, natural gas is increasingly becoming a valuable commodity.[8] Increased demand is also driving an increase in global trade. Gas is transported as LNG, which can be shipped via tankers anywhere in the world. This has allowed gas to be dramatically mobilised and this, in turn, has led to increased globalisation leading to the likelihood of LNG becoming a global commodity.[9]

4.3 The Australian gas market

Australia has an abundance of natural gas and as demand increases – given the fact that gas is a relatively flexible resource and, as a clean fuel, is a preferred fossil fuel within a carbon-restricted economy – total gas consumption is expected to double by 2019–2020 with gas being expected to make a much greater contribution to electricity generation.[10] Gas is the third largest energy resource in Australia after coal and uranium. Most conventional gas resources are located off the northwest coast of Australia and have been developed for domestic use and for export as LNG. Unconventional gas resources also exist in the major coal basins of eastern Australia and are also developed for domestic use and for export as LNG. Australia possesses significant shale and tight gas resources, although exploration for

---

7 See the report by J Podesta and T E Wirth, 'Natural Gas: A Bridge Fuel for the 21st Century' (Center for Am. Progress 10 August 2009) <http://www.americanprogress.org/issues/green/report/2009/08/10/6513/natural-gas-a-bridge-fuel-for-the-21st-century>.

8 See R J Pierce, 'Natural Gas: A Long Bridge to a Promising Destination' (2012) 32 *Utah Environmental Law Review* 245, 251 where the author notes that the economic benefits of natural gas development have the potential to yield major global benefits.

9 See especially S L Sakmar, 'Global Gas Markets: The Role of LNG in the Golden Age of Gas and the Globalization of LNG Trade' (2013) 35 *Houston Journal of International Law* 655, 658 where the author notes that the long-term view is that demand for natural gas will continue to increase as more countries look to meet the rising demand for energy with lower emission fuels. See also International Energy Agency, *World Energy Outlook* (2010) 192 <http://www.worldenergyoutlook.org/mediaweo2010 (ie numerals).pdf.>.

10 See Australian Bureau of Agricultural and Resource Economics, *Australian Energy Projections to 2029–2030* (2010) 44 where it is predicted that gross natural gas production, including LNG, in the western market is projected to grow strongly, at an average rate of 7.1 per cent a year, to reach 4968 petajoules in 2029–2030. This is underpinned by increasing demand in the domestic market and increasing global demand for LNG.

these resources is in the early stage and the exact quantity of available resources is unclear.[11]

There are three primary and quite diverse gas markets in Australia: the western market, the northern market, and the eastern market. The western market covers Western Australia. The eastern market connects Victoria, New South Wales, Queensland, South Australia, and Tasmania. The northern market covers the Northern Territory and parts of South Australia.

Approximately 92 per cent of Australia's conventional gas resources are located in the Carnarvon, Browse, and Bonaparte Basins off the north-west coast. There are also some resources in south-west, south-east, and central Australia. Large coal seam gas ('CSG') resources exist in the coal basins of Queensland and New South Wales, with further potential resources in South Australia. Known tight gas accumulations are located onshore in South Australia, Western Australia, and Victoria, while contingent and potential shale gas resources are located in Queensland, Northern Territory, South Australia, and Western Australia.

Most of the Australian markets are expanding as a result of strong export demand. Growth in LNG exports has been supported by the development of a number of greenfield – that is, previously undeveloped – projects, including the Gorgon LNG, Ichthys, Wheatstone, and Browse projects. It is projected that by 2029–30, LNG exports from the western market have the potential to reach 73 million tonnes (3986 petajoules), reflecting an average annual growth rate over the projection period of 9 per cent.[12]

A significant component of this increase, particularly in the eastern market, has been a consequence of the contribution of unconventional gas – in particular, CSG – to the gas production profile.[13] In 2007–2008, CSG accounted for approximately 6 per cent of total gas consumption in Australia and 80 per cent in Queensland. Increasing production of CSG in both New South Wales and Queensland is projected, with growth likely to increase from 118 petajoules in 2007–2008 to 2507 petajoules by 2029–2030, representing 88 per cent of

---

11 See Australian Government, Department of Resources, Energy and Tourism, Geoscience Australia, Bureau of Resources and Energy Economics, *Australian Gas Resource Assessment 2012* (2012) 1 <http://www.ga.gov.au/webtemp/image_cache/GA21116.pdf>.

12 See Australian Government, Bureau of Resources and Energy Economics, *Resources and Energy Quarterly* (March Quarter 2014) http://www.bree.gov.au/publications/resources-and-energy-statistics. The March Quarter Report for 2014 projects a growth in global gas consumption by 12 per cent over the forecast period 2013–2019 to around 3930 billion cubic metres. The main regions of growth will be in Asia (predominantly China and India), followed by Africa and the Middle East. Coincident with this trend in consumption will be a shift in gas production, with a decline in European production more than offset by significant growth in Australia.

13 Coal seam gas and unconventional gas development and regulation are discussed in Chapter 5.

Figure 4.1 The Australian gas market
Source: Adapted from Geoscience Australia

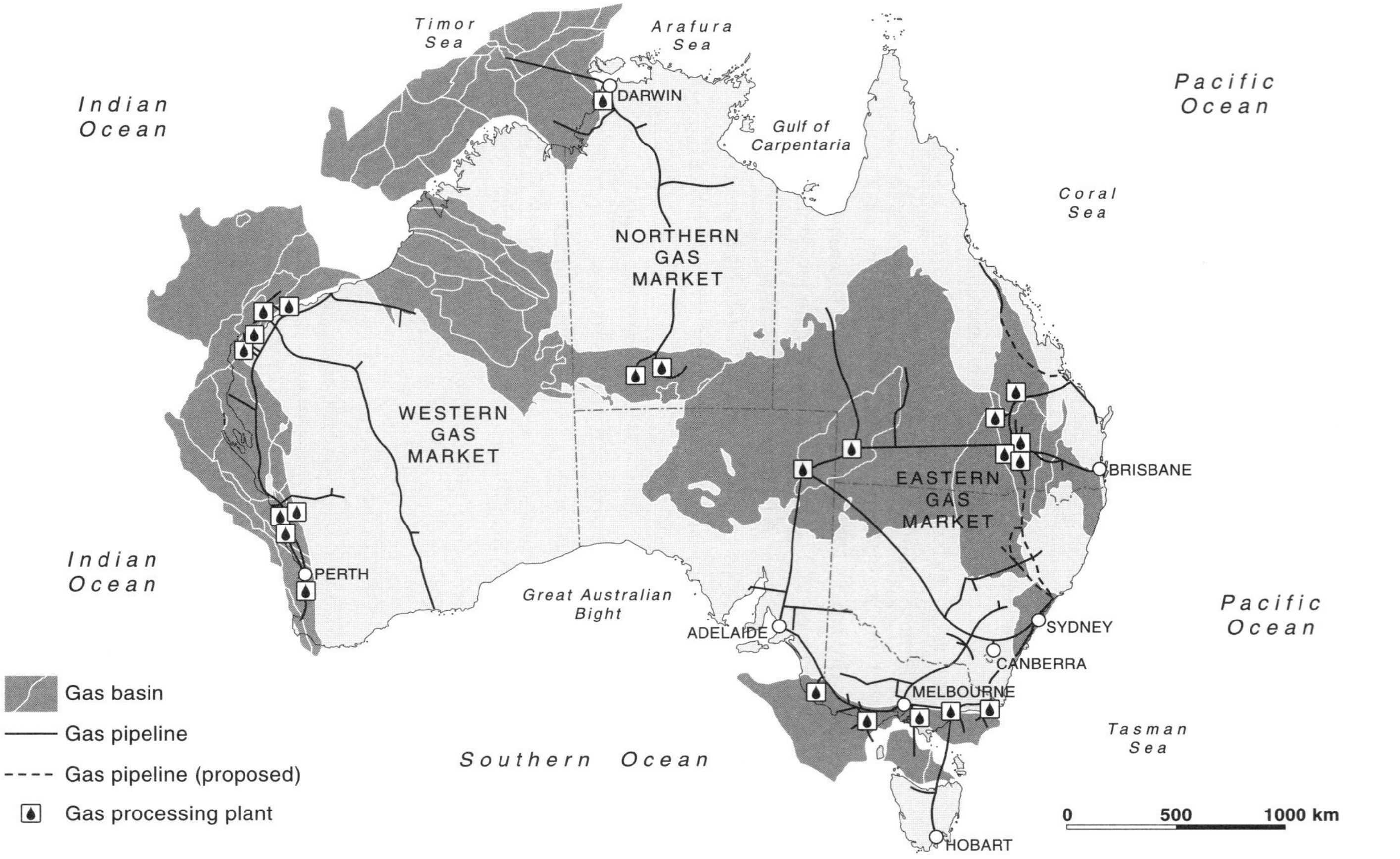

gas production in the eastern gas market.[14] It is also anticipated that from 2015, CSG will be converted to LNG and there are a number of conversion projects being planned in Queensland. If these projects proceed to commercialisation, the potential capacity for CSG-LNG development is 43 million tonnes a year by 2020, which would provide enough gas for both domestic and export demand for at least the next 20 years.

Australian gas consumption has grown by 4 per cent per year over the past decade. In 2009–2010, gas accounted for 23 per cent of Australia's primary energy consumption and 15 per cent of electricity generation. During this time, approximately 48 per cent of Australian gas production was exported as LNG. Higher export volumes combined with international oil prices increased the value of exports in 2010–2011 to $10.4 billion. The primary gas users in Australia are the manufacturing sector (32 per cent), electricity generation (29 per cent), mining sector (23 per cent), and residential sector (10 per cent). No tight or shale gas is currently produced in Australia.[15]

The 2015 Energy White Paper made it clear that despite the strong likelihood of rising gas prices, given the export price, the Federal Government does not support a domestic gas reservation policy. Rather, in line with its productivity and efficiency objectives, it is seeking to reduce regulatory impediments and accelerate gas production.[16]

4.4 The regulatory framework for natural gas

A discussion of natural gas regulation in Australia must necessarily commence with an outline of the energy framework. Part IIIAA of the *Competition and Consumer Act 2010* (Cth) establishes the Australian Energy Regulator (AER). The Australian Energy Market Agreement sets out the legislative and regulatory framework for Australia's energy markets. It provides for national legislation that is then implemented within each participating state and territory. South Australia is the lead legislator. Other jurisdictions have introduced application legislation to give effect to the South Australian legislation.

---

14 See A Syed, J Melanie, S Thorpe, and K Penney, 'Australian energy projections to 2029–2030' (ABARE research report 10.02, Department of Resources, Energy and Tourism, Canberra, March 2010).
15 See T Wood and L Carter, *Getting gas right – Australia's energy challenge* (June 2013) 10 http://grattan.edu.au/wp-content/uploads/2014/04/189_getting_gas_right_report.pdf.
16 The Energy White Paper was released in April 2015. See Australian Government, Department of Industry and Science, *Energy White Paper* (April 2015) http://www.ewp.industry.gov.au.

Key energy market legislation includes:

- The *National Electricity Law* (NEL) which is found in the Schedule to the *National Electricity (South Australia) Act 1996* (SA). The NEL establishes obligations in the national electricity market and for electricity networks. The NEL is supported by the *National Electricity Rules* and *National Electricity (South Australia) Regulations* 2003–2007 and 2010.
- The NGL is a Schedule to the *National Gas (South Australia) Act 2008* (SA). The NGL establishes obligations for gas pipelines, gas wholesale markets, and a gas market bulletin board. The NGL is supported by the *National Gas Rules* and *National Gas (South Australia) Regulations 2008*.
- The *National Energy Retail Law* is a Schedule to the *National Energy Retail Law (South Australia) Act 2011* (SA). The *National Energy Retail Law* (NERL) regulates the supply and sale of energy to retail customers. The NERL is supported by the *National Energy Retail Regulations 2012* (SA).
- The *Australian Energy Market Commission Establishment Act 2004* (SA) establishes the Australian Energy Market Commission (AEMC) and the *Australian Energy Market Commission Establishment Regulations 2005*.

4.4.1 The former natural gas access code

The gas access regime in Australia was previously governed by the Natural Gas Pipelines Access Inter-Government Agreement of 7 November 1997; the respective *Gas Pipeline Access Acts* of the Commonwealth, states and territories; and the Gas Pipelines Access Law (the 'Law') which included the *National Third Party Access Code for Natural Gas Pipeline Systems* (the 'Gas Code'), set up under the *Gas Pipelines Access (South Australia) Act 1997* (SA). Both the Law and the Gas Code were statutory instruments, and comprised Schedule 1 and Schedule 2 to both the *Gas Pipelines Access (South Australia) Act 1997* (SA) and the *Gas Pipelines Access (Western Australia) Act 1998* (WA). This regime no longer exists as it has been replaced by the National Gas Law ('the NGL'), which is discussed below; however, the functionality of the NGL can be better understood following a brief historical analysis of the preceding framework first established under the Gas Code.

The Gas Code was established in November 1997 by agreement between the Commonwealth, state, and territory governments.[17] The Law, including the Gas Code, implemented the access objectives which were agreed by all jurisdictions.

There are four primary elements to the gas sector in Australia: supply, transmission along high-pressure pipelines, distribution, and retailing. The Gas Code

---

17 The Gas Code was established pursuant to the *Gas Pipelines Access (South Australia) Act 1997* (SA), schs 1 and 2.

sought to regulate the downstream sector by implementing principles relevant for access to Australian natural gas transmission and distribution through pipeline services. It also established a national access regime for natural gas pipeline systems. This framework was the direct consequence of the 1994 decision by the Council of Australian Governments ('COAG') to agree to free and fair trade in natural gas at both the state and national levels which followed on from the competition policy review articulated within the Hilmer Report. One of the primary findings of the Hilmer Report was that in the early 1990s, the essential transmission and distribution pipeline facilities were government-owned and enjoyed significant protection from competition. The report suggested implementing a nationwide body to regulate gas pipeline access to prevent anti-competitive behaviour and foster consistency to access issues across the economy. Such a body could then encourage, where appropriate, the sharing of sector expertise to further promote the benefit of expertise, centralisation, and cross-industry experience for the gas pipeline framework.

The idea underlying the inception of the Gas Code was to make the gas industry more responsive to demand and to introduce greater innovation, entrepreneurship and competition on the part of the pipeline owners.[18]

The Gas Code allowed third parties to negotiate access to the pipelines with the owners of those pipelines. This ensured that gas users were able to contract directly for gas supply with an upstream producer who could then access the pipeline to deliver the gas.

The primary objectives of the Gas Code was to establish a framework for third party access to gas pipelines. This was to be achieved through the following measures:

(a) facilitating the development and operation of a national market for natural gas
(b) preventing abuse of monopoly power
(c) promoting a competitive market for natural gas whereby customers could choose suppliers, including producers, retailers and traders
(d) providing a right of access to natural gas pipelines on condition that the right was exercised in a fair and reasonable manner for both Service Providers and Users
(e) providing for the resolution of disputes.[19]

The Gas Code also required the owner or operator of a 'covered' pipeline to lodge an 'access arrangement' with the relevant regulator.

---

18 See the discussion by A Moran, *Natural Gas in Australia after the Hilmer Revolution* (Institution of Public Affairs, 2002) 4.
19 See *National Third Party Access Code for Natural Gas Pipeline Systems* s 3.1, 11, sch 1, cl 1: Introduction. The National Gas Code is set out in *Gas Pipelines Access (South Australia) Act* (SA) 1998, schs 1 and 2 (now repealed).

The primary features of the Gas Code may be summarised as follows:

- **Coverage**: the Code established a mechanism by which pipelines (including distribution systems) became subject to the Gas Code
- **Up-front access arrangement**: the Code relied upon an up-front access arrangement that outlined services and reference tariffs applicable to a covered pipeline
- **Pricing principles**: the Code established structured pricing principles
- **Segregation**: the Code implemented ring fencing so that businesses providing services to covered pipelines were segregated[20]
- **Disclosure**: the Code established information disclosure requirements
- **Arbitration**: the Code set up binding arbitration to determine disputes
- **Timelines**: the Code implemented specific timelines for all processes.

The Gas Code was designed to provide a clear national access regime with consistency between different jurisdictions. The scope of the Gas Code was limited to pipelines used for the haulage of natural gas. The definition of a 'pipeline' in this context included gas transmission pipelines, distribution networks and related facilities, but excluded upstream facilities. The upstream sector in this context refers to facilities that promote the searching for potential underground or underwater natural gas fields, drilling of exploratory wells, and subsequently, drilling and operating the wells that recover and bring the natural gas to the surface.

Schedule A of the Gas Code listed the pipelines that were automatically covered. Other pipelines could become 'covered' under the Code depending upon whether or not they fitted within the specified criteria. For a pipeline to come within the application of the Gas Code it had to be established that:

(a) access (or increased access) to services provided by means of the pipeline would promote competition in at least one market (whether or not in Australia), other than the market for the services provided by means of the pipeline

(b) that it would be uneconomic for anyone to develop another pipeline to provide the services provided by means of the pipeline

(c) that access (or increased access) to the services provided by means of the pipeline can be provided without undue risk to human health or safety

(d) that access (or increased access) to the services provided by means of the pipeline would not be contrary to the public interest.[21]

In 2006, following a report issued by the Productivity Commission, the Ministerial Council on Energy implemented key policy changes to the Gas Code ahead of

---

20 'Ring fencing' refers to segregating a business providing services using a covered pipeline.
21 See *National Third Party Access Code for Natural Gas Pipeline Systems* (WA) s 1.9.

the introduction of a new national gas law.[22] These changes included a new test for coverage under the gas access regime that allowed only those pipelines generating a material increase in competition in a related market to be covered by the Gas Code. Further, a range of pipelines satisfying clear criteria were able to be quarantined from coverage for up to 15 years and international pipelines were exempted from coverage with the aim of promoting investment and regulatory certainty.[23] A new objects clause was also introduced that sought to provide a uniform guiding principle that would promote efficient investment in, and operation and use of, natural gas services for the long-term interests of consumers of natural gas with respect to price, quality, safety, reliability and security of supply'.

4.4.2 The National Gas Law: Functions of the AER and the AEMC

In 2008, the Ministerial Council on Energy Standing Committee of Officials completed consultation on and drafting of a new National Gas Law ('the NGL'). The NGL was developed to reform the governance arrangements for the regulation of national gas pipeline services in Australia. The NGL fully replaces the Gas Pipelines Access Law (including the Gas Code) discussed above, which previously regulated pipeline services throughout Australia.

The NGL introduces new high-level policy direction, economic regulation, rule-making, and rule enforcement. The reforms were intended to encourage efficient investment in gas infrastructure, streamline the rule change process, and increase transparency in the gas market. The NGL is modelled on a similar regime which exists with respect to the National Electricity Law.[24] The increased consistency between electricity and gas regulation resulting from the implementation of the latest NGL is expected to strengthen the national character of the governance and economic regulation of the energy sector.[25]

---

22 Productivity Commission, *Review of the Gas Access Regime* (10 August 2004) http://www.pc.gov.au/inquiry/gas.
23 See the discussion on this by M J F Sweeney, 'Reform of the Gas Access Regime – Decision of the Ministerial Council on Energy' (2006) 25(2) *Australian Resources and Energy Law Journal* 132, 141 where the author notes that the intent of the Ministerial Council on Energy was to introduce changes which 'minimise the potential for regulatory distortions that can deter investment and at the same time, better establish a balance between the long term interests of consumers with the need to promote efficient investment in natural gas services, particularly transmission and distribution pipelines'.
24 The National Electricity Law is set out in the Schedule to the *National Electricity (South Australia) Act 1996* (SA).
25 See generally South Australia, *Parliamentary Debates*, House of Assembly, 12 June 2008, Hon Mr Williams MLA; cl 2.1 of the *Australian Energy Market Agreement 2004* (Cth) as amended in 2006.

The *National Gas (South Australia) Act 2008* (SA) which implemented the National Gas Law commenced on 1 July 2008. Mirror application legislation has been enacted in other jurisdictions implementing the National Gas Law in all states and territories.[26] The NGL reforms the gas access regulation and replaced the *Gas Pipelines Access (South Australia) Act 1997* (SA), which was commonly known as the National Third Party Access Code or the 'Gas Code'.

The NGL is supported by the new *National Gas Rules* ('the NGR') which regulate access to natural gas pipeline services and broader elements of natural gas markets and which have the effect of law.[27] The NGR implements the broader provisions of the NGL. The NGR sets out the processes by which the gas markets are operated and the responsibilities and obligations of the market participants.

The objectives of the NGR are to:

- facilitate efficient, competitive, and reliable gas markets
- regulate the operation and administration of the gas markets
- regulate activities of participants in the gas markets
- provide for access to the DTS (declared transmission system for pipeline access) and to ensure its security.

The NGL established two principal regulatory bodies for the gas and electricity sectors: the Australian Energy Regulator ('the AER') and the Australian Energy Market Commission ('the AEMC'). In Western Australia, it is the Economic Regulation Authority ('the ERA') rather than the AER which regulates gas markets, and this includes pipeline access. In all other states, the AER is the primary regulator.

The AER is responsible for monitoring, investigating, enforcing, and reporting on compliance by regulated entities with obligations under the NGL (as well as the NEL) and their respective Rules. The staffing and resources for the AER are provided by the Australian Competition and Consumer Commission ('the ACCC'). The ERA has equivalent responsibilities under the NGL in Western Australia. The functions and powers of the AER/ERA and the manner in which those functions and powers may be exercised are set out in ss 27 and 28 of the NGL which are extracted below:

---

26 See *National Gas (Queensland) Act 2008* (Qld); *National Gas (New South Wales) Act 2008* (NSW); *National Gas (ACT) Act 2008* (ACT); *National Gas (Victoria) Act 2008* (Vic); *National Gas (Tasmania) Act 2008* (Tas); *National Gas (Northern Territory) Act 2008* (NT); *Australian Energy Market Act 2004* (Cth). Western Australia applied the NGL in its jurisdiction on 1 January 2010 under the *National Gas Access (WA) Act 2009* (WA).

27 The *National Gas Rules* (NGR) that support the new NGL were introduced pursuant to s 294 of the *National Gas (South Australia) Act 2008*. The current consolidated version of the NGR is version 24. The NGR is available at http://www.aemc.gov.au/getattachment/2aacd64d-5572-49d2-b9fc-be3f5879c741/National-Gas-Rules-Version-24.aspx.

National Gas (South Australia) Act 2008 (SA) – Schedule: National Gas Law

27 Functions and powers of the AER

(1) The AER has the following functions and powers:
   (a) to monitor compliance by persons (including AEMO) with this Law, the Regulations and the Rules, including compliance with an applicable access arrangement, an access determination and a ring fencing decision; and
   (b) to investigate breaches or possible breaches of provisions of this Law, the Regulations or the Rules, including offences against this Law; and
   (c) to institute and conduct proceedings in relation to breaches of provisions of this Law, the Regulations or the Rules, including offences against this Law; and
   (d) to institute and conduct appeals from decisions in proceedings referred to in paragraph (c); and
   (e) AER economic regulatory functions or powers; and
   (f) to prepare and publish reports on the financial and operational performance of service providers in providing pipeline services by means of covered pipelines; and
   (g) to approve compliance programs of service providers relating to compliance by service providers with this Law or the Rules; and
   (h) any other functions and powers conferred on it under this Law or the Rules.
(1a) The AER has the following functions and powers in relation to the Procedures:
   (a) to investigate breaches or possible breaches of the Procedures referred to the AER by AEMO; and
   (b) to institute and conduct proceedings in relation to breaches of the Procedures referred to the AER by AEMO; and
   (c) to institute and conduct appeals from decisions in proceedings referred to in paragraph (b); and
   (d) to approve, in consultation with AEMO, compliance programs relating to compliance by Registered participants with the Procedures.
(2) The AER has the power to do all things necessary or convenient to be done for or in connection with the performance of its functions.

28 Manner in which AER must perform or exercise AER economic regulatory functions or powers

(1) The AER must, in performing or exercising an AER economic regulatory function or power –
   (a) perform or exercise that function or power in a manner that will or is likely to contribute to the achievement of the national gas objective; and
   (b) if the AER is making a designated reviewable regulatory decision –
      (i) ensure that –
         (A) the covered pipeline service provider that provides the pipeline services to which the applicable access arrangement decision will apply; and

    (B) users or prospective users of the pipeline services that the AER considers have an interest in the matter; and
    (C) any user or consumer associations or user or consumer interest groups that the AER considers have an interest in the matter,
    are, in accordance with the Rules –
    (D) informed of the material issues under consideration by the AER; and
    (E) given a reasonable opportunity to make submissions in respect of the decision before it is made; and
  (ii) specify –
    (A) the manner in which the constituent components of the decision relate to each other; and
    (B) the manner in which that interrelationship has been taken into account in the making of the decision; and
  (iii) if there are 2 or more possible designated reviewable regulatory decisions that will or are likely to contribute to the achievement of the national gas objective –
    (A) make the decision that the AER is satisfied will or is likely to contribute to the achievement of the national gas objective to the greatest degree (the ***preferable designated reviewable regulatory decision***); and
    (B) specify reasons as to the basis on which the AER is satisfied that the decision is the preferable designated reviewable regulatory decision.

(2) In addition, the AER –
 (a) must take into account the revenue and pricing principles –
  (i) when exercising a discretion in approving or making those parts of an access arrangement relating to a reference tariff; or
  (ii) when making an access determination relating to a rate or charge for a pipeline service; and
 (b) may take into account the revenue and pricing principles when performing or exercising any other AER economic regulatory function or power, if the AER considers it appropriate to do so.

(3) For the purposes of subsection (2)(a)(ii), a reference to a "reference service" in the revenue and pricing principles must be read as a reference to a "pipeline service".

The monitoring powers of the AER are extremely important as they promote compliance with the NGL. The capacity of the AER to institute formal enforcement proceedings in relation to breaches of the NGL, including breaches relating to an access determination, a ring fence determination, or the preparation and publication of reports on the financial and operational performance of service providers is crucial.

Example

In 2014 the AER investigated the claim that Epic Energy submitted incorrect allocation data to the AEMO and therefore was in breach of r 369 of the *National Gas Rules*.

Epic Energy South Australia Pty Ltd (Epic Energy) submitted incorrect Short Term Trading Market (STTM) allocation data to the Australian Energy Market Operator (AEMO) for the Moomba to Adelaide Pipeline (MAP) over 13 days between 29 June and 16 July 2013. On six of these days, Epic Energy's error resulted in the ex post price being set incorrectly. This distorted deviation payments which were payable because demand for gas could not be accurately predicted. On 8 July 2013, demand was under forecast in the Adelaide hub by about 10 terajoule or 10 000 gigajoule (GJ). This meant that some participants collectively paid around $10 000 too much because they paid an incorrect price of $7.42/GJ instead of $6.45/GJ.

The error was the product of Epic Energy's meters; however, even when Epic Energy became aware of the fault, it did not proceed to immediately rectify it. The failure to do so generated incorrect data and, as a result, the gas flow to the market was over-reported. Epic Energy had advised the AER that conditional alarms would be put in place by April 2012 and that erroneous data could be avoided in future by installing conditional alarms to detect for anomalous flow readings from a gas valve. Epic Energy advised the AER in October 2012 that conditional alarms had been installed when, in fact, they had not.

The broader impact of this failing was an erosion in confidence in market data and market outcomes. This, in turn, had an impact upon participation and competition in the market.

The AER found that Epic Energy was required to submit an allocation notice in accordance with the NGR requirements every day to the AEMO in order to assess the amount of gas delivered by trading participants against registered facility services. The delivery amounts are derived from daily meter measurements and from apportioning or allocating gas flows between trading participant offers and from the nomination of gas flows to and from the Adelaide hub. AEMO then uses this allocation data to calculate prices and payments in the short term trading market, which trading participants then become subject to.

Following an investigation, the AER held that Epic Energy was in breach of r 369 of the NGR and that the breach was the product of a failure to implement appropriate governance processes to effectively implement, review, and test the IT systems required for operating in the short term money market. The AER then issued an infringement notice on Epic Energy, specifying an infringement penalty of $20 000 on 20 December 2013.

The NGL confers a broad range of enforcement responses to breaches of the NGL and the NGR upon the AER. In addition to initiating court enforcement proceedings, as the Epic Energy example above illustrates, the AER has the power to issue infringement notices to regulated entities and to seek enforceable undertakings from regulated entities. An infringement notice will always provide the recipient with the option of paying a penalty or electing to have the matter heard in court. The AER's infringement notice regime is set out in div 5 of pt 6 of the NEL, and pt 7 of ch 8 of the NGL.

The enforcement orders that the AER may seek in a court include: orders to pay a civil penalty, orders to cease infringing conduct, orders to remedy infringing conduct, or orders to require the regulated entity to adopt a compliance plan. It has been suggested that these enforcement orders are too restrictive and should include community service and adverse publicity orders as well as orders for the payment of compensation or damages akin to the types of orders that can be made by the ACCC under ss 86C and 86D of the *Competition and Consumer Act 2010* (Cth).[28]

The Australian Energy Market Commission (AEMC) has a more expansive role under the NGL to that which it undertook under the former Gas Code. In addition to making recommendations on coverage and price regulation as well as exemptions for greenfield (undeveloped) pipelines, the AEMC is directly responsible for the classification and reclassification of pipelines, determining the form of regulation that should be applied to covered pipelines, and ensuring that adequate advice is given upon an application for amendment to a pipeline description referred by the relevant minister. The AEMC also evaluates applications for greenfield pipeline incentives (15-year no-coverage determinations and price regulation exemptions) by providing recommendations to ministers.

The full powers and functions of the AEMC are set out pursuant to s 74 of the NGL which is extracted below:

National Gas (South Australia) Act 2008 (SA) – Schedule: National Gas Law

74 Subject matter for National Gas Rules

(1) Subject to this Division, the AEMC, in accordance with this Law and the Regulations, may make Rules, to be known, collectively, as the "National Gas Rules", for or with respect to –

---

28 For a detailed review of the adequacy of the enforcement procedures under the NGL see Allens Linklaters and NERA Economic Consulting, *Review of Enforcement Regimes Under the National Energy Law: A Report Prepared for the Standing Council on Energy and Resources* (15 August 2013) 39–40.

(a) regulating –
  (i) access to pipeline services; and
  (ii) the provision of pipeline services; and
  (iii) the collection, use, disclosure, copying, recording, management and publication of information in relation to natural gas services; and
  (iv) the operation of a regulated retail gas market; and
  (v) AEMO's declared system functions and the operation of a declared wholesale gas market; and
  (va) AEMO's STTM functions and the operation of a short term trading market of an adoptive jurisdiction; and
  (vi) the activities of Registered participants, users, end users and other persons in a regulated gas market; and
  (vii) the safety, security and reliability of pipelines; and
  (viii) the connection of premises of retail customers; and
  (aaa) AEMO's gas trading exchange functions and the operation of a gas trading exchange; and
  (aa) facilitating and supporting the provision of services to retail customers; and
  (b) any matter or thing contemplated by this Law, or necessary or expedient for the purposes of this Law.
(2) Without limiting subsection (1), the AEMC, in accordance with this Law and the Regulations, may make Rules for or with respect to any matter or thing specified in Schedule 1 to this Law.
(3) Rules made by the AEMC in accordance with this Law and the Regulations may –
  (a) be of general or limited application;
  (b) vary according to the persons, times, places or circumstances to which they are expressed to apply;
  (c) confer functions or powers on, or leave any matter or thing to be decided or determined by –
    (i) the AER, the AEMC or AEMO; or
    (ii) any panel or committee established by the AEMC; or
    (iii) any other body established, or person appointed, in accordance with the Rules;
  (d) confer rights or impose obligations on any person or a class of person (other than AEMO, the AER or the AEMC);
  (e) confer a function on the AER, the AEMC or AEMO to make or issue guidelines, tests, standards, procedures or any other document (however described) in accordance with the Rules, including guidelines, tests, standards, procedures or any other document (however described) that leave any matter or thing to be determined by the AER, the AEMC or AEMO;
  (f) empower or require any person (other than a person referred to in paragraph (e)) or body to make or issue guidelines, tests, standards, procedures or any other document (however described) in accordance with the Rules;

(fa) provide for Procedures governing the operation of regulated gas markets;

(g) apply, adopt or incorporate wholly or partially, or as amended by the Rules, the provisions of any standard, rule, specification, method or document (however described) formulated, issued, prescribed or published by any person, authority or body whether –
   (i) as formulated, issued, prescribed or published at the time the Rules are made or at any time before the Rules are made; or
   (ii) as amended from time to time;

(h) confer a power of direction on the AER, the AEMC or AEMO to require a person conferred a right, or on whom an obligation is imposed, under the Rules to comply with–
   (i) a guideline, test, standard, procedure or other document (however described) referred to in paragraph (e) or (f); or
   (ii) a standard, rule, specification, method or document (however described) referred to in paragraph (g);

(i) if this section authorises or requires Rules that regulate any matter or thing, prohibit that matter or thing or any aspect of that matter of thing;

(j) provide for the review of, or a right of appeal against, a decision or determination made under the Rules and for that purpose, confer jurisdiction on the Court;

(k) require a form prescribed by or under the Rules, or information or documents included in, attached to or given with the form, to be verified by statutory declaration;

(l) in a specified case or class of case, exempt –
   (i) AEMO; or
   (ii) a Registered participant or class of Registered participant; or
   (iii) any other person or body performing or exercising a function or power, or conferred a right, or on whom an obligation is imposed, under the Rules or a class of any such person or body

   from complying with a provision, or part of a provision, of the Rules;

(m) provide for the modification or variation of a provision of the Rules (with or without substitution of a provision of the Rules or part of a provision of the Rules) as it applies to–
   (i) AEMO; or
   (ii) a Registered participant or class of Registered participant; or
   (iii) any other person or body performing or exercising a function or power, or conferred a right, or on whom an obligation is imposed, under the Rules or a class of any such person or body;

(n) confer an immunity on, or limit the liability of, any person or body performing or exercising a function or power, or conferred a right, or on whom an obligation is imposed under the Rules;

(na) require a person or body performing or exercising a function or power, or on whom a right is conferred or an obligation is imposed under the Rules, to indemnify another such person or body;

(o) contain provisions of a savings or transitional nature consequent on the amendment or revocation of a Rule.

The AEMC is responsible for creating and updating the NGR that regulate the provision of pipeline services, access to pipeline services, monitoring obligations, and gas market rules. In this respect, the AEMC produces the statutory rules that are enforced by the AER. It also provides expert advice for Commonwealth, state, and territory governments. The AEMC is an independent national body responsible to the Council of Australian Governments through the Standing Council on Energy and Resources.[29] The NGR requires that the AEMC establish and maintain a register of all pipelines that are, or have been, subject to any form of regulation or exemption from regulation under the *National Third Party Access Code for Natural Gas Pipelines* (the 'old scheme'). This is an onerous responsibility and the task of mapping all of these pipelines has been undertaken by the AEMC. Australia has an extensive network of pipelines with three distinct interconnecting networks:

- the eastern Australia network incorporating South Australia, New South Wales, the Australian Capital Territory, Queensland, and Victoria
- the central network incorporating the Northern Territory
- the western network incorporating Western Australia.

The pipeline networks are not presently interlinked. A number of new pipelines and extensions to existing pipelines have been built and a number of major new pipelines projects are planned. New pipeline proposals totalling 11 000 km in length are at various stages of development.[30] The map (Figure 4.2) provided on the Australian Parliament website is extracted below to highlight the range and scope of the pipelines across Australia.

Finally, the Australian Energy Market Operator ('AEMO') was created by the COAG and developed under the guidance of the Ministerial Council on Energy ('the MCE'). The AEMO seeks to improve the national character of energy market governance by bringing together the responsibility for the functionality of both gas and electricity retail and wholesale markets as well as the management of Victoria's gas transmission network and national transmission planning under the one operational framework. As part of its gas market functions, the AEMO is responsible for the establishment of a short-term trading market which sets a daily wholesale price for natural gas.

---

29 The Council of Australian Governments' Standing Council on Energy and Resources (SCER) is the primary body responsible for examining issues of national significance in the energy and resources sectors. The Council aims to encourage a prudent and competitive mineral and energy sector through governance and policy development. See further the COAG Energy Council website at: http://www.scer.gov.au/about-us/.
30 See the discussion by M Roarty, 'Natural Gas: Energy for the New Millennium' (Research Paper No 5, Parliament of Australia, 1999) http://www.aph.gov.au/About_Parliament/Parliamentary_Departments/Parliamentary_Library/pubs/rp/rp9899/99rp05.

Figure 4.2 Pipelines across Australia

Source: Adapted from MapsOfWorld

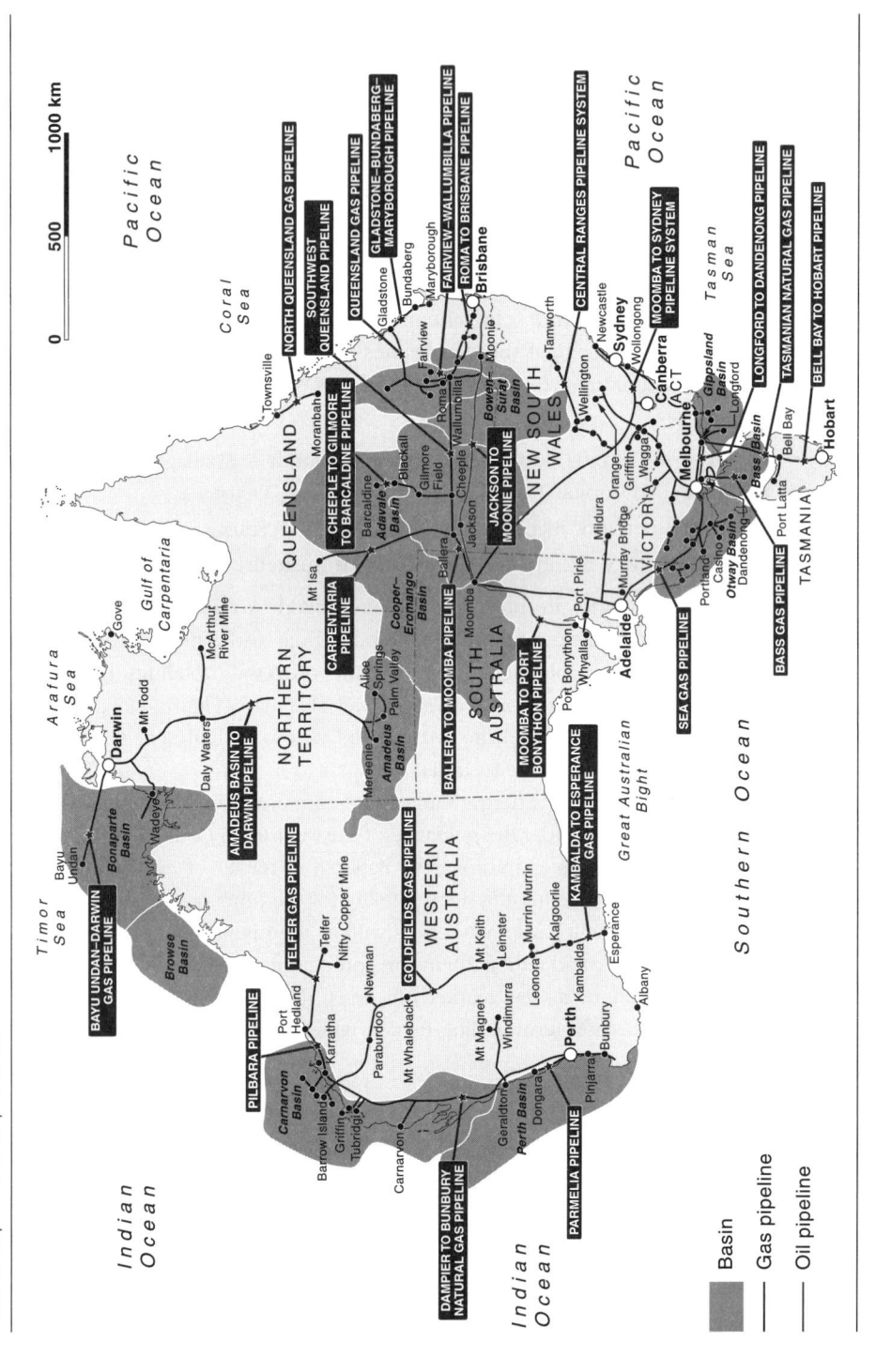

The legislative functions of the AEMO are set out in s 91A of the NGL, which is extracted below:

National Gas (South Australia) Act 2008 (SA) – Schedule: National Gas Law

91A AEMO's statutory functions

(1) AEMO has the following functions:
   (a) to operate and administer markets for natural gas in accordance with this Law, the Rules and the Procedures;
   (b) to promote the development, and improve the effectiveness of the operation and administration of, gas markets;
   (ba) conduct trials relating to the operation and administration of markets, or parts of markets, for natural gas that are or will be governed by this Law, the Rules and the Procedures;
   (c) to register persons as Registered participants;
   (d) to exempt certain persons from being registered as Registered participants;
   (e) to facilitate retail customer transfer, metering and retail competition (including balancing, allocation and reconciliation of gas deliveries and withdrawals to and from subnetworks);
   (f) for an adoptive jurisdiction – the declared system functions or STTM functions (as the case requires);
   (g) to make, amend or revoke Procedures;
   (ga) the gas trading exchange functions;
   (h) to operate and maintain the Natural Gas Services Bulletin Board;
   (i) to prepare, periodically review, revise, and publish the gas statement of opportunities;
   (j) to investigate breaches or possible breaches of the Procedures;
   (k) any functions conferred by jurisdictional gas legislation or an application Act;
   (l) any other functions conferred under this Law, the Rules or the Procedures.
(2) AEMO must, in carrying out functions referred to in this section have regard to the national gas objective.

The primary mandate of the AEMO is to operate both the retail and wholesale gas markets in south-east Australia and to regulate access to the Victorian gas Declared Transmission System ('the DTS'). The DTS transports natural gas to the vast majority of Victorian homes and businesses: it transports gas from Longford in the east to and from Culcairn in the north and Iona in the west. The Declared Transmission System Service Provider ('the DTSSP') maintains the gas transmission system facilities used to transport the gas between production or storage facilities and another transmission pipeline or storage facility, as well as to customers or the distribution networks. Operation of the wholesale market is governed by a set of processes, responsibilities, and obligations that are set out in the *Declared Wholesale Gas Market Rules* ('the *DWGMR*') and these are articulated in pt 19 of the NGR.

The gas market in Australia is regulated via either the short-term trading market ('the STTM'), which exists in New South Wales, Queensland and South Australia, or the wholesale market which exists in Victoria. Both markets are operated by AEMO. Under the STTM, shippers deliver gas to the market to be sold and users buy gas which is then delivered to the consumers and the prices are determined by the daily market price, which reflects the supply and demand context. AEMO has control of the market but is not involved in the operation of production facilities, transmission pipelines, storage facilities, or distribution networks. Further, AEMO does not interfere with the fundamental contractual arrangements that exist between pipeline operators and shippers. The regulations for the STTM are part of the NGR which is set out by the AEMC.

Under the wholesale market in Victoria, participants trade in the imbalance that exists in the declared transmission system between the gas supplied by a participant and the gas consumed by a participant. The wholesale gas market determines the price at which the imbalance may be traded. Based upon hourly aggregate demand forecasts, AEMO will declare a market price at the beginning of each day which is utilised as the transactional cost of gas that is injected into and withdrawn from the declared transmission system. Participants 'offer' gas into the wholesale market via competitive bidding.[31]

4.4.3 The National Gas Law: An overview

The current NGL has introduced a national gas objective with the aim of implementing uniform regulatory objectives.[32] The objective is articulated within s 23 of the NGL as follows:

> ... to promote efficient investment in, and efficient operation and use of, natural gas services for the long term interests of consumers of natural gas with respect to price, quality, safety, reliability and security of supply of natural gas.

The AEMC and the AER are both required to have regard to the national gas objective in performing their functions under the NGL. The national gas objective is primarily an economic concept that requires account to be taken of the long-term time frame, competitive access coverage, safety measures, reliability in investment, and overall promotion of consumer interests.[33] The inclusion of a single, guiding

---

31 See the discussion by AEMO, *Guide to Victoria's Wholesale Declared Gas Market* (February 2012) www.aemo.com.au/~/media/Files/Other/corporate/0000–0359 pdf.pdf
32 See *National Gas (South Australia) Act 2008* (SA), Schedule: National Gas Law, pt 3, div 1, s 23.
33 Western Australia, *Parliamentary Debates*, Legislative Assembly, 18 June 2008 (John Kobelke, MLA).

principle promoting consistency and certainty in regulatory decision-making is regarded as important because it creates a more attractive environment for investment.

In addition to establishing access entitlements, the broader functions of the NGL may be summarised as follows:

- sets out the requirements and duties of the provider of a covered pipeline service, including ring fencing
- sets out the access arrangements for covered pipelines, including price and revenue regulation and the arbitration of access disputes by the regulator
- sets up the Natural Gas Services Bulletin Board and the Short Term Trading Market
- outlines enforcement proceedings where a breach of the NGL occurs
- outlines the way in which decision-making under the NGL is to be carried out
- sets up the scheme register.

4.4.4 Pipeline classification under the National Gas Law

The NGL only has an application to what it describes as 'covered' pipelines.[34] Access to 'uncovered' pipelines remains a matter for private commercial negotiation and arrangements without recourse to a regulator. As with its predecessor, the NGL applies what is known as 'economic regulation' to 'covered pipelines'. Covered pipelines are those which meet the coverage criteria. Pursuant to s 15 of the NGL, coverage is determined according to market threshold criteria. The criteria requires a 'material increase' in at least one market to occur where access to a pipeline services occurs. Further, it must be proven to be uneconomic to develop another pipeline to provide pipeline services and also that access to the pipeline does not cause risk to human health or safety and is not detrimental to the public interest. In *The Pilbara Infrastructure Pty Ltd v Australian Competition Tribunal* the High Court concluded that 'public interest' in this context 'imports a discretionary value judgement that is to be made by reference to undefined factual matters' and, as such, the range of matters the Minister may have regard to when considering whether to be satisfied that access (or increased access) would not be contrary to the public interest is very wide indeed.[35]

---

34 See *National Gas (South Australia) Act 2008*, Schedule: National Gas Law, s 15 which sets out pipeline coverage criteria.
35 (2012) 246 CLR 379, [42].

The economic focus of the new coverage provisions were highlighted in the second reading speech for the National Gas (South Australia) Bill 2008:

> The national gas objective is an economic concept and should be interpreted as such. The long-term interest of consumers of gas requires the economic welfare of consumers, over the long-term, to be maximised. If gas markets and access to pipeline services are efficient in an economic sense, the long term economic interests of consumers in respect of price, quality, reliability, safety and security of natural gas services will be maximised. By the promotion of an economic efficiency objective in access to pipeline services, competition will be promoted in upstream and downstream markets.[36]

A pipeline will be covered by the NGL where the following criteria can be established:

- the relevant minister, taking into account the recommendations of the National Competition Council, determines that the pipeline should be covered on the basis that it meets all four of the 'pipeline coverage criteria' in s 15 of the NGL (which are, as outlined above: that access would promote a material increase in competition in another market, it is uneconomic to duplicate the pipeline, access is consistent with safety, and access is not contrary to the public interest);[37]
- the pipeline is developed following a competitive tender process approved by the AER under the new regime; or
- the service provider voluntarily submits an access arrangement in respect of the pipeline to the AER.[38]

The implementation of a tiered coverage regime under the NGL is crucial because it promotes the use of natural gas and reduces entry barriers. The importance of ensuring that vertically integrated organisations do not control access and limit competition was clearly outlined by the Federal Court in *BHP Billiton Iron Ore Pty Ltd v The National Competition Council*:

> ... it is the very prevention of a vertically integrated organization using its control over access to an essential facility to limit effective competition in dependent markets that is a key activity that the access regime seeks to deal.[39]

All persons are entitled to apply to the National Competition Council ('the NCC') to have a pipeline covered by the NGL regime or, alternatively, to have coverage

---
36 South Australia, *Parliamentary Debates*, Legislative Assembly, 12 June 2008, 2697 (PF Condon).
37 See *National Gas (South Australia) Act 2008* (SA) Schedule: National Gas Law s 15.
38 Ibid ss 15–17.
39 [2007] ATPR 42, 141.

revoked. Coverage will not, however, apply to pipeline which are subject to a 15-year 'greenfields' exemption.[40]

All service providers of fully regulated pipelines are required to submit access arrangements, including the price and other terms and conditions, to the AER for approval. Public consultation must be sought and a draft and final decision is implemented by the AER within prescribed time frames. If the AER decides not to approve the service provider's proposed access arrangement, an alternative access arrangement must be proposed by the AER within two months of the final decision being issued.[41]

The AER must take into account the revenue and pricing principles articulated within s 21 of the NGL when making decisions about tariffs and charges.[42] It must also take full account of the national gas objective. The revenue and pricing principles may be summarised as follows:

- a service provider is entitled to a reasonable opportunity to recover all proper costs incurred in providing reference services and complying with regulatory obligations
- a service provider needs to promote economic efficiency through incentives in the provision of reference services
- the value of the relevant pipeline must be considered
- consideration must be given to any pipeline valuation determined within a previous access arrangement or decision
- consideration should be given to the economic costs involved as well as the potential for under and over investment in the pipeline by a service provider
- consideration must be given to the potential for under and over utilisation of the pipeline.

Under the NGL, pipelines are automatically classified as either 'transmission' pipelines or 'distribution' pipelines by the NCC. The primary function of a transmission pipeline is to convey gas to a market. The primary function of a distribution pipeline is to reticulate gas within a market.

The classification of a pipeline affects the obligations to which the pipeline service provider is subject under the NGL and also which relevant minister has jurisdiction to determine coverage.

Section 98 of the NGL, which is extracted below, governs the making of pipeline classification decisions by the Council.

---

40 Section 151 of the NGL entitles a service provider to apply for a 15-year exemption with respect to a greenfields pipeline project before the pipeline is commissioned.
41 See NGL s 27.
42 See NGL s 28(2).

National Gas (South Australia) Act 2008 (SA) – Schedule: National Gas Law

98 Initial classification decision to be made as part of recommendation

(1) The NCC must, as part of a coverage recommendation, classify the pipeline the subject of an application under section 92 as a transmission pipeline or a distribution pipeline (an *initial classification decision*). In doing so, the NCC must apply the pipeline classification criterion.

(2) The NCC must as part of an initial classification decision –
   (a) if it classifies the pipeline the subject of the application as a transmission pipeline – determine whether the transmission pipeline is also a cross boundary transmission pipeline;
   (b) if it classifies the pipeline the subject of the application as a distribution pipeline – determine whether the distribution pipeline is also a cross boundary distribution pipeline.

(3) The NCC must also determine, as part of an initial classification decision, the participating jurisdiction with which the pipeline the subject of the application under section 92 is most closely connected if the NCC determines the pipeline is also a cross boundary distribution pipeline. In doing so, the NCC must apply the jurisdictional determination criteria.

The pipeline classification criterion for transmission and distribution characterisation is set out in s 13 of the NGL which is extracted below:

National Gas (South Australia) Act 2008 (SA) – Schedule: National Gas Law

13 Pipeline classification criterion

(1) The pipeline classification criterion is whether the primary function of the pipeline is to–
   (a) reticulate gas within a market (which is the primary function of a distribution pipeline); or
   (b) convey gas to a market (which is the primary function of a transmission pipeline).

(2) Without limiting subsection (1), in determining the primary function of the pipeline, regard must also be had to whether the characteristics of the pipeline are those of a transmission pipeline or distribution pipeline having regard to –
   (a) the characteristics and classification of, as the case requires, an old scheme transmission pipeline or an old scheme distribution pipeline;
   (b) the characteristics of, as the case requires, a transmission pipeline or a distribution pipeline classified under this Law;
   (c) the characteristics and classification of pipelines specified in the Rules (if any);
   (d) the diameter of the pipeline;

(e) the pressure at which the pipeline is or will be designed to operate;
(f) the number of points at which gas can or will be injected into the pipeline;
(g) the extent of the area served or to be served by the pipeline;
(h) the pipeline's linear or dendritic configuration.

Section 13(1) makes it clear that pipelines will be classified according to their primary function as a distribution pipeline or a transmission pipeline. Section 13(2) includes the words 'without limiting subsection (1)', meaning that the primary function test remains the main basis for pipeline classification despite the need to examine the physical characteristics of the pipeline including the diameter, the linear configuration, and the pressure at which the pipeline is designed to operate. The factors set out in s 13(2) are therefore intended to be informative rather than determinative.

For cross-boundary distribution pipelines the most important determinant is the jurisdiction with which the pipeline is most closely connected. This is set out in s 14 of the NGL which is extracted below:

National Gas (South Australia) Act 2008 (SA) – Schedule: National Gas Law

14 Jurisdictional determination criteria – cross boundary distribution pipelines

The pipeline jurisdictional determination criteria are –

(a) whether more gas is to be delivered by a cross boundary distribution pipeline in the jurisdictional area of 1 participating jurisdiction than in the jurisdictional area of any other participating jurisdiction;

(b) whether more customers to be served by a cross boundary distribution pipeline are resident in the jurisdictional area of 1 participating jurisdiction than in the jurisdictional area of any other participating jurisdiction;

(c) whether more of the network for a cross boundary distribution pipeline is in the jurisdictional area of 1 participating jurisdiction than in the jurisdictional area of any other participating jurisdiction;

(d) whether 1 participating jurisdiction has greater prospects for growth in the gas market served or to be served by a cross boundary distribution pipeline than any other participating jurisdiction;

(e) whether the regional economic benefits from competition are likely to be greater for 1 participating jurisdiction than for any other participating jurisdiction.

Whatever pipeline classification is made, an applicant is entitled to apply for a reclassification. In deciding whether a pipeline should be reclassified, the NCC must take into account the pipeline classification criterion in ss 13(1) and (2) and should also have regard to the national gas objective. In particular, consideration should be given to the issue of whether changing the obligations to which the pipeline's service provider is subject is likely to have any effect on the efficiency of pipeline access or the operation of gas markets. If it is likely that a change in pipeline classification will diminish the rights of third parties in a manner that is not consistent with the national gas objective, the reclassification may be refused.

4.4.5 Light and fully regulated pipelines

Where a pipeline is covered, two forms of regulation are available: full regulation and light regulation. The implementation of this dual regime is intended to prevent all covered pipelines from necessarily being subjected to 'full' regulation under the access arrangement process. Rather, eligible pipeline service providers operating covered pipelines may apply for a 'light regulation determination' and thereby avoid the upfront setting of reference tariffs (price regulation) under the access arrangement process. To encourage investment, the NGL also provides a 15-year no-coverage exemption for a greenfields pipeline project, where exemption is sought prior to a pipeline being commissioned. The NGL also provides an exemption for a major extension to a covered pipeline where the extension is specifically exempted by the AER.[43]

The primary concern of the NGL in this respect is to ensure that the form of regulation that is appropriate for particular pipeline services is proportionate to the degree of market power that is involved. Expensive, heavy regulation is only really needed where there is the potential for significant inefficiencies flowing from the exploitation of market power.[44]

A fully regulated pipeline requires the service provider to submit an 'access arrangement' – as outlined below – and to have that arrangement approved by the AER.

Light regulation removes price regulation and the requirement for an access arrangement. Light regulation adopts a negotiate/arbitrate model for third party access, with referral to arbitration by the regulator in the event of an access dispute. This process reflects the negotiate/arbitrate model to access which is set out in pt IIIA of the *Competition and Consumer Act 2010* (Cth). Light regulation can be more expedient and efficient where negotiation is successful. Where, however, negotiation is unsuccessful, light regulation may result in additional costs and time delays. Light coverage is generally granted to access holders to secure the competitiveness of investments.[45]

---

43 See NGL s 151(1).
44 This is discussed in National Competition Council, *Guide to Light Regulation of Covered Pipeline Services* (16 February 2010) 13 <http://ncc.gov.au/images/uploads/National_Gas_Law_-_Light_regulation_of_covered_pipeline_services.pdf>.
45 See NGL s 15(a) which expressly sets out that 'that access (or increased access) to pipeline services provided by means of the pipeline would promote a material increase in competition in at least 1 market (whether or not in Australia), other than the market for the pipeline services provided by means of the pipeline'.

The two-tiered market (covered and uncovered and light and full regulation) will only operate efficiently where the regulation is properly monitored. The objectives underpinning the promotion of competitiveness and public welfare, set out in s 15 of the NGL, may only be effectively implemented to accommodate the NGL objective where they are fairly and effectively monitored.[46]

The determination as to whether a covered pipeline is a light or full pipeline must be made in accordance with the requirements specified in s 112(2) of the NGL, which corresponds with s 34 of the NGR. Section 34 of the NGR is set out below:

National Gas Rules

34 Application for light regulation determination (Section 112(2) of the NGL)

(1) An application for a light regulation determination must:
   (a) be in writing; and
   (b) identify the pipeline that provides, or is to provide, the services for which the determination is sought and include a reference to a website at which a description of the pipeline can be inspected; and
   (c) include a description of all pipeline services provided or to be provided by means of the pipeline; and
   (d) include the applicant's reasons for asserting that the pipeline services should be light regulation services; and
   (e) include other information and materials on which the applicant relies in support of the application.
(2) The application must also include the following information:
   (a) the capacity of the pipeline and the extent to which that capacity is currently utilised; and
   (b) for a transmission pipeline, a description of:
      (i) all locations *served* by the pipeline (i.e. all locations at which *receipt or delivery points* for natural gas carried by the pipeline exist); and
      (ii) all pipelines that currently *serve* the same locations; and
      (iii) all pipelines that currently pass within 100 km of any location *served* by the pipeline; and
   (c) for a distribution pipeline, a description of:
      (i) the geographical area *served* by the pipeline; and
      (ii) the points at which natural gas is, or is to be, injected into the pipeline; and
   (d) a description of the pipeline services provided, or to be provided, by the pipeline; and
   (e) an indication of any other sources of energy available to consumers of gas from the pipeline; and
   (f) the identity of the parties with an interest in the pipeline and the nature and extent of each interest; and

---

46 Sweeney, above n 22, 5-6.

(g) a description of the following relationships:
   (i) any relationship between the owner, operator and controller of the pipeline (or any 2 of them);
   (ii) any relationship between the owner, operator or controller of the pipeline and a user of pipeline services or a supplier or consumer of gas in a location or geographical area *served* by the pipeline;
   (iii) any relationship between the owner, operator or controller of the pipeline and the owner, operator or controller of any other pipeline *serving* any one or more of the same locations or the same geographical area; and
(h) an estimate of the annual cost to the service provider of regulation on the basis of light regulation and on the basis of full regulation; and
(i) any other information the applicant considers relevant to the application of the National Gas Objective or the form of regulation factors in the circumstances of the present case.

4.4.6 Access determination disputes

An access dispute may arise where a service provider refuses to provide a requested pipeline service. This may occur as a result of a safety issue whereby, in accordance with s 114 of the NGR, the service provider must lodge a safety of operation notification. An independently appointed expert is then required to make a determination on this issue. In proceedings for the resolution of an access dispute, the relevant dispute resolution body must be bound by the findings of the approved expert. In proceedings for an access dispute, the relevant dispute resolution body must, where relevant, take into account:

- the value of any past capital contribution made by a party to the dispute; and
- the extent the party has re-couped any such past capital contribution.

An access determination may require the service provider to carry out an expansion of the capacity of the access dispute pipeline but may not require the service provider to extend the geographical range of the pipeline. Where such an expansion of capacity is ordered, the access arrangement may be varied to provide for the capital costs of the expansion, consequential adjustments to reference tariffs, and the establishment of a speculative capital expenditure account.

The relevant provisions are set out in ss 118 and 119 of the NGR (which refer back to s 191 of the NGL). Sections 118 and 119 of the NGR are extracted below:

National Gas Rules

118 Access determination requiring expansion of capacity (Section 191 of the NGL)

(1) An access determination:
   (a) may require the service provider to carry out an expansion of the capacity of the access dispute pipeline; but

(b) may not require the service provider to extend the geographical range of the access dispute pipeline. with an access determination unless the service provider agrees.
(2) However:
   (a) the service provider cannot be required to carry out an expansion of the capacity of a *light regulation pipeline* unless the prospective user funds the capacity expansion in its entirety; and
   (b) the service provider cannot be required to fund, in whole or part, an expansion of the capacity of a *full regulation pipeline* unless the extension and expansion requirements of the applicable access arrangement provide for the relevant funding; and
   (c) an expansion of capacity required under an access determination must be:
      (i) technically and economically feasible; and
      (ii) consistent with the safe and reliable operation of the pipeline.
(3) A user or prospective user acquires no interest in a pipeline by funding an expansion of capacity of the pipeline in accordance with an access determination unless the service provider agrees.

119 Variation of applicable access arrangement to accommodate capacity expansion (Section 191 of the NGL)
(1) This rule applies if an access determination:
   (a) requires a service provider to expand the capacity of the access dispute pipeline; and
   (b) requires a prospective user of incremental capacity to contribute some or all the cost.
(2) The access determination may make consequential amendments to the applicable access arrangement.
(3) The consequential amendments must provide for one or more of the following:
   (a) a mechanism to roll some or all the capital costs of the expansion into the capital base;
   (b) consequential adjustments to reference tariffs;
   (c) a surcharge to be levied on users of incremental services;
   (d) the establishment of a speculative capital expenditure account and regulation of its operation.
(4) The access determination (and the consequential amendments to the access arrangement) must set out the terms and conditions of access for a prospective user of incremental capacity who is to contribute some or all the cost of the capacity expansion.

4.4.7 Access arrangement information for light and full regulation pipelines

The NGL mandates that access arrangement information that is reasonably necessary for both users and prospective users be disclosed.[47] The nature and scope

---

47 See NGR s 44 which requires publication of copies of an access arrangement.

of the information to be revealed will depend upon whether the pipeline is covered by the NGL and whether the pipeline is classified as light or fully regulated.

In assessing the access arrangement, the NGL establishes a 'fit for purpose' model.[48] This model differs significantly from the previous Gas Code, which preferred to presume the suitability of a proposed access arrangement, even in circumstances where that arrangement did not represent the optimum or most appropriate outcome, provided it could be shown that the arrangement was reasonable. Under the NGL, no presumptions are made with respect to a proposed access arrangement and the AER is not obliged to approve any element of a proposed access arrangement unless it is clearly found to be fit for the purpose. The NGR sets out that the service provider, when submitting an access arrangement for the AER's approval, must submit information regarding the access proposal.[49]

The nature of access arrangement information is set out in s 42 of the NGR which is extracted below:

National Gas Rules

42 General requirements for access arrangement information

(1) *Access arrangement information* for an access arrangement or an *access arrangement proposal* is information that is reasonably necessary for users and prospective users:
   (a) to understand the background to the access arrangement or the *access arrangement proposal*; and
   (b) to understand the basis and derivation of the various elements of the access arrangement or the *access arrangement proposal*.
(2) *Access arrangement information* must include the information specifically required by the Law.

Within three months of a pipeline becoming a covered pipeline regulated under the NGL, a service provider is required to submit an access arrangement proposal proposing a full access arrangement for a covered pipeline. This is set out in s 132 of the NGL. No such requirement exists for a light regulation pipeline which need only comply with limited access arrangements.

The NGL distinguishes between the requirements for a 'limited access' arrangement for a light regulation pipeline and the full access arrangement requirements for a fully regulated pipeline.

The requirements for a limited access arrangement for a light regulation pipeline are set out in s 116(2) of the NGL, which corresponds with s 45 of the NGR. Section 45 of the NGR is extracted below:

---

48 See NGL s 28.
49 See NGR s 43.

National Gas Rules

45 Requirements for limited access arrangement (and limited access arrangement proposal)

(1) A limited access arrangement for a light regulation pipeline must:
   (a) identify the pipeline and include a reference to a website at which a description of the pipeline can be inspected; and
   (b) describe the pipeline services the service provider proposes to offer to provide by means of the pipeline; and
   (c) state the terms and conditions (other than price) for access to the pipeline services likely to be sought by a significant part of the market; and
   (d) if the access arrangement is to contain queuing requirements – set out the queuing requirements; and
   (e) set out the capacity trading requirements; and
   (f) set out the extension and expansion requirements; and
   (g) state the terms and conditions for changing receipt and delivery points; and
   (h) if there is to be a review submission date – state the review submission date and the revision commencement date; and
   (i) if there is to be an expiry date – state the expiry date.
(2) The access arrangement information for the limited access arrangement must include the following:
   (a) the capacity of the pipeline and the extent to which that capacity is currently utilised;
   (b) the key performance indicators for the pipeline.
(3) This rule extends to an access arrangement proposal consisting of a proposed limited access arrangement for a light regulation pipeline.

An access arrangement for a fully regulated pipeline is far more comprehensive in nature and scope and will necessarily include obligations to specify reference services, reference tariffs, capacity trading requirements, extension and expansion requirements and, where relevant, to review submission dates.

The requirements for a full access arrangement pursuant to a fully regulated gas pipeline are set out in s 48 of the NGR which is extracted below:

National Gas Rules

48 Requirements for full access arrangement (and full access arrangement proposal)

(1) A full access arrangement must:
   (a) identify the pipeline to which the access arrangement relates and include a reference to a website at which a description of the pipeline can be inspected; and
   (b) describe the pipeline services the service provider proposes to offer to provide by means of the pipeline; and
   (c) specify the reference services; and

(d) specify for each reference service:
  (i) the reference tariff; and
  (ii) the other terms and conditions on which the reference service will be provided; and
(e) if the access arrangement is to contain queuing requirements – set out the queuing requirements; and
(f) set out the capacity trading requirements; and
(g) set out the extension and expansion requirements; and
(h) state the terms and conditions for changing receipt and delivery points; and
(i) if there is to be a *review submission date* – state the *review submission date* and the revision commencement date; and
(j) if there is to be an *expiry date* – state the *expiry date*.

(2) This rule extends to be an *access arrangement proposal* consisting of a proposed full access arrangement.

The NGL seeks to optimise economic efficiency whilst at the same time maintaining the safety and reliability of pipeline access and taking account of the long-term interests of consumers. The NGL introduces a range of flexible regulations with the aim of creating a gas marketplace that promotes investment and progression and, at the same time, optimises consumer interests. There remain, however, many areas where improvement is needed under the NGL.

A reform of the NGL was announced in 2012, with the Council of Australian Governments endorsing a comprehensive package of national energy market reforms developed in collaboration with the Standing Council on Energy and Resources to respond to the current challenges of rising energy prices. A key component of the proposed reforms lay in the reforms recommended by the AEMC to promote more efficient outcomes in setting revenues and prices for consumers. It was felt that this was best achieved through the conferral of greater powers to the AER to undertake benchmarking, publish information on the relative efficiency of electricity network businesses, improve the enforcement powers of the AER, and develop enhanced budget and performance reporting.[50]

The new rules proposed by the AEMC are now set out in the NGR pursuant to the *National Gas Amendment (Price and Revenue Regulation of Gas Services) Rule 2012*. In implementing the rules, the AEMC notes that the most significant factors determining the revenues earned and the prices charged by proponents within the

---

50 The AEMC released its final position on new rules to regulate the NGL in 2012. These rules are now set out in the NGR pursuant to the *National Gas Amendment (Price and Revenue Regulation of Gas Services) Rule 2012*. For an outline of the proposed reforms see COAG Energy Council, *Energy Market Reform* http://www.scer.gov.au/workstreams/energy-market-reform/. See also the *Energy White Paper*, above n. 16.

NGL was the rate of return on capital as well as the size of the regulatory asset. The new rules affect the NGL by altering the rate of return. The new rate of return framework requires the regulator to estimate the best possible rate of return, taking into account market circumstances, financial models, and any other relevant information at the point when a regulatory determination has been made. The aim is to ensure that the AER has the capacity to adopt a variety of approaches to estimating the return on debt thereby allowing greater potential for reduced risk in debt financing.[51]

4.5 REVIEW QUESTIONS

1. What is Liquid Natural Gas (LNG) and what is it used for?
2. Briefly outline the current scope of the Australian gas market and its likely growth over the next decade.
3. What is the purpose of the *National Gas Law* (existing as a Schedule to the *National Gas (South Australia) Act 2008* (SA))?
4. What were the objectives of the predecessor to the NGL – the *National Third Party Access Code for Natural Gas Pipeline* – and why were they felt to be important in the context of the Australian domestic gas market? Are these objectives similar to those which have been articulated within the subsequently introduced *National Gas Rules*?
5. What role does the Australian Energy Market Commission have under the new NGL?
6. Consider the following problem:

    In 2014 the AER investigated an allegation that AGX was in breach of the *National Gas Rules* because it failed to adequately provide an outline of the financial and operational performance of its retail gas market. The report provided contained flaws and incorrect data. This data effectively meant that the submissions of AGX to the AEMO were also incorrect. This created errors in the market data and impacted upon market participation and competition. Is it possible for the AER to terminate the operating licence of AGX as a consequence of this activity?

7. Describe the state of the existing gas pipeline infrastructure in Australia.
8. What is the difference between 'covered' and 'uncovered' pipelines under the NGL?
9. Consider the following problem:

    The Moomba to Sydney pipeline (MSP) is utilised by vertically integrated energy companies (AGL, Origin, TRUenergy), energy retailers (Country Energy, EnergyAustralia), and some larger industrial companies (VISY and others). The

---

51 See AEMC, *Information: New Rules for Networks* (15 November 2012) http://www.aemc.gov.au/getattachment/d00de929-2e7d-421e-b082-e40c2e09abb9/Information-sheet-Final-position.aspx. The AER has indicated it will use benchmarking in making its assessments under the revised rules.

integrated energy companies and energy retailers generally use the transport services of the pipeline to ship gas for on-sale. At present only Origin uses gas transported by the MSP for electricity generation. MSP is owned by EAPL. EAPL argue that MSP should be subject to 'light regulation' under the NGL because any market power arising from the operation of the MSP was low. This was a consequence of the substitution threat from the other pipelines that exist in the area and increasingly, other sources of gas supply that do not rely on the MSP preferring to take advantage of swap contracts as an alternative to pipeline transport. Do you agree with EAPL that the pipeline should, in light of the market conditions, only be subject to light regulation? What do you think is the purpose of differentiating between 'light regulation' and 'full regulation' under the NGL?

4.6 FURTHER READING

AEMC, Information: New Rules for Networks (15 November 2012) http://www.aemc.gov.au/getattachment/d00de929-2e7d-421e-b082-e40c2e09abb9/Information-sheet-Final-position.aspx.

Allens Linklater and NERA Economic Consulting, *Review of Enforcement Regimes Under the National Energy Law: A Report Prepared for the Standing Council on Energy and Resources* (15 August 2013).

Australian Bureau of Agricultural and Resource Economics, Australian Energy Projections to 2029–2030 (2010).

Australian Government, Bureau of Resources and Energy Economics, Resources and Energy Quarterly (March Quarter 2014) http://www.bree.gov.au/publications/resources-and-energy-statistics.

Australian Government, Department of Resources, Energy and Tourism, Geoscience Australia, Bureau of Resources and Energy Economics, Australian Gas Resource Assessment 2012 (2012) <http://www.ga.gov.au/webtemp/image_cache/GA21116.pdf>.

Australian Energy Regulator, Weekly Gas Reports www.aer.gov.au/publications.

Australian Energy Regulator, *State of the Energy Market 2012*, Chapter 4: Gas Pipelines. http://www.aer.gov.au/sites/default/files/State%20of%20the%20energy%20market%202012%20%20Chapter%204%20Gas%20pipelines%20%28A4%29.pdf.

V Chandra, *Fundamentals of Natural Gas: An International Perspective* (Penwell Book Publishing, Tulsa, 2006).

D Clough, 'Economic Duplication and Access to Essential Facilities in Australia' (2000) 28 *Australian Business Law Review* 325.

COAG Energy Council, *Gas Transmission Pipeline Capacity Trading* (15 May 2013) http://www.scer.gov.au/workstreams/energy-market-reform/gas-market-development/gtpct/.

COAG Energy Council, *Limited Merits Review* (June, 2013) http://www.scer.gov.au/general-council-publications/regulatory-impact-statements.

COAG Energy Council, *Energy Market Reform* http://www.scer.gov.au/workstreams/energy-market-reform/.

M Groves, 'Energy Rule Changes in the Long Term Interests of Consumers' (2013) *Competition and Consumer Law News* 148.

F H Hilmer, 'The Bases of Competition Policy' (1994) 17 *University of New South Wales Law Journal* [v].

Independent Committee of Inquiry into Competition Policy in Australia, Commonwealth Government, National Competition Policy (1993).

Independent Committee of Inquiry into Competition Policy in Australia, *Hilmer Report, 1993* (AGPS, Canberra, 1993).

International Energy Agency, World Energy Outlook (2010) http://www.worldenergyoutlook.org/mediaweo2010.pdf.

J C Jacobs, 'Unit Operation of Oil and Gas Fields' (1948) 57 *Yale Law Journal* 1207.

L McDonald, 'The National Third Party Access Code For Natural Gas Pipeline Systems: A Balancing Act' (2003) 22 *Australian Resources and Energy Law Journal* 81.

A Moran, *Natural Gas in Australia after the Hilmer Revolution* (Institution of Public Affairs, 2002).

National Competition Council, *The National Access Regime: A Guide to Part IIIA of the Trade Practices Act – Part B Declaration* (NCC, Melbourne, December 2002).

National Competition Council, Guide to Light Regulation of Covered Pipeline Services (16 February 2010) http://ncc.gov.au/images/uploads/National_Gas_Law_-_Light_regulation_of_covered_pipeline_services.pdf.

W Pengilley 'Hilmer and Essential Facilities' (1994) 17 *University of New South Wales Law Journal* 1.

R J Pierce, 'Natural Gas: A Long Bridge to a Promising Destination' (2012) 32 *Utah Environmental Law Review* 245.

J Podesta and T E Wirth, 'Natural Gas: A Bridge Fuel for the 21st Century' (Center for Am. Progress, 10 August 2009) http://www.americanprogress.org/issues/green/report/2009/08/10/6513/natural-gas-a-bridge-fuel-for-the-21st-century.

H P Pressler, 'Legal Problems Involved in Cycling Gas in Gas Fields' (1945) 24 *Texas Law Review* 19.

Productivity Commission, Review of the Gas Access Regime (10 August 2004) www.pc.gov.au/inquiry/gas.

M Roarty, 'Natural Gas: Energy for the New Millennium' (Research Paper No 5, Parliament of Australia, 1999) http://www.aph.gov.au/About_Parliament/Parliamentary_Departments/Parliamentary_Library/pubs/rp/rp9899/99rp05.

S L Sakmar, 'Global Gas Markets: The Role of LNG in the Golden Age of Gas and the Globalization of LNG Trade' (2013) 35 *Houston Journal of International Law* 655.

M F Sweeney, 'Reform of the Gas Access Regime – Decision of the Ministerial Council on Energy' (2006) 25(2) *Australian Resources and Energy Law Journal* 132.

A Syed, J Melanie, S Thorpe, K Penney, 'Australian energy projections to 2029–30' (ABARE research report 10.02, Department of Resources, Energy and Tourism, Canberra, March 2010).

T Wood and L Carter, Getting gas right – Australia's energy challenge (June 2013) http://grattan.edu.au/wp-content/uploads/2014/04/189_getting_gas_right_report.pdf.

Y Wu, 'Gas Market Integration: Global Trends and Implications for the EAS Region' (Discussion Paper 11.20, University of Western Australia, 2011).

G Yarrow, M Egan and J Tamblyn, *Review of The Limited Merits Review Regime: Stage Two Report* (Standing Council on Energy and Resources, 2012).

5

UNCONVENTIONAL GAS REGULATION

5.1	Introduction	176
5.2	What is unconventional gas?	178
5.3	How is unconventional gas extracted?	180
5.4	Environmental and social issues associated with unconventional gas extraction	184
5.5	Regulatory frameworks for unconventional gas: Queensland and New South Wales	188
5.6	Regulatory framework: *Environment Protection and Biodiversity Conservation Act 1999* (Cth)	230
5.7	Review questions	232
5.8	Further reading	233

5.1 Introduction

The International Energy Agency has predicted that resource intensive countries are entering into what they describe as the 'golden age of gas'.[1] This is largely a product of the dramatic global expansion in the production of unconventional gas. In Australia, the expansion of the unconventional gas industry has been concentrated in the eastern states, where the number of drills has increased rapidly over the last five years.[2] Given the significant reserves estimated to exist within Queensland, New South Wales, Victoria, and South Australia, continued expansion is likely as international demand for liquid natural gas (LNG) in the export market increases.[3] The advantages that unconventional gas offers – as a new, abundant and relatively less pollutant form of fossil fuel – are strong factors driving the growth of the industry.[4] Despite this, there are many concerns associated with the progression of the industry. Hydraulic fracturing is a new extraction

---

1 See International Energy Agency, *World Energy Outlook 2011: Are We Entering a Golden Age of Gas?* (2011) <http://www.iea.org/publications/worldenergyoutlook/goldenageofgas/>.
2 In Queensland, it is estimated that there is 39 954 PJ CSG reserves. By 2012, 3500 wells existed in Queensland, with 1070 wells being drilled that year. See Australian Petroleum, Production and Exploration Association, *Economic Significance of Coal Seam Gas In Queensland: Final Report* (June 2012). http://www.appea.com.au/wp-content/uploads/2013/05/120606_ACIL-qld-csg-final-report.pdf.
3 Over the next five years, Australia's LNG exports are projected to increase at a rate of 19 per cent a year, underpinned by a number of new projects under construction. The value of Australia's LNG exports is also expected to more than double from $8 billion in 2009–2010 to $18.5 billion in 2015–2016, an average annual increase of 15 per cent. The Australian LNG industry is aiming to export 60 million tonnes of LNG by 2020. These figures are set out in A Schultz and R Petchey, *Australian Energy Statistics: Energy Update 2011* (Australian Bureau of Agricultural and Resource Economics and Sciences, 29 June 2011) http://data.daff.gov.au/data/warehouse/pe_abares99010610/EnergyUpdate_2011_REPORT.pdf. The Report outlines trends in Australian energy production, consumption and trade between 1973–74 and 2009–2010 across the different jurisdictions and industry sectors.
4 See the discussion by K J Flaherty, 'Quandary or Quest: Problems of Developing Coal Bed Methane as an Energy Resource' (2000) 15 *Journal of Natural Resources and Environmental Law* 71 at 76 where the author outlines the fact that because coal bed methane can be found virtually wherever coal exists, and because of the vast quantities that are estimated to exist within coal rich areas, commercial exploitation has proceeded ahead of regulatory protection. For a more recent discussion on the advantages of CSG development see M Roarty, 'The Development of Australia's Coal Seam Gas Resources' (Parliament of Australia, Science, Technology, Environment and Resources Section, July 2011) http://parlinfo.aph.gov.au/parlInfo/download/library/prspub/957068/upload_binary/957068.pdf;fileType=application/pdf#search=%22background%20note%20(parliamentary%20library,%20australia)%22. This paper discusses how the development of CSG deposits in Queensland and New South Wales will not only enable the supply of natural gas for the growing eastern Australian market but also enable the establishment of major export LNG industries, providing an impetus to employment, infrastructure investment, and Australia's exports.

process that is used consistently with low permeability gases, such as shale and tight gas. Hydraulic fracturing has significant potential to cause environmental damage to the subsurface through chemical contamination, water contamination, water depletion, subsurface fissures causing seismic disruption, subsurface elements becoming airborne and causing air pollution and health hazards, and the potential for infrastructure degradation.

Health hazards can occur through the disturbance of subsurface materials. The National Institute for Occupational Safety in the United States has collected air samples from 11 hydraulic fracturing sites across the US. At 31 per cent of the US sites, silica concentrations exceeded health regulations to such an extent that even if mine workers wore proper respiratory equipment, they may not be adequately protected against lung disease.[5] American studies have also shown that the closer you live to a drill, the greater the risk of adverse health effects, and this generates issues of broader environmental justice for communities impacted by unconventional gas projects.[6]

Chemical additives are also generally added during hydraulic fracturing activities. Each well has the potential to produce enormous quantifies of toxic fluid which contain both chemical additives and other naturally occurring radioactive materials, liquid hydrocarbons, brine water, and heavy materials. Fissures created by the hydraulic fracturing process can create pathways for the release of gases, chemicals, and radioactive materials.

The unconventional gas industry in Australia needs to be supported by a rigorous regulatory framework that implements strong and meticulous environmental assessment. This is particularly important given the new technologies that are associated with the extraction process and the uncertainty regarding how these processes may impact upon the long-term health of the environment. Regulation needs to be implemented promoting independent and transparent review processes as well as strong and effective community consultation programs.

This chapter examines the existing regulation of coal seam gas and shale gas in Australia. The chapter focuses on the regulatory frameworks implemented in New South Wales and Queensland. It evaluates the different types of regulatory initiatives that have evolved in response to the new resource management demands

---

5 See the report by the Department of Health and Human Resources, Cincinnati, *NIOS Field Effort to Assess Chemical Exposure Risks to Gas and Oil Workers* (1998) http://www.cdc.gov/niosh/docs/2010–130/pdfs/2010–130.pdf.
6 See the report by L M McKenzie, R Z Witter, L S Newman, J L Adgate, 'Human Health Risk Assessments of Air from Unconventional Gas' (2012) *Science of the Total Environment* http://cogcc.state.co.us/library/setbackstakeholdergroup/Presentations/Health%20Risk%20Assessment%20of%20Air%20Emissions%20From%20Unconventional%20Natural%20Gas%20-%20HMcKenzie2012.pdf The report suggests that risk management approaches should focus upon reducing exposures to emissions during well completions.

associated with unconventional gas expansion. The discussion includes an examination of the new strategic land use plans, gateway framework, the new CSG fracking and well integrity Codes, the Land Use Access Code, and EPA initiatives that have been introduced in NSW and Queensland, as well as the federal initiatives implemented to address the impact of unconventional gas expansion on water resources in the *Environmental Protection Biodiversity Conservation Act 1999* (Cth).

5.2 What is unconventional gas?

'Unconventional gas' is natural gas that has been extracted from a range of different forms of unconventional reservoirs. In terms of its chemical composition – which is primarily methane – unconventional gas is a resource that is identical to conventional natural gas. It is referred to as 'unconventional' because of its atypical geological locations. Unconventional gas is found in highly compact rock or coalbeds and therefore extraction of the gas requires a specific set of production techniques. There are three different forms of unconventional gas: shale gas, tight-bed gas, and coalbed methane (referred to in Australia as 'coal seam gas').

Hydrocarbons are trapped in subsurface formations that are known as 'reservoir rock'. The type of hydrocarbons that are trapped in these reservoirs are mainly gas; however, oil deposits may also exist. The hydrocarbons that are trapped are not found in large, contained pools but rather in minute pores between the grains which comprise the matrix of the rock. The quality and character of the reservoir rock is determined according to its porosity and permeability. Porosity refers to the empty space that exists between the rock and indicates the capacity of the reservoir rock to contain fluids.

A reservoir rock that is highly porous will be capable of containing a larger volume of gas or oil than one with a lower porosity. Apart from porosity, it is also important for gas or fluids which are contained within a reservoir rock to be capable of moving. In order for this to occur, the empty spaces between the rock need to be interconnected. The level of interconnectedness between the pores in a reservoir and the measurement of the reservoir's ability to transmit the gas or oil is known as its 'permeability'. The permeability measure is known as the 'Darcy'. The Darcy is a unit of permeability – named after the French engineer, Henry Darcy – that is widely used in petroleum, engineering, and geology. The Darcy is the primary means of distinguishing unconventional gas from conventional gas. Low-permeability reservoir rocks will generally only have a permeability of a few dozen micro-Darcy; whereas a good conventional reservoir will generally have a permeability of one Darcy.

Both shale gas and tight gas tend to be trapped within reservoir rocks that have a low permeability. This means that the interconnectedness of the pores in the rock is low and it is difficult for the gas to migrate. Generally, shale and tight gas are found in ultra-compact rock reservoirs. The permeability of shale gas can be as low as one-thousandth (a 'nanoDarcy') of the permeability of tight gas reservoirs.

5.2.1 Shale gas

Shale gas is extracted from a geological layer known as the 'source rock' that is comprised of a clay-rich sedimentary rock. Source rock arises where organic rock sediments deposited on the bottom of oceans or lakes are gradually covered by additional layers of sediment. As the layers become deeply buried, the sediments are consolidated into rock and the organic matter is transformed into hydrocarbons in the form of oil and gas. The transformation of the organic matter into hydrocarbons occurs as a result of differences in temperature as well as subsurface pressure. Ordinarily, newly-formed hydrocarbons migrate up through the pores and cracks of the rock and reach the surface. In the case of shale gas, most of the gas released during the transformation of the organic matter stays in place within the rock.

5.2.2 Tight gas

Tight gas is similar to shale gas in that it amounts to gas trapped within rock. However, tight gas is trapped in ultra-compact reservoirs that are characterised by low porosity and permeability. The rock pores that contain the gas are minuscule and the interconnections between them are so limited that the gas can only migrate through the pores with great difficulty.

5.2.3 Coal seam gas

Coal seam gas is a by-product of coal. Coal forms when plant material is 'coalified' into lignite, sub-bituminous coal, bituminous coal, and anthracite coal.[7] At different stages during the coalification process, biogenic and thermogenic methane forms.[8] Much of this methane will escape to the surface or migrate into the surrounding rock; however, a portion will remain trapped within the micro-pores of the coal in areas known as 'coal cleats' or 'seams'. Naturally-occurring water contained within the coal seams creates pressure that holds the methane gas in place.

---

7 See E A Craig and M S Myers, 'Ownership of Methane Gas in Coalbeds' (1987) 24 *Rocky Mountain. Min. Law Inst.* 767, 782.
8 See I Gray, *Reservoir Engineering in Coal Seams: Part 1—The Physical Process of Gas Storage and Movement in Coal Seams* (Society of Petroleum Engineers, 1987) 28–34 where the formation of methane is discussed.

The methane gas contained within the micro-pores of the coal is regarded as a 'pure gas' in the sense that it is non-toxic and contains few impurities.[9] In order to extract the gas from the coal seam, it is necessary to remove the water that holds the gas in place. Removing the water will drop the pressure in the seam, allowing the gas to be captured. This is achieved by pumping the water out of the aquifer. The extracted water has varying levels of contamination and salinity. This water – known as 'produced water' – is an important aspect of the extraction process and, if not disposed of properly, can create significant environmental concerns.[10]

5.3 How is unconventional gas extracted?

One of the most important differences between conventional and unconventional forms of natural gas lies in the extraction process. The location of unconventional gas within unconventional reservoirs means that ordinary extraction processes associated with conventional forms of natural gas are inappropriate. In many respects, the expansion of the unconventional gas industry has been the product of technological innovation. It has been noted that the use of technology to unlock energy resources contained within shale formations will be responsible for reshaping the United States energy economy.[11] From 2010–2035, the total volume of natural gas produced per year is projected to increase from 21.4 trillion to 31.4 trillion cubic feet per year. By 2035, shale gas will comprise more than half of all domestic production in the United States.[12] The extraction technology has developed to such an extent that the commercial extraction and exploitation of unconventional gas

---

9 See the discussion by D Mathew, 'The Nature of Gas in Coal: Technical Challenges of Co-Location of Coal and Coalbed Methane' (2005) *Australian Mineral and Petroleum Association Yearbook* 368, 369.
10 See the discussion by T Nunan, 'Legal Issues Emerging from the Growth of the Coal Seam Gas Industry in Queensland' (2006) 25 *Australian Resources & Energy Law Journal* 189, 190 where the author notes that unlike water extracted by a landowner via a water bore, 'associated water' is considered to be a regulated waste in Queensland, following the introduction of the *Water Supply (Safety and Reliability) Act 2008* (Qld) s 201A.
11 See Goldman Sachs, Global Market Institute, 'Unlocking the Economic Potential of North America's Energy Resources' (June 2014) 45–7 where the report notes that the expanding technological development for natural gas fired energy and renewable energy will generate continued growth.
12 See the International Energy Agency, above n 1. See also United States Energy Information Administration, *Annual Energy Outlook 2013: With projections to 2040* (2013) 2, 15 <http://www.eia.gov/forecasts/aeo/pdf/0383(2013).pdf>.

has become economically feasible within the current environment.[13] This was not always the case. Scientists and drilling operators have known of the existence of unconventional gas for decades; the issue, however, is that they have been unable to access the resource because the technology to do so did not exist. In particular, the low permeability of shale gas has traditionally made large-scale commercial extraction expensive and difficult.[14]

The development of this technology has prompted concerns regarding its impact upon public and environmental health. Policy-makers are increasingly cognisant of the importance of ensuring that new technological processes are carefully monitored to effectively evaluate the level of risk posed and to ensure that the risk is properly minimised or managed. Apprehension regarding chemical contamination and water depletion from hydro-fracturing, as well as increased fugitive emissions and seismic activity, has prompted a suite of regulatory changes in Australia (see discussion below). To fully understand the nature and scope of these regulatory and policy developments it is crucial to first understand the way in which these new technological processes function.

5.3.1 Hydraulic fracturing and horizontal drilling[15]

The extraction of unconventional gas, particularly from dense rock formations, is the direct result of new extraction technology known as 'hydraulic fracturing' and 'horizontal drilling'. Hydraulic fracturing – or 'fracking' – is the process of using hydraulic pressure to create additional fractures in an underground rock formation in order to encourage the release of trapped, unconventional gas. The low permeability of unconventional gas reservoirs means that oil and gas may not move through the formation quickly enough to justify the expense involved in drilling a well. Where cracks or fractures in the rock formation are created, those fractures may be utilised as pathways to the wellbore, and this increases the rate at which gas may flow to the well. This, in turn, makes drilling low permeability rocks far more economical. Hence, the purpose of hydraulic fracturing is to generate pathways in low permeability formations.[16]

---

13 For an interesting overview on the 'technological revolution' that has occurred in the United States, see R H Pifer, 'A Greener Shade of Blue: Technology and the Shale Revolution' (2013) 27 *Notre Dame Journal of Law Ethics and Public Policy* 131, 133–5.
14 See the discussion by S S Sakmar, 'The Global Shale Gas Initiative: Will the United States be the Role Model for the Development of Shale Gas Around the World?' (2011) 33 *Houston Journal of International Law* 369, 370–1.
15 Hydro-fracturing is also known as: hydraulic fracturing, fracing, and fracking. 'Fracking' is the shortened term that is utilised most frequently. See H Wise, 'Fracturing Regulation Applied' (2012) 22 *Duke Environmental Law and Policy Forum* 316.
16 See US Department of Energy, National Energy Tech. Lab., 'Modern Shale Gas Development in the United States: A Primer' (April 2009) http://www.netl.doe.gov/technologies/oilgas/

Prior to utilising hydraulic fracturing, an operator must drill a well. Once a vertical well has been drilled down to a depth of up to 3.2 km and has reached the layer of deep rock in which the gas resides, a horizontal well is drilled along the layer of deep rock. Once the vertical and horizontal wells have been properly cased in cement and steel to prevent leakage, high-pressure pumps push a fracturing fluid down the horizontal 'fracking' well at high pressure, inducing the formation of fractures within the rock.

Once the fracture has occurred, the high-pressure pumps are turned off and the pressure of the formation pushes the fracturing fluid back through the well and up to the surface. This fluid is known as 'flowback water' and it is recovered at the surface of the drill. It is usual for 30–70 per cent of the fluid used in the fracturing process to be recovered as flowback, with the remainder either returning gradually to the surface along with the oil or gas, or continuing to reside within the fractured pore space.

The fracturing fluid is composed of a base fluid, which contains water and other chemical additives, as well as small particles known as 'proppants'. Proppants are small granules of sand or ceramic particles.[17] The chemical additives include corrosion inhibitors to protect the well's piping, biocides to inhibit microbial growth, friction reducers to diminish the friction between the flowback fluid and the well, and viscosity adjusters to assist the fluid in carrying the proppants into the fissures in the fracking drill.

During hydraulic fracturing, the fracking fluid carries the proppants and deposits them into the newly created fractures. When the fracturing fluid is removed, the proppant remains within the fissures of the rock and props them open to prevent them from closing when the fracking fluid is removed.[18]

5.3.2 Water pumping for coal seam gas extraction

To extract CSG a well needs to be drilled directly into the targeted coal seam. The coal seam is generally located 200 to 1000 m beneath the surface and is therefore reasonably shallow.[19] The hole is drilled using drilling fluids to lubricate and cool the drill rods and the drill bit, remove the rock cuttings, maintain pressure control of the well, and stabilise the hole. Drilling fluid is generally a mixture of water, clays, and additives such as bentonite, cellulose, polymer, barite and guar gum that control fluid loss, density, and viscosity. The drill well is cased with steel and the gap between the steel and the rock is pressure cemented from the coal seam to

---

publications/epreports/shalegasprimer2009.pdf E.S-4, 57. See also H R Williams and C J Meyers, *Manual of Oil and Gas Terms* (LexisNexis, 10th ed, 1997) 775.

17 See R Beckwith, 'Proppants: Where in the World' (2011) 63 *Journal of Petroleum Technology Online* 36–40.
18 See J G Speight, *The Chemistry and Technology of Petroleum* (McGraw-Hill, 2nd ed, 1991) 141.
19 See above n 9, 369.

the ground surface. This is to ensure that all of the formations overlying the coal seam are isolated from fluid and gas passing from the inside of the well. Water is then pumped from the coal seam and this reduction in water pressure releases the gas from the coal.[20]

Production of CSG normally requires the drilling of a number of wells at an increased density of spacing than is normally required for conventional gas production. CSG drilling forms the majority of unconventional gas drills in Queensland.[21] CSG is adsorbed into the coal matrix and is held in place by the pressure of formation water. To extract the gas, a well must be drilled into the coal seam and formation water from the coal cleats and fractures must be pumped from the coal and removed. The removal of water in the coal seam reduces the pressure, enabling the CSG to be released (desorbed) from the coal micropores and cleats, and allowing the gas and 'produced water' to be carried to the surface. In some cases the coal permeability is low which means that the most economically viable option is for coal to be hydraulically fractured.

Produced water – also referred to as 'CSG water' or 'wastewater' – is the water that is pumped out of coal seams so that the CSG can be released. Over time, the volume of produced water normally declines and the volume of produced gas increases.[22] Once the produced water and the gas reach the surface, they are separated. The methane is collected and passed to a central compressor station where it is added to a pipeline network for delivery to users. The produced water is piped for use or further treatment.

The amount of water that may be used within a CSG site varies as water production can change markedly between different coal seams depending on the underground water pressures and the particular geology. It has been roughly estimated by the CSIRO that water production in Queensland averages approximately 20 000 litres per well per day.

The water that is produced from a coal seam has generally been underground for a long time with little fresh water penetration and often has a high saline content that approximates to about one-sixth that of sea water. CSG water contains mainly sodium chloride – varying from 200 to more than 10 000 milligrams per litre – sodium bicarbonate, and traces of other compounds. The high level of salt contained in the water makes effective treatment and disposal of produced water imperative.[23]

---

20 See above n 8, 28–34 where the formation of methane is discussed.
21 It is estimated that today CSG comprises 90 per cent of Queensland gas production. See the report by the Australian Petroleum Production & Exploration Association, *Economic Significance of Coal Seam Gas In Queensland: Final Report* (June 2012) http://www.appea.com.au/wp-content/uploads/2013/05/120606_ACIL-qld-csg-final-report.pdf.
22 Ibid.
23 For a detailed outline of the nature of CSG produced water see CSIRO, *CSG Produced Water and Site Management* (April 2012) http://www.csiro.au/Portals/Publications.aspx.

5.4 Environmental and social issues associated with unconventional gas extraction

The increased commercialisation and extraction of unconventional gas has generated a diverse array of new environmental and social concerns because of: (i) the anticipated impact of the new technologies associated with the extraction process upon the environment; and (ii) the expansion of unconventional gas projects into land not previously connected with the mining industry. The combination of these two issues has created heightened social awareness and mobilised land-owner activism.

Coal seam gas mining involves the removal of vast amounts of water because the extraction of the gas depends upon pumping the water out of the coal seam. The removal of large amounts of groundwater can significantly affect aquifer levels and the capacity of aquifers to recharge. In some cases, the removal of significant volumes of water will deplete the aquifer level completely. The impact that the extraction of unconventional gas might have upon groundwater is one of the most significant concerns connected with unconventional gas development. A clean and safe supply of groundwater is essential for the drinking water needs of country towns, major industries – especially agriculture – and to support groundwater dependent ecosystems. Groundwater quality decline and contamination creates a serious threat to food security, human and animal health and the degradation of wetlands and rivers.[24]

Further, a reduced availability of groundwater can significantly affect agricultural and rural industries that may be reliant upon this water supply. This is a particular concern for agricultural industries within New South Wales and Queensland, where the majority of coal seam gas mining is concentrated.

While an area's natural assimilation capacity may reduce the impact of any leakage, this is variable.[25] Groundwater may be at risk if pathways for the migration of gases – and possibly saline fluids and fracturing chemicals – exist deep underground. This may occur where there are leaky well casings, natural fractures in the rock, and old abandoned wells. Other concerns include how to deal with wastewater disposal, greenhouse gas emissions, and the cumulative effects of the large number of wells and related infrastructure that are required for development; for example, roads,

---

[24] See National Water Commission, *Impact of Groundwater Pumping on Groundwater Quality* (16 May 2012) http://archive.nwc.gov.au/rnws/ngap/groundwater-projects/managing-risks-to-groundwater-quality/impact-of-groundwater-pumping-on-groundwater-quality.

[25] See especially Council of Canadian Academies, 'Environmental Impacts of Shale Gas in Canada: The Expert Panel on Harnessing Science and Technology to Understand the Environmental Impacts of Shale Gas Extraction' (May 2014) http://www.scienceadvice.ca/

compressor stations, pipeline rights-of-way, and staging areas, which can affect agricultural production, destroy the habitat, and result in deforestation.[26]

Additionally, the removed water has a high saline content and may, during the extraction process, have chemical constituents added to it. The danger of toxic or highly saline removed water being released into adjoining waterways and contaminating domestic water supplies is a grave concern given the potentially devastating impacts involved.

Reliable and timely information is crucial if the environmental effects of coal seam gas and shale gas development are to be properly managed into the future. In most instances, unconventional gas development has proceeded in the absence of sufficient environmental baseline data.[27] This makes it difficult to identify and characterise relevant impacts or to dismiss impacts that are inappropriately associated with unconventional gas development. Monitoring that has been undertaken in the past suggests that gas leakage into aquifers is a strong concern, as are fugitive emissions; however, the full environmental impact of coal seam gas and shale gas development may take decades to fully comprehend.[28]

A particular concern associated with the expansion of the unconventional gas industry in Australia has been its social impact. Unconventional gas projects are generally located in smaller, rural communities. The progression of the project generally has a direct impact upon the local economy, the social and physical infrastructure

---

uploads/eng/assessments%20and%20publications%20and%20news%20releases/shale%20gas/shalegas_fullreporten.pdf. The report specifically notes that the impact of fracking, particularly upon groundwater sources, is very site specific.

[26] In the United States, Congress has mandated environmental impact studies to be undertaken. See the report by M Ratman and M Tiemann, 'An Overview of Unconventional Oil and Natural Gas: Resources and Federal Actions' *Congressional Research Services* (21 November 2014) https://www.fas.org/sgp/crs/misc/R43148.pdf. The report notes, at p. 15, that the EPA in the US conducted a study on the relationship between hydraulic fracturing and drinking water to gain a better understanding of potential contamination risks. The EPA is conducting retrospective case studies at five sites to develop information about the potential impacts of hydraulic fracturing on drinking water resources under different circumstances. The case studies include: (i) the Bakken Shale in Dunn County, ND; (ii) the Barnett Shale in Wise County, TX; (iii) the Marcellus Shale in Bradford County, PA; (iv) the Marcellus Shale in Washington County, PA; and (v) coalbed methane in the Raton Basin, CO.

[27] See especially Council of Canadian Academies, above n 25, 4. For an Australian perspective see NSW Chief Scientist and Engineer, *Initial Report on the Independent Review of Coal Seam Gas Activities in NSW* (July 2013), Recommendation 3, vii: 'That a pre-major-CSG whole-of-State subsidence baseline be calculated using appropriate remote sensing data going back, say, 15 years.'

[28] For a discussion on the concertns relating to fugitive emissions see S Day, L Connell, D Etheridge, T Norgate and N Sherwood, 'Fugitive Greenhouse Gas Emissions from Coal Seam Gas Production in Australia' (October 2012) *CSIRO Energy Technology* 22 where the authors note the uncertainties inherent in the current methodologies used for estimating fugitive emissions from many sectors of the gas industry generally, and the lack of specific information regarding emissions from the Australian CSG industry in particular.

of the communities, and the social relations that exist within the communities. The primary reason that local communities embrace unconventional gas development, if they do at all, is the economic benefits that such projects can attract. This is particularly important within smaller rural communities with histories of economic decline.

Rapid mining development of the sort that has been connected with unconventional gas extraction and production can dramatically alter the character and context of local communities. The influx of industry workers generally means that rental units are quickly filled and this has the direct effect of escalating both purchase prices and rental rates and pushing long-term residents out of the housing market. This, in turn, can exacerbate stress on these individuals and families, causing social disruption and changes in patterns of behaviour and culture within communities.[29]

These impacts have prompted significant regulatory changes, particularly in those eastern states where CSG projects are most established; those being New South Wales and Queensland. These changes have particularly focused upon improved access and compensation provisions for affected landholders and more focused environmental assessment procedures.

Two significant examples of CSG production are the Gladstone and Narrabri projects. In Queensland, Santos and Petrona's multi-billion dollar Gladstone LNG (GLNG) project will be the world's first project to process coal seam gas into liquefied natural gas as a cleaner energy source. It involves:

- exploration and production of CSG in the Surat and Bowen Basin gas fields
- construction and operation of a 435 km gas pipeline from the gas fields to Gladstone
- construction and operation of a gas liquefaction and export facility on Curtis Island and associated infrastructure.

The GLNG project involves extraction of CSG from fields around Roma, Emerald, Injune, and Taroom, a 435 km gas transmission pipeline from the gas fields to Gladstone, an LNG facility of approximately 10 million tonnes per annum on Curtis Island near Gladstone, and associated infrastructure including marine facilities, port dredging, and a potential access road and bridge at Gladstone. The LNG facility is proposed to be developed in three stages, the first of which will have a capacity of approximately 3–4 million tonnes per annum. The LNG which is produced will be exported to overseas markets.[30]

---

29 See the discussion by R S Krannich and T Greider, 'Personal Well-Being in Rapid Growth and Stable Communities: Multiple Indicators and Contrasting Results' (1984) 49(4) *Rural Sociology* 541, 541–52.
30 For a full outline of the GLNG Project see the outline on the Santos website at http://www.santosglng.com/the-project.aspx. See also T Hunter, 'Australia's Unconventional Gas Resources' in CEDA, *Australia's Unconventional Energy Options* (September 2012) http://www.ceda.com.au/media/263565/cedaunconventionalenergyfinal.pdf.

In New South Wales, CSG has been developed in the Narrabri area in northwest New South Wales. A comprehensive exploration program has been undertaken in and around the Pilliga region by Santos. The program is designed to gather further information on gas reserves, gas composition, and flow rates. The exploration program includes the drilling of 15 exploration and pilot wells and one core hole. It also includes recommencing pilots inactive since Santos took over the Narrabri operations in November 2011.

In Queensland, the assessment and granting of tenure is not conducted in consultation with the community. However, if tenure is granted, the authority holder may be obliged to consult landowners or occupiers during the tenure period; for example, s 74 of the *Petroleum and Gas (Production and Safety) Act 2004* (Qld) ('*PGPSA*') makes it clear that the holder of an authority to prospect must consult, or use reasonable endeavours to consult, with each owner and occupier of private or public land on which authorised activities for the authority are proposed to be carried out or are being carried out. Hence, although landholders have no statutory right to preclude resource title-holders from accessing land, there is nevertheless an obligation on title-holders to consult with each owner and occupier of public and private land in order to reach an agreement concerning access, the carrying out of authorised activities for the purpose of the resource title, and the title-holder's compensation liability.[31]

The *PGPSA* does not give the community any right to object to the issuance of an authority to prospect and consequently, concerns regarding these titles are generally expressed through environmental review avenues. Landholders do have a right to object to the issuance of a mining lease under the *Mineral Resources Act 1989* (Qld) and objection will be heard before the Land Court of Queensland; however, the usual outcome of these objections is that the landholder is granted compensation and the title-holder is permitted to retain its licence.[32]

In New South Wales, community consultation is a condition of all exploration licences. Prior to the granting of a Petroleum Exploration Licence (PEL) for CSG, local communities have the opportunity to comment on applications through a

---

31 See Queensland Government, *Guide to Queensland's New Land Access Laws* (2010) http://mines.industry.qld.gov.au/assets/land-tenurepdf/6184_landaccesslaws_guide_print.pdf.
32 See *Mineral Resources Act 1989* (Qld) s 269. Note that the proposed Mineral Resources and Energy (Common Provisions) Bill 2014 incorporates changes to the right to object and, in particular, streamlines the objection process with the *Environmental Protection Act 1994* (Qld). If passed, this will mean that only those neighbours with direct boundaries will be entitled to object to the issuance of a mining lease. However, mines that exceed 10 hectares are automatically subject to a site-specific environmental assessment which is subject to public notification, meaning that a broader range of objections may be heard via this avenue.

public comment process.[33] This process helps to ensure that the transparency of the decision-making process is sustained. It also allows the interests of the state to be balanced by the interests of the communities directly affected by the activities. It is a condition of title that exploration licence-holders meet strict publication and consultation requirements in order to involve communities in making decisions that affect them.

5.5 Regulatory frameworks for unconventional gas: Queensland and New South Wales

5.5.1 Queensland: *Petroleum and Gas (Production and Safety) Act 2004*

5.5.1.1 Regulatory requirements for resource titles

Coal seam gas represents a significant component of the resource market in Queensland. Since the mid-1990s, CSG has grown and now supplies approximately 80 per cent of the Queensland market. Gas is a cleaner energy in terms of its usage and the Queensland Gas Scheme, which resulted in 20 per cent of electricity being sourced from gas-fired plants by 2012, but which was closed at the end of 2013, was a significant factor underlying the increase in exploration for natural gas. A further factor has been the expanding export market for LNG and the facilities proposed for the Gladstone project.

---

33 Consultation is available pursuant to the public comment process which only applies to exploration licences. After lodging an exploration licence application the applicant must publish a notification in a major metropolitan newspaper, circulating statewide, and at least one regional newspaper. For exploration licences proposed for tender or released as a result of a part transfer of an existing licence held by the Director General of NSW Trade and Investment, the Department will publish notification of the proposed tender or transfer in a newspaper circulating statewide and in at least one local newspaper. The notification must include the exploration licence application number (where appropriate), location of the proposed exploration area, contact details for the applicant, and reference to the Department's website. Submissions regarding the application must be made within 28 days of the proposal being published. For a further outline on this, see NSW Department of Trade and Investment, Resources and Energy, 'Public Comment Process: For the exploration of coal and petroleum including coal seam gas' (October 2011) http://www.resourcesandenergy.nsw.gov.au/__data/assets/pdf_file/0009/426582/Public-Comment-Process-Document.pdf.

Exploration and production of onshore CSG in Queensland falls within the scope of the petroleum legislation. Both CSG and shale come within the definition of 'petroleum' in s 76K of the *Petroleum Act 1923* and s 299 of the *PGPSA*. Ownership of petroleum is vested in the Crown pursuant to s 10 of the *Petroleum Act 1923* (Qld) and s 26 of the *PGPSA*.

The regulation of CSG, as a component of petroleum, comes within the application of the *PGPSA*. This is the primary Act regulating the issuance of onshore petroleum permits for unconventional gas in Queensland.[34] To conduct CSG mining in Queensland, a petroleum lease must be issued pursuant to pt 2 of the *PGPSA*. Once granted, a petroleum lease allows the leaseholder to explore for petroleum, test for and carry out petroleum production, evaluate underground reservoirs for petroleum storage, and extract or produce gas from coal. Section 109, which outlines the scope of these provisions, is extracted below. Section 110 allows a leaseholder to construct and operate petroleum pipelines within the area of the petroleum lease. Section 111A allows a leaseholder to construct a facility to manage and deal with produced water from CSG operations.

Petroleum and Gas (Production and Safety) Act 2004 (Qld)

109 Exploration, production and storage activities

(1) The lease holder may carry out the following activities in the area of the lease –
   (a) exploring for petroleum;
   (b) subject to section 152 –
      (i) testing for petroleum production; and
      (ii) evaluating the feasibility of petroleum production; and
      (iii) testing natural underground reservoirs for storage of petroleum or a prescribed storage gas;
   (c) petroleum production;
   (d) evaluating, developing and using natural underground reservoirs for petroleum storage or to store prescribed storage gases, including, for example, to store petroleum or prescribed storage gases for others.
   (e) plugging and abandoning, or otherwise remediating, a bore or well the lease holder reasonably believes is a legacy borehole and rehabilitating the surrounding area in compliance with the requirements prescribed under a regulation.
(2) However, the holder must not carry out any of the following –
   (a) extraction or production of a gasification or retorting product from coal or oil shale by a chemical or thermal process;

---

34 See the accompanying *Petroleum and Gas (Production and Safety) Regulations 2004* (Qld).

(b) exploration for coal or oil shale to carry out extraction or production mentioned in paragraph (a);
(c) GHG stream storage.
(3) The rights under subsection (1) may be exercised only by or for the holder.

The *PGPSA* also provides for circumstances where a petroleum title is jointly obtained with the consent of a coal exploration tenement holder. This may obviously be the case where an applicant applies for a petroleum title for a CSG project over land that is already the subject of a coal licence. In such a situation, the *PGPSA* provides for the making of a CSG statement. The statement must include: (i) the likely effect of proposed petroleum production on the future development of coal or oil shale resources from the land; (ii) the technical and commercial feasibility of coordinated petroleum production and coal or oil shale mining from the land; and (iii) include an overview of a proposed safety management plan for all operating plant, or proposed operating plant, for proposed petroleum production under the lease that may affect possible future safe and efficient mining under a coal or oil shale mining lease.[35]

There are two primary forms of petroleum tenure that may be issued in Queensland which have a specific application to onshore unconventional gas projects: an authority to prospect and a petroleum lease. An authority to prospect confers authority to explore for petroleum upon the holder. An authority to prospect will be issued in response to the relevant minister making a call for tenders pursuant to s 35 of the *PGPSA*.

Tender requirements include a statement about: (i) how and when the tenderer proposes to consult with, and keep informed, each owner and occupier of private or public land on which authorised activities for the proposed authority are, or are likely to be, carried out; and (ii) a proposed work program that complies with the initial work program requirements.

An authority to prospect will continue for no longer than 12 years. Section 42 of the *PGPSA* makes it clear that the issuance of the authority is dependant upon a holistic review that includes, in particular, consideration of the financial and technical resources of the applicant, the ability of the applicant to manage petroleum exploration and production, and the applicant's proposed work program.

The work program that accompanies an authority to prospect is an extensive document whose primary objective is to overview the activities proposed within the authorised area, overview the extent and nature of petroleum exploration and petroleum production to be carried out during the following year, the precise location where exploration and production activities are proposed to be carried out, and the anticipated costs of those activities.[36]

---

[35] See *PGPSA* (Qld) s 306(1).
[36] See *PGPSA* (Qld) s 48(1).

The holder of an authority to prospect may, following exploration activities, apply to the Minister to indicate that all or a specific area within the authority is a potential commercial area and has market opportunities for either petroleum production or storage.[37]

Once issued, an authority to prospect will be subject to a range of requirements and prohibitions. Of particular relevance for unconventional gas is the restriction on the holder of an authority to prospect on flaring or venting petroleum in a gaseous state unless the flaring or venting is authorised.[38]

In addition to acquiring an authority to prospect, amendments to the *PGPSA* introduced in 2012 now require land that is likely to be highly prospective for coal, petroleum, and gas to be subject to a competitive cash bid tender process for the allocation of exploration rights. Pursuant to this new process, a cash bid component is required to be included as part of a tender application for an authority to prospect for petroleum and gas. The cash bid should take into consideration the potential in-ground value of the resources for the respective land area(s) specified in any official state 'call for tenders' document.[39]

The preferred holder of the tender must meet stringent environmental and tenure approval requirements before exploration tenure is granted. Similarly, all existing approval requirements associated with protecting landholders' rights, covering land access, conduct, and compensation will continue to apply as part of the permit assessment.

Once an authority to prospect has expired, the applicant may seek permission from the Minister to apply for a more expansive resource title in the form of a petroleum lease. A petroleum lease may be issued where an applicant is seeking to explore for petroleum, to evaluate the feasibility of petroleum production, or to test underground reservoirs for storage of petroleum. A petroleum lease gives its holder the right to produce petroleum within the area of the petroleum lease. The maximum area that may be granted pursuant to a petroleum lease is 75 sub-blocks and the maximum term is 30 years.[40]

The holder of a petroleum lease cannot, however, carry out any exploration, extraction, or production of a gasification or retorting product from coal or oil shale by a chemical or thermal processor or any greenhouse gas stream storage.[41]

The primary distinction between an authority to prospect and a petroleum lease is that the holder of a petroleum lease has surpassed the prospecting stage and is

---

37 See *PGPSA* (Qld) s 89.
38 See *PGPSA* (Qld) s 72.
39 See the *Mining and Other Legislation Amendment Act 2013* (Qld) which amended the *Mineral Resources Act 1989* (Qld) s 136E to allow for the inclusion of cash-bid component in determining the tender. See also the amendments to the *PGPSA* (Qld) in ch 2, div 2.
40 See *PGPSA* (Qld) s 109.
41 See *PGPSA* (Qld) s 109(2).

ready to produce petroleum. An application for a petroleum lease must include a proposed 'initial development plan' outlining the relevant information about the nature and extent of activities that are proposed to be carried out pursuant to the petroleum lease. Further, the Minister must be of the opinion that the applicant is capable of carrying out the activities described in the development plan, taking account of their financial and technical resources as well as their ability to manage petroleum exploration and production.[42] Generally, an initial development plan will cover a five-year period of activities proposed for the petroleum lease. Once the five-year period expires, or in circumstances where there is a significant change to an activity, the holder of the petroleum lease will be required to lodge a proposed 'later development plan'.[43]

The proposed 'later development plan' gives detailed information regarding the nature and extent of activities proposed to be carried out under the petroleum lease, and also highlights any significant change (and reasons for the change) from activities provided for under the previously approved development plan. Generally, a proposed later development plan will also cover a five-year period.

The *PGPSA* gives the holder of a petroleum title extensive rights to use and take underground water and these rights form what are referred to in the legislation as 'underground water rights'.[44] The legislation makes it clear that the holder of such a title may take or interfere with underground water if the taking or interference occurs during the course of authorised activities and may also use water for carrying out authorised activities and such rights are taken to be authorised by the *Water Act 2000* (Qld). No specific limit is placed upon the volume of water that may be utilised for the purpose of exercising underground water rights.[45]

Specific provisions exist in ch 3 of the *PGPSA* for CSG mining. The primary purpose of these express provisions is to clarify rights to explore for and extract CSG and to address issues arising from exploration and production including upholding safety standards; optimising the development and use of Queensland's coal, oil, shale, and petroleum resources; providing certainty of tenure for future investments relating to coal; and ensuring that, provided it is commercially and technically feasible, petroleum leases for CSG are issued in a safe and efficient manner.[46]

These objectives clearly prioritise the commercial and investment imperatives connected with the development of the CSG industry in Queensland and whilst safety and efficiency are mentioned within the legislation, these factors will only

---

42 See *PGPSA* (Qld) s 121.
43 See *PGPSA* (Qld) div 4.
44 See *PGPSA* (Qld) s 185(2)(a).
45 See *PGPSA* (Qld) s 185.
46 See *PGPSA* (Qld) s 295.

qualify the development of a CSG project where to do so would be consistent with commercial feasibility.

The provisions in ch 3 of the *PGPSA* impose additional requirements upon the application for an authority to prospect that involves the extraction of CSG or shale oil. In particular, an applicant is required to provide a CSG statement, which must accompany other application documents for both an authority to prospect and a petroleum lease.[47] The CSG statement must outline: (i) the likely effect of proposed petroleum production on the future development of coal or oil shale resources from the land; and (ii) the technical and commercial feasibility of coordinated petroleum production and coal or oil shale mining from the land; and must include an overview of a proposed safety management plan for all operating plant for proposed petroleum production that may affect safe and efficient mining.[48]

Given the expansion and scope of the CSG industry in Queensland, a significant concern associated with the issuance of a resource title for unconventional gas is the possibility of overlap, particularly in areas where the potential for gas reserves is significant. It is, however, possible for an authority to prospect for CSG resources to be issued over land which is already subject to an existing coal or oil shale exploration tenement. In such circumstances, an applicant must make a special request and the Minister may require the applicant to conduct negotiations with the existing coal or oil shale tenement holder in order to come to an arrangement that provides the best resource use outcomes without affecting the existing rights or interests of the parties.[49] It is not possible, however, for an overlapping authority to prospect to be issued where a petroleum lease has already been issued for CSG or oil shale or, where an environmental impact assessment has been approved for a petroleum lease for CSG or oil shale.[50]

Further, a work plan lodged in support of a petroleum lease for CSG or oil shale must include a statement of how the effects on, and the interests of, any relevant overlapping or adjacent coal or oil shale mining tenement holder have, or have not, been considered and this must also be outlined within the CSG statement.[51] Further, consistent with the CSG statement, the initial development plan should seek to optimise petroleum production in a safe and efficient manner.

In addition to providing a CSG statement that addresses the specific criteria, a petroleum lease for CSG and oil shale must also provide a safety management plan. The safety management plan must outline, in detail, the safety procedures to be implemented during the process of exploring for and extracting CSG

---

47 See *PGPSA* (Qld) ss 305, 328.
48 See *PGPSA* (Qld) s 306.
49 See *PGPSA* (Qld) s 313.
50 See *PGPSA* (Qld) ss 315–16.
51 See *PGPSA* (Qld) s 381.

and oil shale. The PGPSA prescribes the content of the safety management plan. Section 675(1) prescribes what must be included and is extracted below:

Petroleum and Gas (Production and Safety) Act 2004 (Qld)

675 Content requirements for safety management plans

(1) A safety management plan for an operating plant must include details of each of the following to the extent they are appropriate for the plant –
   (a) a description of the plant, its location and operations;
   (b) organisational safety policies;
   (c) organisational structure and safety responsibilities;
   (ca) for an operating plant, other than a coal mining–CSG operating plant – the operator of the plant;
   (d) each site at the plant for which a site safety manager is required;
   (e) a formal safety assessment consisting of the systematic assessment of risk and a description of the technical and other measures undertaken, or to be undertaken, to control the identified risk;
   (f) if there is proposed, or there is likely to be, interaction with other operating plant or contractors in the same vicinity, or if there are multiple operating plant with different operators on the same petroleum tenure, geothermal tenure or GHG authority[52] –
      (i) a description of the proposed or likely interactions, and how they will be managed; and
      (ii) an identification of the specific risks that may arise as a result of the proposed or likely interactions, and how the risks will be controlled; and
      (iii) an identification of the safety responsibilities of each operator;
   (g) a skills assessment identifying the minimum skills, knowledge, competencies and experience requirements for each person to carry out specific work;
   (h) a training and supervision program containing the mechanism for imparting the skills, knowledge, competencies and experience identified in paragraph (g) and assessing new skills, monitoring performance and ensuring ongoing retention of skill levels;
   (i) safety standards and standard operating and maintenance procedures applied, or to be applied, in each stage of the plant;
   (j) control systems, including, for example, alarm systems, temperature and pressure control systems, and emergency shutdown systems;

---

[52] See *PGPSA* (Qld) s 675(1). See also *PGPSA* (Qld) s 388 (additional content requirements) and s 705B (content requirements for principal hazard management plans) *Petroleum and Gas (Production and Safety) Regulation 2004* reg 59. The *PGPSA* (Qld) and its Regulations also impose other obligations that may be audited by the Petroleum and Gas Inspectorate of the Department of Natural Resources and Mines.

(k) machinery and equipment relating to, or that may affect, the safety of the plant;
(l) emergency equipment, preparedness and procedures;
(m) communication systems including, for example, emergency communication systems;
(ma) a process for managing change including a process for managing any changes to plant, operating procedures, organisational structure, personnel and the safety management plan;
(n) the mechanisms for implementing, monitoring and reviewing and auditing safety policies and safety management plans;
(p) key performance indicators to be used to monitor compliance with the plan and this Act;
(q) mechanisms for –
   (i) recording, investigating and reviewing incidents at the plant; and
   (ii) implementing recommendations from an investigation or review of an incident at the plant;
(r) record management, including, for example, all relevant approvals, certificates of compliance and other documents required under this Act;
(s) to the extent that, because of the *Work Health and Safety Act 2011,* schedule 1, part 2, division 1, that Act does not apply to a place or installation at the plant, details, including codes and standards adopted, addressing all relevant requirements under that Act that would, other than for that section, apply;
(t) if the operating plant is, under the NOHSC standard, a major hazard facility – each matter not mentioned in paragraphs (b) to (r) that is provided for under chapters 6 to 10 of that standard;

Note– For what is a major hazard facility under the NOHSC standard, see chapter 4, definition major hazard facility and chapter 5 (Identification and classification of a major hazard facility), section 5.6.

Section 674 of the *PGPSA* requires the operator of an operating plant to have, for each stage of the plant, a safety management plan that complies with s 675 of the *PGPSA*. A safety management plan may be a stand-alone document, but more often the elements of the plan can be or are already incorporated into the organisation's existing documented safety management system.[53] The safety management plan must be kept available for public inspection.[54]

---

[53] For a discussion regarding the preparation of safety management plans see Department of Natural Resources and Mines, *SafeOP for Petroleum and Gas: A Guide to the Legislative Requirements for Operating Plant, Part A Explanatory Guide* (2013) http://mines.industry.qld.gov.au/assets/petroleum-pdf/safeop.pdf.

[54] See *PGPSA* (Qld) s 675.

5.5.1.2 Access and compensation framework

The *PGPSA* introduced a land access framework with the specific aim of reducing land and resource conflicts directly associated with the expansion of unconventional gas projects.[55] The key features of the access framework include a specific requirement that all resource authority-holders must comply with a single land access code. Further important requirements include: an entry notice requirement that is mandated for 'preliminary' activities – that is, those having a minor impact on landholders; a conduct and compensation agreement to be negotiated before a resource authority holder comes onto a landholder's property to undertake 'advanced activities' – that is, those likely to have a significant impact on a landholder's business or land use and a graduated process for negotiating and resolving disputes about agreements which ensures that matters are only referred to the Land Court as a last resort. These requirements ensure stronger compliance and enforcement powers for government agencies where the *Land Access Code 2010* (Qld) is breached.

In addition to the laws and the access code, the Queensland Government has introduced a range of materials aimed at assisting both petroleum title-holders and landholders. These materials include standard agreements that may be entered into between title-holders and landholders regarding access arrangements, an outline of the obligations upon title-holders regarding conduct and compensation negotiations, a general guide to land access laws, and tips for negotiating with resource companies.[56]

The starting point for the land access framework in Queensland lies in the provision under the *PGPSA* which reserves upon the state an exclusive right to enter and carry out any petroleum-related activity or to authorise others to carry out a petroleum-related activity over land which, immediately prior to the resource title being issued, was owned by the state.[57] Significantly, this statutory reservation

---

[55] See the discussion by N Swayne, 'Regulating Coal Seam Gas in Queensland: Lessons in an Adaptive Management Approach' (2012) 29(2) *Environmental and Planning Law Journal* 163, 178 where the author notes that the 'mere presence of CSG projects on local land will cause disruptions to the landholder including as a result of the location of infrastructure on the land including drill sites, well heads, gathering lines, compressor stations, fluid storage and treatment facilities, and access roads. These, in conjunction with noise impacts and impacts on visual amenity will affect practices on the land such as locations of stock, pasture and crops'.

[56] See Queensland Government, Department of Natural Resources and Mines, 'Queensland Government Response to the Report of the Land Access Review Panel' (December 2012) http://mines.industry.qld.gov.au/assets/native-title-pdf/qg-response-land-access-framework.pdf.

[57] The reservation applies to each land grant, issued under another Act, relating to land that, (a) immediately before the grant, was unallocated state land as defined under the *Land Act 1994*; and (b) that is, or was, issued on or after the commencement of the 1923 Act: *PGPSA* (Qld) s 27(1).

allows the state to authorise others to carry out a petroleum activity but does not expressly reserve any right in the state to authorise others to enter private land for any non-petroleum activity.[58]

Further, the *PGPSA* sets out that the holder of a resource title cannot enter private land to undertake 'advanced activities' unless they have entered into a conduct and compensation agreement, or a deferral agreement has been entered into with the affected landholder, or the matter has been referred to the Land Court for determination.[59] Entering into a conduct and compensation agreement is crucial for all parties involved as it promotes negotiation and discussion between the parties which provides a greater chance of core rights and obligations being properly articulated.[60] This reduces the prospect of conflict because it ensures that the parties have coordinated the framework for an interactive relationship.

The *PGPSA* draws a distinction between preliminary activities and advanced activities. Preliminary activities are authorised activities with no impact, or a minor impact, on the business or land use activities of any landowner: such activities may include taking soil samples and surveying. Advanced activities are all other activities and include activities such as road construction, drilling, and earth removal etc.[61] A conduct and compensation agreement must be entered into with each owner or occupier who, as an 'eligible claimant', is entitled to seek compensation for the effects of the activities before the holder of an authority to prospect can enter land inside the petroleum authority.[62] The holder of the authority to prospect must also comply with other statutory prerequisites for entry onto the land, including providing the landowner with an entry notice. The *PGPSA* exempts the holder of an authority to prospect from the requirement to enter into a conduct and compensation agreement in circumstances where the holder already has a right to enter the land, the entry is to preserve life or property, because of an emergency that exists or may exist, or the landowner is an applicant or respondent to a Land Court application to determine compensation.[63]

---

58 See *PGPSA* (Qld) s 27(2)(b)(ii).
59 See *PGPSA* (Qld) s 500 which expressly requires each eligible claimant to the land to be a party to a conduct and compensation agreement. There is also provision for a 'deferral agreement' which is an agreement that the title-holder can enter the land conditional upon the entry into a conduct and compensation agreement at a later date: *PGPSA* (Qld) ss 500A(e)(i), 500B.
60 See *PGPSA* (Qld) s 495. If a deferral agreement has been entered into between the landowner and the petroleum tenure holder or a Land Court exemption to the entering into of an access and compensation agreement, a written notice of entry must be provided to each and every owner and occupier of the land at least 10 days prior to entry.
61 See *PGPSA* (Qld) sch 2, Dictionary.
62 See *PGPSA* (Qld) ss 500, 532(1).
63 See *PGPSA* (Qld) s 500A(a), (e), (f).

Prior to accessing the authorised land which is subject to the resource title in order to undertake either preliminary or advanced activities, the holder of an authority to prospect must generally give the landowner an entry notice describing the land to be entered, the activities to be carried out, and when and where these activities will be carried out.[64]

An entry notice will not be required where there is a conduct and compensation agreement relating to the land that contains a waiver of the entry notice requirement.[65] If an entry notice has not been served and a conduct and compensation agreement containing a waiver of entry notice has not been entered into prior to entry, the entry will be unlawful. In *O'Connor v Arrow (Daandine) Pty Ltd* the Court issued a declaration that entry onto land was unlawful and the construction of a pipeline was restrained until a valid entry notice was served.[66] The holder of an authority to prospect must wait 10 business days after the entry notice is provided before they are able to enter the land.[67]

The *PGPSA* distinguishes between private land inside the area of a petroleum authority – 'authority land' – and private land outside the petroleum authority's area that the petroleum title-holder may need to cross over in order to enter the authority land – 'access land'. A right to enter 'access land' is conferred by s 502 of the *PGPSA*.

The right to access private land outside the authority's area exists where it is reasonable and necessary. Section 502 gives the holder of a petroleum title rights to cross the land, enter the area of the authority, and to carry out activities on that land which are reasonably necessary and dependent upon access rights. Prior to any access right being exercised by the holder of a petroleum title, an access agreement must be entered into with each owner and occupier of the land where the exercise of such rights is likely to have a permanent impact on the land.[68]

An access agreement may authorise access to land, outside of the petroleum authority, in order for the title-holder to carry out reasonably necessary activities. The agreement will establish where, when, and how often the holder of the authority to prospect may cross the land, the method of crossing, and other matters including any construction of roads, tracks, gates, or fences. The *PGPSA* does not specifically delineate the contents of an access agreement, other than to state

---

64 See *PGPSA* (Qld) ss 495, 496.
65 See *PGPSA* (Qld) ss 497(1)(c), 498.
66 [2009] QSC 432, [48], [49].
67 See *PGPSA* (Qld) s 495(2).
68 'Permanent impact' on the land means a continuing effect on the land or its use or a permanent or long-term adverse effect on its current lawful use by an occupier of the land; for example, building a road: *PGPSA* (Qld) s 503(3).

that it may include a waiver of entry notice and may also include a compensation arrangement.[69]

By contrast, a conduct and compensation agreement seeks to outline how and when the land contained within the authority to prospect is to be accessed and further, how authorised activities are to be conducted. A conduct and compensation agreement should also include provisions articulating the particulars of any access arrangement. The primary purpose, however, of the conduct and compensation agreement is to address the compensation liability of the affected landholder and define the agreed basis for compensation.[70] The *PGPSA* does not restrict the matters that may be included within a conduct and compensation agreement although its terms must not be inconsistent with the provisions of the *PGPSA* and will often include provisions already contained within the *Land Access Code 2010* (Qld).

A conduct and compensation agreement must address the compensation liability that is owed by the holder of the petroleum authority towards the landowner. A petroleum authority holder must compensate the landowner for any compensatable effect he or she suffers and which is caused by relevant authorised activities.[71] Compensatable effects come under five broad heads:

- deprivation of possession of the land's surface
- diminution of the land's value
- diminution of the use made of the land or any improvement on it
- severance of any part of the land from other parts of the land or from other land of the landowner
- any cost, damage, or loss arising from the carrying out of authorised activities on the land.

Compensatable effects may also cover consequential damages the landowner incurs as well as accounting, legal, or valuation costs necessarily and reasonably incurred by the landowner in negotiating the agreement.[72] The agreement may cover both monetary and non-monetary forms of compensation.[73]

The *Land Access Code 2010* (Qld) was established under the *PGPSA* pursuant to s 24A. Section 24A(1) sets out that 'a regulation may make a single code for all resource Acts (the Land Access Code) that –

---

69 See *PGPSA* (Qld) ss 503(2), 506(2), (3).
70 See the discussion by S Christensen, P O'Connor, W D Duncan and A Phillips, 'Regulation of Land Access for Resource Development: A Coal Seam Gas Case Study from Queensland' (2012) 21(2) *Australian Property Law Journal* 110, 117.
71 See *PGPSA* (Qld) s 532(2).
72 See *PGPSA* (Qld) s 532(4)(b).
73 See *PGPSA* (Qld) s 534(2)(b)(i).

(a) states best practice guidelines for communication between the holders of authorities and owners and occupiers of private land; and

(b) imposes on the authorities mandatory conditions concerning the conduct of authorised activities on private land.

The *Land Access Code 2010* (Qld) establishes best practice benchmarks for landholder and petroleum title-holder communications which must be complied with, and which have a particular cogency to unconventional gas projects given the fact that many of these projects impact private, agricultural landholdings.

The general principles articulated in the *Land Access Code* for all holders of petroleum titles are as follows:

- liaise closely with the landholder in good faith
- advise the landholder of the holder's intentions relating to authorised activities well in advance of them being undertaken
- advise the landholder of any significant changes to operations or timing
- minimise damage to improvements, vegetation, and land
- respect the rights, privacy, property, and activities of the landholder
- rectify, without undue delay, any damage caused by the authorised activities
- promptly pay compensation agreed with the landholder once the agreed milestones are reached
- abide by this Code before, during and after undertaking activities
- be responsible for all authorised activities and actions undertaken by employees and contractors of the resource authority
- regard as confidential information obtained about the landholder's operations.[74]

In order to ensure proper and efficient communications, the holder of a petroleum title should make early contact with the landholder to visit and inspect the property. Further, to facilitate efficient communication in the initial stages of the process, both the petroleum title-holder and the landholder should each appoint a responsible person to negotiate the agreement and to undertake all communications in relation to land access.

The *Land Access Code* specifically mandates a range of conditions that relevant parties must adhere to. These are:

- oral or written notice to the landholder
- training for all persons acting for the holder of a petroleum title of obligations under the *Land Access Code*

---

[74] See *Land Access Code 2010* (Qld), 4. Available at http://mines.industry.qld.gov.au/assets/land-tenure-pdf/land_access_code_nov2010.pdf.

- existing access points, tracks or roads must be used where practicable
- existing access points, tracks or roads must be kept in good repair
- new access points, roads or tracks must be located at a place that minimises the impact upon the landholder's business or land use activities
- damages to land, access points, roads or tracks must be notified to the landholder as soon as possible and the damage must be repaired
- private land must be used in a way that minimises disturbances to people, livestock and property
- all relevant steps must be taken to prevent the spread of the reproductive material of a declared pest
- the location and plan of any camp established on land must be organised and agreed upon with the landholder
- rubbish or waste accumulating on the land must be suitably disposed of
- gates, grids and fences damaged or removed must be repaired or replaced as soon as practicable.[75]

These mandatory provisions seek to identify key conflict areas that are connected, in particular, with the expansion of unconventional gas projects in Queensland. By anticipating potential areas where conflict may arise and introducing a range of compulsory interface obligations for both petroleum title-holders and landholders, the opportunity for conflict is significantly diminished.

The Land Access Framework that has been implemented in Queensland is one of the most progressive in the country. It provides a sound basis for negotiation between mining proponents and landholders regarding access to private land and compensation for damage to that land. In this regard, the framework has significantly reduced the incidence and potential for protracted conflict and disruption within the CSG industry in Queensland and generated greater engagement between title-holders and affected landowners.[76]

---

75 Ibid.
76 A review of the Queensland Land Access Framework was conducted by the Land Access Review Panel in 2012 and, in response, the Queensland Government released its Six Point Action Plan. This plan is outlined by the Department of Natural Resources and Mines in the 'Queensland Government Response to the Report of the Land Access Review Panel' (December 2012) https://www.dnrm.qld.gov.au/__data/assets/pdf_file/0014/193100/qld-gov-response-land-access-framework.pdf. Pursuant to this plan, the following recommendations were suggested:
   - an expansion of the heads of compensation to include legal costs incurred by landholders in seeking advice regarding a Conduct and Compensation Agreement;
   - an expanded Land Court jurisdiction to incorporate dispute resolution processes; and
   - Conduct and Compensation Agreements to be noted on affected land titles.

5.5.2 New South Wales

The progression of CSG mining in New South Wales has been subjected to strong and consistent community objection. This has resulted in the introduction of a suite of focused regulatory developments that include: a 2 km exclusion zone around residential and village areas, an Aquifer Interference Policy, a Code of Practice for Well Integrity and fracture stimulation, and a ban on harmful BTEX chemicals and evaporation ponds.[77]

The core regulatory framework for mining in New South Wales consists of four key acts: the *Petroleum (Onshore) Act 1991* (NSW), the *Environmental Planning and Assessment Act 1979* (NSW), the *Protection of the Environment Operations Act 1997* (NSW), and the *Water Management Act 2000* (NSW). The operation of these Acts to CSG mining is now subject to the changes articulated within the New South Wales Gas Policy 2014 ('Gas Policy'), which was implemented following an extensive review of the industry undertaken by the NSW Chief Scientist.

The comprehensive report by the NSW Chief Scientist gave particular focus to the potential risks associated with CSG extraction given the uncertain impact that new technology might have upon the environment and landscape and how those risks might be appropriately managed. The report concluded that:

> Management of potential risks associated with CSG, as with other industries, requires effective controls; high levels of industry professionalism; systems to predict, assess, monitor and act on risks at appropriate threshold conditions; legislation; regulation; research; and commitment to rapid remediation, continuous improvement and specialist training. The Review studied the risks associated with the CSG industry in depth and concludes that – provided drilling is allowed only in areas where the geology and hydrogeology can be characterised adequately, and provided that appropriate engineering and scientific solutions are in place to manage the storage, transport, reuse or disposal of produced water and salts – the risks associated with CSG exploration and production can be managed. That said, current risk management needs improvement to reach best practice. In particularly sensitive areas, such as in and near drinking water catchments, risk management needs to be of a high order with particularly stringent requirements on companies operating there in terms of management, data provision, insurance cover, and incident-response times.[78]

---

77 At the date of writing, 30 new regulatory measures for CSG mining had been introduced in NSW.
78 See the NSW Chief Scientist and Engineer, M O'Kane, *Final Report of the Independent Review of CSG Activities in NSW* (September 2014) 10 http://www.chiefscientist.nsw.gov.au/__data/assets/pdf_file/0005/56912/140930-CSG-Final-Report.pdf.

The Gas Policy adopts all of the 16 recommendations made for improving CSG projects in the final report by the NSW Chief Scientist, which was released on 14 September 2014.[79] The recommendations from this report include:

- implementing new data-gathering processes and a comprehensive, publicly-available information portal
- developing a centralised risk management and impact prediction resource to improve the quality and consistency of environmental assessments for CSG projects
- creating an expert scientific and engineering advisory body to advise the government on matters such as CSG industry impacts, science and technology developments, and the assessment of NSW sedimentary basins
- full cost recovery for the regulation and support of the CSG industry
- a plan to manage legacy matters (such as abandoned CSG wells and incomplete compliance checking)
- one single regulatory regime for all onshore subsurface gas resources (excluding water) in the state
- a separation of the process for allocating rights to exploit subsurface resources (excluding water) from the process of regulating the activities giving effect to that exploitation – that is, exploration and production activities – but both processes still to be implemented by a single independent regulator.[80]

Exclusions zones on the issuance of new petroleum exploration licences for CSG projects are in force in New South Wales.[81] Further, there has been a freeze on the issuance of new petroleum titles for CSG development until the different elements of the Gas Policy have been fully implemented.[82] New licences are unlikely to be issued until the new framework is fully operational. All existing applications for petroleum exploration licences are to be extinguished and the costs associated with application fees are refundable.[83] Legislation effecting this extinguishment was introduced to complement the announcement of the new Gas Policy.[84] Further, existing issued petroleum licences may take advantage of the buy-back offer being put forward by the New South Wales Government which allows a

---

79 Ibid.
80 Ibid 12–15.
81 The exclusion zones apply to 2.7 million hectares of existing and future residential land across NSW and the equine and viticulture critical industry clusters in the Upper Hunter and are an important component of the strategic regional land use policy.
82 The NSW Government has not issued any new petroleum exploration licences for CSG since April 2011. The freeze will continue until mid-2015.
83 At the date of writing there were 16 petroleum exploration licences pending which covered 43 per cent of the state.
84 See *Mining and Petroleum Legislation Amendment Bill 2014* (NSW) schs 1, 5.

holder to exchange the exploration licences for a limited compensatory payment. All new petroleum exploration licences issued following the implementation of the new Gas Policy are to be subject to new conditions and codes. These changes are unlikely to impact extensively upon the major existing exploration projects in Camden, Gloucester, and Narrabri which have already been issued, although the licence conditions imposed on these titles may be expected to meet the criteria of new, higher standards.

Further, a strategic release framework will set out the areas where CSG gas exploration may occur in NSW and any determination will involve an assessment of economic, environmental, and social factors. Exclusion zones were also introduced in 2012 to prevent approval of new CSG activities within 2 km of residential zones and within critical industry clusters. The NSW Government also put a hold on CSG activities in the special areas of the Sydney water catchment. Once the strategic release framework is implemented, CSG exploration will only be allowed to be conducted in authorised areas although the hold on the special areas of the Sydney water catchment will endure until further findings are released.[85]

The aim of the strategic release framework is to ensure that focused CSG development occurs but also to promote the implementation of strategic gas projects. Given that NSW currently imports approximately 95 per cent of its gas interstate, it recognised a strong need to increase domestic fuel security. Hence, where a project is deemed to be one that has the capacity to provide a significant amount of gas for the NSW market, it may be characterised as a 'strategic energy project' and, in such a situation, is to be overseen by a case manager. This effectively means that the approval process for the project is to be coordinated by government, although it will not be immune from the obligation to comply with environmental standards and satisfy community consultation requirements. It is also clear that the government will actively investigate the possibility of gas pipelines being extended into the NSW region to ensure greater gas accessibility for the state.

Another important proposed amendment under the new Gas Policy is that petroleum licences are to be subject to public expressions of interest to ensure that the most suitable and capable proponents acquire titles. The proposed test – which will replace the interim 'public interest' test which was implemented in accordance with the findings of a 2013 ICAC Report – will require decisions regarding current or proposed mining and exploration licences and mining/production leases to be

---

[85] This is a component of the new Strategic Release Framework which is effective from July 2015. See NSW Government, *NSW Gas Plan: Common Questions and Answers* https://www.nsw.gov.au/sites/default/files/miscellaneous/sc000218_nsw_gas_plan_questions_and_answers_web.pdf.

based upon whether the current or proposed holder is a 'fit and proper' person.[86] The focus of the provision is upon the suitability and character of the applicant, with consideration being given to the past criminal conduct, if any, of the applicant, as well as the nature and scope of their technical expertise and capability.

The implementation of this provision flows from the previous findings in the NSW ICAC Report which indicated that a regulatory environment lacking in clear criteria and direction had the potential to generate improper incentives with respect to the issuance of resource titles. The Report stated:

> Where the environment is opaque, uncertain and discretionary, incentives may be created to lobby and persuade decision-makers to achieve favourable outcomes; the greater the value of the coal resource to be transferred from government to a private entity, the greater the incentive. Similarly, the greater the private investment put at risk by uncertain policy and regulatory decisions, the greater the incentive to improperly lobby and manipulate government decision-making and policy.[87]

Section 24 of the Mining and Petroleum Legislation Amendment Bill 2014 (NSW) seeks to amend the existing s 380A of the *Mining Act 1992* (NSW) and replace it with a provision which specifically articulates the nature and scope of a 'fit and proper' applicant and sets out the range of factors that need to be taken into account. The proposed s 380A from the Mining and Petroleum Legislation Amendment Bill 2014 (NSW) is extracted below:

Mining and Petroleum Legislation Amendment Bill 2014 (NSW)

380A Fit and proper person consideration in making certain decisions about mining rights

(1) Despite anything to the contrary in this Act, any of the following decisions under this Act may be made on the ground that, in the opinion of the decision-maker, a relevant person is not a fit and proper person (without limiting any other ground on which such a decision may be made):

---

86 In 2013, following findings about the improper grant of two coal mining exploration licences by the Independent Commission Against Corruption (ICAC), a 'public interest' test was implemented. A public auction method was recommended as the preferred method of allocation for exploration licences in place of direct allocation. ICAC also recommended that irrespective of the means of allocation, an exploration licence should not be granted until the state has undertaken due diligence on the applicant and is satisfied that the applicant has the technical expertise and financial capacity to fund a satisfactory level of exploration activity. See the Independent Commission Against Corruption, *Reducing the Opportunities and Incentives for Corruption in the State's Management of Coal Resources* (2013) <http//www.icac.nsw.gov.au>.

87 See ICAC, above n 87, 12.

(a) a decision to refuse to grant or renew a mining right (a relevant person in such a case being an applicant for the grant or renewal of the mining right),

(b) a decision to refuse to transfer a mining right (a relevant person in such a case being the proposed transferee),

(c) a decision to cancel a mining right or to suspend operations under a mining right (in whole or in part), a relevant person in such a case being a holder of the mining right,

(d) a decision to restrict operations under a mining right by the imposition or variation of conditions of a mining right (a relevant person in such a case being a holder of the mining right).

(2) For the purpose of determining whether a person is a fit and proper person, the decision-maker may take into consideration any or all of the following matters (but without limiting the matters that can be taken into consideration for that purpose):

(a) whether the person or (in the case of a body corporate) a director of the body corporate or of a related body corporate has compliance or criminal conduct issues (as defined in this section),

(b) in the case of a body corporate, whether a director of the body corporate or of a related body corporate is or has been a director of another body but only if the person was a director of that other body corporate at the time of the conduct that resulted in the compliance or criminal conduct issues,

(c) the person's record of compliance with relevant legislation (established to the satisfaction of the decision-maker),

(d) in the case of a body corporate, the record of compliance with relevant legislation (established to the satisfaction of the decision-maker) of any director of the body corporate or a related body corporate,

(e) whether, in the opinion of the decision-maker, the management of the activities or works that are or are to be authorised, required or regulated under the mining right are not or will not be in the hands of a technically competent person,

(f) whether, in the opinion of the decision-maker, the person is not of good repute,

(g) in the case of a body corporate, whether, in the opinion of the decision-maker, a director of the body corporate or a related body corporate is not of good repute,

(h) whether, in the opinion of the decision-maker, the person is not of good character, with particular regard to honesty and integrity,

(i) in the case of a body corporate, whether, in the opinion of the decision-maker, a director of the body corporate or a related body corporate is not of good character, with particular regard to honesty and integrity,

(j) whether the person, during the previous 3 years, was an undischarged bankrupt or applied to take the benefit of any law for the relief of bankrupt or insolvent debtors, compounded with his or her creditors or made an assignment of his or her remuneration for their benefit,

(k) in the case of an individual, whether he or she is or was a director of a body corporate that is the subject of a winding up order or for which a controller or administrator has been appointed during the previous 3 years,

(l) in the case of a body corporate, whether the body corporate or a related body corporate is the subject of a winding up order or has had a controller or administrator appointed during the previous 3 years,
(m) whether the person has demonstrated to the decision-maker the financial capacity to comply with the person's obligations under the mining right,
(n) whether the person is in partnership, in connection with activities that are subject to a mining right or proposed mining right, with a person whom the decision-maker considers is not a fit and proper person under this section,
(o) whether the person has an arrangement (formal or informal) in connection with activities that are subject to a mining right or proposed mining right with another person whom the decision-maker considers is not a fit and proper person under this section, if the decision-maker is satisfied that the arrangement gives that other person the capacity to determine the outcome of decisions about financial and operating policies concerning those activities,
(p) any other matters prescribed by the regulations.

5.5.2.1 The ownership framework under the *Petroleum (Onshore) Act 1991* (NSW)

The *Petroleum (Onshore) Act 1991* (NSW) ('the *POA*') is administered by the New South Wales Office of Coal Seam Gas ('OCSG'). The *POA* regulates the issuance and administration of petroleum titles. Petroleum is defined broadly to include any naturally occurring hydrocarbon or mixture of hydrocarbons, whether in a gaseous, liquid or solid state.[88] The breadth of this definition includes shale, tight gas, and CSG. All unconventional gas licences in New South Wales will be issued pursuant to the *POA*.

The ownership framework set out under the *POA* is based upon the public ownership of petroleum (the core principles of a public ownership framework are outlined in Chapter 1). As Chapter 1 sets out, under a public ownership framework, ownership of petroleum is vested in the Crown, rather than the surface estate owner, and all Crown grants, leases, and licenses are treated as having impliedly reserved ownership of petroleum in the Crown.[89] The Crown retains ownership over minerals that exist in a natural state on or below the surface of the land.[90] This means that it is an offence to mine for minerals in the absence of a petroleum title.[91]

---

[88] See *POA* (NSW) s 3.
[89] See *POA* (NSW) s 6(1), (2). This vesting provision alters the common law position under the maxim: *Cujus est solum, ejus est usque ad coelum et ad inferos* (whoever owns the soil, it is theirs all the way up to heaven and down to hell).
[90] See *POA* (NSW) s 6.
[91] See *POA* (NSW) s 7.

The public ownership structure of the *POA* is to be contrasted with the private, accession framework that underpins mineral ownership within other countries around the world, including the United States.[92] As Chapter 1 sets out, under the public ownership framework, the state in right of the Crown is responsible for determining a framework for suitable applicants to apply for a statutory licence. This system has been described as a 'concession system'.[93]

Once issued, the licence will enable the holder to explore for or produce unconventional gas.

5.5.2.2 Petroleum licences and conditions under the *Petroleum (Onshore) Act 1991* (NSW)

The licensing framework in New South Wales is grounded in the core assumption that subsurface resources, including unconventional gas, have a separate 'ownership' capacity to the strata in which they reside despite their corporeal integration.[94] Once subsurface minerals have been extracted, and the licence-holder acquires control of the mineral, ownership in the mineral passes from the Crown to the licence-holder, subject to the entitlement of the state to receive royalty payments in accordance with revenue agreements.[95]

The provisions in the *POA* provide little substantive guidance as to how minerals are to be disaggregated from the land and this means that the exact scope of mineral ownership is not entirely clear. To overcome this issue, the *POA* delineates the particular rights and obligations that attach to each form of petroleum title.[96] Once advertised, the validity of the rights attaching to a petroleum title cannot be challenged after the expiration of three months.[97]

---

92 See the outline of this in Chapter 1. For a more general discussion on the operation of a state-based mineral ownership framework see Y Omorogbe and P Oniemola, 'Property Rights in Oil and Gas Under Dominial Regimes' in A McHarg, B Barton, A Bradbrook and L Godden (eds) *Property and the Law in Energy and Natural Resources* (Oxford University Press, 2010) 115, 118.
93 The concession system arises where the state, as owner of the mineral resource, grants rights to explore or exploit the mineral pursuant to an objective and defined statutory licensing framework. See J K Boyce, 'From Natural Resources to Natural Assets' in J K Boyce and B G Shelley (eds), *Natural Assets: Democratizing Environmental Ownership* (Island Press, Washington DC, 2003) 7.
94 For a discussion of the *ad coelum* principle under common law see J Sprankling, 'Owning the Center of the Earth' (2008) 55 *UCLA Law Review* 979, 1000–1.
95 This is discussed in Chapter 10.
96 For a discussion on the tensions associated with the establishment of statutory vesting provisions upon landholders see S Christensen, P O'Connor, W D Duncan and A Phillips, 'Regulation of Land Access for Resource Development: A Coal Seam Gas Case Study from Queensland' (2012) 21 *Australian Property Law* 110, 112.
97 See *POA* (NSW) s 25.

The rapid expansion of petroleum titles for unconventional gas within areas in New South Wales which have been traditionally associated with farming and tourism, combined with the uncertainty regarding the scope of ownership rights and how they will interact with surface ownership, has generated significant community tension.[98] In responding to this, the NSW Gas Policy seeks to ensure that all future issued licences are subject to new conditions which incorporate best practice requirements and which are specifically tailored to respond to the risks associated with different regions and activities. Further, any environmental conditions imposed upon issued licences are to be fully monitored by the Environmental Protection Authority (discussed below).

The *POA* outlines a range of different forms of petroleum titles, and each will constitute personal rather than real property.[99] The *POA* provides for the issuance of explorations licences.[100] An exploration licence for unconventional gas under the *POA* amounts to a permissory right which cannot exceed six years in duration. It authorises the holder to access the land, and conduct exploratory drills and these rights are exclusive.[101] A 'special prospecting authority' may also be issued under the *POA*. This authority confers on the holder an exclusive right to conduct speculative geological, geophysical, or geochemical surveys or scientific investigations on the land.[102]

The *POA* also provides for the issuance of an assessment lease that confers rights of retention upon the holder in an area where a significant petroleum deposit is identified and mining that deposit is not deemed to be commercially viable in the short-term, although there is a reasonable prospect it will become so in the future. The holder of an assessment lease retains an exclusive right to prospect for and assess petroleum deposits.[103]

A production lease issued under the *POA* confers a right upon a holder to extract and produce petroleum, including the right to conduct mining operations on the land and to construct and maintain mining operations.[104] The rights conferred by the production lease are exclusive in nature. The range of entitlements encompassed by the production lease include: the right to conduct mining operations; the right to construct buildings; the right to plan waterways, roads, pipelines, dams,

---

98 For an outline of landholder issues see T Marshall, 'Coal Seam Gas (CSG) Licences and Landholders' Rights' (2011) 26(9) *Australian Environment Review* 226, 234–6.
99 See *POA* (NSW) s 26. Note that petroleum titles do not constitute personal property for the purposes of the *Personal Property Securities Act 2009* (Cth). This is set out in the *POA* (NSW) s 126B.
100 A full discussion of the nature and scope of an exploration licence is outlined in Chapter 1.
101 See *POA* (NSW) s 29.
102 See *POA* (NSW) s 38.
103 See *POA* (NSW) s 33.
104 See *POA* (NSW) s 41.

reservoirs, tanks, pumping stations and other structures; and the right to bring in equipment necessary for the full enjoyment of the lease.[105]

The *POA* makes a fundamental distinction between high and low-impact titles. An exploration licence or a special prospecting authority will be characterised as 'low-impact' where it is found to be unlikely to have a significant impact on the land over which the title is granted.[106] Low-impact exploration licences and special prospecting authorities are collectively described within the *POA* as 'prospecting titles'.[107]

Under the Gas Policy, all gas exploration licences will be assessed and determined by the Minister for Resources and Energy and all gas production is to be assessed and determined by the Minister for Planning or the independent NSW Planning Assessment Commission.

Where a petroleum title is issued for an unconventional gas project, it may be subject to *POA* conditions.[108] If the licence-holder does not comply with the attached conditions, a pecuniary penalty may be awarded or the Minister may, in special circumstances, consider whether the licence should be suspended or cancelled.[109] The Minister may, at his or her discretion, request a cancellation if a condition of the title is contravened, or if the land subject to the title is required for a public purpose.[110] Significantly, all breaches of the *POA* are regarded as criminal offences.[111]

One of the standard title conditions for licences under the *POA* is an environmental management obligation. This effectively requires the holder of the licence to maintain ongoing environmental standards and to carry out progressive environmental assessment. This is an extremely important requirement for unconventional gas projects given the fact that the full environmental impact of exploring for and extracting unconventional gas upon the surrounding environment is not clear.

Where it is established that an environmental management condition is breached, the Minister has a discretion to determine whether the breach should result in a suspension or cancellation of title. Ordinarily, a breach of this type of condition will attract a penalty and will not result in a licence being cancelled or

---

105 See *POA* (NSW) s 41. For a discussion of the expansive scope of statutory entitlements for production leases in Queensland see M Weir and T Hunter, 'Property Rights and Coal Seam Gas Extraction: The Modern Property Law Conundrum' (2012) 2 *Property Law Review* 71, 77–80.
106 See *POA* (NSW) s 45C.
107 See *POA* (NSW) s 45B(1).
108 See *POA* (NSW) s 23.
109 See *POA* (NSW) s 22.
110 See *POA* (NSW) s 22.
111 See *POA* (NSW), s 22(1). See also *Connell v Santos NSW Pty Ltd* [2014] NSWLEC 1, [67]–[68] where the Court held that a breach of a condition on title frustrates the statutory scheme and the imposition of criminal offences upholds the efficacy and credibility of the legislative scheme.

suspended.[112] The difficulty, however, with imposing financial sanctions, is that they do not have a strong deterrence factor where the licence-holder is easily able to make payment.[113]

A cancellation of title for a breach of the *POA* is only possible once the Minister has first reviewed the option of suspension.[114] The approach is well-outlined by the New South Wales Land and Environment Court in the decision of *Connell v NSW Pty Ltd*.[115] On the facts, Santos NSW Pty Limited ('Santos') pleaded guilty to four charges of committing offences pursuant to s 136A(1) of the *POA* in that, without reasonable excuse, it failed to comply with conditions of a petroleum title issued over land in the Narrabri region of New South Wales. The condition required Santos to report any incidents causing or threatening material harm to the environment. This incident was the result of multiple leaks at the water treatment plant. Attempts were made to repair those leaks. Despite this, a spill of water at the water treatment plant resulted in a burst pipeline and this in turn resulted in 7000 litres of production water escaping from the water treatment plant.

This incident occurred prior to Santos taking over the site. Santos pleaded guilty to breaching the environmental management condition by not accurately reporting the spill in its report and detailing the problem in the petroleum operations plan. The Land and Environment Court was then asked to issue a sanction against Santos for these offences.

Preston CJ took into account the objective circumstances of the offences, as mitigated by the subjective circumstances of the defendant. His Honour also considered the important of imposing a sentence that achieves the objective of:

> denouncing the conduct of the defendant, ensuring that it is adequately punished for the offences it has committed, making it accountable for its actions, recognising the harm done to the environment by commission of the offences, and preventing crime by deterring other operators in the coal seam gas industry from committing similar offences of failing to comply, without reasonable excuse, with the conditions of a petroleum title it holds.[116]

After evaluating the circumstances, Preston CJ imposed a range of minimal pecuniary sanctions:

---

112 See *POA* (NSW) s 22(3A).
113 See the discussion by C Clifford, 'Conservationists say Santos fine for Pilliga spill too small', *ABC* (online), 12 January 2014 http://www.abc.net.au/news/2014–01–11/conservationists-say-santos-fine-for-pilliga-spill-too-small/5195618.
114 See *POA* (NSW) s 22.
115 [2014] NSWLEC 1.
116 [2014] NSWLEC 1, [171].

(a) In relation to the first charge, a fine of $30 000 for the offence forming the basis of this charge. This figure was to be discounted by 30 per cent for the assistance provided by the defendant to the authorities which resulted in a fine of $21 000; and

(b) In relation to the second, third, and fourth charges, a fine of $15 000 for each offence forming the basis of these charges. This figure was discounted by 30 per cent for the assistance provided by the defendant to the authorities for each offence, which resulted in a fine of $10 500 for each offence.[117]

During the course of his judgment, Preston CJ specifically held that the use of criminal law sanctioning mechanisms under the *POA* 'ensures the credibility of the legislative scheme.'[118] His Honour also noted that the defendant's efforts to mitigate the harm to the environment meant that any fine should fall 'at the lower end of the scale of seriousness.'[119] Amendments to the *POA* now provide for up to $1 100 000.00 in penalties for a breach of a petroleum title condition.[120]

5.5.2.3 Land access, compensation and access disputes

Significantly, for the purposes of unconventional gas licensing, prospecting in NSW may only be carried out pursuant to a prospecting title where a land access arrangement has been entered into between the title-holder and the landholder.[121] The access arrangement must include provisions relating to the nature of the prospecting, the means by which access will be obtained, the areas of land affected, and the compensation payable.[122]

Access arrangements are not mandated for high-impact exploration or special prospecting titles, or for assessment or production leases, which seems an unusual anomaly given the potential for surface estate disturbance posed by the issuance of such titles. However, proposed amendments to the *POA* would allow the Minister to cancel a title should compensation not be paid in accordance with a land access arrangement.[123]

In an attempt to increase the rigour of ongoing environmental management obligations, the Petroleum (Onshore) Amendment Bill 2013 (NSW) has proposed the

---

117 [2014] NSWLEC 1, [172].
118 [2014] NSWLEC 1, [68].
119 [2014] NSWLEC 1, [113]. The Court mitigated the sanction in accordance with the new s 136A(3).
120 See *POA* (NSW) ss 136A, 136A(3).
121 See *POA* (NSW) s 69C(1).
122 See *POA* (NSW) s 69D.
123 *Petroleum (Onshore) Amendment Bill* 2013 (NSW) cl 22(3A)(d)

introduction of a new 'environmental assessment permit'.[124] This permit would allow entry onto land covered by a petroleum title for the environmental assessment of petroleum title activities.[125] This type of ongoing assessment is likely to result in the detection of environmental management breaches thereby reducing the need to remediate the impact of such breaches after they have occurred. It is also likely to encourage full and accurate reporting of non-compliance with environmental management conditions.

The regulation of access rights remains, however, a source of significant conflict between landholders and CSG proponents in New South Wales. The General Standing Committee Report on CSG has recommended strengthening landholder rights for land access.[126] In response, the Petroleum (Onshore) Amendment Bill 2013 seeks to introduce a Land Access Code for petroleum exploration. At the date of writing, the Bill (which lapsed on prorogation), remains with the Legislative Council after reaching the second reading.[127] Given the widespread community concern regarding coal seam gas expansion, the process of passing the Bill has been delayed by opposition and debate.

The proposed NSW Code is akin to the *Land Access Code 2010* that exists in Queensland.[128] Such a code has the potential to significantly improve landholder entitlements. The Bill enables the regulations to prescribe a land access code for petroleum exploration.[129] The draft NSW Land Access Code contains 'mandatory provisions', which are taken to be included in an access arrangement (and which would be required for all petroleum exploration on the surface of land) except to the extent that the agreement expressly excludes or varies them as well as 'aspirational' practices.[130]

---

124 *Petroleum (Onshore) Amendment Bill* 2013 (NSW) cl 104R. The holder of a permit under this section, and any employee or agent of the holder, may, in accordance with the permit: (a) enter the land to which the permit relates; and (b) do on that land all such things as are reasonably necessary to carry out the assessment to which the permit relates.
125 Ibid.
126 See Parliament of New South Wales, *New South Wales Inquiry into Coal Seam Gas*, (Legislative Council General Purpose Standing Committee No 5, May 2012) 137–46. Changes to the *Petroleum (Onshore) Act 1991* (NSW) were recommended including developing an access code covering both the exploration and production phases.
127 See NSW Draft Code of Practice for Land Access for CSG and Petroleum Exploration (November 2013) issued pursuant to s 69DB *Petroleum (Onshore) Act* 1991 (NSW) at: http://www.haveyoursay.nsw.gov.au/assets/premier-and-cabinet/cal-gray/Code-of-Practice-for-Land-Access-1.pdf. At the date of writing, the Draft Code of Practice for Land Access for CSG and Petroleum Exploration was still open for public discussion and feedback.
128 See *Land Access Code 2010* (Qld) at <http://mines.industry.qld.gov.au/assets/land-tenure-pdf/land_access_code_nov2010.pdf>.
129 The Bill also extends the amount of costs which a landholder can recover from a mining or petroleum exploration title holder in connection with an access arrangement for that title holder, to include costs associated with negotiating, and entering into, the arrangement.
130 Petroleum (Onshore) Amendment Bill 2013 s 69DC.

The draft Land Access Code also extends the amount of costs which a landholder may recover from a mining or petroleum exploration title to include costs associated with negotiating, and entering into an access arrangement. A significant limitation of the draft NSW Land Access Code however, lies in the fact that it only has an application to prospecting titles.[131]

Some of the draft provisions of the proposed Land Access Code include:

3.6 The Explorer must:
- **(a)** keep the Landholder informed of the progress of, and variations to, an exploration activity on their land;
- **(b)** minimise potential for any damage to the Landholder's property during the exploration activity;
- **(c)** be responsible for all actions undertaken by employees and contractors of the explorer;
- **(d)** notify the Landholder of any damage to the Landholder's land, livestock or property, caused or contributed to by the carrying out of the exploration activity; and
- **(e)** rectify without undue delay any damage to the Landholder's property to the extent caused or contributed to by the carrying out of the exploration activity.

3.9 The Explorer must provide the Landholder with a written, plain language explanation of the way in which the water regulatory requirements are being implemented in relation to the title.

3.10 The Explorer must provide any available monitoring/testing results carried out under the water regulation requirements listed in 3.9 to the Landholder and/or his or her representative and the Landholder's request.

3.12 The Explorer must use the Landholder's land in a way that minimises disturbance to existing land uses (including crops), livestock and property, including roads, gates, grids and fences.[132]

The *POA* sets out that compensation must be paid to interest holders injuriously affected by the issuance of a petroleum title, and this includes native title-holders.[133] Unlike Queensland, the *POA* does not require the parties to enter into access and compensation agreements prior to the issuance of a petroleum title and, in the absence of such an agreement, the affected party will need to rely upon the separate compensation provisions.[134] The *POA* restricts compensation to surface loss flowing from prospecting operations carried out pursuant to a 'prospecting title'. It does not

---

131 Petroleum (Onshore) Amendment Bill 2013 s 69DB.
132 See the Draft Code of Practice for Land Access NSW made under s 69DB of the *Petroleum (Onshore) Act 1991* (NSW) at: http://www.haveyoursay.nsw.gov.au/assets/premier-and-cabinet/cal-gray/Code-of-Practice-for-Land-Access-1.pdf.
133 See *POA* (NSW) s 107.
134 This position may be contrasted with Queensland where the parties are required to enter into a land access agreement setting out access, conduct and compensation prior to the issuance of title: see *PGPSA* (Qld) ss 500, 503.

include any injury or loss resulting from unauthorised land access by a petroleum title-holder.[135] The amount of compensation payable will depend on the circumstances. In the context of unconventional gas, consideration may be given to the loss caused or likely to be caused to the surface of the land, including crops, trees, grasses, or other vegetation on the land, or damage to buildings and improvements on the land which have been caused by prospecting or petroleum mining operations. If water depletion, occurring as a consequence of CSG mining, can, for example, be shown to result in crop deterioration, compensation may be available.

Compensation is also payable for the deprivation of possession or use of the surface of the land or easement rights and also for the destruction of stock.[136] Compensation is not payable where the operations do not affect 'any portion of the surface of the land'.[137] Hence, where an unconventional gas project causes minimal surface damage but the extraction has a damaging impact upon the subsurface, which may result in long-term damage but which is not immediately apparent on the surface, compensation will not be available.

Under the Gas Policy, the NSW Government will commission the Independent Pricing and Regulatory Tribunal to benchmark compensation rates annually. The Tribunal will then be asked to consider both fixed rate compensation and compensation that takes into account the economic benefits of exploration and production over the expected life of the wells. These benchmarks may then be used as a guide by landowners in their compensation agreements with industry.

Further, under the Gas Policy, the NSW Government proposes to establish a Community Benefits Fund, with voluntary contributions from both gas companies and the NSW Government. The Fund will finance local projects in communities where gas exploration and production occurs. Companies may choose to make upfront payments which may then be credited against future royalty payments.

Disputes regarding land access under the *POA* were the sole subject of an extensive review conducted by Mr Bret Walker SC in 2014. The Land Access Arbitration Framework Review – known as the 'Walker Review' – made 31 recommendations to improve the arbitration land access framework. The NSW Government, in accordance with the Gas Policy, endorsed all the recommendations set out in the Walker Review.[138] The aim of the Gas Policy is to ensure that the arbitration process is efficient, balanced, and transparent, and that the reasonable cost of

---

135 See *POA* (NSW) s 69D(1)(f).
136 Ibids 109.
137 Ibids 107. The Petroleum (Onshore) Amendment Bill 2013 cl 107(1) sets out that compensation is payable by all petroleum title-holders.
138 Department of Resources and Energy, New South Wales, Bret Walker, *Examination of the Land Access Arbitration Framework: Mining Act 1992 and Petroleum (Onshore) Act 1991* (20 June 2014) http://www.resourcesandenergy.nsw.gov.au/__data/assets/pdf_file/0018/527112/Brett-Walker-Examination-of-the-Land-Access-Arbitration-Framework.pdf.

negotiating an access arrangement is borne by gas companies, rather than landholders. Key recommendations from the Walker Review are extracted below:

Walker Review Recommendations

Recommendation 1

It is recommended that the *Mining Act* and the *Petroleum (Onshore) Act* be amended to provide for appointment of Panel Arbitrators for a maximum of three years.

Recommendation 2

Includes minimum qualifications for all land access arbitrators including members of the Arbitration Panel.

Recommendation 7

It is recommended that a person be eligible to be appointed to the Panel if they:

> are an accredited arbitrator through a recognised body such as the NSW Law Society, the Institute for Arbitrators and Mediators or the National Mediator Accreditation System;
>
> and have extensive arbitration experience.

In addition, a person must either:

> have extensive resources or agricultural industry experience;
>
> or be a legal practitioner who is eligible for appointment to the Supreme Court, with considerable litigation experience.

Recommendation 8

It is recommended that the Division of Resources and Energy induct Panel Arbitrators and provide annual seminars to provide updates on the *Mining Act* and the *Petroleum (Onshore) Act* and any related policy initiatives.

Recommendation 9

It is recommended that Panel Arbitrators maintain consistent minimum qualifications and be required to undertake continuous education and training consistent with accreditation requirements

5.5.2.4 Codes of practice for CSG: Fracture stimulation and well integrity

The Office of Coal Seam Gas in New South Wales ('OCSG') is responsible for the administration of CSG titles and activity approvals granted under the *POA* as well as any associated assessments under the *Environmental Planning and Assessment Act 1979* (NSW) ('the *EPAA*'). The OCSG monitors auditing and title compliance and is also responsible for the application of workplace health and safety requirements for CSG operations under both the *POA* and the *Work Health and Safety Act 2011* (Cth).

The OCSG is also responsible for administering two Codes of Practice introduced pursuant to the *POA* which impact upon unconventional gas mining. The first is the Fracture Stimulation Code and the second is the Well Integrity Code. The codes follow on from the decision to become a signatory to the National Partnership Agreement on Coal Seam Gas ('NPACSG').[139] The release of the codes follows a 12-month moratorium on the use of hydraulic fracturing for CSG drilling in New South Wales.

The provisions of each Code are recognised as conditions of title and any CSG proponent breaching the Code will be subject to a range of remedies including, at the discretion of the Minister, cancellation of the title.

Each Code is divided into three sections: principles, mandatory requirements, and aspirational standards. Under the Fracture Stimulation Code, all CSG explorers engaging in hydraulic fracturing must have an approved fracture stimulation management plan ('FSMP') in place prior to commencing hydraulic fracturing.[140] The FSMP must summarise the nature, location, scale, timing, and duration of the activity. It must also demonstrate that all risks to the environment are properly managed through an effective risk management process and a review of environmental factors must also be included.[141] The review must address, inter alia, public safety, chemical use, impact upon water resources, air pollution, loss of well integrity, and induced seismicity.[142]

All chemicals to be utilised during the fracture stimulation activity must be outlined in the FSMP.[143] In early 2012, the NSW Government banned the use of BTEX chemical compounds in all drilling and hydraulic fracturing activities.[144]

---

139 See Council of Australian Governments, *National Partnership Agreement on Coal Seam Gas and Large Coal Mining Development: Intergovernmental Agreement on Federal Financial Relations*. Entered into between the Commonwealth and NSW, Victoria, Queensland, South Australia and the Northern Territory. The Agreement is available for viewing at: https://www.ehp.qld.gov.au/management/impact-assessment/pdf/partnership-agreement.pdf.
140 NSW Government, Department of Trade and Investment Resources and Energy, *Code of Practice for CSG: Fracture Stimulation Activities* (September 2012) 1.2 (a).
141 Ibid 4.2 (d).
142 Ibid 1.2 (a).
143 Ibid 6.2.
144 State Environmental Planning Policy, 2007 (Mining, Petroleum, Production and Extractive Industries) (Mining SEPP), T1-O-120. The use of the volatile organic compounds benzene, toluene, ethylbenzene and xylene (BTEX) in CSG drilling additives and CSG fracture stimulation additives is prohibited.
  1. All CSG drilling additives and CSG fracture stimulation additives must be tested by a NATA-certified laboratory and demonstrated to meet Australian drinking water health guideline values, which are currently:
     a. mg/L for benzene
     b. 0.8 mg/L for toluene
     c. 0.3 mg/L for ethylbenzene
     d. 0.6 mg/L for xylene

Henceforth, all chemical fluids used during hydraulic fracturing must be disclosed to and approved by the Office of Coal Seam Gas.[145] Any risks to human health arising from the usage of such chemicals, the risks and consequences of surface spills and the manner in which the chemicals will be stored must be outlined.[146] Water resources impacted by the fracture stimulation activity, the possibility of cross-contamination between coal-bed waters and shallower water sources, changes in groundwater pressure, changes in surface water levels, and changes to water quality characteristics must also be outlined.[147]

CSG explorers engaging in hydraulic fracturing must consult with stakeholders prior to undertaking the activity.[148] The exact nature of the consultation is not, however, outlined by the Code and provisions relating to the nature, location, scale, time, and duration of the operation are included as aspirational benchmarks rather than mandatory conditions.[149] All CSG explorers engaging in hydraulic fracturing must also implement an SMP for the safety of workers and to ensure the safety of operational procedures being conducted at the site.[150]

The Code of Practice for CSG Well Integrity applies to the construction, production, maintenance, and abandonment of CSG wells in New South Wales. The mandatory requirements under this Code include the requirement that no leaks occur through or between any casing strings in order to avoid contamination of groundwater or other aquifer resources, and avoiding leakage.[151]

The Well Integrity Code mandates the use of appropriate well production materials to prevent contamination of groundwater.[152] The Code does not seek to impose rigorous obligations regarding water usage and waste disposal. Hence the use of sustainable construction practices and operating procedures for the conservation of water usage and the minimisation of waste are only included as aspirational rather than mandatory requirements.[153] The Well Integrity Code effectively requires CSG wells which are constructed in New South Wales to have three layers of concrete and steel to reduce as much as possible the risk of gas or fluids escaping from the well. This also protects groundwater and soils from the risk of contamination.

---

145 'BTEX' stands for volatile organic compounds of benzene, toluene, ethylbenzene and xylene which had been utilised internationally and in Queensland. See above n 145.
146 Ibid.
147 NSW Government, Department of Trade and Investment Resources and Energy, *Code of Practice for CSG: Fracture Stimulation Activities* (September 2012) 7.2 (c).
148 Ibid 2.2 (a).
149 Ibid 2.3 (a).
150 Ibid 5.2 (c).
151 NSW Government, Department of Trade and Investment Resources and Energy, *Code of Practice for CSG: Well Integrity* (September 2012) 4.1.1.
152 Ibid 4.1.2.
153 Ibid 4.1.3.

5.5.2.5 The gateway process

The planning framework in New South Wales has undergone significant reform. The new system is to be implemented via two Acts: the first Act will set up the operational components and the subsequent Act will establish the administrative and compliance arrangements.[154]

The broader reform framework of this new planning law has no explicit impact upon unconventional gas; however, increased focus has been given to sustainable development initiatives, increased review of local plans (every four years), and the implementation of new planning policies directed at addressing issues of state significance – especially the way in which agricultural and rural resources are managed. The indirect impact of this new framework upon the administration of all petroleum titles in New South Wales, particularly those associated with CSG, is significant. The new gateway process is also incorporated within the ambit of this new framework and this process has direct and explicit application to CSG project appraisals.

The most conspicuous change introduced has been the implementation of strategic regional land use policies ('SRLUPs').[155] The primary aim of the SRLUP is to map strategic agricultural land across NSW in order to provide enhanced protection against the impacts of state significant mining and CSG activity and, in particular, to determine whether agriculture or mining should be given priority within the mapped areas.[156]

To date, no petroleum titles have been issued in New South Wales for shale gas. The primary form of unconventional gas mining is therefore CSG. Mining for CSG in NSW has, to date, largely centred around the Hunter region, the Gloucester Basin, the Gunnedah Basin, the Southern Coalfield (near Camden) and the Clarence Moreton Basin in north-eastern NSW. More limited exploration activity has also occurred in Illawarra, Central Coast, and Sydney.[157]

At the time of writing, the only SRLUPs that had been issued were for the Upper Hunter Valley and New England North West although mapping has commenced for the Central West and the Southern Highlands.[158]

The approach adopted by the SRLUPs in the Upper Hunter and New England North West to the management of CSG expansion has been to identify land known

---

[154] See the Planning Bill 2013 (NSW) and the Planning Administration Bill 2013 (NSW).
[155] See K Owens, 'Strategic Regional Land Use Plans: Presenting the Future for Coal Seam Gas Projects in NSW' (2012) 29 *Environmental and Planning Law Journal* 113, 115.
[156] See T Poisel, 'Coal Seam Gas Exploration and Production in New South Wales: The Case for Better Strategic Planning and More Stringent Regulation' (2012) 29 *Environmental Planning Law Journal* 129, 149.
[157] See Parliament of New South Wales, above n 127, 32.
[158] The Upper Hunter and New England North West areas were priorities because these are areas where CSG activity has, to date, been concentrated in NSW.

as 'biophysical, strategic, agricultural land' ('BSAL') and to set out that where a CSG mining application applies to a BSAL area, the application must proceed through what is known as the 'gateway process'.[159]

A development application for a CSG project, or a state significant mining project, whether stand alone (Greenfield) or an extension (Brownfield) project, which applies to BSAL may not be lodged unless a gateway certificate has been issued or the land is verified as not containing BSAL. A site will contain BSAL where it has been mapped under the relevant SRLUP or where an individual application is found to meet the criteria. If an application does not contain BSAL it may proceed straight to the development application stage.

The gateway process has three key elements to it: first, the gateway panel which is established must determine whether the project would significantly affect BSAL by reducing the agricultural productivity of the land due to impacts on the land through surface area disturbance and subsidence, impacts on the soil profile, fertility and rooting depths, soil salinity, changes to soil pH, and impacts on highly productive groundwater. Second, the gateway panel must consider whether the proposal would lead to significant impacts on what has been described as the 'critical industry clusters'.[160] The criteria included for this analysis is: surface area disturbance, subsidence, reduced access to agricultural resources,

---

159 See, eg, NSW Government, Department of Planning and Infrastructure, *Strategic Regional Land Use Plan Upper Hunter* (September 2012) 21–2 <http://www.nsw.gov.au/sites/default/files/initiatives/upperhunterslup_sd_v01.pdf>. The definition of 'strategic agricultural land' covers 'biophysical, strategic, agricultural land' and 'critical industry clusters'. These definitions include land with high rainfall, land used for agricultural product that provides significant employment, land used for equine activities and land used for viticulture.

160 See State Environmental Planning Policy (Mining, Petroleum Production and Extractive Industries) Amendment (Coal Seam Gas) 2014, cl 4 which amends cl 9A(5) to include critical industry clusters within the scope of CSG exclusion zones. Critical industry clusters are geographically identified within the Upper Hunter Strategic Regional Land Use Plan. A critical industry cluster (CIC) is a localised concentration of interrelated productive industries based on an agricultural product that provides significant employment opportunities and contributes to the identity of the region. The CIC also needs to have potential to be substantially impacted by coal seam gas or mining proposals. CICs must meet the following criteria:
- There is a concentration of enterprises that provides clear development and marketing advantages and is based on an agricultural product
- The productive industries are interrelated
- It consists of a unique combination of factors such as location, infrastructure, heritage and natural resources
- It is of national and/or international importance
- It is an iconic industry that contributes to the region's identity, and
- It is potentially substantially impacted by coal seam gas or mining proposals.

Two CICs have been identified in the Upper Hunter Valley: an equine CIC and a Viticulture CIC.

reduced access to support services and transport routes, and loss of scenic and landscape values.[161]

Prohibitions on the expansion of CSG development in NSW have been largely introduced via amendments to the State Environmental Planning Policy (Mining, Petroleum Production and Extractive Industries) 2007 (Mining SEPP). CSG exclusions zones now include: land within a residential zone; designated 'village areas', which are not zoned residential but are considered to have rural village characteristics (there are currently seven of these areas); and future residential growth areas, including a 2 km buffer around those areas. These areas are physically identified by reference to planning instruments (including local environment plans) and SRLUPs. Pipelines associated with CSG development have also been banned within the exclusions zones although they are permissible within the 2 km buffer zone.[162] Once the strategic release plan is fully implemented and accords with the NSW Gas Policy, CSG development will only be permitted in approved and pre-assessed areas and in accordance with strategic imperatives. This effectively means that specifically designated exclusion zones will become redundant.

Finally, the gateway panel must evaluate any advice it receives from the Commonwealth Independent Expert Scientific Commission on CSG and Large Coal Mining Development (IESC) and take into account any issues relevant to the new Aquifer Interference Policy.[163]

The IESC is in the process of preparing bioregional assessments for all areas where CSG development is planned. These assessments will evaluate the ecology, hydrology, and geology of the areas to determine the potential risks of CSG mining with respect to water resources. At the time of writing, the IESC had commissioned an assessment into the Gloucester Basin. Further assessments of other important water resources have also been planned. Once the Gas Policy has been implemented, an online portal is to be developed that will draw together data from environmental regulators in NSW, including expanded assessments by the NSW Office of Water, monitoring surface water quality and groundwater impacts. The idea is to ensure that information is available in a single online portal for any interested parties.

Where a CSG proposal passes the gateway analysis and is approved by the panel, an unconditional certificate will be issued; once issued, the CSG proposal will then proceed to a full merit evaluation.[164] Where, however, the CSG proposal does not pass the gateway analysis, it is not automatically rejected. Rather, a conditional

---

161 Ibid.
162 Ibid.
163 The Aquifer Interference Policy is outlined in detail further in this chapter.
164 See NSW Government, above n 160, 80.

certificate will be issued requiring the applicant to address the conditions of the gateway certificate in a subsequent development application.

The aim of the new gateway process is to filter applications for CSG and state significant mining projects that have an impact upon vital agricultural land in NSW. The absence of any legal entitlement by landholders to veto the issuance of CSG titles has effectively meant vigorous environmental review processes operate as de facto processes for landholder objection. In this way, the new gateway review process performs both a public and a private role in ensuring not only the proper evaluation and assessment of multi-variate environmental risks but also provides an avenue of redress for CSG projects that are proposed over private land interests.[165]

The difficulty with the new gateway process is that it relies heavily upon research modelling, particularly with respect to groundwater impacts in BSAL areas, and this modelling has not yet been finalised. As a consequence, any environmental review process must necessarily be predictive in nature.[166] Arguably, the incorporation of such speculative impact assessment merely produces an operationally redundant environmental assessment framework.[167] Despite these endemic issues, the gateway framework represents a significant improvement because of the increased focus it gives to CSG appraisals. The impact upon communities of CSG expansion has been both cumulative and synergistic. CSG projects affect the ownership spectrum of landholders and, despite fiscal compensation, significantly diminish the scope of permissible activities. Setting up an independent expert framework for the review of environmental impacts and land management improves what can be a disengaging and uncertain process for affected landholders.

---

165 See the discussion by D Lloyd, H Luke and W E Boyd, 'Community perspectives of natural resource extraction: coal seam gas mining and social identity in Eastern Australia' (2013) 10 *Collabah* 1 which evaluates the way in which the community is involved in the expansion of coal seam gas development. See also S Petrova and D Marinova, 'Social Impacts of Mining: Changes within the Local Social Landscape' (2013) 22.2 *Rural Society* 153 which examines the increasing tendency to conflate ownership objection into environmental objection based upon social-ecological transformations.

166 See Geoscience Australia, *Summary of Advice in Relation to the Potential Impacts of CSG Extraction in the Surat and Bowen Basins, Queensland, Phase 1, Final Report* (2010) 54 where the report notes 'the current groundwater modelling is inadequate in terms of scale and detail to identify the impacts of multiple CSG developments on groundwater interactions'. See also D Barrett, N Kunz, C Moran C and S Vink, *Scoping Study: Groundwater Impacts of Coal Seam Gas Development – Assessment and Monitoring* (Centre for Water in the Minerals Industry, 2008) 9.

167 See Owens, above n 156, 118 where the author notes that SRLUPs will not be 'able to take us any further in strategically assessing the likely cumulative effects of CSG developments across regions' and will be unlikely to add anything over and above what should already be required by way of environmental assessment.

5.5.2.6 *Environmental Planning and Assessment Act 1979* (NSW); *Protection of the Environment Operations Act 1997* (NSW)

Until SRLUPs have been fully implemented and the gateway process is completely operational under the new Gas Policy framework, the planning and development of land in New South Wales continues to be regulated by three primary sources: the *Environmental Planning and Assessment Act 1979* (NSW) (*EPAA*), the *Environmental Planning and Assessment Regulations 2000* (NSW) and the relevant environmental planning instrument, whether that be a state environmental planning policy (SEPP) or a local environmental policy (LEP). Environmental planning instruments provide for the protection of the environment, including the preservation of trees, native flora, fauna and animals, and the preservation of vulnerable ecological communities.[168] Environmental planning instruments also prescribe the circumstances where 'development consent' for a particular CSG project is required.

The *EPAA* has been amended to include two new categories of development: state significant developments (SSD) and state significant infrastructure (SSI).[169] Neither the SSD nor the SSI are specifically defined and will therefore exist where so declared by a state environmental planning policy.[170] SSD applications are assessed by the NSW Department of Planning and Infrastructure. Development and consent for these projects may only be granted by the Minister pursuant to the provisions of pt 4, div 4.1 and pt 5.1, div 1 of the *EPAA*.[171]

In the context of CSG development, the categories of SSD and SSI specifically include: development for the purpose of petroleum production, drilling or operating petroleum exploration wells in an environmentally sensitive area of state significance, and development for the purpose of petroleum related works that is ancillary to another SSD or with a capital investment value of more than $30 million.[172]

Where a CSG proposal is characterised as being of low scale (five or fewer wells) or where it involves strati-graphic boreholes or monitoring wells, it may not

---

168 See *EPAA* (NSW) s 26.
169 The categories of SSD and SSI are set out in the *State Environmental Planning Policy (State and Regional Development) 2011* (NSW). State significant development is assessed and determined under div 4.1 of pt 4 of the *EPAA* (NSW) and state significant infrastructure is assessed and determined under pt 5.1 of the *EPAA* (NSW). See T Poisel, 'Coal Seam Gas Exploration and Production in New South Wales: The Case for Better Strategic Planning and More Strategic Regulation' (2012) 29 *Environmental and Planning Law Journal* 129, 138. This discussion will focus upon state significant development under pt 4, div 4.1.
170 See the definition set out in the *State Environmental Planning Policy (State and Regional Development) 2011*, schs 1, 2, 3 for a detailed outline.
171 See *EPAA* (NSW) ss 89C, 115U.
172 See State Environmental Planning Policy (State and Regional Development) 2011 cl 5 of sch 1 and sch 3.

constitute a SSD.[173] In practice, most large-scale CSG projects will come within the application of div 4.1 of the *EPAA* and will therefore need to obtain development consent.[174] Significantly, however, an SSD will not generally require approvals from other major governmental agencies, as the process is consolidated into a single evaluation process.[175]

Once a project is found to constitute an SSD, a full merit evaluation of the development application must be conducted. An environmental impact statement must accompany the application.[176] Where a project has no application to environmentally sensitive land or where it is exploratory and smaller in scale, it may not be regarded as an SSD. Where this occurs, the Minister, in consultation with the Planning Assessment Commission, may declare the project an SSD if it is clear that it has the potential to deliver major public benefits and it is a significant and complex project.[177]

The *EPAA* framework anticipates, however, that smaller-scale projects which do not affect areas defined as environmentally sensitive and which have less of a public and environmental impact will not be subject to development consent. Where a CSG proposal does not constitute an SSD or an SSI and therefore does not trigger div 4.1 or pt 5.1, pt 4 of the *EPAA* may have an application.[178]

A CSG project will require development consent under pt 4 where it comes within the application of the *State Environmental Planning Policy (Mining, Petroleum and Extractive Industries)*.[179] All petroleum production development on:

- land where agriculture or industry may be carried out
- land which is already the subject of a production lease under the *POA*
- land in any part of a waterway or a coastal estuary

which involves drilling or exploration in an environmentally sensitive area of state significance other than strati-graphic boreholes or monitoring wells, must obtain development consent under pt 4 where div 4.1 or pt 5.1 does not apply.

Where the project constitutes a 'designated development' and satisfies the petroleum works threshold, the development application must also be accompanied

---

173 See State Environmental Planning Policy (State and Regional Development) 2011 cl 6 of sch 1.
174 See *EPAA* (NSW) s 89D.
175 Hence, the Office of Environmental Heritage, the Environmental Protection Authority or the Office of Water do not need to provide approval to an SSD that is independent to the development consent evaluation.
176 See *Environmental Planning and Assessment Regulation 2000* (NSW) sch 1, pt 1, reg 2(1)(e).
177 See *EPAA* (NSW) s 89C(3).
178 This is set out in the *EPAA* (NSW) s 76A. The relevant State Environmental Planning Instrument for pt 4 is the *State Environmental Planning Policy (Mining, Petroleum Production and Extractive Industries) 2007* ('Mining SEPP').
179 See the Mining SEPP s 7(2).

by an EIS.[180] The *EPAA* will require all decisions that 'significantly affect the environment' to be subject to an EIS.[181]

An EIS must include a general description of the environment affected by the development, the likely impact on the environment, the measures proposed to mitigate any adverse effects of the development, and a justification for the project having regard to biophysical, economic and social considerations.[182] Given the detailed nature of the EIS, it is usually prepared by an expert, on behalf of a CSG proponent.

The *EPAA* also specifically imposes a duty upon all determining authorities to consider, to the fullest possible extent, all matters likely to affect the environment when determining whether to approve an activity.[183]

Exactly what type of project will attract an EIS is debatable. In *Fullerton Cove Residents Action Group Inc v Dart Energy Ltd (No 2)* the New South Wales Land and Environment Court concluded that an EIS was not required for a pilot CSG program because there was no evidence that the potential cumulative impact of the project would have a significant impact on the environment.[184] The Court reached this conclusion despite the fact that the department had ranked the project, in accordance with specifically prepared EIS guidelines for mining and exploratory projects, as a 'medium risk'.[185]

The Court further concluded that a failure to obtain a baseline groundwater assessment for the pilot CSG program did not infringe the mandated duty in the *EPAA* to consider the environmental impact of activities to the fullest possible extent.[186]

---

180 Section 77A(2) of the *EPAA* (NSW) sets out that a designated development does not include a state significant development despite any declaration to the contrary. The petroleum works threshold is defined in the *Environmental Planning and Assessment Regulation 2000* sch 3, reg 27 to include: works that produce more than 5 petajoules of natural gas per year; that store natural gas products with an intended storage capacity in excess of 200 tonnes for liquefied gases; that are located within 40 metres of a natural water body, or wetland, an area of high water table or highly permeable soils, a drinking water catchment, or a floodplain. The requirement for an EIS is set out in the *EPAA* (NSW) s 78A(8)(a) and the *Environmental Planning and Assessment Regulation 2000* (NSW) cl 50 and sch 1, pt 1, reg 2 (1)(c).
181 See *EPAA* (NSW) s 112.
182 See *Environmental Planning and Assessment Regulations 2000* sch 2, reg 7.
183 See *EPAA* (NSW) s 111.
184 There are two aspects to this duty: a duty to examine the environmental impact and a duty to take account of existing knowledge, information, and assessments: *Fullerton Cove Residents Action Group Inc v Dart Energy Ltd (No 2)* [2013] NSWLEC 38, [142] (Pepper J).
185 [2013] NSWLEC 38, [308].
186 The Environmental Impact Assessment Guidelines for NSW ('ESG2 guidelines') come under the Mineral Resources Branch of the NSW Department of Trade and Investment which is responsible for the administration of authorisations under the *Mining Act 1992* (NSW) and petroleum titles under the *Petroleum (Onshore) Act 1991* (NSW). The ESG2 guidelines are available at: <http://www.resourcesandenergy.nsw.gov.au/__data/assets/pdf_file/0007/427444/2012-03-05-ESG2-Environmental-Impact-Assessment-Guidelines-FINAL.pdf>.

Whilst compliance with this duty was found to be 'pivotal' to the objectives of the *EPAA*, it argued that there was no 'standard of absolute perfection' which meant that it would be sufficient for the purposes of this duty if the determining authority considered the impact of the project upon groundwater without necessarily having to commission and evaluate a specific groundwater assessment.[187]

Where a CSG project does not require development consent under either pt 4 or div. 4.1, it may be subject to a less rigorous environmental assessment process. A CSG project outside the application of pt 4 or div. 5.1 only needs to provide the determining authority with a Review of Environmental Factors (REF). Factors relevant to a REF include: impact on an ecosystem; transformation of the aesthetic, recreational or environmental value of an area; long-term environmental degradation; pollution; and environmental problems associated with the disposal of waste.[188]

The *EPAA* makes it clear that neither an SSD nor an SSI will be exempted from the need to obtain an aquifer interference approval under the *Water Management Act 2000* (NSW).[189] Section 89J(1)(g) of the *EPAA* is significant because it confirms the need for a CSG proponent to undergo separate aquifer interference approval where the activities of the project involve interference, penetration or obstruction of water from an aquifer.[190]

The *Protection of the Environment Operations Act 1997* (NSW) ('the *PEOA*') now includes CSG operations as 'scheduled activities'.[191] This means that all CSG activities in NSW must now hold an environmental protection licence (EPL) issued by the Environmental Protection Authority (EPA) in addition to the relevant approved petroleum title. As such, the EPA will become the lead regulator for compliance and enforcement of conditions of approval for gas activities, including consent conditions and activity approvals.

This is an important development because the EPA is an independent, credible, and publicly accessible body, with a high level of transparency and public

---

187 [2013] NSWLEC 38, [147]–[148] where the Court concluded that the duty set out in s 111 of the *EPAA* (NSW) was not infringed by the absence of a separate groundwater assessment given that Hunter Water had provided a 'letter' indicating that the risk to groundwater pressure and contamination was not extensive.
188 See *EPAA* (NSW) s 111. The factors relevant to an REF are set out in the *Environmental Planning and Assessment Regulations 2000* (NSW) reg 228. The Department of Industry and Investment has also issued a guideline on the preparation of a REF which specifically requires the potential impacts on surface and groundwater to be evaluated.
189 See *EPAA* (NSW) s 89J(1)(g) which sets out that activities under s 91 of the *Water Management Act 2000* (NSW) apart from the aquifer interference approval will be exempted where a development consent for a SSD exists; for SSI a similar provision exists in s 115G(1).
190 The provisions of the *Water Management Act 2000* (NSW) are discussed in more detail in the next section. Section 89J(1)(g) of the *EPAA* (NSW) appears to be contrary to the provisions of the AIP.
191 See *Protection of the Environment Operations Act 1997* (NSW) sch 1.

accountability. Its environmental assessment process is broadly similar to the environmental impact assessment connected with the *EPAA*, although there are a number of key differences. First, a breach of an EPL is more likely to result in suspension or revocation of the licence by the EPA.[192] Second, the *PEOA* incorporates provisions for mandatory environmental audits. An environmental audit is essentially a survey of the environmental impact of the ongoing activities of a corporation and extends beyond the date of initial assessment.

The *PEOA* provides for mandatory environmental audits where the EPA suspects that the holder of the EPL has contravened the *PEOA*, its regulations or the conditions of the EPL, and that the contravention has caused, is causing or will cause harm to the environment.[193] The information acquired under an environmental audit is one of the primary means of overseeing the environmental management conditions attached to a petroleum title. The improved transparency, accountability, and independence of the *PEOA* framework increases the prospect of the EPA discovering environmental damage or risk from CSG projects in breach of environmental conditions.

5.5.2.7 The *Water Management Act 2000* (NSW) and the Aquifer Interference Policy

The reforms implemented to the environmental review process in New South Wales have been reinforced by strategic reforms to the water access licensing laws. The *Water Management Act 2000* (NSW) ('the *WMA*'), in combination with the water sharing plans, govern water allocation and usage in NSW. Water entitlements under the *WMA* may be acquired through the issuance of water access licences, which apply to all water entitlements other than basic landholder entitlements which exempt stock and domestic rights, harvestable rights, native title rights, and other exempted activities such as bushfire fighting.[194]

The Aquifer Interference Policy (AIP), which was introduced as a component of the Strategic Regional Land Use Policy, further reinforces the provisions of the *WMA*.[195] The stated purpose of the AIP is to 'explain the role and requirements of the *WMA* in the water licensing and approval processes for aquifer interference

---

192 The provisions relating to suspension and revocation of an EPL are set out in *Protection of the Environment Operations Act 1997* (NSW) s 79. There are many more instances of the EPA revoking an EPL than of the Minister cancelling a petroleum title.
193 See N Gunningham and J Prest, 'Environmental Audit as a Regulatory Strategy: Prospects and Reform' (1993) 15 *Sydney Law Review* 492, 494 where the authors refer to the audit as a 'survey of the activities by a corporation or other entity to assess the extent of that corporations current and potential impact on the environment'.
194 See ss 52–55.
195 See NSW Government, Department of Primary Industries, Office of Water, *Aquifer Interference Policy* (September 2012) http://www.water.nsw.gov.au/Water-management/Law-and-policy/Key-policies/Aquifer-interference/Aquifer-interference.

activities'.[196] The AIP seeks to outline the specific requirements that CSG proponents must satisfy when removing water for the purpose of CSG extraction in order to determine their potential impacts on water resources. It also outlines the information and modelling that proponents need to provide to enable the impacts to be assessed.

Water access licences (WALs) entitle licence-holders to specified shares in the available water within a particular water management area and allow for water to be taken at specified times, rates, or circumstances from those locations. WALs may be granted to provide access to the available water governed by a water-sharing plan pursuant to the *WMA*.

All CSG mining involves the removal of significant amounts of groundwater. Consequently, a CSG proponent must obtain a water access licence in order to commence operations. This requirement is anticipated in both the *EPAA* and the *WMA*.[197]

The *WMA* allows an application to be made for a WAL where a person has acquired the right to apply for the licence pursuant to the issuance of a controlled allocation order. The Minister or delegate can make a controlled allocation order that provides for a small volume of additional aquifer access licences to be issued over specific groundwater sources with unassigned water.[198] The order may not include water sources where current water requirements have either reached or are close to reaching the long-term average annual extraction limit.

All CSG proponents must hold a water access licence authorising the taking of water during the course of a mining activity.[199] In addition to holding an access licence, a CSG proponent must apply for an aquifer interference approval because CSG mining also comes within the definition of an 'aquifer interference activity' under the *WMA*.[200] Aquifer interference is defined broadly in the *WMA* to include the taking of water from an aquifer during the course of CSG mining.[201] The definition also includes the extraction of sand and other base materials.[202]

Applications for access licences and aquifer interference approvals over biophysical strategic agricultural land are evaluated during the course of the gateway

---

196 See NSW Government, above n 196, 1.1.
197 See *EPAA* (NSW) s 89J(1)(g). An SSD is exempted from other authorisations under different legislation including: *Coastal Protection Act 1979* (NSW); *Fisheries Management Act 1994* (NSW); *Heritage Act 1977* (NSW); *National Parks and Wildlife Act 1974* (NSW) and the *Rural Fires Act 1997* (NSW).
198 See *WMA* (NSW) s 65.
199 See *WMA* (NSW) s 60I.
200 See *WMA* (NSW) s 91F(2).
201 See *WMA* (NSW) s 4.
202 See *Water Management (General) Regulation 2011* reg 22.

process.[203] There is no need for a CSG proponent who has undergone gateway assessment to apply separately under the *WMA*, unless the removal of water induces flow from another water source located beyond the strategic agricultural land zone.[204]

An aquifer interference approval will only be granted if the Minister is satisfied that adequate arrangements are in force to ensure that no more than minimal harm will be done to the aquifer or its dependent ecosystems. The AIP sets out a range of minimal impact considerations; these considerations may also be used for any determination which is made by the gateway panel.

Two levels of minimal impact considerations are set out in the AIP. The AIP draws a fundamental distinction between high-productive and low-productive groundwater sources and threshold minimal impact considerations are developed for each source.[205] The factors relevant to impact include: acidity concerns, waterlogging or water table rises affecting land use or dependent ecosystems, hydraulic connection between aquifers, whether hydraulic fracturing will modify the hydraulic connection between aquifers, and method of disposing of associated waters.[206]

If the impacts are less than the prescribed Level 1 minimal impact considerations, they are acceptable. If the impacts exceed the prescribed Level 1 minimal impacts by no more than the accuracy of an otherwise robust model, the project will be 'within the range of acceptability'.[207] In such a situation, extra monitoring and mitigation or remediation may be imposed via adaptive management conditions.

If the predicted impacts are greater than the Level 1 minimal impact considerations by more than the accuracy of an otherwise robust model, additional assessment may be required. If it reveals that the predicted impacts do not prevent the long-term viability of the relevant water-dependent asset, the impacts will be acceptable.[208]

The minimum harm strategy underpinning the aquifer interference approvals in the *WMA* and the AIP is based upon the premise that CSG proponents need only prove that adequate arrangements are in place to ensure that the minimum harm that does occur to an aquifer does not exceed a minimum standard. The overarching strategy of the AIP is not to preclude approving projects, even where they do exceed the minimum impact thresholds.

---

203 See NSW Government, above n 196, 3.1. The Mining SEPP s 7(2) requires the Minister to provide advice to on aquifer impacts at either (a) the gateway stage or (b) during the development application. Advice must be based upon the requirements set out in 3.2 of the AIP.
204 See NSW Government, above n 196, pt 2.1.
205 See NSW Government, above n 196, cl 3.2. Highly productive groundwater sources contain water supply works that can yield water at a rate greater than 5 litres per second.
206 See NSW Government, above n 196, cl 3.2.2.
207 See NSW Government, above n 196, cl 3.2.1.
208 Ibid.

The AIP mandates the adoption of a risk management approach to assessing the potential impacts of aquifer interference activities and the level of detail that the proponent must provide will be proportional to a combination of the likelihood of impacts occurring on water sources, users, and dependent ecosystems and the potential consequences of these impacts.[209]

The minimum harm strategy works from a basic assumption of harm. The concern is not whether harm should be precluded but, rather, how the environmental risks associated with that harm might be best managed. Arguably, this approach is inconsistent with the fundamental international principle of ecological sustainable development, the precautionary principle, which imposes a positive obligation to avoid potentially irreversible environmental harm, despite the absence of scientific certainty.[210]

5.6 Regulatory framework: *Environment Protection and Biodiversity Conservation Act 1999* (Cth)

The *Environment Protection and Biodiversity Conservation Act* 1999 (Cth) ('the *EPBCA*') has an application to matters of national environmental significance.[211] The *EPBCA* aims to achieve ecologically sustainable development by applying an environmental assessment regime to matters of 'national environmental significance'.[212]

Where a CSG project falls within the scope of the *EPBCA*, the assessment regime functions as an additional regime to state-based evaluation processes.[213]

---

209 See NSW Government, above n 196, cl 3.2.2.
210 The precautionary principle was set out in Principle 15 of the United Nations' *Rio Declaration on Environment and Development 1992*. Most legislation in Australia endorses this definition and mandates its application via: (i) careful evaluation to avoid, wherever practicable, serious or irreversible damage to the environment; and (ii) an assessment of the risk-weighted consequences of various options.' See the Australian Government, Department of the Environment, *Intergovernmental Agreement on the Environment*, May 1992, para 3.5. <http://www.environment.gov.au/about-us/esd/publications/intergovernmental-agreement>.
211 The broader scope and aims of the *EPBCA* (Cth) is discussed in Chapter 9.
212 See *EPBCA* (Cth) s 3(1)(b).
213 See Australian Government, Department of the Environment, C McGrath, *Review of the EPBC Act* (2006) http://www.environment.gov.au/node/22544.

The *EPBCA* will, however, only apply to reinforce existing state regulatory requirements in circumstances where it can be established that the application has or will have a significant impact upon a matter of national environmental significance.[214] Matters of national environmental significance are specifically outlined within the *EPBCA* and include a CSG project that has a significant impact upon a water resource.[215]

Section 24D of the *EPBCA* gives CSG projects a national environmental focus. It requires all CSG projects (or large coal mining developments) that have a significant impact on a water resource to be referred to the federal Minister.[216] The federal Minister must make a determination as to whether the project in issue constitutes a 'controlled action'. If the Minister determines that the project is likely to have a significant impact on a water resource, it will be characterised as a controlled action and assessed and approved in accordance with the environmental assessment provisions of the *EPBCA*.[217]

'Coal seam gas development' is defined broadly to cover any activity involving CSG extraction that has, or is likely to have, a significant impact on water resources, including any direct or cumulative impacts connected with salinity and salt production.[218] A water resource is defined broadly to include surface or ground water, a watercourse, lake, wetland or aquifer and will include water, organisms, and other components and ecosystems that contribute to the water resource.[219]

The *EPBCA* does not expressly define what amounts to a 'significant impact' on a water resource. Hence, impacts have been held to include changes in the quantity, quality or availability of surface or ground water, alteration in ground pressure, alteration in drainage patterns, or any substantial reduction in the availability of water for the environment and the impact can be direct or indirect.[220]

Assessment under the *EPBCA* can range from preliminary documentation to a public environment report or a full-scale public inquiry. Most assessments require production of an environmental impact assessment and a period for public comment.[221]

Most CSG projects in Australia fall within the scope of the water resources trigger. As such, all CSG projects should first be referred to the federal Minister for

---

214 See *EPBC* (Cth) pt 3, div 1.
215 Ibid s 24D.
216 Ibid s 24D(1).
217 Ibid pt 7, div 1.
218 Ibid s 528.
219 Ibid.
220 See Australian Government, Department of the Environment, *Significant Impact Guidelines 1.3: Coal Seam Gas and Large Coal Mining Developments – impacts on water resources* (2013) http://www.environment.gov.au/resource/significant-impact-guidelines-13-coal-seam-gas-and-large-coal-mining-developments-impacts.
221 See *EPBC* (Cth) pt 8, div 3.

approval and, subject to a determination that the project amounts to a controlled action, environmental assessment.

The possibility of overlap and inconsistency between state and Commonwealth legislation is mitigated by the existence of bilateral agreements that the Commonwealth and the state may enter into.[222] Bilateral agreements allow the Commonwealth to 'accredit' particular state or territory assessment processes and approval decisions and delegate to the states and territories the responsibility for conducting environmental assessments under the *EPBCA*.

The Commonwealth government has announced a reform of the *EPBCA* and a taskforce has been established.[223] Proposed environmental reforms include the introduction of a new National Centre for Cooperation on Environment and Development that would bring together scientists, industry, non-government organisations, and governments to create national standards for accrediting environmental impact assessments, developing a national biodiversity policy to protect ecosystems, and the introduction of a national productivity compact on regulatory and competition reform involving the Commonwealth, states and territories aimed at expediting approval processes.[224] For a further and more detailed discussion of the operative dimensions of the Commonwealth *EPBC Act* see Chapter 9.

5.7 REVIEW QUESTIONS

1. What is the difference between shale gas mining and coal seam gas mining?
2. What are some of the social and environment concerns associated with hydraulic fracturing for shale gas mining?
3. Describe the primary environmental impact of CSG mining upon water resources and explain how some of the risks might be mitigated?
4. Explain the circumstances in which the gateway framework in NSW comes into operation?
5. Is the issuance of an exploration licence for CSG mining in Queensland subject to the precondition that community consultation be sought?
6. Santos Mining Pty Ltd decide to seek an exploratory licence for CSG mining in rural land not far from a university campus. The nearby landowners are farmers and they

---

222 Ibid ch 3, pt 5.
223 See A Hawke, *The Australian Environment Act: Report of the Independent Review of the Environment Protection and Biodiversity Conservation Act 1999* (October 2009) http://www.environment.gov.au/resource/australian-environment-act-report-independent-review-environment-protection-and.
224 See the discussion on the proposed reform by the National Compact on Regulatory and Competition Reform at: https://www.coag.gov.au/sites/default/files/Annex-A-and-B-and-C-to-the-Compact.pdf.

are concerned. They found out about the application by reading the local newspaper but they were angry not to have been personally informed by representatives from Santos Mining Pty Ltd. Further, Santos has not sought to consult with the landowners at all and has entered farmland and commenced drilling within 500 metres of a domestic residence. A small group of local farmers come to you for advice. Advise the farmers on the basis that the farms are located: (i) in New South Wales; and (ii) in Queensland.

7. What additional environmental and safety management obligations does the Queensland legislation require for the issuance of permits to progress a CSG project and are they sufficient?

8. Can the breach of an environmental condition on an exploration licence issued in New South Wales result in the termination of that licence?

9. Consider the following problem:

> In Queensland, BG Group seek an exploration licence over privately owned farmland in order to develop a CSG project. They ask the farmer to enter into an access and compensation agreement and request that he brings a lawyer with him to a meeting so that they are able to 'negotiate' the terms. The farmer does not want to incur the expense of the lawyer and feels that this should be paid for by BG Group. Is this possible?

10. What role does the New South Wales Environmental Protection Agency play in the approval of licences to progress CSG projects and how does that role intersect, if at all, with the requirements of the Commonwealth to oversee matters of national environmental significance in the *Environment Protection and Biodiversity Conservation Act 1999* (Cth)?

5.8 FURTHER READING

Australian Government, Department of the Environment, C McGrath, *Review of the EPBC Act* (2006) http://www.environment.gov.au/node/22544.

Australian Government, Department of the Environment, *Significant Impact Guidelines 1.3: Coal Seam Gas and Large Coal Mining Developments – impacts on water resources* (2013) http://www.environment.gov.au/resource/significant-impact-guidelines-13-coal-seam-gas-and-large-coal-mining-developments-impacts.

Australian Government, National Water Commission, *Position Statement: Coal Seam Gas and Water* (December 2010) http://nwc.gov.au/__data/assets/pdf_file/0003/9723/Coal_Seam_Gas.pdf.

Australian Petroleum, Production and Exploration Association, *Economic Significance of Coal Seam Gas In Queensland: Final Report* (June 2012) http://www.appea.com.au/wp-content/uploads/2013/05/120606_ACIL-qld-csg-final-report.pdf.

Baker & McKenzie, Shale Gas – Global Environmental Law and Regulation (June 2013) http://www.bakermckenzie.com/NLEnvironmentSGELRJun13/.

D Barrett, N Kunz, C Moran and S Vink, *Scoping Study: Groundwater Impacts of Coal Seam Gas Development – Assessment and Monitoring* (Centre for Water in the Minerals Industry, 2008).

R Beckwith, 'Proppants: Where in the World' (2011) 63 *Journal of Petroleum Technology Online* 36.

J K Boyce, 'From Natural Resources to Natural Assets' in J K Boyce and B G Shelley (eds) *Natural Assets: Democratizing Environmental Ownership* (Island Press, Washington DC, 2003).

W Brady, 'Hydraulic Fracturing Regulation in the United States: The Laissez-Faire Approach of the Federal Government and Varying State Regulations' (2012) *University of Denver Publication* 1.

E Burleson, 'Cooperative Federalism and Hydraulic Fracturing: A Human Right to a Clean Environment' (2012) 22 *Cornell Journal of Law and Public Policy* 289.

S Christensen, P O'Connor, W D Duncan and A Phillips, 'Regulation of Land Access for Resource Development: A Coal Seam Gas Case Study from Queensland' (2012) 21(2) *Australian Property Law Journal* 110.

C Clifford, 'Conservationists say Santos fine for Pilliga spill too small', ABC (online), 12 January 2014 http://www.abc.net.au/news/2014–01–11/conservationists-say-santos-fine-for-pilliga-spill-too-small/5195618.

Council of Australian Governments, *National Partnership Agreement on Coal Seam Gas and Large Coal Mining Development: Intergovernmental Agreement on Federal Financial Relations*. https://www.ehp.qld.gov.au/management/impact-assessment/pdf/partnership-agreement.pdf.

Council of Canadian Academies, 'Environmental Impacts of Shale Gas in Canada: The Expert Panel on Harnessing Science and Technology to Understand the Environmental Impacts of Shale Gas Extraction' (May 2014) http://www.scienceadvice.ca/uploads/eng/assessments%20and%20publications%20and%20news%20releases/shale%20gas/shalegas_fullreporten.pdf.

E A Craig and M S Myers, 'Ownership of Methane Gas in Coalbeds' (1987) 24 *Rocky Mountain. Min. Law Inst.* 767.

CSIRO, *CSG Produced Water and Site Management* (April 2012) http://www.csiro.au/Portals/Publications.aspx.

J Dammel, 'Notes from Underground: Hydraulic Fracturing in the Marcellus Shale' (2011) 12 *Minnesota Journal of Law, Science and Technology* 773.

S Day, L Connell, D Etheridge, T Norgate and N Sherwood, 'Fugitive Greenhouse Gas Emissions from Coal Seam Gas Production in Australia' (October 2012) *CSIRO Energy Technology*.

Department of Health and Human Resources, Cincinnati, *NIOS Field Effort to Assess Chemical Exposure Risks to Gas and Oil Workers* (1998) http://www.cdc.gov/niosh/docs/2010-130/pdfs/2010-130.pdf.

Department of Resources and Energy, New South Wales, Bret Walker, *Examination of the Land Access Arbitration Framework: Mining Act 1992 and Petroleum (Onshore) Act 1991* (20 June 2014) http://www.resourcesandenergy.nsw.gov.au/__data/assets/pdf_file/0018/527112/Brett-Walker-Examination-of-the-Land-Access-Arbitration-Framework.pdf.

Desheng Hu, Shengqing Xu, 'Opportunity, challenges and policy choices for China on the development of shale gas' (2013) 60 *Energy Policy* 21.

J Entin, 'The Law and Policy of Hydraulic Fracturing: Addressing the Issues of the Natural Gas Boom' (2013) 63 *Case Western Reserve Law Review* 965.

K J Flaherty, 'Quandary or Quest: Problems of Developing Coal Bed Methane as an Energy Resource' (2000) 15 *Journal of Natural Resources and Environmental Law* 71.

K Galloway, 'Landowners vs Miners Property Interests' (2012) 37(2) *Alternative Law Journal* 77.

Geoscience Australia, *Summary of Advice in Relation to The Potential Impacts of CSG Extraction in the Surat and Bowen Basins, Queensland, Phase 1, Final Report* (2010).

Goldman Sachs, Global Markets Institute, 'Unlocking the Economic Potential of North America's Energy Resources' (June 2014) http://www.goldmansachs.com/our-thinking/our-conferences/north-american-energy-summit/unlocking-the-economic-potential-of-north-americas.pdf.

I Gray, *Reservoir Engineering in Coal Seams: Part 1—The Physical Process of Gas Storage and Movement in Coal Seams* (Society of Petroleum Engineers, 1987).

N Gunningham and J Prest, 'Environmental Audit as a Regulatory Strategy: Prospects and Reform' (1993) 15 *Sydney Law Review* 492.

T Hunter, 'Australia's Unconventional Gas Resources' in CEDA, *Australia's Unconventional Energy Options* (September 2012) http://www.ceda.com.au/media/263565/cedaunconventionalenergyfinal.pdf.

Independent Commission Against Corruption, *Reducing the Opportunities and Incentives for Corruption in the State's Management of Coal Resources* (2013) http://www.icac.nsw.gov.au.

International Energy Agency, *World Energy Outlook 2011: Are We Entering a Golden Age of Gas?* (2011) http://www.iea.org/publications/worldenergyoutlook/goldenageofgas/.

R S Krannich and T Greider, 'Personal Well-Being in Rapid Growth and Stable Communities: Multiple Indicators and Contrasting Results' (1984) 49(4) *Rural Sociology* 541.

L Letts, 'Coal Seam Gas Production – Friend or Foe of Queensland's Water Resources?' (2012) 29 *Environmental Planning Law Journal* 101.

D Lloyd, H Luke and W E Boyd, 'Community perspectives of natural resource extraction: coal seam gas mining and social identity in Eastern Australia' (2013) 10 *Collabah* 1.

R Lyster, 'Coal Seam Gas in the Context of Global Energy and Climate Change Scenarios' (2013) (13/28 Legal Studies Research Paper, Sydney Law School).

T Marshall, 'Coal Seam Gas (CSG) Licences and Landholders' Rights' (2011) 26(9) *Australian Environment Review* 226.

D Mathew, 'The Nature of Gas in Coal: Technical Challenges of Co-Location of Coal and Coalbed Methane' (2005) *Australian Mineral and Petroleum Association Yearbook* 368.

A McHarg, B Barton, A Bradbrook and L Godden (eds), *Property Rights in Oil and Gas Under Dominial Regimes* (Oxford University Press, 2010).

L M McKenzie, R Z Witter, L S Newman, J L Adgate, 'Human Health Risk Assessments of Air from Unconventional Gas' (2012) *Science of the Total Environment* http://cogcc.state.co.us/library/setbackstakeholdergroup/Presentations/Health%20Risk%20Assessment%20of%20Air%20Emissions%20From%20Unconventional%20Natural%20Gas%20-%20HMcKenzie2012.pdf.

T Measham, F Haslam McKenzie, K Moffat and D M Franks, 'An Expanded Role for the Mining Sector in Australian Society?' (2013) 22(2) *Rural Society* 184.

L Mullins, 'The Equity Illusion of Surface Ownership in Coalbed Methane Gas; The Rise of Mutual Simultaneous Rights in Mineral Law and The Resulting Need for Dispute Resolution in Split Estate Relations' (2009) 16 *Missouri Environmental Law and Policy Review* 109.

National Water Commission, *Impact of Groundwater Pumping on Groundwater Quality* (16 May 2012) http://archive.nwc.gov.au/rnws/ngap/groundwater-projects/managing-risks-to-groundwater-quality/impact-of-groundwater-pumping-on-groundwater-quality.

NSW Chief Scientist and Engineer, *Initial Report on the Independent Review of Coal Seam Gas Activities in NSW* (July 2013).

NSW Chief Scientist and Engineer, M O'Kane, *Final Report of the Independent Review of CSG Activities in NSW* (September 2014) http://www.chiefscientist.nsw.gov.au/__data/assets/pdf_file/0005/56912/140930-CSG-Final-Report.pdf.

NSW Department of Trade and Investment, Resources and Energy, 'Public Comment Process: For the exploration of coal and petroleum including coal seam gas' (October 2011) http://www.resourcesandenergy.nsw.gov.au/__data/assets/pdf_file/0009/426582/Public-Comment-Process-Document.pdf.

NSW Government, Department of Primary Industries, Office of Water, *Aquifer Interference Policy* (September 2012) http://www.water.nsw.gov.au/Water-management/Law-and-policy/Key-policies/Aquifer-interference/Aquifer-interference.

T Nunan, 'Legal Issues Emerging from the Growth of the Coal Seam Gas Industry in Queensland' (2006) 25 *Australian Resources & Energy Law Journal* 189.

Y Omorogbe and P Oniemola, 'Property Rights in Oil and Gas Under Dominial Regimes' in A McHarg, B Barton, A Bradbrook and L Godden (eds) *Property and the Law in Energy and Natural Resources* (Oxford University Press, 2010).

K Owens, 'Strategic Regional Land Use Plans: Presenting the Future for Coal Seam Gas Projects in NSW' (2012) 29 *Environmental and Planning Law Journal* 113.

S Petrova and D Marinova, 'Social Impacts of Mining: Changes within the Local Social Landscape' (2013) 22(2) *Rural Society* 153.

R H Pifer, 'A Greener Shade of Blue: Technology and the Shale Revolution' (2013) 27 *Notre Dame Journal of Law Ethics and Public Policy* 131.

T Poisel, 'Coal Seam Gas Exploration and Production in New South Wales: The Case for Better Strategic Planning and More Stringent Regulation' (2012) 29 *Environmental Planning Law Journal* 129.

Queensland Government, *Guide to Queensland's New Land Access Laws* (2010) http://mines.industry.qld.gov.au/assets/land-tenurepdf/6184_landaccesslaws_guide_print.pdf.

Queensland Government, Department of Natural Resources and Mines, 'Queensland Government Response to the Report of the Land Access Review Panel' (December 2012) http://mines.industry.qld.gov.au/assets/native-title-pdf/qg-response-land-access-framework.pdf.

Queensland Government, Department of Natural Resources and Mines, *SafeOP for Petroleum and Gas: A Guide to the Legislative Requirements for Operating Plant, Part A Explanatory Guide* (2013) http://mines.industry.qld.gov.au/assets/petroleum-pdf/safeop.pdf.

M Ratman and M Tiemann, 'An Overview of Unconventional Oil and Natural Gas: Resources and Federal Actions' Congressional Research Services (21 November 2014) https://www.fas.org/sgp/crs/misc/R43148.pdf.

M Roarty, 'The Development of Australia's Coal Seam Gas Resources' (Parliament of Australia, Science, Technology, Environment and Resources Section, July 2011) http://parlinfo.aph.gov.au/parlInfo/download/library/prspub/957068/upload_binary/957068.pdf;fileType=application/pdf#search=%22background%20note%20(parliamentary%20library,%20australia)%22.

A Ross and P Martinez-Santos, 'The Challenge of Groundwater Governance: Case Studies from Spain and Australia' (2010) 10 *Reg Environ Change* 299.

S S Sakmar, 'The Global Shale Gas Initiative: Will the United States be the Role Model for the Development of Shale Gas Around the World?' (2011) 33 *Houston Journal of International Law* 369.

J Schremmer, 'Avoidable Fraccident An Argument Against Strict Liability for Hydraulic Fracturing' (2013) 60 *University of Kansas Law Review* 1215.

A Schultz and R Petchey, *Australian Energy Statistics: Energy Update 2011* (Australian Bureau of Agricultural and Resource Economics and Sciences, 29 June 2011) http://data.daff.gov.au/data/warehouse/pe_abares99010610/EnergyUpdate_2011_REPORT.pdf.

B Sovacool, 'Who Shale Regulate the Fracking Industry' (2013) 24 *Villanova Environmental Law Journal* 189.

J G Speight, *The Chemistry and Technology of Petroleum* (McGraw-Hill, 2nd ed, 1991).

J Sprankling, 'Owning the Center of the Earth' (2008) 55 *UCLA Law Review* 979.

D Steinway, 'Hydraulic Fracturing and the Shale Gas Boom' (2012) 5 *International Environmental Law Review* 180.

N Swayne, 'Regulating Coal Seam Gas in Queensland: Lessons in an Adaptive Management Approach' (2012) 29(2) *Environmental and Planning Law Journal* 163.

United States Energy Information Administration, *Annual Energy Outlook 2013: With projections to 2040* (2013) http://www.eia.gov/forecasts/aeo/pdf/0383(2013).pdf.

US Department of Energy, National Energy Tech. Lab., 'Modern Shale Gas Development in the United States: A Primer' (April 2009) http://www.netl.doe.gov/technologies/oilgas/publications/epreports/shalegasprimer2009.pdf

M Weir and T Hunter, 'Property Rights and Coal Seam Gas Extraction: The Modern Property Law Conundrum' (2012) 2 *Property Law Review* 71.

H R Williams and C J Meyers, *Manual of Oil and Gas Terms* (LexisNexis, 10th edition, 1997)

H Wise, 'Fracturing Regulation Applied' (2012) 22 *Duke Environmental Law and Policy Forum* 316.

6

RENEWABLE ENERGY: REGULATION, THE RET, WIND ENERGY, AND THE MARKET FRAMEWORK

6.1	Introduction	240
6.2	What is renewable energy?	241
6.3	The renewable energy market	244
6.4	Statutory regulation: *Renewable Energy (Electricity) Act 2000* (Cth)	246
6.5	Background to the RET	250
6.6	Australian Renewable Energy Agency (ARENA), the Clean Energy Regulator (CER), and the Clean Energy Finance Corporation (CEFC)	252
6.7	The Large-scale Renewable Energy Target	254
6.8	The SRES and the Solar Credits Scheme	257
6.9	The economics of renewable energy	258
6.10	National electricity market	260
6.11	Wind energy: Regulatory and practical issues	262
6.12	Review questions	273
6.13	Further reading	274

6.1 Introduction

This chapter deals with the regulation of renewable energy. Renewable energy is crucial to climate change mitigation and a shift towards realising sustainable, low-carbon societies. There has been a global divestment from carbon intensive, fossil fuels across all sectors of the investment community and a shift towards investment in the clean energy sector. For example, by 2030 China plans to increase the non-fossil fuel share of all energy to approximately 20 per cent.[1] This objective is to be achieved through the efficient use of fossil fuels and the expansion of renewable energy sectors to generate low-carbon electricity. It will also include the widespread implementation of carbon capture technology (CCS). These industries are rapidly progressing. Solar panels cost 90 per cent less than they did in 1980 and 60 per cent less than they did in 1998.[2] Decarbonisation of the atmosphere – with the aim of preventing global temperature increases – is achievable provided there is a large scale shift to renewable energy. The primary pathway for this will be asset transformation, the progression of research and development, and the facilitation of financing mechanisms.[3]

This chapter examines the broader regulatory frameworks that subsidise renewable energy at both the state and federal levels and has a specific focus on solar and wind energy. Whilst an overview of the nature and forms of renewable energy have been dealt with in Chapter 1, the purpose of this chapter is to provide an outline of the renewable energy sector in Australia and to review the regulatory frameworks supporting the progression of this sector. This chapter provides a detailed outline of the *Renewable Energy (Electricity) Act 2000* (Cth) and examines the operative elements of the renewable energy target and the role that the various clean energy bodies have assumed in implementing this framework. Consideration is given to some of the economic implications underlying renewable energy progression as well as some of the existing market impediments. This chapter concludes by focusing upon one of the most significant emergent renewable energy industries in Australia; namely,

---

1 On 11 November 2014, US President Obama announced a new target to cut net greenhouse gas emissions 26–28 per cent below 2005 levels by 2025. At the same time, President Xi Jinping of China announced targets to peak CO_2 emissions around 2030, with the intention to try to peak early and to increase the non-fossil fuel share of all energy to around 20 per cent by 2030. See The White House, Office of the Press Secretary, Fact Sheet: U.S.-China Joint Announcement on Climate Change and Clean Energy Cooperation (11 November 2014) http://www.whitehouse.gov/the-press-office/2014/11/11/fact-sheet-us-china-joint-announcement-climate-change-and-clean-energy-c.
2 The reducing cost of solar panels is discussed by N Mee, 'Here Comes the Sun: Solar Power Parity with Fossil Fuels' (2012) 36 *William and Mary Environmental Law and Policy Review* 119, 131.
3 See Climate Works, *Pathways to Deep Carbonisation in 2050: How Australia Can Prosper in a Low Carbon World* (September 2014) <http://www.climateworksaustralia.org/project/current-project/pathways-deep-decarbonisation-2050-how-australia-can-prosper-low-carbon>.

wind energy. The nature of wind farming, the regulatory principles, and best practices standards are reviewed and the social and community issues associated with the progression of this important renewable industry for Australia are examined.

6.2 What is renewable energy?

Renewable energy is generally defined as energy that is generated from resources that may be naturally replenished, such as solar, wind, rain, tides, waves and geothermal heat. For the purposes of this chapter, the term 'renewable energy' is used broadly to incorporate those energy sources found to produce little or no greenhouse gases and therefore excludes conventional nuclear and fossil fuel energy. Unlike non-renewable energy – or 'fossil fuels' – renewable energy is not depleted when used because it is not derived from the burning of physical minerals or hydrocarbons. Rather, renewable energy is a product of regeneration because it is derived from natural resources that lack a corporeal identity and may be continually replenished.[4]

There are five main sources of renewable energy: wind, water, solar, geothermal, and biomass. The renewable energies derived from these sources include: wind energy using wind turbines; hydroelectric energy, tidal energy or wave energy; solar energy using photovoltaic cells or solar ovens (among others); geothermal energy, usually in the form of steam; biogas (methane) from waste (or landfill gas); biofuel, such as ethanol; and biomass. Renewable resources are regarded as fully sustainable; however, there are some environmental impacts: geothermal exploration may cause loss of thermophile biodiversity, seismic effects, and stream pollution; hydro-electric generation may, in some cases, have a detrimental impact on river ecology; and wind turbines may cause bird fatalities.[5]

To date, Australia's energy needs have been largely met by fossil fuels. Australia's abundant and low-cost coal resources have been used to generate three-quarters of domestic electricity and our transport system is heavily dependant upon oil, some of which is imported.[6] However, global awareness of climate change is causing a fundamental restructuring of the energy market. In an effort to reduce emissions, governments are heavily subsidising the development of renewable

---

[4] For an excellent discussion on the nature of renewable energy see E T Freyfogle, *Natural Resources Law* (Thomson/West, 2007) 2–5. The author describes natural resources law as 'the expansive body of rules and processes governing the ways people interact with nature' as well as ownership of land and discrete components of nature.

[5] See especially the discussion by D J Kochan and T Grant, 'In the Heat of the Law, It's Not Just Steam: Geothermal Resources and The Impacts on Thermophile Biodiversity' (2007) 13 *Hastings Journal Environmental Law & Policy* 35, 37–40.

[6] See Australian Government, Department of Resources, Energy and Tourism, Geoscience Australia and Australian Bureau of Agricultural and Resource Economics (ABARE), *Australian Energy Resource Assessment Report* (2010), 2 <http://www.ga.gov.au/image_cache/GA16725.pdf>.

energies. This has caused transformational change to the energy market and the Australian economy.

Renewable energy resources are expected to play an increasingly important role in Australia's energy mix over the next two decades, especially with respect to electricity generation. Renewable energy resources are diverse in nature. They include geothermal, hydro-electricity, wind, solar, ocean, and bio-energy sources.[7] It has been predicted that the global fossil fuel industry may face a $30 trillion loss in revenues over the next two decades if the world takes action to address climate change and decarbonise the global energy system.[8] The global investment banking company, Citigroup, has noted that renewable energy sources are reaching a point of disruption and have described this point as the start of the 'age of renewables', arguing that renewable energy is becoming increasingly cost competitive.[9]

This is particularly true in Australia given the likelihood of a significant rise in the cost of fossil fuel consumption and the rapid development and expansion of renewable energy technology. Renewable energy is not subject to the same global market volatility as traditional fossil fuel resources and this, combined with a decrease in the cost of renewable energy generation, has made renewable energy sources an increasingly preferable and safer investment option.[10] Renewable energy is also a much safer option for global energy security.[11]

About 16 per cent of global energy consumption is derived from renewable resources with approximately 10 per cent of all energy from traditional biomass and 3.4 per cent from hydro-electricity. New renewables (small hydro-electricity projects, modern biomass, wind, solar, geothermal, and biofuels) account for another 3 per cent and this percentage is growing rapidly. National renewable energy markets are projected to continue strong growth in the near future given that, at the end of 2014, it was apparent that wind power was growing at the rate of 30 per cent annually. In Australia, the share of wind energy in total electricity generation is projected to increase from 1.5 per cent in 2007–2008 to 12.1 per cent in 2029–2030.[12]

---

7 See G Czisch, *Global Renewable Energy Potential—Approaches to its Use* (2001) <http://www.iset.uni-kassel.de/abt/w3-w/folien/magdeb030901/>.
8 See G Parkinson, 'Fossil fuels face $30 trillion losses from climate, renewables', (28 April 2014) *RenewEconomy*.
9 See G Parkinson, 'Citigroup says the 'Age of Renewables' has begun', (27 March 2014) *RenewEconomy*.
10 This is discussed by M Wilder, 'The Global Economic Environment and Climate Change' in Committee for Economic Development Australia, *The Economics of Climate Change* (June 2014) 26.
11 See the discussion in the *New York Times* by the President of the United Nations Conference on Climate change, Laurent Fabrius, on 24 April 2015: Laurent Fabius, 'Our Climate Imperatives', *New York Times* (online), 24 April 2015 <http://www.nytimes.com/2015/04/25/opinion/laurent-fabius-our-climate-imperatives.html>.
12 See Australian Government, above n 6, 239.

There are core differences between renewable and non-renewable energy or 'fossil fuel' resources. Fundamentally, renewable energy resources are not derived from fuels, such as coal or gas, which have formed over millions of years by the natural anaerobic decomposition of dead organisms buried deep within the earth. Renewable energy is not a derivative of the subsurface strata of the earth but is regenerative and process-based and in this respect, unlike fossil fuels, its scope is immeasurable. Renewable energy does not depend upon extraction technology and does not need to be physically removed from the strata and 'consumed', but is created and constantly recreated. These core differences mean that renewable energy is subject to fundamentally different ownership and control principles. The natural, kinetic process of renewable energy necessarily operates beyond the parameters of private control.[13] On the other hand, there are also broad similarities.

Renewable energy resources can exist over wide geographical areas, in contrast to fossil fuel resources which are concentrated in location as they are dependant upon the existence of relevant reserves. In light of this, the deployment of renewable energy is largely sourced in broader control principles connected to surface estate ownership. The development of renewable energy projects depends upon the construction of an energy processing plant, which must be physically located in an area suitable for renewable energy generation. It is not feasible to impose ownership rights upon emancipated renewable energy resources – such as wind and solar – which lack any tangible or defined presence. Rather, the generation of such resources needs to be protected through the firm recognition of access entitlements.

Access entitlements for both solar and wind energy are crucial because they form the foundation for the unfettered production of energy. Solar and wind access may be protected by a range of property rights including express easements, implied easements, prescription, or restrictive covenants. These rights provide the energy proponent with intangible ownership rights that support the unfettered access of solar light and the free and undisturbed flow of wind. Unfortunately, unlike the United States, easement rights of this nature have not been endorsed under either common law or statute in Australia. The principal argument against solar and wind easements relates to the perception that the acceptance of these would restrict the development of the servient tenement (i.e. the burdened land), combined with

---

[13] This does not mean that renewable energy projects cannot be controlled. For example, in Denmark, the three wind turbines at the Hvide Sande Harbour were set up in December 2011. Eighty per cent of the wind farm is owned by the Holmsland Klit Tourist Association foundation, a local business fund which initiated and financed the project. Hvide Sande's North Harbour Turbine Society I/S pay an annual rent of €644 000 to the local harbour. The other 20 per cent is owned by local residents living within a 4.5 km radius, as per the guidelines set out by the Danish *Renewable Energy Act*. This wind cooperative has 400 local stakeholders and, with an annual return of 9–11 per cent, the turbines are expected to pay for themselves in 7–10 years. The fund is used to initiate new business initiatives for the benefit of the harbour and local municipality.

the fact that the easement right itself has an intangible and amorphous character. Despite this, the Law Reform Commission of Victoria recommended that solar access easements be created, stating:

> The potential gains in facilitating the development of solar energy collection given the current concern in relation to fossil fuel use, both as to stocks and effects, more than justify facilitating the means by which solar energy easements may be acquired. It is thought that any difficulties that might arise in relation to property development raise issues that relate to the removal of these easements rather than their creation.[14]

The absence of developed access protection in the form of easement rights for renewable energy projects in Australia has meant that the industry has had to stimulate investment through the implementation of strong sector regulation and market subsidies.[15]

6.3 The renewable energy market

With the exception of crude oil, Australia's fossil fuel resources are expected to last for many more decades, even with increased levels of production.[16] Coal remains the dominant emission source for greenhouse gas within the electricity sector and coal is Australia's largest energy resource. Australia has abundant, high-quality coal resources (black and brown) and gas (conventional, coal seam gas and potentially tight gas) resources, which are widely distributed across the country. Resources of oil (crude oil, condensate, and LPG) are more limited and Australia relies heavily on imports to meet demand for transport fuels.

The existence of a strong fossil fuel industry in Australia is, at first glance, suggestive of institutionalised economic impediments to the growth of renewables.

---

14 See Victoria Law Reform Commission, *Easements and Covenants*, Report No 41 (1992) 27–8. See also Tasmania Law Reform Institute, *Law of Easements in Tasmania*, Final Report No 12 (March 2010) 35 http://www.utas.edu.au/__data/assets/pdf_file/0011/283835/EasementsFinalReportA4.pdf.
15 For an excellent discussion of the core ownership and operational differences connected with renewable energy resources see A B Klass, 'Property Rights on the New Frontier: Climate Change, Natural Resource Development, and Renewable Energy' (2011) 38 *Ecology Law Quarterly* 63. See Freyfogle, above n 4, 57–9 discussing the role of private property rights within natural resources law. For a comparison of the approaches in the United States and Australia, see A J Bradbrook, 'Australian and American Perspectives on the Protection of Solar and Wind Access' (1988) 28 *Natural Resources Journal* 229.
16 See generally the discussion by CSIRO in 'Change and Choice: The Future Grid Forum's Analysis of Australia's potential electricity pathways to 2050' (December 2013) which explores the challenges of the electricity framework for the future.

However, the imperatives associated with climate change mitigation and the crucial need to address greenhouse gas emissions from the electricity power sector has stimulated strong growth. The implementation of low-carbon electricity through renewable energy sources represents one of the key factors for future decarbonisation processes.[17] This, in combination with a carbon price, is the most effective means of transitioning Australia to a low emissions economy. As outlined by Professor Garnaut:

> The evolution of the electricity sector under carbon pricing should not cause the community anxiety. Australia has an incomparable range of emissions-reducing options. The early stages of the transition will see expansion of gas at the expense of coal alongside the emergence of a range of renewable energy sources. The carbon price will arbitrate between the claims of different means of reducing emissions as the profitability of each is affected by many domestic and international developments. Whether or not coal has a future at home and as an export industry depends on the success of technologies for sequestration of carbon dioxide wastes. There is little reason for concern about the physical security of energy supply during the transition to a low-emissions economy, but I propose some cost-effective measures to ease anxieties in parts of the community.[18]

Australia has five electricity systems and numerous stand-alone, remote electricity systems. The largest is the National Energy Market in eastern Australia. This is followed by the south-west and the north-west interconnected systems in Western Australia and the Darwin-Katherine and Alice Springs system in the Northern Territory. The implementation of new sources of electricity from renewable energy means that it is increasingly important to ensure an effective expansion of the renewable energy grid. In 2009, the Australian Energy Market Commission (AEMC) acknowledged the significant impact that government policies had made in incentivising renewable energy; it went on to outline the importance of developing new connection hubs or scale efficient network extensions to properly manage this expansion.[19]

As in many other countries, the progression of renewable energy markets in Australia has been encouraged by government policy, implemented in response

---

[17] The concept of 'deep decarbonisation' depends upon a virtually complete decarbonisation of the electricity grid using renewables or a mix of renewables combined with either CCS or nuclear. See Climate Works, above n 3.
[18] R Garnaut, *Garnaut Climate Change Review, Update 2011: Australia in the Global Response to Climate Change*, Summary, 11 http://www.garnautreview.org.au/update-2011/garnaut-review-2011.html.
[19] See AEMC (Australian Energy Market Commission), *Review of Energy Market Frameworks in light of Climate Change Policies: Final Report* (30 September 2009).

to concerns regarding climate change, energy independence, and economic stimulus. Both the federal and state governments have focused on creating incentive programs and directing funding to encourage the development of the renewable energy sector. As outlined by ABARE:

> The energy sector currently accounts for more than half of Australia's net carbon dioxide (CO_2) emissions. The move to a lower emissions economy requires a shift from the current heavy dependence on fossil fuels to a greater use of energy sources and technologies that reduce carbon emissions, such as renewable energy and carbon capture and storage. At present renewable energy sources account for only modest proportions of Australia's primary energy consumption (around 5 per cent) and electricity generation (7 per cent), although their use has been increasing strongly in recent years. Proposed developments in Australia's energy policy will seek to significantly boost the role that renewable energy plays over the next two decades.[20]

6.4 Statutory regulation: *Renewable Energy (Electricity) Act 2000* (Cth)

A key component of the renewable energy framework in Australia has been the introduction at the Commonwealth level of the *Renewable Energy (Electricity) Act* (Cth) *2000* ('the *REEA*'). This legislative framework introduced a mandatory renewable energy target. The core aim of the renewable energy target is to encourage the additional generation of electricity from renewable sources.[21] The renewable energy legislation in Australia represents an important component in the process of transforming an emissions intensive electricity generation sector to a zero carbon energy source.[22] This is particularly the case in Australia in 2015, where the current absence of any mandated carbon price means that the

---

20 See Australian Energy Resource Assessment Report, above n 6, 2.
21 For an excellent discussion on the framework for renewable energy in Australia see A Kent and D Mercer, 'Australia's Mandatory Renewable Energy Target (MRET): An Assessment' (2006) 34 *Energy Policy* 1046; J Prest, 'A Dangerous Obsession with Least Cost? Climate Change, Renewable Energy Law and Emission Trading' in D Gumley and D Winterbottom (eds) *Climate Change Law: Comparative, Contractual And Regulatory Considerations* (Thomson Reuters, 2009) 179.
22 See especially the Intergovernmental Panel on Climate Change, *Contribution of Working Groups I, II and III to the Fourth Assessment Report of the Intergovernmental Panel on Climate Change* (Intergovernmental Panel on Climate Change, 2007) 64.

renewable energy sector plays a far more significant role in emission reductions. It is impossible for Australia to limit emissions unless our heavy reliance on coal and natural gas for electricity generation is replaced by a dramatic increase in renewable energy. Whilst the RET has been criticised on the grounds that it is unlikely to lead to further emission reductions in a capped stationary energy sector and may in fact impose additional costs on consumers, these arguments are offset by the fact that the RET is more likely to drive infrastructure investment than pure market alternatives.[23] Indeed, the RET has been a crucial mechanism supporting the development of renewable energy within a difficult environment, given that Australia's self-sufficiency in coal and gas has traditionally given us a greater sense of energy security. Lack of government intervention has, traditionally, retarded the use of renewable energy and the development of a renewable energy industry. By promoting a sustained increase in electricity generation from renewable sources under the RET scheme, Australia is far better equipped to meet its rising energy demand and reduce its reliance on fossil fuel energy sources and establish a more balanced energy mix.[24]

The Warburton Review of the RET, conducted in August 2014, concluded that because the renewables industry is now well established in Australia, the main rationale for the RET depends on its capacity to contribute towards emissions reductions in a cost effective manner. The review concluded that the RET was, however, a high cost approach to reducing emissions because, rather than focusing directly upon emissions, it focuses upon electricity generation and therefore promotes activity in renewable energy ahead of alternative, lower cost options for reducing emissions that exist elsewhere in the economy.[25]

Various stakeholders in the renewables sector have expressed strong concern that any repeal of the RET would affect existing and future investment and that this would amount to what might best be described as 'sovereign risk'. The Warburton Review acknowledged that a repeal would have adverse financial implications and therefore came up with two proposals. The first being to allow the RET to continue until 2030 for existing and committed renewable generators but closing it to new entrants. The second being to modify the RET, so that it increases in proportion with growth in electricity demand. Modification should occur by setting targets one year in advance that correspond to a '50 per cent share of new growth'.[26] The panel felt this would protect investors in existing renewable generators and

---

23 P Christoff, 'Aiming High: On Australia's Emissions Reductions Target' (2008) 31 *University of New South Wales Law Journal* 861, 876.
24 See A J Bradbrook, 'Government Initiatives Promoting Renewable Energy for Electricity Generation in Australia' (2002) 25 *University of New South Wales Law Journal* 124.
25 See Renewable Energy Target Scheme, *Report of the Expert Panel* (August 2014), 6 https://retreview.dpmc.gov.au/sites/default/files/files/RET_Review_Report.pdf.
26 Ibid 7.

support additional renewable generation when demand increased. Under this option, targets would not be mandated for future years, exposing renewable energy investors to the same market risk as other energy investors face.[27]

The Clean Energy Council has argued that the RET is important because it has attracted $20 billion in new investment in Australia and a further $14.5 billion in new investments from projects outlined for the years leading up to 2020.[28] Without the support of the RET, the renewable energy sector in Australia is less commercially viable. When the target was first conceived, it was envisaged that 20 per cent of total power production by 2020 would equate to 41 000 gigawatt/hours of renewable energy produced each year. If the RET were completely abolished, it would reduce renewable energy production in 2020 to 16 000 gigawatt/hours per year.

The aim of the RET is to support renewable energy production and development, increase consumer and industrial access to renewable energy, and improve economic productivity. In a contemporary energy framework, the use of subsidies is crucial for the implementation of broader environmental goals associated with the promotion of sustainable energy markets. As is apparent from the Warburton Review, the subsidies may be criticised on the grounds that they are unfair and economically inefficient, especially when fossil fuels are cheaper. However, such arguments generally overlook the fact that fossil fuels have costs that are not fully internalised, thus making their market cost cheaper despite the fact that their actual cost, when the global impact of GHG emissions is accounted for, is much higher.[29] The broader justification for using a primarily economic instrument such as the RET to promote renewable energy is derived from two core factors: externality costs and market failure. The full cost of fossil fuel energy production is not incorporated into energy costing because externalities such as environmental damage, climate change impact, or impact on human health are not taken into account. This is effectively a market failure because if such costings were properly taken into account the renewable energy sector would be more competitive.[30] Hence, to properly accommodate this market failure, economic instruments grounded in the 'polluter must pay' principle become important. Where industries incur extra costs for pollution activities they are more likely to deviate from established practices in order to reduce costs. In this sense, economic instruments promoting renewable energy industries perform an important market adjustment function.

---

27 Ibid 7.
28 See the Clean Energy Council, Roam Consulting, *Renewable Energy Target policy analysis* (23 May 2014) http://www.cleanenergycouncil.org.au/policy-advocacy/renewable-energy-target/ret-policy-analysis.html.
29 See J P Fershee, 'Energy Subsidies' in K Bosselman et al (eds), 'The Law and Politics of Sustainability' *Berkshire Encyclopaedia of Sustainability* (Bershire Publishing Group, 2012) 158–60.
30 See the discussion by A J Bradbrook, 'Creating Law for the Next Generation Energy Technologies' (2011) 2 *George Washington Journal of Energy and Environmental Law* 17, 19–20.

Further, the argument that the abolition of the RET would create a more efficient free market is inaccurate given the subsidies that fossil fuel sources already receive and these subsidies are supporting long established market participants that already have proven consumer bases and distribution systems.[31] The only way to achieve a perfect free market would be to eliminate *all* energy subsidies, including those associated with fossil fuels.

The primary aim of the RET in Australia is to reduce emissions of greenhouse gases in the electricity sector by ensuring that renewable energy sources remain ecologically sustainable. The Renewable Energy Target Scheme is regulated by the following Acts: *Renewable Energy (Electricity) Act 2000* (Cth); *Renewable Energy (Electricity) (Small-Scale Technology Shortfall Charge) Act 2010* (Cth); *Renewable Energy (Electricity) (Charge) Act 2000* (Cth); *Renewable Energy (Electricity) Regulations 2001* (Cth).

The *REEA* seeks to achieve these aims through the creation of online certificates by eligible renewable authorities. These certificates are based on the amount of electricity in megawatt hours (MWh) generated by a renewable energy power station, or a small-scale solar panel, wind, or hydro system that is displaced by a solar water heater or heat pump. Every year, electricity retailers must purchase and surrender a defined number of certificates. The financial incentive for investing in renewable energy industries and for installing solar water heaters, heat pumps and, small-scale solar panels and wind systems is derived from trading in these certificates. The REC Registry, which is managed by the Clean Energy Regulator, creates the certificates and manages their trade.

The implementation of this scheme has increased the number of installations of small-scale renewable energy systems and stimulated investment in renewable energy power stations. In the period from 2001–2011, nearly 1.4 million small-scale installations, such as solar panels and solar water heaters, had certificates created and validated against them in the REC Registry.

The objective of the RET is to ensure that 20 per cent of Australia's electricity is produced from renewable energy sources by 2020. Despite the differences in climate change policy in Australia, there is at least ostensible bi-partisan commitment to reduce Australia's emissions between 5 per cent and 25 per cent below 2000 levels by 2020. In order to achieve this, annual targets are set for every year of the scheme. Australian electricity retailers and large wholesale purchasers of electricity are required to demonstrate that they meet these annual targets. Compliance is demonstrated by surrendering renewable energy certificates (RECs). One REC is equivalent to one additional MWh of electricity generated from renewable

---

31 See J P Fershee, 'Promoting an All of the Above Approach or Pushing (Oil) Addiction and Abuse: The Curious Role of Energy Subsidies and Mandates in U.S. Energy Policy' (2012) 7 *Environmental Energy and Policy Law Journal* 125, 143.

energy sources (above a benchmark set in 1997). A failure to surrender adequate RECs may lead to a shortfall charge.

Electricity retailers and wholesale buyers – referred to as 'liable entities' – can choose to either generate the electricity from renewable energy sources themselves or purchase the RECs from others that have done so. This effectively creates a market for RECs. In this way, the RET creates demand for additional renewable energy generation through the imposition of an obligation on entities that purchase wholesale electricity to surrender a certain number of renewable energy certificates each year.

The *REEA* establishes two types of RECs: 'large-scale generation certificates' (LRECs) and 'small-scale generation certificates' (SRECs). The Large-scale Renewable Energy Target ('LRET') has a target of 41 000 gigawatt hours (GWh) by 2020 and only large-scale renewable energy projects are eligible. The Small-scale Renewable Energy Scheme ('SRES') targets 4000 GWh annually and is eligible only to small-scale or household installations.

The LRET is fixed annually according to the Renewable Power Percentage (RPP). The RPP determines how many RECs each liable entity needs to surrender to discharge their liability.[32] For example, the RPP for 2011 was 5.62 per cent of total estimated electricity consumption, which is equivalent to 10.6 million RECs.

The prices of LRECs fluctuate on the market and can vary with any number of external and internal factors up to the level of the shortfall charge. Liable entities must surrender a fixed proportion of LRECs annually. If a wholesale purchaser of electricity fails to surrender the appropriate number of RECs, a shortfall of $65 per MWh is enforced for each outstanding REC.[33] It is possible for a partial exemption to be taken into account in determining the renewable energy shortfall charge of a liable entity with respect to one or more emissions-intensive trade-exposed industries. The renewable energy shortfall charge is defined in s 46 of the *REEA* to include either a large-scale or a small-scale technology shortfall charge and refers to a penalty that is payable to the Commonwealth.

6.5 Background to the RET

The RET scheme began with the Commonwealth Government's Mandatory Renewable Energy Target (MRET), which targeted the generation of 9500 GWh of extra renewable electricity per year by 2010. Interim annual targets were set to ensure consistent progress towards achieving the 9500 GWh target and these have been applied to calendar years up to and including 2020. In 2009, the *REEA* was

---

32 See *REEA* (Cth) s 39.
33 See *REEA* (Cth) s 6.

amended and the MRET was replaced with the RET.[34] This altered the target from 9500 GWh by 2010 to 45 000 GWh by 2020 and introduced what is now known as the Solar Credits Scheme (discussed below).

In contradistinction to the MRET, the RET is not linked to a percentage of total energy generation but is expressed as a quantity of additional renewable energy. Hence, the additional 45 000 GWh/pa, which is to be achieved by 2020, may not accurately represent 20 per cent of Australia's total electricity supply if electricity demand rises in the decade preceding 2020.[35]

Shortly after the passage of the 2009 legislation, REC prices fell. The new framework was criticised in the media for promoting domestic solar installations over large-scale renewable energy projects.[36] These factors generated significant uncertainty on the market and threatened to deter potential investment in future large-scale renewable energy projects. In response, the government announced a series of reviews and consultation papers and ultimately amended the legislation in 2010 to create a dichotomy or 'split' between the LRET and SRES. Additionally, the amendments allow the renewable energy regulator to adjust the RECs multiplier to benefit domestic solar systems. Annual LRET targets were adjusted where the number of RECs exceeded a value of $34.5 million by the end of the 2010 period – the aim being to provide a guaranteed demand for RECs from large-scale projects.[37]

The primary objective of the both the LRET and the SRES is to encourage additional investment in renewable energy generation and, in so doing, to reduce emissions of greenhouse gases in the electricity sectors.

The explicit objective is set out in s 3 of the *Renewable Energy (Electricity) Act 2000* (Cth):

Renewable Energy (Electricity) Act 2000 (Cth)

3 Objects/outline

The objects of this Act are:

(a) to encourage the additional generation of electricity from renewable sources; and
(b) to reduce emissions of greenhouse gases in the electricity sector; and
(c) to ensure that renewable energy sources are ecologically sustainable.

---

34 See Renewable Energy (Electricity) Bill 2009 (Cth).
35 It is interesting to note that ABARE has predicted that Australian energy demand may rise by up to 25 per cent by 2020: ABARE, *Australian Energy Predictions to 2030*, (2010) 28.
36 ABC Television, 'Renewable Energy on the Brink', *7:30 Report*, 8 December 2009 (P McCutcheon); ABC Radio National, 'Renewable Energy Reform Pacific Hydro', *Breakfast*, 1 March 2010 (F Kelly); ABC Radio National, 'Renewable Energy Reform – Keppel Prince Engineering', *Breakfast*, 1 March 2010 (F Kelly).
37 Renewable Energy (Electricity) Amendment Bill (Cth) 2010 s 124.

The *REEA* seeks to achieve this through the issuance of certificates for the generation of electricity using eligible renewable energy sources and requiring certain purchasers – known as 'liable entities' – to surrender a specified number of certificates for the electricity that they acquire during a year. If a liable entity does not have enough certificates to surrender, the liable entity will be required to pay what is known as the renewable energy shortfall charge.

There have been a number of significant policy changes since the MRET was first introduced. In particular, a carbon pricing mechanism was introduced, which has subsequently been replaced by an emission reduction fund. Under a carbon pricing mechanism, emitting greenhouse gases costs fossil fuel companies and this gives them an incentive to reduce emissions. Compensation is then paid to taxpayers to help mitigate any price increases, such as the cost of electricity, and this is then passed on to consumers.

By way of contrast, under the newly-implemented emission reduction fund, businesses compete to win tenders and be paid to undertake emission reduction projects. The White Paper that outlines the operation of the emission reduction fund indicates that the fund will function alongside existing programs that are already working to offset Australia's emissions growth, such as the Renewable Energy Target and energy efficiency standards on appliances, equipment, and buildings.[38]

6.6 Australian Renewable Energy Agency (ARENA), the Clean Energy Regulator (CER), and the Clean Energy Finance Corporation (CEFC)

In addition to the implementation of greenhouse gas reduction frameworks, the Commonwealth Government has introduced the Australian Renewable Energy Agency (ARENA), the Clean Energy Regulator (CER), and the Clean Energy Finance Corporation (CEFC). Each of these organisations supports the future development and expansion of the renewable energy sector. ARENA is an independent agency established in 2012 with the objective of improving the competitiveness

---

38 See Emission Reduction Fund, *White Paper* (April 2014) 3 http://www.environment.gov.au/system/files/resources/1f98a924-5946-404c-9510-d440304280f1/files/emissions-reduction-fund-white-paper_0.pdf.

of renewable energy technologies and increasing the supply of renewable energy in Australia.[39] All of ARENA's funding is devoted to supporting renewable energy projects and research activities to capture and share knowledge regarding renewable energy projects.

The Clean Energy Regulator (CER) is an independent federal statutory authority. It was created in 2012 pursuant to the *Clean Energy Regulator Act 2011* (Cth). One of the portfolio of responsibilities of the CER is to administer the RET and this involves assessing emissions to determine the liability of an eligible entity,[40] allocating carbon units, and operating the Australian National Registry of Emissions Units.[41] The CER has the ability to track and locate ownership of certificates issued pursuant to the RET via the REC Registry. The responsibilities of the CER extend well beyond the administration of the RET and include:

- providing education on the carbon pricing mechanism and how it works
- assessing emissions data to determine each emitting entity's liability
- operating an emissions unit registry
- monitoring, facilitating and enforcing compliance with the carbon pricing mechanism
- allocating units, including freely allocated units, fixed price units and auctioned units
- administering the RET and the Carbon Farming Initiative (CFI)
- accrediting auditors for the CFI, the carbon pricing mechanism and the NGER scheme
- working with other national law enforcement and regulatory bodies, including the Australian Securities and Investments Commission, the Australian Competition and Consumer Commission, the Australian Transaction Reports and Analysis Centre, the Australian Federal Police and the Commonwealth Director of Public Prosecutions.[42]

The CEFC works collaboratively with co-financiers and project proponents to seek ways to secure financing solutions for the clean energy sector. The CEFC focuses upon ensuring that renewable energy projects are effectively commercialised and deployed. The CEFC was set up and operates pursuant to the *Clean Energy*

---

39 See *Australian Renewable Energy Agency Act 2011* (Cth) ss 7, 8.
40 See *Clean Energy Act 2011* (Cth) pt 5.
41 See *Australian National Registry of Emissions Units Act 2011* (Cth) pt 2.
42 See Clean Energy Regulator, *About the Clean Energy Regulation: What we do* (2014) http://www.cleanenergyregulator.gov.au/About-us/our-work/Pages/default.aspx. The *National Greenhouse and Energy Reporting Act 2007* (Cth) requires entities emitting GHG to report what are defined as 'scope 1 direct GHG emissions' and 'scope 2 indirect emissions' from electricity and energy to the Clean Energy Regulator.

Finance Corporation Act 2012 (Cth) whose object, as set out in s 3, is to facilitate increased flows of finance into the clean energy sector. The functions of the CEFC are set out in s 9, which is extracted below:

Clean Energy Finance Corporation Act 2012

9 Corporation's functions

(1) The Corporation has the following functions:
   (a) its investment function (see subsection 58(1));
   (b) to liaise with relevant persons and bodies, including ARENA, the Clean Energy Regulator, other Commonwealth agencies and State and Territory governments, for the purposes of facilitating its investment function;
   (c) any other functions conferred on the Corporation by this Act or any other Commonwealth law;
   (d) to do anything incidental or conducive to the performance of the above functions.
(2) In performing its functions, the Corporation must act in a proper, efficient and effective manner.

6.7 The Large-scale Renewable Energy Target

The Large-scale Renewable Energy Target (LRET) was created with the aim of generating strong financial incentive to develop and expand the renewable energy sector, such as wind and solar farms or hydro-electric power stations. As outlined above, the primary mechanism for achieving this is by the creation of a statutory demand for certificates. In this context, the certificates are known as Large-scale Generation Certificates (LGCs). In effect, the LGC is a form of electronic currency created by the REC Registry. An LGC is available for every megawatt hour of net renewable energy that is transmitted to electricity consumers. The LGC can only be created over electricity generated by renewable energy delivered to an electricity network and it can only relate to renewable electricity which is generated above the baseline of the relevant power station.[43]

A full list of eligible renewable energy sources is included in s 17 of the *Renewable Energy (Electricity) Act 2000* (Cth) ('*REEA*'). There are currently more than 15 different types of renewable energy sources being used in accredited

---

[43] For a full discussion of the nature of the LGC see Clean Energy Regulator, Renewable Energy Target, *Large-scale Renewable Energy Target* (2014) http://ret.cleanenergyregulator.gov.au/About-the-Schemes/lret.

power stations. Electricity generated from fossil fuels, or waste products derived from fossil fuels, are not eligible for LGCs. Section 17 of the *REEA* is extracted below:

Renewable Energy (Electricity) Act 2000

17 What is an eligible renewable energy source?
(1) The following energy sources are eligible renewable energy sources:
    (a) hydro;
    (b) wave;
    (c) tide;
    (d) ocean;
    (e) wind;
    (f) solar;
    (g) geothermal-aquifer;
    (h) hot dry rock;
    (i) energy crops;
    (j) wood waste;
    (k) agricultural waste;
    (l) waste from processing of agricultural products;
    (m) food waste;
    (n) food processing waste;
    (o) bagasse;
    (p) black liquor;
    (q) biomass-based components of municipal solid waste;
    (r) landfill gas;
    (s) sewage gas and biomass-based components of sewage;
    (t) any other energy source prescribed by the regulations.
(2) Despite subsection (1), the following energy sources are not eligible renewable energy sources:
    (a) fossil fuels;
    (b) materials or waste products derived from fossil fuels.

The core aim of the *REEA* is to encourage the generation of renewable energy and, in so doing, promote the existing renewable energy sector and encourage future growth. The Large-scale Renewable Energy Target determines the amount of renewable energy that is to be created by renewable energy power stations for every year up to 2030. Eligible power stations create LGCs according to the additional renewable electricity they produce above their baseline. The Clean Energy Regulator determines the baseline which, as a general rule, represents the average amount of electricity that has been generated during the period

1994–1996. Those power stations that have generated electricity for the first time after 1 January 1997 will commence with a baseline of zero.[44]

All LGCs must be generated by renewable energy power stations through the online REC Registry and must comply with the requirements for creation before they may be traded. The LGC is sold on an open market and the price will vary according to supply and demand. The LRET places a legal requirement on liable entities to purchase a particular number of LGCs each year. The LRET is calculated using the Renewable Power Percentage ('RPP'). The RPP takes into account the required amount in the RET for the year, the estimated amount of electricity acquired by liable entities for the year, as well as any under or over surrender of LGCs against the annual targets of previous years. It also takes account of the approximate amount of partial exemptions – known as Partial Exemption Certificates ('PECs') – to eligible emissions-intensive, trade-exposed ('EITE') industries that are expected to be claimed for the year.[45]

A liable entity must apply the annual RPP to the total megawatt hours of relevant electricity that they acquire from relevant electricity grids in order to ascertain how many LGCs need to be purchased or surrendered for the coming year. The LGC may be purchased directly from eligible renewable power stations or from authorised agents and ownership will be transferred following registration with the REC Registry. The price of the LGC is highly variable and in the past it has fluctuated between $10 to $60.

The Clean Energy Regulator requires all liable RET entities to surrender the required number of LGCs to meet their annual liability as determined by the RPP. Once surrendered, a certificate becomes invalid and may not be sold, traded, or purchased. If a RET liable entity does not surrender the required number of LGCs it will be required to pay what is known as a 'shortfall charge' which is currently $65 for every non-surrendered LGC.

The RET and the LGC have been instrumental in promoting the renewable energy sector in Australia. They represent the core building blocks for generating a strong renewable sector and are particularly crucial components for climate change mitigation. These economic instruments provide financial support to stimulate demand and attract investment capital; they also facilitate a flexible market currency and promote market transparency and integrity.[46]

---

[44] For a full discussion of the objectives of the *REEA* (Cth) see Clean Energy Regulator, Renewable Energy Target, *How the Renewable Energy Target Works* (2014) http://ret.cleanenergyregulator.gov.au/About-the-Schemes/How-the-RET-works/How-the-Renewable-Energy-Target-works.

[45] See the outline of the LRET by J Prentice, 'Making Effective Use of Australia's Natural Resources – The Record of Australian Renewable Energy Law Under the Renewable Energy (Electricity) Act 2000 (Cth)' (2011) *Renewable Energy Law and Policy Review* 5.

[46] See the discussion by M McDonnell, K Engel and A Barnhart, 'The Potential and Power of Renewable Energy Credits to Enhance Air Quality and Economic Development in Arizona' (2011) 43 *Arizona State Law Journal* 809. See also P K Oniemola, 'Case for the Promotion of Renewable Energy Through the Use of Economic Instruments' (2011) 1 *Dublin Legal Review Quarterly* 34, 43.

6.8 The SRES and the Solar Credits Scheme

The Small-scale Renewable Energy Scheme ('SRES') functions by reference to measurements of renewable energy known as a REC ('renewable energy certificate'). A REC is a measurement of renewable energy, which may be traded for cash. One REC represents one megawatt hour ('MWh') of electricity. RECs can be generated from a number of systems, including solar hot water, heat pumps, solar PV, and small wind generation units, provided they meet certain conditions.

At the start of 2011, the RET scheme changed for the SRES so that RECs for small systems changed to effectively be renamed Small-scale Technology Certificates ('STCs'). The price of an STC is set by the government and may be subject to change over time. Initially, however, it has been fixed at $40 when traded through the government-managed STC Clearing House. The actual amount of electricity generated under the SRES may or may not exceed the 4000 GWh estimate. This price fix was an attempt to reduce some of the volatility that the REC market had previously experienced. An important point is 1 STC = 1 REC, as they effectively represent the same power value.

The SRES is only open to small-scale or household schemes. In order to be eligible for the RET scheme, the scheme must establish it complies with the following criteria:

- the system must be new
- its components must be listed in the Clean Energy Council list of accredited components
- it must be installed correctly by a Clean Energy Council accredited installer
- it must be installed on an eligible premises (this is only required if Solar Credits are claimed)
- it must be compliant with all local, state, and federal requirements for its type of installation
- documentation for small-scale systems, demonstrating compliance with the legislated requirements, must be completed and signed by the owner, installer, and/or Registered Agent. These documents must be produced, if requested by the Clean Energy Regulator[47]
- a Registered Agent is eligible to create STCs if the owner has correctly signed over the certificates to them and this signed documentation can be provided to the Clean Energy Regulator.

---

47 See Clean Energy Regulator, Renewable Energy Target, *The Small-scale Renewable Energy Scheme* (2014) <http://ret.cleanenergyregulator.gov.au/About-the-Schemes/sres>.

A significant component of the SRES was the program known as the Solar Credits Scheme. Solar credits were introduced to provide an additional financial incentive for solar panel installations by multiplying the number of certificates these systems could create under the scheme. The Solar Credits Scheme applied to the initial 1.5 kW of installed renewable energy capacity of an eligible small generation unit. Bigger units with extra capacity were eligible for the standard 1:1 rate of STC creation. The STC 'multiplier' was set at five up until 2011 and then declined to three on 1 July 2011 and it expired in 2013. All systems installed after this date that attract the SRES are entitled to STCs; however, no solar credit multiplier is now available.

The solar credit system allowed the SRES to rely heavily upon the capacity of the solar energy system and on the predicted amount of electricity it was capable of generating over a deemed period. The SRES has the capacity to contribute significantly to market growth in the renewable sector because of the strong incentives it provides for individual users.[48]

6.9 The economics of renewable energy

Since its inception, the renewable energy regulatory framework has experienced a broad range of impediments, many of which are systemic to the renewable energy sector. Renewable energy projects are costly and generally involve a significant amount of capital investment. The pay-back periods from renewable energy technologies can be lengthy and the immature nature of the technology means that renewable energy is often lacking in competitiveness, especially compared to fossil fuel markets. This is the case despite the fact that the Stern Review on Climate Change made it clear that the costs involved in shifting to a zero carbon environment are minimal when compared with the costs of delay or inaction on climate change.[49]

Promoting market access to renewable energy resources is critical in order to ensure that renewable energies gains a foothold in the energy market where the costs of development and production may otherwise be limited.[50] Fossil fuel markets have tended to be cheaper than renewables because the markets have failed

---

[48] See especially the discussion by P Salkin, 'Key to Unlocking the Power of Small-Scale Renewable Energy: Local Land Use Regulation' (2012) 27 *Journal Of Land Use and Environmental Law* 339.

[49] Sir N Stern, *Review on the Economics of Climate Change* (Cambridge University Press 2007) (i).

[50] See especially the discussion by A L Carleton, 'Mandating Market Access for Renewable Energies in Australia' (2008) 26 *Journal of Energy and Natural Resources Law* 402, 404 where the author notes that global awareness of climate change is causing a restructuring in the energy market.

to internalise the external cost of carbon pollution.[51] The commercial competitiveness of renewable energies is impacted by their comparative cost to fossil fuel energies. Two market distortions affect these costs: (i) the hidden environmental costs, which are externalised from market prices by fossil fuel energies but incorporated into the costs; and (ii) the subsidy preference given to fossil energy sources in Australia, particularly coal and, increasingly, natural gas.

Internalising the environmental costs for non-renewable energies would provide a more accurate comparison between renewable and non-renewable energy markets.[52] The reason for this is because it would illustrate the true ecological costs connected with the production of both resources. Many developed countries support fossil fuel subsidies, along with energy taxes. Their presence necessarily reduces the ability of renewable energies to compete in the market. This reduces the ability of the market to respond effectively to energy diversity as preference is often given to resources that have government support.

There is, however, no doubt that these market distortions are gradually changing. Concerns associated with environmental and economic risks of carbon intensive fossil fuel industries have directly influenced the strategies of major international lending institutions. At the 2014 World Economic Forum, the President of the World Bank, Jim Yong Kim, supported divestment from carbon intensive assets and suggested that pensions funds should increasingly recognise their future responsibilities towards renewable investment projects.[53] Similarly, Christine Lagarde, Managing Director of the International Monetary Fund, called for governments to phase out energy subsidies and spur investment in low-carbon technologies of the future.[54]

Divestment away from carbon intensive investments – particularly those connected with fossil fuels – is a growing occurrence across many sectors of the investment community with 'climate smart' investment increasingly becoming a global preference.[55] In this respect, despite the implementation of the RET and other schemes, Australia continues to lag a long way behind given the developments

---

51 See the discussion by J Prentice, 'Making Effective Use of Australia's Natural Resources – The Record of Australian Renewable Energy Law under the Renewable Energy (Electricity) Act 2000 (Cth) (2011) *Renewable Energy Law And Policy Review* 5, 10–15 where the author discusses the problems with the RET scheme, noting that 'decreased demand for RECs meant there were growing concerns over the viability of investment in large-scale RE projects'.
52 See the discussion by E Rotenberg, 'Energy Efficiency in Regulated and De-Regulated Markets' (2006) 24 *UCLA Journal of Environmental Law and Policy* 259.
53 Yong Jim, 'World Bank President Jim Yong remarks at Davos Press Conference' (Speech delivered at the World Economic Forum, Davos, 23 January 2014).
54 C Lagarde, 'A New Multilateralism for the 21st Century' (Speech delivered at The Richard Dimbleby Lecture, London, 3 February 2014).
55 Climate Bonds Initiative, *Bonds and Climate Change: The State of the Market in 2013* (2013) http://www.ieefa.org/wp-content/uploads/2014/02/Thermal-Coal-Outlook-February-2014-Observations-IEEFA.pdf.

and initiatives which are occurring in the rest of the world. In the United States, for example, companies like Apple are protecting themselves from future energy price rises with data centres and manufacturing facilities that are powered by 100 per cent renewable energy.[56]

Focusing research and development on enabling technologies for renewable energy projects combined with targeted and strategic marketing is critical if the high technology and capital costs traditionally associated with renewable energy production are to be effectively addressed and investment opportunities for the Australian renewable energy sector improved. A rapid transition to a renewable energy mix in Australia requires fundamental structural changes that are best implemented not only by market mechanisms such as the RET but also through core shifts in economic policy. Ensuring market access for renewable energies is essential but it is also important to facilitate ongoing competitiveness. As outlined by one commentator:

> Regulation is required to ensure that competition and choice are maintained. Particularly in the energy sector, where monopolies once controlled the market and still retain a dominant and potentially exclusive market share, support from the legislature is required to ensure new market entrants (namely, renewable energies) gain access to the electricity market and consumers have a choice as to which energy products they purchase.[57]

6.10 National electricity market

Market enforcement is the responsibility of the Australian Energy Regulator (AER) which is charged with the obligation of developing rules and progressing the electricity market on a national level. This includes the economic regulation of the wholesale electricity market. The Australian Energy Market Operator (AEMO) operates the energy market, which is also responsible for the National Electricity Market (NEM). The NEM provides the regulatory framework for grid access, grid updates, and strategic transmission planning.

The NEM was implemented prior to the progression of the renewable industry and the introduction of the RET. The primary objective of the NEM is economic

---

56 Apple, *Environmental Responsibility,* (2014) http://apple.apple.com/au/environment/. See the discussion by M Wilder 'The Global Economic Environment and Climate Change' in Committee for Economic Development Australia, *The Economics of Climate Change* (June 2014).
57 A L Carleton, 'Mandating Market Access for Renewable Energies in Australia' (2008) 26 *Journal of Energy and Natural Resources Law* 402, 403.

efficiency. The objective of the national electricity law has been expressed as follows:

> The objective of this law is to promote efficient investment in and efficient operation of electricity services for the long term interests of consumers of electricity with respect to: price, quality, safety, reliability and security of supply of electricity; and the reliability safety and security of the national electricity system.[58]

The National Electricity Law does not incorporate a clear recognition of public objectives, such as climate change mitigation or sustainable development. This has produced a regulatory environment where decision-makers have not been specifically obligated to take these issues into account, despite the fact that the market growth of the renewable sector is expanding and electricity represents the main energy source.[59]

There are some significant structural barriers connected with the expansion of renewable energy projects under the existing NEM framework. The cost of updating the grid to include large-scale renewable power is predicted to be enormous, with constraints on grid capacity being forecast particularly in high wind areas such as western Victoria and South Australia.[60] This means that in order to connect more renewable energy generation within these areas, the transmission grid must be upgraded. In Victoria, for example, most of the electricity grid infrastructure will need to be replaced within the next 10 years.[61]

The cost associated with financing grid connection and updating transmission lines to allow for increased capacity from new renewable energy projects are, in Victoria, currently borne by the generator wanting to be connected. The National Electricity Regulator has not made it clear who should pay for the augmentations needed to facilitate network connection in Victoria.[62] Service and price levels for new investments into transmission services and for transmission revenue are set by the AER for five years following applications from the transmission businesses.[63]

---

58 *National Electricity (South Australia) Act* 1996, (Schedule) – National Electricity Law s 7.
59 See A Kallies, 'Impact of Australian Market Design on Access to the Grid and Transmission Planning for Renewable Energy in Australia: Can Overseas Examples Provide Guidance?' (2011) *Renewable Energy Law and Policy Review* 147.
60 See the Australian Energy Regulator Report, *State of the Energy Market* (2009) 140 https://www.aer.gov.au/node/6313 which notes that there are 'physical limits on the amount of power that can flow over any one part or region of the network'.
61 This was the conclusion of the Victorian Government, 2009 Victorian Bushfires Royal Commission, *Final Report* (2009) vol 11 [4.3].
62 AEMO, *Victorian Electricity Transmission Network Connection Augmentation Guidelines* (Issue 2, March 2007) 4.
63 Ibid.

It is also likely that any future grid upgrade will incorporate the new technology associated with micro-grids: this means that instead of having one centralised generating station serving as the supply centre, smaller electricity grids, with access to all the essential assets of a larger grid – such as generators, transmission lines, substations, and switchgear – will be utilised. The significant advantage of the micro-grid lies in the fact that in the event of a catastrophic failure, the electricity framework will not be fully disabled.[64]

6.11 Wind energy: Regulatory and practical issues

In addition to the existing economic impediments associated with the progression of the renewable energy sector in Australia, there are many practical barriers. Technology advancements and breakthroughs have mitigated many of these concerns; however, social licensing and community engagement issues remain an important concern. Of particular interest in this regard are the problems experienced with the progression of the wind industry.

The wind energy industry has been the fastest growing renewable energy source in Australia; this is largely because it is a proven technology and has relatively low operating costs and environmental impacts. A wind farm is constituted by a group of wind turbines that are located in the same area and utilised to produce electric power. A large wind farm may consist of several hundred individual wind turbines and will have the capacity to cover hundreds of square kilometres across an extended area. This area does not need to be completely devoted to the wind turbines, as the land between those turbines may be used for agricultural or other purposes. Wind turbines are also increasing in size and can be up to 150 m in diameter.

Wind farms may be located on or offshore. Offshore winds can also be quite powerful, so a lot of energy is generated; however, these farms can be expensive to construct. Additional costs come from infrastructure, maintenance, and oversight. Offshore wind farms are more commonly located in Europe where there are 55 offshore wind farms across 10 European countries that generate energy for five million homes.[65] In Australia, there are currently no offshore wind farms.

---

64 See S Suryanarayanan and E Kyriakides, *Microgrids: An Emerging Technology to Enhance Power System Reliability* (2012) http://smartgrid.ieee.org/march-2012/527-microgrids-an-emerging-technology-to-enhance-power-system-reliability.

65 See European Wind Energy Association (EWEA), *Deep Water: The Next Step for Offshore Wind Energy* (2013) http://www.ewea.org/fileadmin/files/library/publications/reports/Deep_Water.pdf.

As of 2009, however, there were 85 onshore wind farms: 57 of which were in Victoria, South Australia, and Western Australia (19 in each state). The capacity of all these installations amounted to 1703 MW, with 48 per cent of total capacity in South Australia.[66] Significant future growth in the solar and wind energy sector is predicted. The Bureau of Resources and Energy Economics' (BREE) Australian Technology Energy Assessment, which was created in 2012, predicted that solar photovoltaics and onshore wind will have the lowest levelised cost of electricity of all of the renewable options in Australia leading up to 2030. There are a number of reasons for this. First, technology associated with solar photovoltaic has significantly reduced in cost as a result of strong global production. Further, wind and solar energy have a much greater capacity to respond to off-grid power stations.[67]

Wind energy accounts for almost one-quarter of Australia's clean energy generation (22.9 per cent). In 2011, approximately 5000 GWh of electricity (powering over 700 000 homes) was generated by 1052 wind turbines across 52 operating wind farms.

This means that wind energy is the primary form of renewable energy in Australia and is therefore directly affected by the RET framework. As outlined by the Clean Energy Council:

> Wind power is likely to be the dominant technology during the early years of the [RET]. It is currently the least expensive form of renewable energy and has a proven track record of being rolled out on a large scale.[68]

Each state and territory is constitutionally responsible for energy matters that come within its own jurisdiction. National energy policy is mainly implemented at the state and territory level through existing planning systems. The planning systems in each state, as they pertain to wind farms, are largely similar. All require the provision of detailed environmental assessments of wind farm proposals, all require a public consultation process to be undertaken, and all contain provision for public submissions to be made on the proposal.

The system in New South Wales provides a good example. Like most states, NSW does not have a specific regulatory framework dealing with wind farm proposals. Rather, proposals are assessed pursuant to a range of environmental planning Acts. In NSW these are: the *Environmental Planning and Assessment Act 1979* (NSW) ('the

---

66 See Senate Community Affairs References Committee, *The Social and Economic Impact of Rural Wind Farms* (2011) 2 http://www.pacifichydro.com.au/files/2012/06/Senate-Enquiry-The-Social-and-Economic-Impacts-of-Rural-Wind-Farms-report.pdf.
67 See the report by AECOM for Austrade, *Australian Remote Renewables: Opportunities for Investment* (2013) 5. The report noted at p.6 that it is 'currently cheaper to source electricity from a hybridised solar photovoltaic-diesel or wind turbine-diesel system than with existing diesel-fired generation alone in many off-grid regions of Australia'.
68 Clean Energy Council, *Clean Energy Australia 2010* (December 2010) 8.

EPAA'); *Environmental Planning and Assessment Regulation 2000*; State Environmental Planning Policies (SEPPs); and Local Environmental Plans (LEPs). Renewable energy proposals are now considered under pts 3A, 4 and 5 of the *EPAA* (NSW).[69]

Part 3A of the *EPAA* sets out the planning approvals regime for major infrastructure and other projects, including 'critical infrastructure' projects. Renewable energy proposals with a capital cost of more than $30 million (or $5 million in an environmentally sensitive area of state significance) are considered a major project. 'Critical infrastructure' projects are a type of major project deemed by the Minister for planning to be essential to the state for economic, social, or environmental reasons. The NSW Department of Planning regards only a minority of major projects covered by pt 3A of the Act to constitute 'critical infrastructure'. Renewable energy projects with the capacity to produce at least 30 MW of electricity are classified as 'critical infrastructure' and this means that a large number of wind farm proposals must be formally assessed by the Minister for Planning rather than individual councils.

A similar situation exists under the regulatory framework in Victoria, which, in 2014, had 12 operational wind farms, with wind energy comprising 40 per cent of renewable energy production. Until 2010, the Planning Minister was the sole responsible authority for wind farm proposals where the capacity would exceed 30 MW. However, following the *Inquiry into the Approvals Process for Renewable Energy Projects in Victoria* by the Victorian Environment and Natural Resources Committee, this process was found to be outdated, complex, and time-consuming. Consequently, municipal councils were given increased responsibilities with respect to wind farm proposals. The primary goal was to speed up the process given that, previously, renewable energy companies had been waiting between eight months and four-and-a-half years for projects to be approved, with an average waiting time of almost one-and-a-half years and the possibility of waiting for up to two-and-a-half years.

In NSW, pursuant to pt 4 of the *EPAA,* which deals with small-scale developments, local government councils are the responsible authority for all wind farm approvals. The relevant SEPP is the *State Environment Planning Policy (Infrastructure) 2007* (NSW), which defines small wind turbines (a wind turbine with a generating capacity of less than 100 kW), small wind turbine systems (a system comprising one or more small wind turbines each of which feed into the same grid or battery bank), and wind monitoring towers. A council's LEP and/or DCP might also contain additional planning controls.

The Commonwealth Government is responsible for particular aspects of wind farm development, particularly air safety. The reason for this is because the development of wind projects may impact upon a matter of national environmental

---

69 See the outline for renewable energy proposals in NSW by the NSW Department of Planning and Environment, *Renewable Energy* (September 2014) <http://www.planning.nsw.gov.au/StrategicPlanning/RenewableEnergy/tabid/394/language/en-US/Default.aspx>. Part 5 of the *EPAA* (NSW) sets out environmental assessment requirements.

significance under the *Environment Protection and Biodiversity Conservation Act 1999* (Cth) ('the *EPBCA*') where, for example, a wind farm project has the potential to affect world heritage properties, national heritage properties, Ramsar Convention wetlands of international importance, listed threatened species and communities, protected migratory species, or the Commonwealth marine environment. Pursuant to the *EPBCA*, where a matter is referred to it, the responsible minister may declare the proposed development of a wind farm constitute to a controlled action. The *EPBCA* has issued a specific policy statement regarding when wind farm proponents may need to refer the project to the Commonwealth Minister in accordance with the *EPBCA* provisions. Policy Statement 2.3 states:

Indirect and off-site impacts

When considering whether an action requires approval under the EPBC Act the Minister must consider all adverse impacts resulting, either directly or indirectly, from the action, regardless of whether the impacts are within the control of the person proposing to take the action. Consequently, when undertaking an assessment of the potential impacts, all potential adverse impacts must be considered, including indirect and offsite impacts. Indirect and off-site impacts include:

- 'downstream' or 'downwind' impacts, such as impacts on wetlands or ocean reefs from sediment washed or discharged into river or creek systems, or disturbance of fauna off-site by noise or blade glint
- 'upstream' impacts, such as those associated with the production of energy used to undertake the action, and
- 'facilitated' impacts which result from further actions which are made possible or facilitated by the action, such as the installation of power lines, access roads or power stations.[70]

The *EPBCA* will generally apply to a wind farm project where the development impacts upon the mortality of birds and bat species as a consequence of collision with the wind turbines.[71] Other impacts include the clearance or disturbance of native vegetation for wind turbines as well as direct and indirect impacts upon World Heritage or National Heritage properties. Where a matter is referred to the Minister pursuant to the *EPBCA*, the Minister retains discretion to determine whether and how the project should proceed.[72] Further, where a proposal for a wind farm includes development on Crown Land, the provisions of the *Native Title Act 1993* (Cth) will

---

70 Department of Environment, Water, Heritage and the Arts, *Environment Protection Biodiversity Conservation Act 1999: Policy Statement 2.3: Wind Farm Industry* (2009) http://www.environment.gov.au/resource/epbc-act-policy-statement-23-wind-farm-industry.
71 Ibid 9.
72 The environmental review process relevant to the *EPBCA* (Cth) is discussed in greater detail in Chapter 9.

continue to apply.[73] Where a proposed wind farm project does impact land, which is either subject to established native title claims, or land which is registered on the Register of Native Title Claims, an Indigenous land use agreement will generally need to be entered into. This occurred in Hughenden, a town which is 376 km southwest of Townsville and 519 km east of Mt Isa, where Windlab Development and the Yirendali people entered into an Indigenous land use agreement in 2011. This agreement authorised the creation of the Kennedy wind farm – which consisted of 300 turbines and a capacity of 750 MW – within an area subject to registered native title claims. The Kennedy wind farm is now Queensland's largest renewable energy project.[74]

The Commonwealth Government may also become involved in the development of wind farms through the Council of Australian Governments (COAG). The Environment Protection and Heritage Council (EPHC) of COAG has released draft *National Wind Farm Development Guidelines* which seek to complement and add to existing planning and development processes at the state level.[75] These guidelines are best practice rather than mandatory and follow on from an Environmental Protection and Heritage Council Report, issued in 2012 in response to growing community concerns, which examined some of the impediments to responsible wind farm development.

Best Practice Guidelines for Implementation of Wind Energy Projects in Australia

2.4 Community consultation

The proponent should provide accurate and timely information to the public and community groups regarding the proposed development. A key feature of a well planned project is the provision of opportunities for timely and meaningful public involvement.[76]

There will inevitably be a diverse range of public attitudes towards any wind energy development, and these views should be considered in the design and development of wind energy projects. Community and other stakeholder consultation should continue throughout the life of the wind farm until it is decommissioned. Wind companies recognise that community engagement is one of the foundations for the success of wind farms, therefore separate community engagement guidelines, known as the Community Engagement Guidelines for the Australian Wind Industry (2012), have been prepared. These Guidelines are available on the CEC website and the environmental considerations are extracted below:

---

73 See *Native Title Act 1993* (Cth) s 34EB(1)(b)(i).
74 *Windlab Development Pty Ltd and Yirendali People Indigenous Land Use Agreement (ILUA)* (7 February 2012) http://www.atns.net.au/agreement.asp?EntityID=5672.
75 See Clean Energy Council, *Best Practice Guidelines for the Implementation of Wind Energy Projects in Australia* (2013): http://www.cleanenergycouncil.org.au/technologies/wind-energy/best-practice-guidelines.html.
76 Ibid 12.

3.3.2 Environmental considerations

Consultation with local or state authorities during feasibility studies will have identified the scope for environmental assessment required to progress a planning application. Where authorities consider that the proposed wind farm could have significant effects on the environment or amenity due to factors such as its nature, size or location, a proponent will be required to submit a formal environmental assessment. In some cases, proponents may be required to carry out a full Environmental Impact Assessment (EIA) or equivalent, depending on the state environmental and/or planning laws. This process may be overseen by either a local or state authority (refer to the relevant state regulatory authority website to understand the framework for an EIA in that jurisdiction). Risk assessment and management via a well structured risk management framework (i.e. AS/NZS 31000) will prepare the proponent for subsequent legislated assessment processes.

An impact assessment will be required to identify all relevant environmental, social and economic effects associated with the proposal. It should be noted that during the course of detailed evaluation of these issues, it may be necessary for the proponent to amend the proposed wind project design, including the number and position of wind turbines. In addition the whole of the development life needs to be considered, from construction right through operation to decommissioning and rehabilitation or reuse.

Typically, the range of issues to be investigated for the development of a large wind farm will include the following:

- **Landscape and visual assessment**

The existing landscape must be described, and the potential landscape and visual impact of the proposed wind farm assessed and evaluated. It is important that visual amenity is always considered in the context of the existing environment, particularly regarding the value that the local community puts on landscape character and attributes. Further detail of visual assessment is provided in Appendix 5.

- **Noise assessment**

A noise assessment must demonstrate that the operational wind farm (including substation) is able to comply with the relevant noise guidelines in the location where it is to be constructed. The assessment should provide details of the methods used to model predicted noise outputs. More detailed information on wind farm operation noise assessment is provided in Appendix 6.

- **Shadow flicker**

The predicted duration of shadow flicker at all nearby houses (relevant receivers) must be determined and assessed against relevant guidance. More detail on the assessment of shadow flicker is provided in Appendix 5.

- **Flora and fauna**

The impact of the wind farm on the ecological values present within the disturbance footprint of the development should be assessed. Ecological values may include flora, fauna (including those that may utilise the site only intermittently), vegetation or fauna habitat. More information on the assessment of ecological values is provided in Appendix 7.

- **Socio-economic**

The impact of the proposed wind farm on local infrastructure, such as health and emergency services, accommodation and community facilities, should be addressed. In addition, an assessment of the economic impact of the proposal on the local and regional economy may also be completed.

- **Heritage assessment**

An assessment of the heritage values present at site should be completed. Heritage may include Aboriginal sites or artefacts and sites of historic heritage. Consultation with the Aboriginal community should be undertaken during Aboriginal heritage assessments and may include engaging endorsed Aboriginal representatives to accompany the field surveys. Further discussion on how such studies might be approached is given in Appendix 2.

- **Transport impact assessment**

An assessment of the type and volume (number of movements per day) of traffic associated with the construction and operation of the wind farm should be completed. The assessment should include consideration of the potential impacts on the local and regional road network and any modifications to the road network that may be required (e.g. widening). The assessment should be undertaken in consultation with the relevant state transport department and local authority. In some cases a program of road maintenance or improvement may be agreed with the local authority to address any potential impacts caused by the movement of heavy vehicles on local roads.

- **Electromagnetic interference assessment**

An assessment of the potential for communication and/or radar interference should be completed and should include discussion with communication system providers. A more detailed discussion of electromagnetic interference is provided in Appendix 9.

- **Aircraft safety assessment**

An assessment of aircraft safety should be completed. Further information on aircraft safety including consultation with the Civil Aviation Safety Authority and other aviation agencies is provided in the Appendix 4.

- **Hydrological assessment**

An assessment of the impact of the proposed development on nearby surface and ground water systems may be appropriate dependent on the location of the wind farm. The assessment should include the potential impacts of erosion and sedimentation of nearby water courses (including potential impacts on riparian vegetation) and potential contamination of groundwater.

- **Emergency and incident management**

The planned management of potential emergencies and incidents should be addressed. Development of emergency response plans should be undertaken in consultation with relevant local and regional emergency services and any contractors working on the site, if a contractor is to manage the site they may develop the emergency response plan although it is recommended that this is reviewed by the proponent. More guidance on the management of fire is provided Appendix 8.

- **Cumulative impacts**

Consideration of the cumulative impacts of the wind farm together with other development in the area may also be appropriate although this can be difficult in practice. Cumulative impacts can refer to landscape and visual effects, and a wide range of other environmental, social and economic impacts, both positive and negative. While many regulators require the assessment of cumulative impacts as part of a development application, few give guidance on how this should be undertaken. It can also be difficult for proponents to access information on other developments in the area. The best approach is to understand as far as possible how various impacts may theoretically change with the addition of another wind development in the area, and proactively discuss with regulators to work towards a positive outcome.

These studies will need to investigate the potential impact of the development (based on a common layout and project description), identify opportunities to avoid or minimise the impacts and specify mitigation measures or offsets for any residual impacts.

The mitigation measures will provide the basis for the environmental management framework (see Appendix 11), to be provided as part of the development application, which will articulate the translation of impacts into mitigation through design, through construction or through operation or identify offsets. At this stage, dependent upon the issue and the potential need for a management plan to be lodged as part of a secondary consent process (such as a Construction Environmental Management Plan, or CEMP), the mitigation measures should look at the performance outcomes rather than detail the method of mitigation. This is where consultation with the regulators can assist to determine the level of detail that they are expecting.

On the completion of the detailed assessment the proponent should be in a position to submit a development application and, if required, a report on the Environmental Impact Assessment (EIA). The EIA will provide:

- a source of information from which individuals and groups may gain an understanding of the proposal, the need for the proposal, the potential environmental impacts and the measures taken to mitigate any adverse effects
- a base for public consultation and informed comment on the project
- a framework against which decision makers can consider the proposal and determine the conditions under which any approval may be given.

Discussion of the assessment processes for many of the environmental issues relevant to wind energy projects are provided in other appendices. From these assessments, a final project design will result. Typically with wind farms, a degree of flexibility will be sought, particularly around micro-siting the turbines. Commonly this is achieved by specifying a development area and cable corridor, within which the development may occur with a final location to be determined following detailed design.[77]

The Victorian Government has introduced amendments to the planning controls that imposed strict noise conditions and prohibited the development of wind farms within 2 km of an existing dwelling in the absence of written consent. These amendments were designed to prohibit the development of wind farms in areas of high conservation and landscape values, including national and state parks and locations featuring a high degree of amenity, environmental value, or significant tourist destinations, such as the Yarra Valley and Dandenong Ranges, Mornington Peninsula, Bellarine Peninsula, Macedon and McHarg Ranges, Bass Coast, and the Great Ocean Road region. Prohibitions have also been placed on locations identified for future urban growth, including land in the Urban Growth Zone and designated regional population corridors specified in the Regional Victoria Settlement Framework Plan in the State Planning Policy Framework.[78] This gives the owners of any dwelling within 2 km of a proposed wind farm the power to decide whether or not the wind farm development should proceed. A July 2012 amendment clarifies that these changes to the planning controls are targeted at wind farms generating electricity for supply to the grid, not for on-site use. The amendments do not impact upon wind farms which are already operational; however, they do apply to any amendments made to permits issued after 15 March 2012.

One of the most significant concerns associated with the development of wind farms in Australia has been the social and community apprehension regarding possible health impacts, particularly those connected with low-frequency noise and vibrations. As outlined by the Committee for Rural Affairs:

> Wind farms introduce into rural environments sounds, and levels of sound, that were not present in the environment before the advent of the wind farms. These sounds may be perceived as intrusive and detrimental to the amenity of people affected by them. The Sustainable Energy Association of Australia has stated that '… much of the significance of this issue…appears to arise from a change in the noise environment and this change has had some amenity impact'.[79]

The Committee recommended that noise standards be adopted by all states and territories and that the planning and operation of rural wind farms include appropriate

---

[77] See Clean Energy Council, *Best Practice Guidelines for the Implementation of Wind Energy Projects in Australia* (2013) 3.3.2.
[78] See Victoria Planning Provisions: ss 78, 82 and 91.
[79] See Community Affairs Reference Committee, *The Social and Economic Impact of Rural Wind Farms* (June 2011) [2.18].

measures to calculate the impact of low frequency noise and vibrations indoors at impacted dwellings. This recommendation has produced a range of changes to regulatory controls.

In New South Wales, Planning Guidelines for Wind Farms were proposed in 2011. These guidelines, inter alia, mandated noise control, which was defined to include a 'single, repeated or sustained exceedance', to be 'properly' controlled.[80] Where a particular 'exceedance', is identified, it may become necessary for the proponent to specifically outline the meteorological conditions pursuant to which the exceedance has occurred and to take all reasonable and feasible measures to mitigate that exceedance. Measures may include sector management to eliminate the occurrence of exceedances where they are a product of meteorological conditions and/or individual negotiation with the affected resident/s. Where a compliance issue may not be resolved in the short term, the draft guidelines authorise the cessation of wind farm operations until the regulator is fully satisfied that acceptable standards have been demonstrated.[81]

Some of these issues were explored by the Victorian Civil and Administrative Tribunal (VCAT) in the decision of *Cherry Tree Wind Farm Pty Ltd v Mitchell Shire Council*.[82]

> The Cherry Tree Wind Farm Pty Ltd proposed the development of a 16-turbine wind farm along the ridge of the Cherry Tree Range, just out of Seymour, Victoria. This proposal was contested by a number of parties: the Mitchell Shire Council, a number of local residents who formed a group known as the 'Trawool Valley and Whiteheads Creek Landscape Guardians', individual residents, and the Waubra Foundation (an organisation whose main objective is to investigate health problems experienced by people living in proximity to wind farms). The parties objected to the proposal on the basis that: it was contrary to the Mitchell Planning Scheme because of its visual impact and because of its effect on local fauna, flora, and the habitat of wildlife. Further, the parties objected on the basis of health and included a detailed submission regarding the alleged adverse health implications for residents living within a reasonably close proximity to the proposed development.

---

[80] NSW Government, Department of Planning and Infrastructure, *Draft NSW Planning Guidelines for Wind Farms* (2011) 34–35 http://www.planning.nsw.gov.au/Portals/0/PolicyAndLegislation/NSW_Wind_Farm_Guidelines_Web_Dec2011.pdf.

[81] Ibid 50 where the guidelines propose that compliance and enforcement of breaches be in accordance with the policy guidelines relevant to the Department of Planning and Infrastructure. The Department may: conduct inspections and audits of approved projects, respond to reports and complaints received from local councils, members of the public and other state agencies, investigate potential breaches and carry out enforcement action where breaches are confirmed.

[82] [2013] VCAT 521 ('*Cherry Tree*').

> **Held**:
> - The Mitchell Planning Scheme did not prohibit the proposed development.
> - The visual impact of the proposal will not be unacceptable.
> - The development will, to a large extent, be screened from the public realm by the roadside vegetation along the Goulburn Valley Highway.
> - The proposal will not have an unacceptable impact on local flora and fauna.
> - The wind farm will not materially affect wildlife habitat and the removal of vegetation need not be permanent.
> - The proposed development will comply with the noise standard prescribed by the Planning Scheme as a decision guideline.
> - The development will not give rise to problems relating to bushfire, salinity, erosion or aviation.
> - Issue of Health Impacts: Tribunal found a paucity of scientific evidence and adjourned for six months until the SA Report on Health Effects of Wind Farms was released.

The conclusions of VCAT in the first *Cherry Tree* determination deferred a full decision on the health impacts of wind farms until the release of further reports aimed at fully reviewing the health implications of wind farms. In the subsequent *Cherry Tree* decision,[83] the Tribunal made it clear that the role of a planning decision-maker was not to set standards in relation to public health or to second guess the considered statements of bodies or authorities which are expert in the area and which carry a statutory responsibility for regulating the area. The Tribunal therefore accepted the statements of health authorities and the conclusions of the issued reports that indicated that the 2 km buffer zone, which had been imposed by the Mitchell Planning Scheme, was in accordance with the requirements of the precautionary principle and constituted a sufficient safety measure.[84] Whilst the Tribunal did not make any specific comments regarding the report, the following observations were set out:

---
83 *Cherry Tree Wind Farm Pty Ltd v Mitchell Shire Council* [2013] VCAT 1939 ('*Second Cherry Tree*').
84 See Victorian Department of Health, *Wind Farms, Sound and Health* (2013) <http://docs.health.vic.gov.au/docs/doc/03C56A16FC34F658CA257B5E00164599/$FILE/1212016_wind_turbine_community_WEB.pdf> which concluded at p.14 that: 'The evidence indicates that sound can only affect health at sound levels that are loud enough to be easily audible. This means that if you cannot hear a sound, there is no known way that it can affect health'. See also the report issued by the National Health and Medical Research Council, *Wind Turbines and Health—A Rapid Review of the Evidence* (2010) http://www.nhmrc.gov.au/_files_nhmrc/publications/attachments/new0048_evidence_review_wind_turbines_and_health.pdf. The report concluded at p.2 that: 'There are no direct pathological effects from wind farms and that any potential impacts on humans can be minimised by following existing planning guidelines'.

The Tribunal has no doubt that some people who live close to a wind turbine experience adverse health effects, including sleep disturbance. The current state of scientific opinion is that there is no causal link of a physiological nature between these adverse health effects and the operation of the wind turbine. The totality of material before the Tribunal suggests, but does not conclusively prove, that adverse health effects are suffered by only a small proportion of the population that surround a wind farm.

Objections to the development of wind farms have also been raised extensively in other states. For example, in *King v Minister for Planning; Parkesbourne-Mummel Landscape Guardians v Minister for Planning; Gullen Range Wind Farm Pty Ltd v Minister for Planning*, the New South Wales Land and Environment Court held that despite the variety of objections raised to the development of a wind farm, the broader social desirability of promoting this type of renewable energy project should not be ignored.[85] The Court concluded that the various objections to the proposal, including one where it was argued that wind farms constituted a 'cumulative visual impact', should be rejected.[86] The Court ultimately held that there was no visual constraint, nor any 'shadow flicker or unacceptable noise on the public domain' imposed as a result of the operational wind turbines.[87]

6.12 REVIEW QUESTIONS

1. What is the definition of a renewable form of energy and how does renewable energy differ from non-renewable energy?

2. Why are access entitlements so important for the development of renewable energy projects?

3. What is the core aim of the Renewable Energy Target and why do you think it might be important to sustain this target in the absence of a carbon tax or emission trading scheme?

4. Explain the rationale in the Warburton Review underlying the conclusion that the RET was a 'high cost' approach to reducing emissions?

5. What are the arguments for and against retaining the RET as a means of promoting the renewable energy sector in Australia?

6. Consider the following problem:

    Origin Energy ('Origin') has a RET liability per year which it meets by purchasing LGCs from a range of eligible renewable energy companies. Origin then acquits

---

85 [2010] NSWLEC 1102
86 [2010] NSWLEC 1102, [649]
87 Ibid.

the annual LGCs to the regulator. In 2013, Origin did not surrender the required number of LGCs and is uncertain what to do. Origin comes to you for advice.

7. What are the primary elements of the Small Scale Renewable Energy Scheme?
8. Why is wind energy an important component of the renewable energy sector in Australia?
9. What are some of the impediments to the progression of wind farms in Australia?
10. Consider the following problem:

    In *Cherry Tree Wind Farm Pty Ltd v Mitchell Shire Council* [2013] VCAT 521, the Tribunal concluded that wind farms were consistent with sustainability objectives and that health impacts were not properly established. Local farmers in central Victoria have argued, however, that any proposal to establish a wind farm will cause medical issues and therefore should be automatically rejected on these grounds. To what extent does the decision in *Cherry Tree* suggest that the broader social benefits associated with the promotion of renewable energy projects outweigh unsubstantiated allegations of health risks?

6.13 FURTHER READING

ABARE, *Australian Energy Predictions to 2030* (2010).

ABC Radio National, 'Renewable Energy Reform Pacific Hydro', *Breakfast*, 1 March 2010 (F Kelly).

ABC Radio National, 'Renewable Energy Reform – Keppel Prince Engineering', *Breakfast*, 1 March 2010 (F Kelly).

ABC Television, 'Renewable Energy on the Brink', *7:30 Report*, 8 December 2009 (P McCutcheon).

AECOM for Austrade, *Australian Remote Renewables: Opportunities for Investment* (2013).

AEMC, *Review of Energy Market Frameworks in light of Climate Change Policies: Final Report* (30 September 2009).

AEMO, *Victorian Electricity Transmission Network Connection Augmentation Guidelines* (Issue 2, March 2007).

Apple, *Environmental Responsibility* (2014) http://apple.apple.com/au/environment/.

Australian Energy Regulator Report, *State of the Energy Market* (2009) https://www.aer.gov.au/node/6313.

Australian Government, Department of Resources, Energy and Tourism, Geoscience Australia and Australian Bureau of Agricultural and Resource Economics (ABARE), *Australian Energy Resource Assessment Report* (2010) http://www.ga.gov.au/image_cache/GA16725.pdf.

A J Bradbrook, 'Australian and American Perspectives on the Protection of Solar and Wind Access' (1988) 28 *Natural Resources Journal* 229.

A J Bradbrook, 'Government Initiatives Promoting Renewable Energy for Electricity Generation in Australia' (2002) 25 *University of New South Wales Law Journal* 124.

A J Bradbrook, 'Creating Law for the Next Generation Energy Technologies' (2011) 2 *George Washington Journal of Energy and Environmental Law* 17.

A L Carleton, 'Mandating Market Access for Renewable Energies in Australia' (2008) 26 *Journal of Energy and Natural Resources Law* 402.

P Christoff, 'Aiming High: On Australia's Emissions Reductions Target' (2008) 31 *University of New South Wales Law Journal* 861.

Clean Energy Council, *Clean Energy Australia 2010* (December 2010).

Clean Energy Council, *Best Practice Guidelines for the Implementation of Wind Energy Projects in Australia* (2013): http://www.cleanenergycouncil.org.au/technologies/wind-energy/best-practice-guidelines.html.

Clean Energy Council, Roam Consulting, *Renewable Energy Target policy analysis* (23 May 2014) http://www.cleanenergycouncil.org.au/policy-advocacy/renewable-energy-target/ret-policy-analysis.html.

Clean Energy Regulator, *About the Clean Energy Regulation: What we do* (2014) http://www.cleanenergyregulator.gov.au/About-us/our-work/Pages/default.aspx.

Clean Energy Regulator, Renewable Energy Target, *Large-scale Renewable Energy Target* (2014) http://ret.cleanenergyregulator.gov.au/About-the-Schemes/lret.

Clean Energy Regulator, Renewable Energy Target, *How the Renewable Energy Target Works* (2014) http://ret.cleanenergyregulator.gov.au/About-the-Schemes/How-the-RET-works/How-the-Renewable-Energy-Target-works.

Clean Energy Regulator, Renewable Energy Target, *The Small-scale Renewable Energy Scheme* (2014) <http://ret.cleanenergyregulator.gov.au/About-the-Schemes/sres>.

Climate Bonds Initiative, *Bonds and Climate Change: The State of the Market in 2013* (2013) http://www.ieefa.org/wp-content/uploads/2014/02/Thermal-Coal-Outlook-February-2014-Observations-IEEFA.pdf.

Climate Works, *Pathways to Deep Carbonisation in 2050: How Australia Can Prosper in a Low Carbon World* (September 2014) <http://www.climateworksaustralia.org/project/current-project/pathways-deep-decarbonisation-2050-how-australia-can-prosper-low-carbon>.

CSIRO, Change and Choice: The Future Grid Forum's Analysis of Australia's potential electricity pathways to 2050 (December 2013).

G Czisch, *Global Renewable Energy Potential–Approaches to its Use* (2001) http://www.iset.uni-kassel.de/abt/w3-w/folien/magdeb030901/.

B Deninger, 'Twenty-First Century Offshore Wind Boom: Why Texas is Leading the Way' (2014) 44 *Texas Environmental Law Journal* 81.

Department of Environment, Water, Heritage and the Arts, *Environment Protection Biodiversity Conservation Act 1999: Policy Statement 2.3: Wind Farm Industry* (2009) http://www.environment.gov.au/resource/epbc-act-policy-statement-23-wind-farm-industry.

Emission Reduction Fund, *White Paper* (April 2014) http://www.environment.gov.au/system/files/resources/1f98a924-5946-404c-9510-d440304280f1/files/emissions-reduction-fund-white-paper_0.pdf.

European Wind Energy Association (EWEA), *Deep Water: The Next Step for Offshore Wind Energy* (2013) http://www.ewea.org/fileadmin/files/library/publications/reports/Deep_Water.pdf.

F A Felder, 'Climate Change Mitigation and the Global Energy System' (2014) 25 *Villanova Environmental Law Journal* 89.

J P Fershee, 'Energy Subsidies' in K Bosselman et al eds, 'The Law and Politics of Sustainability' *Berkshire Encyclopaedia of Sustainability* (Berkshire Publishing Group 2012).

J P Fershee, 'Promoting an All of the Above Approach or Pushing (Oil) Addiction and Abuse: The Curious Role of Energy Subsidies and Mandates in U.S. Energy Policy' (2012) 7 *Environmental Energy and Policy Law Journal* 125.

A C Fink, 'Securitise me: Stimulating Renewable Energy Financing by Embracing the Capital Markets' (2014) 12 *University of New Hampshire Law Review* 109.

E T Freyfogle, *Natural Resources Law* (Thomson/West, 2007).

R Garnaut, *Garnaut Climate Change Review, Update 2011: Australia in the Global Response to Climate Change* (2011) http://www.garnautreview.org.au/update-2011/garnaut-review-2011.html.

C Gottlieb, 'Regional Land Use Planning: A Collaborative Solution for the Conservation of Natural Resources' (2014) 29 *Journal of Environmental Law and Litigation* 35.

Intergovernmental Panel on Climate Change, *Contribution of Working Groups I, II and III to the Fourth Assessment Report of the Intergovernmental Panel on Climate Change* (Intergovernmental Panel on Climate Change, 2007).

A Kallies, 'Impact of Australian Market Design on Access to the Grid and Transmission Planning for Renewable Energy in Australia: Can Overseas Examples Provide Guidance?' (2011) *Renewable Energy Law and Policy Review* 147.

A Kent and D Mercer, 'Australia's Mandatory Renewable Energy Target (MRET): An Assessment' (2006) 34 *Energy Policy* 1046.

A B Klass, 'Property Rights on the New Frontier: Climate Change, Natural Resource Development, and Renewable Energy' (2011) 38 *Ecology Law Quarterly* 63.

D J Kochan and T Grant, 'In the Heat of the Law, It's Not Just Steam: Geothermal Resources and The Impacts on Thermophile Biodiversity' (2007) 13 *Hastings Journal Environmental Law & Policy* 35.

C Lagarde, 'A New Multilateralism for the 21st Century' (Speech delivered at The Richard Dimbleby Lecture, London, 3 February 2014).

M McDonnell, K Engel and A Barnhart, 'The Potential and Power of Renewable Energy Credits to Enhance Air Quality and Economic Development in Arizona' (2011) 43 *Arizona State Law Journal* 809.

N Mee, 'Here Comes the Sun: Solar Power Parity with Fossil Fuels' (2012) 36 *William and Mary Environmental Law and Policy Review* 119.

National Health and Medical Research Council, *Wind Turbines and Health—A Rapid Review of the Evidence* (2010) http://www.nhmrc.gov.au/_files_nhmrc/publications/attachments/new0048_evidence_review_wind_turbines_and_health.pdf.

NSW Government, Department of Planning and Infrastructure, *Draft NSW Planning Guidelines for Wind Farms* (2011) http://www.planning.nsw.gov.au/Portals/0/PolicyAndLegislation/NSW_Wind_Farm_Guidelines_Web_Dec2011.pdf.

NSW Department of Planning and Environment, *Renewable Energy* (September 2014) <http://www.planning.nsw.gov.au/StrategicPlanning/RenewableEnergy/tabid/394/language/en-US/Default.aspx>.

P K Oniemola, 'Case for the Promotion of Renewable Energy Through the Use of Economic Instruments' (2011) 1 *Dublin Legal Review Quarterly* 34.

S J Panarella, 'For the Birds: Wind Energy, Dead Eagles and Unwelcome Surprises' (2014) 20 *Hastings Journal of Environmental Law and Policy* 3.

G Parkinson, 'Citigroup says the "Age of Renewables" has begun' (27 March 2014) *RenewEconomy*.

G Parkinson, 'Fossil fuels face $30 trillion losses from climate, renewables', (28 April 2014) *RenewEconomy*.

A K Prasad, 'Alternative Energy Technologies for Transportation' (2014) 25 *Villanova Environmental Law Journal* 107.

J Prentice, 'Making Effective Use of Australia's Natural Resources – The Record of Australian Renewable Energy Law Under the Renewable Energy (Electricity) Act 2000 (Cth)'(2011) *Renewable Energy Law and Policy Review* 5.

J Prest, 'A Dangerous Obsession with Least Cost? Climate Change, Renewable Energy Law and Emission Trading' in D Gumley and D Winterbottom (eds) *Climate Change Law: Comparative, Contractual And Regulatory Considerations* (Thomson Reuters, 2009).

Renewable Energy Target Scheme, *Report of the Expert Panel* (August 2014) https://retreview.dpmc.gov.au/sites/default/files/files/RET_Review_Report.pdf.

E Rotenberg, 'Energy Efficiency in Regulated and De-Regulated Markets' (2006) 24 *UCLA Journal of Environmental Law and Policy* 259.

P Salkin, 'Key to Unlocking the Power of Small-Scale Renewable Energy: Local Land Use Regulation' (2012) 27 *Journal Of Land Use and Environmental Law* 339.

Senate Community Affairs References Committee, *The Social and Economic Impact of Rural Wind Farms* (2011) http://www.pacifichydro.com.au/files/2012/06/Senate-Enquiry-The-Social-and-Economic-Impacts-of-Rural-Wind-Farms-report.pdf.

Sir N Stern, *Review on the Economics of Climate Change* (Cambridge University Press, 2007).

C R Sunstein and L A Reisch, 'Automatically Green: Behavioural Economics and Environmental Protection' (2014) 38 *Harvard Journal of Environmental Review* 127.

S Suryanarayanan and E Kyriakides, *Microgrids: An Emerging Technology to Enhance Power System Reliability* (2012) http://smartgrid.ieee.org/march-2012/527-microgrids-an-emerging-technology-to-enhance-power-system-reliability.

Tasmania Law Reform Institute, *Law of Easements in Tasmania*, Final Report No 12 (March 2010) http://www.utas.edu.au/__data/assets/pdf_file/0011/283835/EasementsFinalReportA4.pdf.

Victoria Law Reform Commission, *Easements and Covenants*, Report No 41 (1992).

Victorian Department of Health, *Wind Farms, Sound and Health* (2013) <http://docs.health.vic.gov.au/docs/doc/03C56A16FC34F658CA257B5E00164599/$FILE/1212016_wind_turbine_community_WEB.pdf.

Victorian Government, *2009 Victorian Bushfires Royal Commission, Final Report* (2009).

The White House, Office of the Press Secretary, Fact Sheet: U.S.-China Joint Announcement on Climate Change and Clean Energy Cooperation (11 November 2014) http://www.whitehouse.gov/the-press-office/2014/11/11/fact-sheet-us-china-joint-announcement-climate-change-and-clean-energy-c.

M Wilder, 'The Global Economic Environment and Climate Change' in Committee for Economic Development Australia, *The Economics of Climate Change* (June 2014).

H J Wiseman, 'Remedying Regulatory Diseconomies of Scale' (2014) *Boston University Law Review* 235.

Yong Jim, 'World Bank President Jim Yong remarks at Davos Press Conference' (Speech delivered at the World Economic Forum, Davos, 23 January 2014).

7

CARBON CAPTURE SEQUESTRATION

7.1	Introduction	280
7.2	What is carbon capture and storage?	282
7.3	Why do we need CCS?	286
7.4	Capturing CO_2	288
7.5	Transporting CO_2	291
7.6	Storing CO_2	292
7.7	International CCS projects	292
7.8	CCS in Australia	294
7.9	Regulating the storage of CO_2 in Australia	296
7.10	Review questions	310
7.11	Further reading	311

7.1 Introduction

In a carbon-constrained world, the progression of technologies focused upon reducing emissions has become a strong priority. Carbon capture and storage (CCS) has been internationally recognised as an essential part of cost-effectively reducing global greenhouse gas emissions, especially for heavily fossil fuel reliant countries, such as Australia. A number of CCS demonstration projects have been completed, are underway or planned, and several major commercial CCS projects are proposed for Australia.[1]

CCS refers broadly to the process of transforming carbon emissions into a liquid form in order to inject them into subsurface reservoirs and thereby avoid CO_2 emissions.[2] The technology facilitating CCS has progressed enormously over the last few years and the Global Carbon Capture and Storage Institute has indicated that Australia, as the ninth largest energy producer, has demonstrated its support for CCS through a number of strategic initiatives. These initiatives may be summarised as follows:

- the implementation of the CCS Flagships Program, designed to accelerate the development and demonstration of CCS technologies
- the release of greenhouse gas acreage for commercial exploration of potential CCS storage sites
- small to large-scale demonstration projects
- the implementation of the National Low Emissions Coal Fund
- the implementation of the Low Emissions Technology Demonstration Fund
- state-based initiatives that include the creation of the Queensland Clean Coal Council, NSW Clean Coal Council, and Victoria's Energy Technology Innovation Strategy.

Furthermore, a number of CCS projects are currently operative in Australia for either commercial, research and development, or demonstration purposes. Existing projects include:

---

1 For example, the CarbonNet Project, which is currently in the feasibility stage, is investigating the potential to set up a large-scale CCS network in Victoria's Gippsland region. The project received $70 million in funding from the federal government and $30 million from the state government in 2012.
2 For a detailed outline of the nature of CCS, see Global CCS Institute, *Moving Below Zero: Understanding Bio-energy with Carbon Capture and Storage* (9 April 2014) where the authors note that 'with appropriate storage site selection, 99 per cent of the CO_2 is likely to be locked up for over 1000 years, becoming more secure over time. This is particularly important given that climate change itself will have major impacts on biological systems turning them from carbon sinks to sources'.

- **Callide Oxyfuel Project, Queensland.** This demonstration plant was fully commissioned in December 2012 and will operate for a two-year period. The Project has undertaken a number of studies of the potential for CO_2 storage in the Bowen Basin and Surat Basin for the Global CCS Institute. The overall budget for the Project, including operating costs to December 2014, is A$244 million. The Project has been implemented under a Joint Venture between CS Energy, IHI, J-Power, Mitsui, Glencore, and Schlumberger, with additional funding from the Australian Government, Japanese Government, and ACALET (Australian Coal Association Low Emission Technology).
- **CarbonNet Project, Victoria.** This Project is currently in feasibility stage and is investigating the potential for establishing a large-scale, multi-user carbon capture and storage network. The network could integrate multiple CO_2 capture projects in Victoria's Gippsland region, transporting the CO_2 via a common-use pipeline and injecting it into deep offshore underground storage. In February 2012, the CarbonNet Project received an additional A$70 million from the Australian Government and A$30 million from the State of Victoria.
- **South West Hub Project, Western Australia.** This Project aims to initially store up to 2.4 Mtpa of CO_2 captured from industry and power plants southwest of Perth. The Federal Government has also allocated A$48 million for the National Geosequestration Laboratory (NGL) in Perth, which will be equipped with state-of-the-art infrastructure to research all facets of CCS storage. The Project has recently completed a major 3D seismic survey and is currently preparing a further drilling campaign.
- **The CO_2 CRC Otway Project, Victoria.** This represents Australia's only operational storage demonstration project: 65 000 tonnes of CO_2-rich gas has been injected into a 2 km-deep depleted gas field from 2008–2009 and a major program of monitoring and verification was implemented. This $60 million project, which is supported by 15 companies and 7 government agencies, involves researchers from around the world and its partners include gas, coal and power companies, research organisations, and governments.
- **The Gorgon Project, Western Australia.** The Gorgon Project is unique in that the Gorgon Joint Venture is in the process of constructing the Gorgon Carbon Dioxide Injection Project as an integral component of the larger Gorgon Project. It is anticipated that between 3.4 and 4.0 million tonnes of reservoir CO_2 per year will be injected into the Dupuy Formation over 2 km below the surface of Barrow Island. All government approvals have been granted and construction of the overall Gorgon Project is approximately 75 per cent complete. Injection operations are due to commence following the commissioning of the second LNG processing train in 2015.

As discussed in Chapter 8, longer-term monitoring has revealed that the increasing amount of CO_2 in the atmosphere due to human activity – and this particularly includes the production and use of fossil fuel energy – is causing the earth to warm and the oceans to become more acidic. Scientists have predicted that the temperature will continue to increase and that this will cause the climate to change, sea levels to rise, and ocean and land environments to be adversely affected.[3]

CCS seeks to reduce the amount of CO_2 in the atmosphere by putting it into the ground and, in so doing, implement a strong form of climate change mitigation. Modelling used by the Intergovernmental Panel on Climate Change's Fifth Assessment Report has highlighted the importance of CCS in meeting climate goals. It was outlined that the prospect of avoiding 2°C by 2100 may not actually be possible without CCS and the global macroeconomic costs for those scenarios where CCS was not incorporated were substantially higher. It was also noted that, in the long-term, the largest market for CCS systems is most likely found in the electric power sector, where the cost of deploying CCS will be much higher and, as a result, will be done solely for the purpose of isolating anthropogenic CO_2 from the atmosphere. The IPCC found that this was unlikely to occur without sufficiently stringent limits on greenhouse gas (GHG) emissions that make it economic to incur these additional costs. This is best achieved through regulatory mandates requiring the use of CCS – for example, on new facilities – and/or adequate direct or indirect financial support.[4]

7.2 What is carbon capture and storage?

Carbon capture and storage (CCS) – also referred to as 'bio-CCS' – constitutes a relatively new form of technology which has the capacity to prevent large quantities of the greenhouse gas carbon dioxide (CO_2) from being released into

---

3 See Intergovernmental Panel on Climate Change, T F Stocker, D Qin, G K Plattner, M Tignor, S K Allen, J Boschung, A Nauels, Y Xia, V Bex and P M Midgley (eds), 'Summary for Policymakers' in *Climate Change 2013 The Physical Science Basis: Working Group I Contribution to the Fifth Assessment Report of the Intergovernmental Panel on Climate Change* (Cambridge University Press, 2013) 2.
4 Intergovernmental Panel on Climate Change, O Edenhofer, R Pichs-Madruga, Y Sokona (et al), 'Chapter 7: Energy Systems' in *Climate Change 2014 Mitigation of Climate Change: Working Group III Contribution to the Fifth Assessment Report of the Intergovernmental Panel on Climate Change* (Cambridge University Press, 2014) 28.

the atmosphere. In effect, CCS amounts to the process of 'capturing' carbon and then sequestering that carbon in large underground reservoirs.[5]

As the name implies, CO_2 is captured typically from large industrial processes prior to it being emitted into the atmosphere. Captured CO_2 is then transported to a carefully selected and safe storage site, where it is injected deep into a rock formation for permanent storage away from the atmosphere. The end-to-end process has been described by the IPCC in the following way:

> A 'complete end-to-end CCS system' captures CO_2 from large (e.g., typically larger than 0.1 Mt CO_2/year) stationary point sources (e.g., hydrocarbon-fuelled power plants, refineries, cement plants, and steel mills), transports and injects the compressed CO_2 into a suitable deep (typically more than 800 m below the surface) geologic structure, and then applies a suite of measurement, monitoring, and verification (MMV) technologies to ensure the safety, efficacy, and permanence of the captured CO_2's isolation from the atmosphere.[6]

Given that CCS has the potential to achieve significant emission reductions, it is considered to be a key technology to reduce fossil fuel production. CCS is a significant component of most decarbonisation pathways because it has the capacity to be incorporated into the infrastructure of emergent fossil fuel projects. It also has the capacity to be retrofitted into existing fossil fuel plants. In the United States, for example, the *Clean Air Act* through the Environmental Protection Authority has introduced rules that effectively require all new coal-fired plants to incorporate CCS technology because coal-burning power plants would have to limit carbon emissions to 1100 pounds per MWh over a 12-month operating period; and natural gas plants, meanwhile, would have to stay below 1000 pounds of carbon per MWh. Whilst modern natural gas plants meet the standard, so far even the most efficient and operational coal plants are well above the EPA's proposed threshold.[7]

Some established coal-fired plants are also retrofitting CCS technology. For example, in Canada, the SaskPower's Boundary Dam Unit, a 43-year-old coal-fired generating plant has been retrofitted with CCS technology. The cost was

---

5 There is a large amount of literature available discussing the bio-CCS process. Some of the best include: Global CCS Institute, *The Global Status of CCS: Released by the Global Carbon Capture Institute* (5 November 2014) <http://www.globalccsinstitute.com/publications/global-status-ccs-2014>; Global CCS Institute, *Report to the Global CCS Institute on Legal and Regulatory Developments Related to Carbon Capture Storage between November 2010 and June 2011* (13 December 2011) <http://www.globalccsinstitute.com/publications/legal-and-regulatory-developments-related-carbon-capture-and-storage-between-november-2>.
6 See Intergovernmental Panel on Climate Change, above n 4, 26.
7 See United States Environmental Protection Agency, *Standards of Performance for Greenhouse Gas Emissions From New Stationary Sources: Electric Utility Generating Units; Proposed Rule* (2014) 40 CFR pts 60, 70, 71, 79 and 98.

extensive, at US$1.3 billion, making the project significantly less competitive than conventional power plants. The CCS technology also uses energy produced by the plant which means that the overall energy production is reduced. Similarly, in the US, Mississippi Power's Kemper County energy facility, which is currently under construction, will use coal gasification TRIG technology to turn lignite coal into gas while capturing 65 per cent of CO_2 produced.[8]

Retrofitting existing plants with CO_2 capture is expected to lead to higher costs and significantly reduce overall efficiencies in existing power stations as opposed to newly constructed power stations which may be more efficient.[9]

There are three major stages involved in CCS technology:

1. **Capture** – carbon dioxide must be separated and 'captured' in contradistinction to other gases which may be present within industrial emissions coming from large emitters, such as coal-fired plants or natural gas producers.

2. **Transport** – as soon as CO_2 is separated, it must then be compressed into a form that is suitable for transportation. Transportation will generally occur via pipelines, vehicles, or ships to the suitable geological storage site. Once the carbon dioxide arrives at the site, it is ready for injection.

3. **Storage** – the separated and transported CO_2 is then injected into deep underground storage reservoirs. These reservoirs generally occur at depths of at least 1 km, but sometimes even lower.

CCS – especially bio-CCS – offers a range of significant potential benefits, provided the technology is exercised with full regard to the principles of ecological sustainability. CCS offers an enhanced degree of permanence because it is not dependant upon the biological properties of the biomass to store the CO_2. Indeed, most bio-CCS projects expect to be able to lock the carbon up for over 1000 years and this level of permanence is highly desirable in the context of climate change.[10]

---

8 See the outline of these projects: Power Technology, *Leading the Way: CCS Fitted Coal-Fired Stations Now a Reality* (14 July 2014) <http://www.power-technology.com/features/featureleading-the-way-ccs-fitted-coal-fired-power-stations-now-a-reality-4317055/>.

9 The costs associated with retro-fitting CCS into existing power stations was outlined by the IPCC in its special report on CCS: Intergovernmental Panel on Climate Change, *Carbon Dioxide Capture and Storage* (2005) 22 http://www.ipcc.ch/pdf/special-reports/srccs/srccs_wholereport.pdf.

10 For a discussion on the scientific issues surrounding the 'permanence ' of CCS storage see generally Organisation for Economic Co-operation and Development, International Energy Agency, *Carbon Capture and Storage in the CDM* (November 2007) <http://www.iea.org/publications/freepublications/publication/CCS_in_CDM.pdf>. The report indicates that provided the geological formation is properly chosen, underground formations can retain CO_2 for long periods and that the risk of CO_2 leaks is higher during and shortly after the injection phase, when the gas pressure is high.

Figure 7.1 The carbon capture and storage process
Source: Reproduced with permission of the Global Carbon Capture and Storage Institute

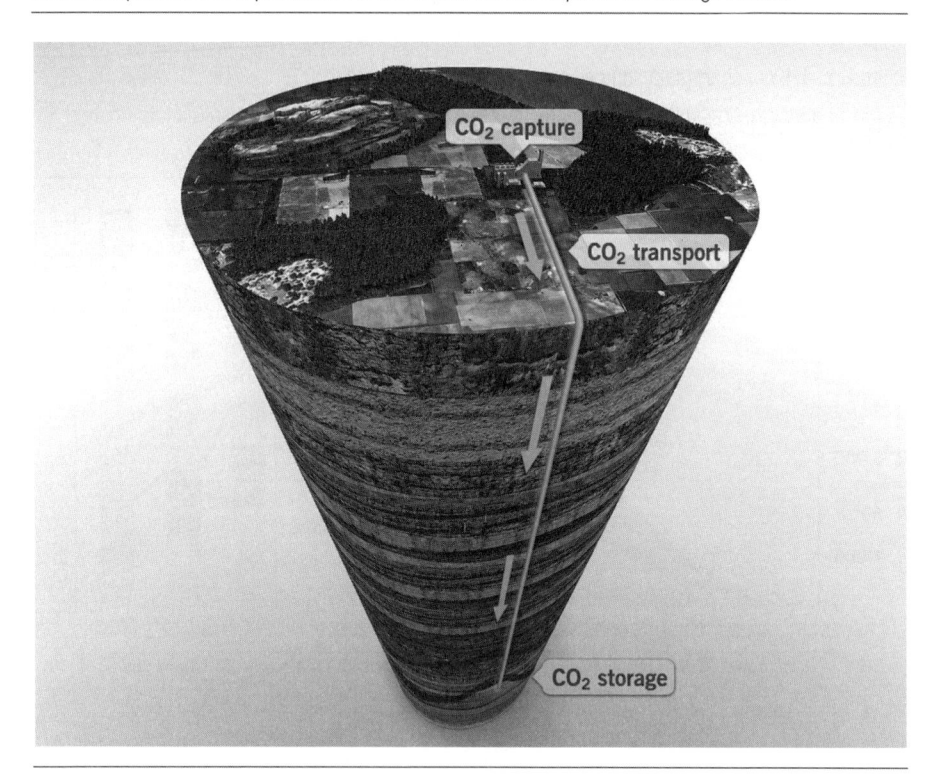

Second, bio-CCS has the capacity to be utilised in industries – such as biofuel or steel production – and also to generate power. This means that the CCS process not only removes CO_2 from the atmosphere through the CCS process, it is also utilised to displace high-emitting fossil fuel energy resources.[11]

Finally, CCS has the potential to remove enormous volumes of CO_2 from the atmosphere. The International Energy Agency ('IEA') has found that CCS may remove up to 10 billion tonnes of CO_2 per year by 2050. The following is an extract from the IEA report detailing the importance of CCS and the need to ensure that the rate of CCS increases in order for it to be effective as a climate change mechanism:

> IEA analysis highlights the importance of CCS to emission reduction. Ultimately, to effectively address climate change, the effectiveness of CCS

---

11 See the discussion in the report published by the Global CCS Institute, *Moving Below Zero: Understanding Bio-energy with Carbon Capture and Storage* (9 April 2014) <http://www.globalccsinstitute.com/publications/moving-below-zero-understanding-bioenergy-carbon-capture-storage>.

depends upon the capture and storage rate. This will need to increase from the tens of megatonnes of capture carbon dioxide in 2013 to the thousands of megatonnes of captured carbon dioxide in 2050.[12]

Figure 7.2   IPCC chart (data sourced from IPCC Special Report 2005)
Source: The information in this chart comes from the Intergovernmental Panel on Climate Change (IPCC), 2005: IPCC Special Report on Carbon Dioxide Capture and Storage.

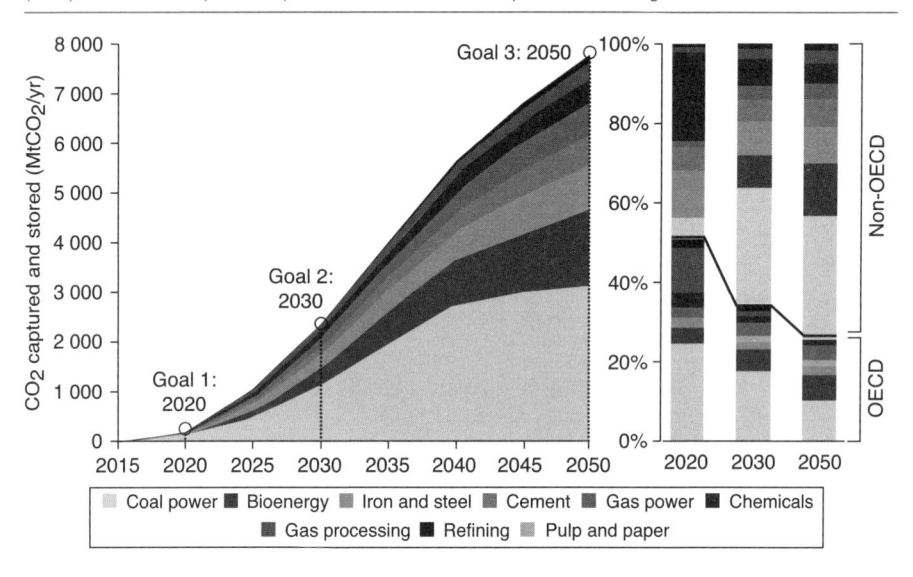

7.3 Why do we need CCS?

As the world faces the challenges associated with climate change, the need for a revolutionary scale of CO_2 mitigation has become apparent. The *Intergovernmental Panel on Climate Change Special Report for CCS* indicated that CCS has the capacity to contribute between 15 and 55 per cent of the required abatement leading up to the year 2100.[13] One of the major sources of CO_2 emissions stems from the electricity that is generated from fossil fuels. In 2011, the IEA predicted that in excess of 40 per cent of the world's energy-related CO_2 emissions stem from fossil fuel combustion.[14]

---

12 See International Energy Agency, *Technology Roadmap; Carbon Capture and Storage* (2013) <http://www.iea.org/publications/freepublications/publication/technology-roadmap-carbon-capture-and-storage-2013.html>.
13 See IPCC, above n 9, 55.
14 See International Energy Agency, *CO₂ Emissions from Fuel Combustions: Highlights* (2013) 9 http://www.iea.org/publications/freepublications/publication/co2emissionsfromfuelcombusti

Despite this, demand for fossil fuels is on the rise, especially in developing countries, where a significant percentage of the population currently has reduced access to electricity and global energy demand is increasing. Within this complex framework, despite developments in renewable energy markets, CCS represents an attractive and viable option. Indeed, given the market impediments connected with the renewable sector for many developing countries, CCS may actually be the only realistic option for reducing emissions from large-scale point sources. Whilst decentralised renewable energy has expanded in recent years to meet the demands of developing countries – including wind, solar, hydro-power and biomass options – there are still significant efficiency and market impediments.[15]

The status of CCS and its economic imperative in the context of a carbon economy has been outlined by a range of authorities which have been summarised in the 2014 *Global Status of CCS Report* issued by the International Energy Agency which states:

> Carbon capture and storage (CCS) has been identified as an essential technology to meet the internationally agreed goal of limiting the temperature increase to 2°C. Deploying CCS technologies and retrofitting fossil fuel plants with CCS avoids the need to retire large parts of this fleet prematurely. This improves the economic feasibility of attaining the climate objective ... However, progress in developing CCS has been disappointingly slow ... Beyond 2020, when demonstrated and deployed at new high efficiency plants, or retrofitted at suitable existing plants, CCS may play a key role in curbing CO_2 emissions from coal-based power generation and industry ... potentially reducing the overall cost of power sector decarbonisation by around US$1 trillion between 2012 and 2035.[16]

The Energy Modeling Forum's (EMF) *EMF 27: Study on Global Technology and Climate Policy Strategies* notes:

> A robust finding is that the unavailability of carbon capture and storage and limited availability of bioenergy have the largest impact on feasibility and macroeconomic costs for stabilizing atmospheric concentrations at low levels ... a substantial number of models were not able to produce 450 ppm without CCS. Indeed, the vast majority of situations in which models could not produce scenarios were those in which CCS was assumed to be unavailable ... unlike other technologies assessed in this study, it is a very versatile technology that has the potential to contribute to decarbonization

---

onhighlights2013.pdf where the report notes that coal accounted for 44 per cent of the global CO_2 emissions due to its heavy carbon content per unit of energy released. Coal is twice as emission-intensive as gas.

15 See IPCC, above n 4, 26.
16 International Energy Agency, *World Energy Outlook 2014* (13 November 2013) 42.

via different processes, such as electricity generation and synthetic fuel production from different feedstock and in industry.[17]

These studies indicate the critical importance of CCS as an adaptive technology capable of curbing greenhouse gas emissions from fossil fuel power stations. CCS provides a means by which the impact of carbon intensive energy production can be mitigated whilst the economic impact of asset retirement is gradually absorbed. However, the deployment of CCS requires significant infrastructure, suitable for the long-term storage of waste products, including carbon dioxide storage, transport, measurement, monitoring, and verification. In this respect, CCS technologies in the mining and energy sector will only become competitive with unabated technologies if the costs associated with the additional equipment attached to the power plant and their corresponding decrease in efficiency is compensated by either a high carbon price or direct financial support.

7.4 Capturing CO_2

The scope and range of carbon capture processes is diverse. The concept of bio-energy and carbon storage – or 'bio-CCS' – refers to the removal of CO_2 from the atmosphere via biomass conversion technologies. Biomass binds carbon from the atmosphere as it grows; with the conversion of the biomass, this carbon is subsequently released as CO_2. If, rather than being released, CO_2 is captured and transported to a storage site and permanently stored deep underground, it would result in net removal of CO_2 from the atmosphere. Bio-CCS may therefore be defined as processes by which CO_2, originating from biomass, is captured and stored. These may be energy production processes or any other industrial processes with CO_2-rich process streams originating from biomass feedstocks.[18]

The process of capturing CO_2 and separating it from a carbon emission source, either before or after it has been combusted in order to produce energy or other products, such as cement and steel, is outlined below. In certain industrial processes, such as some biofuels production routes, the separation of high purity CO_2 is

---

17 See E Kriegler, J P Weyant, G J Blanford, V Krey, L Clarke, J Edmonds, A Fawcett, G Luderer, K Riahi, R Richels, S K Rose, M Tavoni and D P Van Vuuren, 'The role of technology for achieving climate policy objectives: overview of the EMF 27 study on global technology and climate policy strategies' (2014) 123 *Climate Change* 353, 355. The authors conclude that the EMF 27 Study indicates that: 'Non-electric energy end use is hardest to decarbonize, particularly in the transport sector. Technology is a key element of climate mitigation. Versatile technologies such as CCS and bioenergy are found to be most important, due in part to their combined ability to produce negative emissions'.
18 See European Biofuels Technology Platform, *Bio-Mass with CO_2 Capture and Storage (Bio-CCS): The way forward for Europe* (20 June 2012) http://www.biofuelstp.eu/downloads/bioccsjtf/EBTP-ZEP-Report-Bio-CCS-The-Way-Forward.pdf.

already an established part of the process. In such situations, capture for storage will usually only require some dehydration before the gas can be compressed and transported to a storage site.[19]

There are three different capture processes that may be involved with CCS capture:

1. pre-combustion technology
2. post-combustion technology
3. oxyfuel combustion.

7.4.1 Pre-combustion technology

Pre-combustion technology involves the capture and separation of carbon dioxide prior to the combustion of the fossil fuels. The fossil fuels are converted to hydrogen and CO_2 prior to combustion. This separation enables the CO_2 to be captured and stored. The hydrogen is then combusted to generate energy. Pre-combustion technology may only be conducted in new, fully-equipped fossil fuel plants because the process of capturing the CO_2 must be directly linked to the combustion process. Most older or established fossil fuel plants do not incorporate this technology.

7.4.2 Post-combustion technology

Post-combustion carbon capture is probably the most widely utilised process for implementing CCS. This technology involves an absorption-based process. It relies upon the chemical reaction between CO_2 and a solvent. The flue gas from the fossil fuel is brought into contact with the solvent, resulting in a reaction with the CO_2 which allows the CO_2 to be separated. Post-combustion technology is well established, having been in operation for many years.[20]

7.4.3 Oxyfuel combustion

Oxyfuel combustion is the process of burning a fossil fuel using oxygen instead of air. Where fuel is burnt with pure oxygen, the contaminants are removed from the emissions and this makes the capture and separation of CO_2 much easier.

The outline that follows represents the different forms of pre-combustion (enhanced oil recovery), post-combustion (cement industry), and oxyfuel combustion (cement industry) processes.

---

19 See especially the discussion in European Biofuels Technology Platform, above n 18, 8.
20 For a full discussion of post-combustion technology see M Wang, A Lawal, P Stephenson, J Sidders, C Ramshaw and H Yeung, 'Post-Combustion CO_2 with Chemical Absorption: A State of the Art Review' (2011) 89(9) *Chemical Engineering and Design* 1609, 1609–24.

7.4.3.1 Enhanced oil recovery operations

In certain conditions, CO_2 may be used to 'enhance' the recovery of oil. Pursuant to enhanced oil recovery ('EOR'), CO_2 is injected into oil wells under high pressure and subject to low temperatures. This injection process has the effect of producing an overall increase in recoverable oil. The reasons for this may be summarised as follows:

1. The CO_2 cleans the oil that is trapped in the microscopic pores of the reservoir rock and therefore acts like a solvent.
2. The CO_2 simultaneously acts as a pressurising agent to effectively push additional oil out of the reservoir rock.
3. The CO_2 injection reduces the overall viscosity of the oil and this has the effect of promoting greater flow.

EOR has the capacity to generate a market for CO_2 that could play a vital role in contributing to the large capital investment that is required for CCS deployment.[21]

The injection of CO_2 into mature oil reservoirs is an established and effective process for CCS that combines carbon mitigation with improved oil production. It may be applied to a variety of different oil reservoirs in a range of different geological settings. The injected carbon is retained within the reservoir as an intrinsic component of the CO_2 enhanced oil recovery process. This means that all injected CO_2 will ultimately remain stored within the oilfield at the end of enhanced oil recovery operations.[22] The storage aspect has provided significant interest in CO_2 EOR as a potential method of CCS. Currently, about 130 commercial CO_2 EOR operations – also called 'CO_2 floods' – have been deployed around the world, although the vast majority are in the United States.

7.4.3.2 CCS operations in the cement industry

The cement industry is a major source of industrial greenhouse gas emissions and accounts for around 5 per cent of global anthropogenic greenhouse gas emissions. The opportunities for developing CCS operations in this industry revolve primarily around what is referred to as 'post-combustion' capture or 'oxy-fuel'. In essence, post-combustion capture of CO_2 in the cement industry incorporates processes which include chemical absorption, adsorption, membrane, mineralisation, and calcium looping technologies.[23]

---

21 See Global CCS Institute, S Whittiker and E Perkins, *Technical Aspects of CO_2 Enhanced Oil Recovery and Associated Carbon Storage* (2013) 3 http://decarboni.se/sites/default/files/publications/118946/technical-aspects-co2-enhanced-oil-recovery-and-associated-carbon-sto.pdf.
22 See Global CCS Institute, S Whittiker and E Perkins, above n 21, 4.
23 See the discussion of the different forms of CCS technologies that have been and are being developed in this industry in IEAGHG, *Deployment of CCS in the Cement Industry* (December 2013) http://decarboni.se/sites/default/files/publications/162743/deployment-ccs-cement-industry.pdf.

Oxyfuel technology, on the other hand, relies on the combustion of pure oxygen and a recirculation of flue gas in order to enrich CO_2 to a level which allows a relatively easy purification by liquefaction systems. Using oxygen instead of air for the combustion process allows a rich CO_2 flue gas to be obtained which is mainly composed of CO_2 and this gas can be readily cleaned and compressed. An adequate purity of CO_2 must be obtained in the flue gas in order to reach a composition suitable for transport and then subsequent storage. The process requires the integration of two additional units to those deployed at a conventional coal-fired power plant:

- an oxygen production unit known as an air separation unit (ASU) which extracts oxygen out of the surrounding air; and
- a CO_2 purification unit to clean and compress the CO_2 out of the flue gas known as a gas processing unit.[24]

Implementing the full oxyfuel concept means that almost all generated CO_2 can theoretically be captured.[25]

As with enhanced oil recovery, the difficulty with progressing CCS operations within the cement industry lies in the fact that further research and development is needed to improve existing technology and, in light of this, the cost barriers need to be addressed.

7.5 Transporting CO_2

Carbon dioxide is usually transported via a pipeline from a capture location to a specific storage site. It is, however, possible for small volumes of CO_2 to be transported in ships, tanker trucks, or rail tankers. CO_2 pipelines are similar in design to those that are used to transport natural gas and oil. However, in order to be transported, CO_2 must be dehydrated and compressed to great pressures so that it may reach what is described as a 'dense phase'. This effectively means that the CO_2 gas is transformed into something akin to a liquid.[26]

The scale of transportation required for the widespread deployment of CCS is much more significant than any existing small-scale transport because it involves transferring dense, concentrated CO_2 across vast areas of land and sea. This makes road and rail transportation inadequate so pipelines are the usual method for domestic transportation.

---

[24] See Global CCS Institute, *CO₂ Capture Technologies: Oxy Combustion with CO₂ Capture* (January 2012) http://www.globalccsinstitute.com/publications/co2-capture-technologies-oxy-combustion-co2-capture.

[25] See the discussion by K Vatopoulos and E Tzimas, 'Assessment of CO₂ capture technologies in cement manufacturing process' (2012) *Journal of Cleaner Production* 32, 251–61.

[26] See Global CCS Institute, *How CCS Works – Transport* http://www.globalccsinstitute.com/content/how-ccs-works-transport.

The properties of CO_2 vary considerably from other fluids that are commonly transported by pipeline, such as natural gas. In the majority of CO_2 pipelines, the flow needs to be driven by compressors at the pipeline source that ensure that the necessary temperature and pressure are consistently maintained. This means that careful attention must be given to the quality and integrity of the pipeline. It also means that pipelines that are commonly utilised to transport other substances, such as gas, may not be suitable for the transportation of CO_2. Current regulatory frameworks will need to be adapted to accommodate the increase in the volumes of CO_2. This will necessarily incorporate legislative amendments to government planning permissions, environmental impact assessment, and rights of way.

7.6 Storing CO_2

The compressed CO_2 needs to be stored in a suitable geological storage site or reservoir. These reservoirs are generally found deep underground in the rock strata and are natural formations. The compressed CO_2 is injected into these reservoirs, which are generally at located at a depth of 1 km or more below the ground. Once injected, the CO_2 moves slowly through the porous rock, filling the tiny spaces which are described as 'pore space'.

Possible storage sites include depleted oil or gas fields as well as rocks that previously contained saline water formations. These sites generally contain an impermeable trap above them in order to prevent the CO_2 from migrating to the surface. Sites that are compliant with these characteristics have the capacity to retain fluids and gases for millions of years.

Factors which may influence the optimisation of a storage basin include cost, minimising risk, access to a range of uses of the basin, ground surface and seabed, and the value of the resource. Such factors must also be considered within the framework of government energy policies.[27]

7.7 International CCS projects

In the Netherlands, the implementation of CCS is a consequence of both climate change imperatives and the economic benefits connected with the development of CCS technology in association with the energy sector. The benefits of CCS have resulted in a framework where the government works in association with a privately

---

27 For a discussion on CCS storage sites, see T M Kerr, 'Carbon Dioxide Capture and Storage: Priorities for Development' (2008) 4 *Carbon and Climate Law Review* 335 where the author notes that thorough planning and geological analysis of the storage site using established site selection processes and regulatory mechanisms is vital.

run CCS market to supervise the implementation of CCS projects.[28] The *Spatial Planning Act* ('SPA 2008') introduced by the Netherlands allows the government to adapt spatial planning by district/local governments for projects of national importance. The government is also seeking to utilise the large number of depleted gas fields that exist in the region for the storage of CO_2. CCS storage capacity in the Netherlands is estimated to be 11 266 Mt, of which 1150 Mt capacity exists in depleted gas fields and 715 Mt in subsurface aquifers. The largest storage potential exists onshore, although offshore remains a viable option. In February 2011, the government announced that it was not prepared to permit onshore geological storage and is currently only permitting CCS projects for offshore storage.[29]

The Dutch national research and development project for CCS is known as 'CATO-2'. CATO-2 simulates all of the stages relevant to the CCS chain and seeks to promote continued research and development at each level. The project began in 2001 and incorporates a consortium of around 40 partners. The overriding aim is to determine whether – from economical, technical, social and ecological perspectives – CCS may be able to contribute to sustainable energy systems in the Netherlands. The previous research project, CATO-1, provided a range of innovations which have placed the Netherlands in a leading position within the CCS community. In light of this, the focus of CATO-2 is specifically upon facilitating these innovations and promoting integrated development. Research and development is being conducted in advanced capture technologies, storage processes, monitoring techniques, regulation, and public interest concerns.[30]

The Rotterdam Climate Initiative ('RCI') was launched in 2006 with the aim of reducing CO_2 emissions by 50 per cent by 2025 while promoting economic development in the Rotterdam region. This specific target is particularly ambitious and higher than the national target of the Netherlands which is only a 30 per cent reduction by 2020. The RCI includes plans for what is currently the most developed CCS network project in Europe. It involves 18 major companies working collaboratively to provide feasibility studies for CCS projects and a business case. One study has examined the feasibility of carrying captured CO_2 from sites to an offshore storage location, by pipeline or ship, to be stored by either EOR operators or within deep geological formations under the North Sea. The timescale for this CCS project to commence is the middle of 2015 with capture and storage of 20 Mtpa of CO_2 to be achieved by 2025.[31]

---

28 See P J Vergragt, *CCS: The Next Technological Lock-in? A Case study from The Netherlands* (Tellus Institute, Boston, 2008).
29 See the report on CCS in the Netherlands at Zero CO_2NO, *The Netherlands* http://www.zeroco2.no/projects/countries/the-netherlands.
30 For a detailed outlined of CATO-2 see Cato 2, *The CATO-2 programme* http://www.co2-cato.org/cato-2.
31 See the discussion by the Global CCS Institute, *Rotterdam CCS Network* (2013) http://www.globalccsinstitute.com/projects/12661.

7.8 CCS in Australia

Carbon Capture and Storage (CCS) is one of the significant methods of mitigating greenhouse gas emissions. Australia has a number of CCS demonstration projects underway as well as a number of major commercial CCS projects planned. The Australian National Carbon Storage Council was implemented with the aim of bringing together key stakeholders to advise the Australian Government on the accelerated development and deployment of CCS projects throughout Australia.

Most of the storage potential for CCS is located offshore in Australia. The greatest potential for CCS storage exists in north-western Australia. Sometimes, however, areas assessed to have good storage potential do not align with core electricity demand areas. There is little potential for CCS to be implemented in depleted oil and gas fields and, given that most of these fields are still productive in Australia, it will be difficult for CCS to be effectively implemented within these core electricity demand areas. The geological storage sites themselves must be suitable. The reservoirs need to have adequate storage capacity, a sealing caprock so that injected carbon does not leak, and a stable geological environment. Where CO_2 is injected into problematic geological reservoirs, it may actually exacerbate climate change concerns by escaping back into the atmosphere and this potential must be minimised.[32]

Given the paucity of suitable sites, there is a significant potential for conflict in Australia between different 'users' in both on and offshore sites. Offshore development of CCS sites may generate difficulties for existing offshore users, such as fisheries, shipping routes, infrastructure, etc. Probably the greatest issue of overlap lies with the petroleum industries because of the high possibility that large-scale commercial processing of CCS will compromise petroleum production by existing petroleum title-holders.

The Australian Government has developed a number of specific on and offshore policy and regulatory responses to the issue of CCS storage management. Offshore CCS is primarily governed by provisions of the *Offshore Petroleum and Greenhouse Gas Storage Act 2006* (Cth) ('the *OPGGSA*') and its associated regulations. The core object of the *OPGGSA* is to provide an effective regulatory framework for petroleum exploration and recovery and for the injection and storage of greenhouse gases.[33]

The *OPGGSA* provides clear security of title for CCS operators and seeks to clarify longer-term liability issues that may be connected with CCS. Within the

---

[32] See the discussion by S Som, 'Creating Safe and Effective Carbon Sequestration' (2008) 17 *New York University Environmental Law Journal* 961, 964.
[33] *OPGGSA* (Cth) s 3.

offshore zone, the *OPGGSA* allows petroleum title-holders to apply to undertake what it describes as 'greenhouse gas activities'.[34] The holder of a petroleum title may explore for GHG storage formations within their authorised work area.[35] This type of exploration comes within the scope of authorised activities. The fact that such exploration is conducted does not mean, however, that the petroleum title automatically acquires a separate GHG title. A separate GHG licence must still be applied for. Petroleum retention lessees and petroleum production licencees may apply for what is known as a GHG holding lease or a GHG injection licence over a declared GHG storage formation which is located wholly within their title area.

Where a petroleum title-holder takes up this opportunity, they are required to obtain a declaration in accordance with the *OPGGSA*. To give petroleum lessees and licence-holders an opportunity to exercise this option, lessees and licencees will have at least 60 days to provide notice of any intention to release a whole or part of the lease or licence area to bids for a GHG assessment permit.[36] Where a petroleum title-holder does apply for a GHG title during this period, the area may not be offered for other bids until the application by the petroleum title-holder lapses, is withdrawn, or is refused.

Petroleum production licencees may inject a GHG substance, provided it is clearly for purposes relating to the recovery of petroleum (or enhanced hydrocarbon recovery), without the need for a GHG injection licence. In this case, the injection operation will be approved and administered as a component of the field development plan. Where the injection of the GHG substance is carried out for the direct purpose of GHG disposal, the petroleum title-holder will be subject to the GHG injection and storage provisions of the *OPGGSA* which means that a separate GHG injection licence must be obtained.

State governments are also actively seeking to facilitate onshore CCS projects and Victoria, Queensland, and South Australia have all enacted legislation.[37] The regulatory and policy regimes which have been adopted by the state governments have addressed the issue of overlapping tenure and competing/conflicting land use; however, there are also additional problems of overlap and resource conflict with established agricultural industries. Groundwater impacts

---

34 For a full discussion of the application of the *Offshore Petroleum and Greenhouse Gas Storage Act 2006* (Cth) see A Thorpe, 'Too Little, Too Soon: An Assessment of Australian Carbon Capture and Storage Legislation Against the New Standards set for the Clean Development Mechanism' (2012) 3 *Climate Law* 139, 142–4.

35 These regulations have been made and appear as pt 6 of the *Offshore Petroleum and Greenhouse Gas Storage (Resource Management and Administration) Regulations 2011* (Cth).

36 See *OPGGSA* (Cth) ss 297, 304.

37 See *Offshore Petroleum and Greenhouse Gas Storage Act 2010* (Vic); *Greenhouse Gas Storage Act 2009* (Qld); *Petroleum and Geothermal Energy Act 2000* (SA).

are also a potential difficulty that may generate community concerns and the possibility of conflict with unconventional gas operations is significant given the fact that there is a strong coincidence between the unconventional gas resources and potential CCS storage sites.[38]

Some of the different aspects of the Australian regulatory framework for storage, licensing, conflict, and liability are discussed below.

7.9 Regulating the storage of CO_2 in Australia

In Australia, the new regulatory framework for offshore geological storage was implemented at the federal level through detailed amendments to the *Offshore Petroleum and Greenhouse Gas Storage Act 2006* (Cth) ('the *OPGGSA*') which came into effect in 2008. The Victorian Government has implemented mirror legislation in the *Offshore Petroleum and Greenhouse Gas Storage Act 2010* (Vic). Onshore regulation remains the subject of state jurisdiction. In Queensland the relevant Act is the *Greenhouse Gas Storage Act 2009* (Qld).[39] In Victoria the relevent Act is the *Greenhouse Gas Geological Sequestration Act 2008* (Vic) ('*GGGSA*').

In vesting subsurface storage rights in the State, the Victorian and Queensland Acts have displayed a clear preference for state control in the development of subsurface geological storage projects. These statutory initiatives represent strong incentives for the progression of the CCS market.

The legislative framework adopted within Victoria and Queensland vests ownership of subsurface storage space in the state. State ownership of subsurface geological surface formations confers control and, potentially, liability for the process and ongoing management of CCS upon the state government. This vesting may constitute a deprivation of interest for the surface owner thereby raising the possibility of surface owner compensation. The details of each framework are explored below.

---

38 For a discussion on this see Global CCS Institute, *Comparing Different Approaches to Managing CO_2 Storage Resources in Mature CCS Futures* (March 2014) http://decarbonise/sites/default/files/publications/162748/comparing-different-approaches-managing-co2-storage-resources-mature-ccs-futures.pdf.

39 Other Australian states are yet to follow. In 2012, however, Western Australia introduced amendments to the *Petroleum and Geothermal Energy Resources Act 1967* (WA) providing for onshore geological storage of greenhouse gas. Prior to the introduction of these amendments, the injection and storage of greenhouse gases in WA was not regulated, apart from the Gorgon Project, which is the largest commercial carbon capture and storage project in the world and is regulated according to the state agreement set up pursuant to the *Barrow Island Act 2003* (WA).

7.9.1 Victoria: *Greenhouse Gas Geological Sequestration Act 2008*

Victoria was a pioneer state in enacting legislation dealing with onshore geological sequestration.[40] The *GGGSA* was created as a separate Act to the *Petroleum Act 1998* (Vic) because geological storage formations capable of storing greenhouse gases were intended to be treated as separate resources. The main objective of the *GGGSA* is to facilitate and regulate the injection of greenhouse gas substances into underground geological formations for the purpose of permanent storage.[41] In constructing a legal regime for CCS, the *GGGSA* seeks to provide certainty for investors and other affected interest holders with respect to their legal rights and obligations as well as promoting community confidence that CCS projects will be undertaken in a manner that minimises risks to public health and the environment.[42]

Geological storage formations are defined in the *GGGSA* to include 'any seal or reservoir of an underground geological formation' and also 'any associated geological attributes or features of an underground geological formation'.[43] This definition mandates the inclusion of a storage area as well as geological formations that are connected to that storage area. A storage area that has additional non-geological attributes will not be excluded from the definition provided the base requirements can be established. The *GGGSA* makes no distinction between natural and artificial storage areas. As such, any subsurface storage areas with geological attributes that exist within the boundaries of Victoria will come within the scope of the statutory definition.

The pivotal ownership section of the *GGGSA* is s 14, which provides that the Crown retains ownership over all underground geological storage formations, which are located on private land.[44] Prior to the introduction of the *GGGSA* the Crown held no ownership rights in geological storage formations located underneath private land. It did, however, own geological storage formations in Crown land pursuant to the provisions of the *Land Act 1958* (Vic).[45]

---

40 See Minister for State and Energy Resources, State of Victoria. 'Victoria Leads Australia with Carbon Capture and Storage Bill' (Press Release, 9 September 2008 1 <http://www.premier.vic.gov.au/component/content/article/4704/html>.
41 Victoria, *Parliamentary Debates,* Legislative Assemb. October 16, 2008, 3674 (Peter Batchelor, Minister for Energy and Resources).
42 These objectives were explicitly outlined in the explanatory memorandum of the Bill. See *Greenhouse Gas Geological Sequestration Bill* 2008, Explanatory Memorandum.
43 *GGGSA* (Vic) s 3.
44 *GGGSA* (Vic) s 14(1).
45 This is the effect of the *Land Act 1958* (Vic) s 339(1) which applies to Crown land as well as 'depth-limited' freehold grants, issued after 29 December 1891 which are not for the purpose of mining.

The vesting provisions in the *GGGSA* therefore significantly extend the scope of Crown ownership in subsurface storage spaces within 15.24 metres of the surface of the land.[46] The effect of the vesting provision is that in Victoria geological storage spaces coming within the scope of the Act are no longer treated as a component of the land but are recognised as an independent resource, owned by the Crown, in right of the State.[47]

Greenhouse Gas Geological Sequestration Act 2008 (Vic)

14 Underground geological storage formation is the property of the Crown

(1) The Crown owns all underground geological storage formations below the surface of any land in Victoria.
(2) Subsection (1) does not apply in relation to any land (other than Crown land) to the extent that the underground geological storage formation is within 15·24 metres of the surface of the land.
(3) Subsection (1) applies despite any prior alienation of Crown land.
(4) The Crown is not liable to pay any compensation in respect of a loss caused by the operation of this section.

A direct consequence of the vesting provision has been the deprivation of the common law entitlements previously retained by the surface estate owner. The removal of rights of control over subsurface storage areas amounts to a significant loss for surface estate owners, particularly given the increasing value of subsurface storage areas within a carbon economy. This deprivation is not addressed by the *GGGSA*, which explicitly sets out that the Crown is not responsible for the payment of compensation for any loss caused by the operation of these provisions.[48] It is unlikely that the effect of these provisions will offend the *Victorian Charter of Human Rights and Responsibilities* because the deprivation of property to the surface estate owner has occurred 'in accordance with the law' as set out in the *GGGSA*.[49] The *GGGSA*

---

46 *GGGSA* (Vic) s 14(2). Presumably this depth restriction seeks to preclude overlap with deeper storage spaces that may be connected to subsurface mining activities.
47 See T J Logan, 'Carbon Down Under: Lessons From Australia: Two Recommendations for Clarifying Sub-Surface Property Rights to Facilitate Onshore Geologic Carbon Sequestration in the United States' (2010) 11 *San Diego International Law Journal* 561, 579.
48 *GGGSA* (Vic) s 14(4). See also State Government of Victoria, Department of Energy and Earth Resources, *A Regulatory Framework for the Long-Term Underground Geological Storage of Carbon Dioxide in Victoria* (January 2008) http://www.dpi.vic.gov.au/energy/about/legislation-and-regulation/ccs-regulations/a-regulatory-framework-for-the-long-term-underground-geological-storage-of-carbon-dioxide.
49 See *Charter of Human Rights and Responsibilities 2006* (Vic) s 20 which sets out that 'A person must not be deprived of his or her property other than in accordance with law.' See also Victoria, *Parliamentary Debates*, Legislative Council, 16 October 2008, 4475 (Peter Batchelor, Minister for Energy and Resources).

does attempt to alleviate this to some extent by setting out that a CCS operation cannot commence on private land until the surface estate owner has consented or the parties have entered into a compensation agreement.[50]

The amount of compensation payable under any such agreement to a surface estate owner or native title-holders must, however, be referable to a direct loss that is a natural and reasonable consequence of the development of a CCS project.[51] This will include damage to the surface, damage to improvements, deprivation of possession, loss of amenity, or any decrease in market value. Significantly, it will not include any deprivation in the commercial value of the subsurface storage area.[52]

The regulatory framework of the *GGGSA* is founded upon the licensing system set out in the *Petroleum Act 1998* (Vic). The rationale for adopting a similar model is that CCS uses many of the same technologies as the petroleum industry and the industry is familiar with the framework.[53] As such, applicants seeking to develop a CCS project will be required – in the same way as if they were developing a subsurface mining project – to apply for an exploration permit authorising the holder to carry out greenhouse gas sequestration exploration.[54] Where the holder of an exploration permit discovers an underground geological storage formation they will then be entitled to apply for either an injection and monitoring licence or a greenhouse gas sequestration retention lease in order to facilitate progression of the project to the injection phase.[55]

An injection licence authorises the applicant to carry out greenhouse gas substance injection and monitoring within the licence area.[56] A greenhouse gas sequestration retention lease authorises the holder to carry out exploration and formation activities for greenhouse gas sequestration within the lease area.[57]

The *GGGSA* also mandates the creation of a greenhouse gas sequestration register to formalise the creation, dealing, modification and exemption of licences and to record the volume of greenhouse gas substances permitted to be injected into a subsurface geological storage formation.[58] The creation of the register is important because it ensures a permanent record of CCS activities for future reference.

The vesting provision separates subsurface geological formations in private land below a certain depth from the web of corporeal resources coming within the

---

50 *GGGSA* (Vic) s 200.
51 *GGGSA* (Vic) s 201.
52 *GGGSA* (Vic) s 201(1)(a)-(g). The *GGGSA* (Vic) explicitly sets out that compensation is not payable for the value of subsurface storage areas: s 203.
53 See Victoria, *Parliamentary Debates*, above n 49.
54 *GGGSA* (Vic) s 19(1).
55 *GGGSA* (Vic) s 19(2).
56 *GGGSA* (Vic) s 71.
57 *GGGSA* (Vic) s 59.
58 *GGGSA* (Vic) s 281.

common law concept of land ownership. The surface estate owner loses control over this separate resource and ownership is transferred to the State. This confers on the State the power to licence out rights to utilise the storage space for CCS projects. In this situation, the State retains ultimate ownership of the storage space and assumes ownership of the injected greenhouse gases upon the expiration or surrender of the injection licences.[59]

The verification of subsurface geological storage formations under the *GGGSA* lacks proper articulation.[60] The vesting provisions confer title to subsurface storage formations and presume that these 'spaces' have been independently verified. There is, however, no separate clause in the *GGGSA* articulating the nature, scope, or proprietary identity of subsurface storage spaces. As such, it is unclear whether the interest acquired by the Crown is corporeal or incorporeal, whether it is a land interest or a separate resource interest, or whether the ownership relates to the space itself or merely the right to utilise the space.[61]

There are also potential difficulties with superimposing a new resource interest upon a pre-existing subsurface landscape. The excision of subsurface usage rights from a surface owner's entitlements may generate unfairness where owners have previously relied upon the existence of this space. It may also create overlapping ownership regimes where different holders are required to compete for usage which in turn may encourage an 'anti-commons' scenario.[62] The issue of surface access to vested subsurface storage spaces is also problematic. State vested ownership under the *GGGSA* does not expressly include access rights. This means that if a surface estate owner refuses access, the holder of a geological storage licence may be unable to utilise the subsurface resource. In such a scenario, a CCS licence-holder may raise the 'incidental rights' provisions in the *GGGSA*, although

---

59 See *GGGSA* (Vic) s 16 which sets out that upon the cancellation or surrender of an injection licence, the Crown becomes the owner of any greenhouse gas which has been injected into the storage formation.
60 See the discussion by N Swayne and A Phillips, 'Legal Liability for Carbon Capture and Storage in Australia: Where should the losses fall?' (2012) 29 *Environmental Planning Law Journal* 189, 194 where the authors consider ownership issues in the context of longer-term liability.
61 This is a significant issue because if it is the space that is owned, then does ownership cease once the space has been filled? See the discussion by T Lloyd, 'Carbon Capture and Sequestration: Removing The Legal and Regulatory Barriers' (2014) 4 *Climate Law* 187. For a broader discussion about the difficulty of propertising space, see C Rose, 'Property in all the Wrong Places?' (2005) 114 *Yale Law Journal* 991, 1003–7.
62 The concept of anti-commons property was discussed by M Heller, 'The Tragedy of the Anti-Commons: Property in the Transition from Marx to Markets' (1998) 111 *Harvard Law Review* 621, 660–9. Private property rights distribute ownership vertically but problems may be generated where horizontal divisions are created and governments must take care to avoid accidentally creating anti-commons property when they define new property rights.

the scope of these provisions is unclear as they only mandate the carrying out of any act within the licence area which is 'incidental to or necessary for' the purpose of the licence.[63]

A further concern with the state-based focus of the *GGGSA* framework is that it only applies to subsurface storage spaces located within Victoria. The absence of legislative uniformity between the states creates the potential for a patchwork of competing jurisdictional claims.[64] Most subsurface storage areas will not conform to state boundaries, making it likely that competing – and possibly conflicting – legislative and common law measures will apply to a single subsurface storage space.[65]

The implementation of non-uniform, surface-specific legislation is ultimately detrimental to longer-term climate change objectives because the overlap and complexity increases capital and operational costs and generates investment resistance.[66] These concerns must be addressed if a more efficient, transparent, and credible regime for the assessment, approval, and commercial deployment of CCS projects in Australia is to emerge.

7.9.2 Queensland: *Greenhouse Gas Storage Act 2009*

The *Greenhouse Gas Storage Act 2009* (Qld) ('the *GGSA*') was introduced shortly after the Victorian Act with similar objectives. It aims to facilitate greenhouse gas geological storage and reduce the impact of greenhouse gases on the environment by providing a regulatory system for GHG activities.[67]

The *GGSA* vests ownership of all GHG storage reservoirs in the State, expressly setting out that all 'GHG storage reservoirs in land are, and are taken always to have been, the property of the State'.[68]

---

[63] The 'incidental rights' regime for permits, licences and leases is set out in the *GGGSA* (Vic) ss 19(1)(b), 59(b), 71(c).

[64] See also W Buzbee, 'Recognising the Regulatory Commons: A Theory of Regulatory Gaps' (2003) 89 *Iowa Law Review* 1, 10–15 where the author examines the difficulties of overlapping state-based regulation and examines the concept of a 'regulatory commons'. See also B Wiener, 'Global Environmental Regulation: Instrument Choice in Legal Regulation' (1999) 108 *Yale Law Journal* 677.

[65] See W Buzbee, 'State Greenhouse Gas Regulation, Federal Climate Change Regulation and the Preemption Sword' (2009) 1 *San Diego Journal of Climate and Energy Law* 23, 36 where the author argues that overlapping regulatory requirements can create significant hurdles for market development.

[66] See N Swayne and A Phillips, above n 60, 191 where the authors note that in order to encourage private investment in CCS technology, the capital and operational costs need to be carefully monitored.

[67] *GGSA* (Qld) s 3.

[68] *GGSA* (Qld) s 27(1). This does not include any of the adjacent area under the *Petroleum (Submerged) Lands Act 1982* (Qld).

Greenhouse Gas Storage Act 2009 (Qld)

27 GHG storage reservoirs the property of the State

(1) All GHG storage reservoirs in land in the State are and are taken always to have been the property of the State.

(2) To remove any doubt, it is declared that–

   (a) a person does not acquire any property in a GHG storage reservoir or petroleum in it only because the person creates or discovers the reservoir; and

   Note– For other provisions about the ownership of petroleum, see the P&G Act, sections 26 to 28 and Chapter 2, part 6, division 3.

   (b) subsection (1) applies whether or not the land is freehold or other land.

(3) This section applies despite any other Act, grant, title or other document in force from the commencement of this section.

(4) In this section– the State does not include any of the adjacent area under the Petroleum (Submerged Lands) Act 1982.

A GHG storage reservoir is defined in the *GGSA* as 'the spatial extent of an underground geological formation'. This definition focuses upon the subsurface space rather than any right of usage or definitive physical attributes of the storage reservoir. Additional explanatory provisions set out that 'a person does not acquire any property in a GHG storage reservoir or petroleum only because the person creates or discovers the reservoir'.[69] Further, the State will acquire ownership of GHG storage reservoirs 'whether or not the land is freehold or other land'.[70]

Like the Victorian *GGGSA*, the Queensland *GGSA* segregates subsurface geological storage reservoirs from the land and vests title to this newly independent resource in the State. The *GGSA* contains no depth limitations so that all storage reservoirs within Queensland boundaries become the property of the State and are deemed to have always been so.

The deeming provision suggests that GHG storage reservoirs are not a newly verified resource, but rather a resource that the State has always owned. This is a questionable assumption because until the implementation of the *GGSA*, GHG storage reservoirs were assumed to constitute a part of the surface owner's estate under common law.[71] A vesting provision cannot function retrospectively when the resource to which it relates had no pre-existing independent identity.

---

69 *GGSA* (Qld) s 27(2)(a).
70 *GGSA* (Qld) s 27(2)(b).
71 The distinction between 'geological storage formations' and subsurface minerals is discussed by O L Anderson, 'Geologic CO_2 Sequestration: Who Owns the Pore Space?' (2009) 9 *Wyoming Law Review* 97, 101 who notes that the 'mineral owner' owns the minerals as well as rights to use the land for the purpose of extracting the minerals but does not own the geological storage space.

The *GGSA* establishes a framework for GHG permits and for GHG injection and storage leases in a similar manner to the *GGGSA*. Once a storage lease has been utilised, safety checks have been carried out, and the appropriate monitoring done, ownership and responsibility of the stored GHG automatically passes over to the State.[72] The titles issued under the licensing framework may be subject to a range of mandatory conditions; for example, GHG storage leases may only carry out GHG injection in compliance with a test plan that has been approved by the Minister.[73] Further, a GHG storage lease-holder may only use a GHG stream for GHG storage injection or for GHG stream storage where the stream consists of CO_2 or any substance incidentally derived from the process of carbon capture.[74]

The *GGSA* also sets out that all land grants issued in Queensland, whether before or after the commencement of the Act, are now taken to include a reservation to the State of all GHG storage reservoirs, including the exclusive right to enter and carry out any GHG storage activity and the right to authorise or regulate others to carry out GHG activities.[75] This provision has an extensive application as it qualifies the common law rights of surface estate owners not just by divesting them of their common law entitlements to subsurface storage facilities, but through the deemed implementation of incorporeal rights of access and entry which are now taken to encumber the surface estate.[76]

The validity of this retrospective vesting provision is unclear. Where a land grant is issued, the underlying radical title of the Crown is elevated to full beneficial ownership and the grantee derives a tenured estate from the Crown.[77] Whilst the Crown may regulate its underlying title, it may not derogate from grants already issued and only authorised reservations, consistent with the legislation upon which the grant is based, will be valid unless it can be established that the derogation is for the public benefit or the nation as a whole.[78]

---

72 *GGSA* (Qld) s 181(2).
73 *GGSA* (Qld) s 161.
74 *GGSA* (Qld) s 164.
75 *GGSA* (Qld) s 28.
76 By contrast, the Victorian *Greenhouse Gas Geological Sequestration Act 2008* (Vic) sets out that rights which are 'incidental' to exploration permits, injection licences or GHG formation retention leases will be included and presumably this will include access rights. See ss 19(1)(b), 59(b), 71(c).
77 See *Mabo v Queensland [No 2]* (1992) 175 CLR 1, [50]–[51].
78 See *O'Keefe v Williams* (1910) 11 CLR 171, 190 where Griffith CJ stated: 'The obligation of the Crown may also, I think, be put in another way, as an obligation not to do anything in derogation of the rights conferred by the statutory contract'. See also *Wik Peoples v Queensland* (1996) 187 CLR 1, [47] (Brennan CJ): 'the conditions which entitle a person to the grant of a freehold estate under a conditional purchase are prescribed by statute'. An implied covenant not to derogate against the grant will be read down to exclude those measures affecting the nation as a whole which the Crown took for public good: *Commissioners of Crown Lands v Page* [1960] 2 QB 274.

7.9.3 The Australian offshore statutory regime

The Australian statutory framework for offshore geological storage formations has been implemented in accordance with international law. The 1996 *Protocol to the Convention on the Prevention of Marine Pollution by Dumping of Wastes and Other Matter* and the 1992 *Convention for the Protection of the Marine Environment of the North-East Atlantic* ('OSPAR convention') both include CO_2 as a category of waste or other matter which may be considered for dumping in the sub-seabed where particular conditions are met.[79]

In accordance with these international conventions, the subsurface capture of CO_2 in the seabed is exempted from the dumping prohibitions. The Commonwealth and Victorian Governments have introduced offshore legislation to regulate this process. Unlike the onshore position – where ownership principles for subsurface storage facilities are critical – the absence of underlying ownership in the seabed means that the only issue relevant for offshore regulation concerns the nature, scope, and jurisdictional authority applicable to CCS projects arising in these regions.[80]

The Commonwealth and the Victorian offshore geological storage legislation is consistent with the constitutional arrangement that the states regulate the territorial sea out to three miles and the Commonwealth legislation has an application beyond this and in certain designated jurisdictional zones of the high seas.[81]

---

[79] *Protocol to the Convention on the Prevention of Marine Pollution by Dumping of Wastes and Other Matters*, adopted 7 November 1996 (entered into force 24 March 2006); *Convention for the Protection of the Marine Environment of the North-East Atlantic*, adopted 22 September 1992, 2345 UNTS 67 (entered into force 25 March 1998). There are currently 42 parties to the *Protocol* including Australia. Guidelines on how to store CO_2 in sub-seabed geological formations were also determined by parties to the *Protocol*. These are set out in the *2012 Specific Guidelines for the Assessment of Carbon Dioxide for the Disposal into Sub-Seabed Geological Formations* http://www.imo.org/blast/blastDataHelper.asp?data_id=31124.

[80] For a discussion of the ownership principles relevant to Australian territorial waters, see *Yarmirr v Northern Territory* (2001) 208 CLR 1, [70] where Gleeson CJ, Gaudron, Gummow and Hayne JJ concluded that any provisions purporting to vest property in the seabed beneath the territorial sea could not be characterised as conferring 'full ownership' because any such interpretation would be inconsistent with the recognition and enforcement of established public and international fishing, navigation, and free passage rights.

[81] This jurisdictional arrangement is encapsulated in the *Seas and Submerged Lands Act 1973* (Cth), an Act which also gives effect to the articles implemented in the 1982 *United Nations Convention on the Law of the Sea*, opened for signature 10 December 1982, 1833 UNTS 3, entered into force 16 November 1994. Under both the *Protocol to the Convention on the Prevention of Marine Pollution* (above n 79) and the *Convention for the Protection of the Marine Environment of the North-East Atlantic* (above n 79) the CO_2 must form part of a CO_2 stream from CO_2 capture processes for sequestration and may only be considered for dumping if (a) disposal is into a sub-seabed geological formation; (b) the stream consists overwhelmingly of CO_2, although they may contain incidental associated substances derived from the source material and the capture and sequestration processes used, and (c) no wastes or other matter are added for the purpose of disposing of those wastes or other matter. In addition, Annex II and III of the *OSPAR Convention* also provide that CO_2 streams may only be disposed of into sub-seabed geological formations where they are intended to be retained in these formations permanently and will not lead to significant adverse consequences for the marine environment, human health and other legitimate uses of the maritime area.

7.9.3.1 Commonwealth: *Offshore Petroleum and Greenhouse Gas Storage Act 2006*

The *Offshore Petroleum and Greenhouse Gas Storage Act 2006* (Cth) ('*OPGGSA*') was significantly amended in 2008 to accommodate offshore geological storage provisions.[82] This Act and the mirror legislation introduced by the Victorian Government is the product of an ongoing agreement between the Commonwealth and state governments to develop complementary legislative frameworks to regulate offshore greenhouse gas storage.[83]

The *OPGGSA* does not incorporate any vesting provision because the Crown does not have any underlying title in the seabed. The high seas are reserved for the 'common heritage of mankind' and are therefore incapable of being owned or apportioned otherwise than in accordance with rules that promote the common interest of all nations.[84] The Crown has no radical title in the territorial seas because they are not the 'dominion of the common law' and ownership may be inconsistent with established international rights of free passage, navigation, and fishing. As outlined by Gleeson CJ, Gaudron, Gummow and Hayne JJ in *Commonwealth v Yarmirr*, there are 'altogether different rights and interests which arose from the assertion of sovereignty over the territorial sea'.[85] The absence of radical title was therefore consistent with the fact that 'at no time before federation did the Imperial authorities assert any claim of ownership to the territorial seas or sea-bed'.[86] See also the discussion in Chapter 3.

The absence of core, derivative ownership does not preclude the exercise of state and Commonwealth power in the offshore region. As outlined in Chapter 3, the *OPGGSA* was introduced pursuant to the sovereign power of the Commonwealth to regulate territorial waters extending beyond the three-mile limit.[87] The greenhouse gas licences which are issued pursuant to the *OPGGSA* are – unlike onshore CCS licences – likely to be characterised as defeasible, statutory interests because of the absence of an underlying radical title. As regulatory interests, they are subject to the vicissitudes of subsequent

---

82 The Act was previously known as the *Offshore Petroleum Act 2006* (Cth).
83 This was discussed by the Council of Australian Governments meeting where the importance of progressing the collaborative framework between the Commonwealth and the states with respect to Carbon Capture and Storage Flagship programs was highlighted. See Council of Australian Governments, *Communique* (7 December 2009) 9 http://www.coag.gov.au/node/475.
84 See generally C Brennan, 'The Common Heritage of Mankind Principle in International Law' (1983) 21 *Columbia Journal of Transnational Law* 305.
85 *Commonwealth v Yarmirr* (2001) 208 CLR 1, [50].
86 *R v Keyn* (1876) 2 Ex D 63, 239 (Lush J). For a further discussion on this by the Australian High Court see *Commonwealth v Yarmirr* (2001) 208 CLR 1, [50].
87 For a discussion on the distinction between sovereignty of title and sovereignty of power see *Mabo v Queensland [No 2]* (1992) 175 CLR 1, [50]–[51].

modification and amendment and are therefore unlikely to attract the just terms protection in s 51(xxxi) of the *Constitution*.[88]

The regulatory framework of the *OPGGSA* deals with 'eligible' greenhouse gas storage formations. An 'eligible' storage formation is one that is suitable, without engineering enhancements, for the permanent storage of at least 100 tonnes of greenhouse gas injected over a period of time.[89] An applicant may store injected greenhouse gas in an eligible greenhouse gas formation where they have obtained the requisite licence.[90] Greenhouse gas storage formations that do not comply with this storage volume, or which have been artificially enhanced – including those created either directly or indirectly for the purpose of petroleum drilling – will not be subject to the *OPGGSA*.

A greenhouse gas holding lease entitles the holder to explore in the leased area for a potential greenhouse gas storage formation and to inject and store greenhouse gas, on an appraisal basis, in a geological formation located within the leased area.[91] A greenhouse gas injection licence authorises the licencee to inject and permanently store a greenhouse substance into an identified greenhouse gas storage formation located in the licence area.[92] The greenhouse gas injection licence also allows the holder to explore the area for geological storage formations and to inject and store greenhouse gases on an appraisal basis.[93]

The licensing framework of the *OPGGSA* is intended to operate consistently with the petroleum licensing framework. The Act sets out that the issuing of a greenhouse gas storage interest is not precluded from areas where a petroleum exploration licence, production licence, retention lease, special prospecting authority, or access authority has already been issued and vice versa.[94]

The interface between petroleum and storage rights is, however, unclear. For example, where a petroleum exploration licence covers a location that includes a greenhouse gas storage right – and the injection of greenhouse gas into the area may have a 'significant adverse impact' upon either the recovery of petroleum or its commercial viability – the Minister has the power to 'eliminate the risk' by either suspending or imposing specific directions on the storage licence.[95] The *OPGGSA* does not articulate the circumstances that would raise a 'significant adverse impact'

---

88 For further discussion on the regulatory nature of offshore property see K Gray, 'Regulatory Property and the Jurisprudence of Quasi-Public Trust' (2010) 32 *Sydney Law Review* 221, 224–8; S Hepburn, 'Native Title Rights in the Territorial Sea and Beyond: Exclusivity and Commerce in the *Akiba* Decision' (2011) 34(1) *University of New South Wales Law Journal* 159, 162–5.
89 *OPGGSA* (Cth) s 21(1).
90 *OPGGSA* (Cth) pts 3.3, 3.4.
91 *OPGGSA* (Cth) s 319.
92 *OPGGSA* (Cth) s 357.
93 *OPGGSA* (Cth) s 357(1)(c)–(h).
94 *OPGGSA* (Cth) s 458.
95 *OPGGSA* (Cth) s 383.

so that it remains unclear whether the future potential for CO_2 to migrate from the storage area would come within the scope of this provision.[96]

7.9.3.2 Victoria: *Offshore Petroleum and Greenhouse Gas Storage Act 2010*

Like its Commonwealth counterpart, the Victorian *OPGGSA* does not incorporate any vesting provisions for offshore subsurface storage spaces. Rather, it provides for the issuing of assessment permits, greenhouse gas holding leases, greenhouse gas injection licences, and greenhouse gas special authorities.[97]

A gas injection licence is expressly mandated to continue 'indefinitely' and this aligns the statutory grant with a freehold equivalent. The licence may, however, be terminated by the Minister if no operations have been carried on for a continuous period of five years.[98]

If a greenhouse gas interest overlaps with a petroleum interest, and there is a risk that operations will have a significant adverse impact upon either the recovery or commercial viability of petroleum, the Victorian *OPGGSA* will allow the Minister to 'eliminate the risk' by either suspending or imposing specific directions on the storage licence.[99] Unlike the Commonwealth Act, the Victorian *OPGGSA* provides further guidance on how a 'significant adverse impact' should be assessed.[100] The expense of a potential impact must be evaluated in terms of the effect it has upon the operating costs of a petroleum plant and any ongoing costs associated with a reduction in the rate of recovery or quantity of petroleum.[101]

7.9.4 Legal liability for carbon capture in Australia

7.9.4.1 Environmental concerns

The primary concern with CCS lies in the probability of leakage, leakage volume, and the consequences of such leakage. Existing risk modelling regarding the storage of CO_2 raises a number of concerns including a failure to accurately predict the geo-mechanical response of injection, a failure to adequately predict

---

96 This is discussed by S Hazeldine, 'Geological Factors in Framing Legislation to Enable and Regulate Storage of Carbon Dioxide Deep in the Ground' in I Havercroft, R Macrory and R Steward (eds), *Carbon Capture and Storage: Emerging Legal and Regulatory Issues* (Hart Publishing, 2011) 14, 17. It should be noted that the Victorian legislation does address this issue. The *Offshore Petroleum and Greenhouse Gas Regulations 2011* (Vic) regs 157–161 specifically focus upon different types of 'significant adverse impacts'.
97 *OPGGSA* (Vic) pt 3.
98 *OPGGSA* (Vic) s 378. The equivalent provisions in the *OPGGSA* (Cth) is ss 359–360.
99 *OPGGSA* (Vic) s 383.
100 *Offshore Petroleum and Greenhouse Gas Storage Act Regulations 2011* (Vic) regs 157–161.
101 Ibid reg 158(2).

the location of precipitation or dissolution and a failure to predict the risk of induced seismicity.

During the injection and operation phases, the injected liquid CO_2 poses the highest risk of escape and must be physically contained. Where injection occurs into saline formations, the injected CO_2 will tend to migrate upwards and laterally owing to the density of the CO_2 being less than the density of the saline water within the formation. The Intergovernmental Panel on Climate Change (IPCC) has stated that the fraction of CO_2 retained in appropriately selected and managed geological reservoirs is 'very likely' to exceed 99 per cent over 100 years and 'likely' to exceed 99 per cent over 1000 years.[102] Total security and a 100 per cent risk against leakage cannot be guaranteed. The potential damage from unexpected migration includes changes in subsurface pressure generated by the fracturing of the reservoir seal, groundwater contamination, and alternations to the pH of subsurface water in the vicinity of the sequestration site.

7.9.4.2 Tortious actions

Legal liability may arise where the CO_2 trespasses through the injection site into the surface or subsurface of adjacent land. Wrongful entry into the subsoil of land in the possession of another is trespass, including where entry is affected through a natural aperture on the defendant's own land pursuant to excavation or tunnelling.[103] For a trespass to be successful it must be established there has been a direct interference with land that is the fault of the defendant. This means that surface estate owners who are directly impacted by the leakage of CO_2 may be entitled to bring an action in trespass.[104]

An action may also be brought in private nuisance for the 'unreasonable interference with a person's use or enjoyment of land, or some right over or in connection with it. The Attorney General may bring an action in public nuisance for interference to the public at large on public lands. The interference must be sufficiently substantial to cause material injury to the land or property or substantial interference with the comfort and convenience of the occupier of the land. A defendant may argue that he or she had statutory authorisation for the activities which gave rise to the interference, although it is unlikely that harm caused by CO_2 which has migrated beyond the boundaries of the authorised CCS licence will attract this defence.

---

102 See Bert Mertz et al (eds), IPCC Special Report on Carbon Dioxide Capture and Storage (IPCC, 2005) 34. Available at: <http://www.ipcc.ch/report/srccs/>.
103 See *Edwards v Sims* (1929) Ky 24 SW (2d) 619; *Stoneman v Lyons* (1975) 133 CLR 550, 561–2 (Stephen J).
104 See especially the discussion by N Swayne and A Phillips, 'Legal Liability for Carbon Capture and Storage in Australia: Where should the losses fall?' (2012) 29 *Environmental Planning Law Journal* 189. See also G Campbell, 'Carbon Capture and Storage: Legislative Approaches to Liability – Managing Long-Term Obligations and Liabilities' (2009) *AMPLA Yearbook* 324, 344.

Where it can be established that a duty of care is owed by the CCS operator to the plaintiff (who may, for example, be the surface estate owner) and it is breached due to a failure to comply with reasonable safety standards, damages for personal injury and economic loss may be recoverable in negligence. Significantly, Queensland completely exempts surface estate owners from any liability regarding CCS projects that occur upon their land setting out that the owner or occupier is not civilly liable to anyone else for a claim based in tort for damages.[105] Victoria does not provide complete immunity, although it does make it clear that the holder of a greenhouse gas authority will, for the purpose of establishing any common law liability, be regarded as the owner of the land.[106] The risk of leakage from the CCS project and the likely damage such leakage might cause to the environment and the atmosphere will be relevant to determining whether a duty of care has been breached. In particular, the issue of foreseeability is relevant. If a CCS operator could be taken to 'foresee' the likelihood of damage or harm flowing from leakage – even if the possibility of leakage is remote – a failure to take appropriate action – including, for example, selecting a suitable storage site, adopting appropriate injection processes, and monitoring the site – may constitute a breach. Obviously it will be difficult for the courts to ascertain what constitutes a 'reasonable' standard of care given the fact that the technology with CCS is new and monitoring will need to occur over hundreds of years. However, as outlined by Swayne and Phillips, early CCS projects, in determining a consistent management practice, must be aware that 'the fact that a risk of harm could have been avoided by doing something in a different way will not of itself give rise to or affect liability for the way in which the thing was done'.[107]

7.9.4.3 Regulatory standards

Regulatory standards, common industry standards, and expert practice will all be relevant in this context. In Victoria and Queensland the CCS operator must comply with an approved plan which determines the rate and volume of CO_2 injection and the composition of the greenhouse gas that is permitted to be injected.[108] Any amendments to the plan must be appropriately approved by the Minister. In Queensland, the Minister will consider the reasonableness of the variation.[109] In Victoria or Western Australia, the Minister may actually mandate change if they believe it is necessary for a more effective injection process or in order to maximise

---

105 *GGSA* (Qld) s 338A.
106 *GGGSA* (Vic) s 187.
107 Swayne and Phillips, above n 103, 200.
108 See *GGSA* (Qld) s 141(c)(i) injection and monitoring plan; *GGGSA* (Vic) s 93; Petroleum and Geothermal Energy Legislation Amendment Bill 2013 (WA) cl 77 which seeks to introduce s 66(2); *OPGGSA* (Cth) s 358(3)(d), (e), (f), (j); *OPGGSA* (Vic) s 374(1).
109 *GGSA* (Qld) s 158(c).

the volume of substance capable of being stored.[110] Where a 'serious situation' exists, the Minister has power to issue directions to a CCS operator to take any action and they do have the power to prohibit action.[111] Injection may be suspended or terminated, remediation may be required, or changes to the management process may be necessary.[112] A serious situation is defined broadly to include any situation where CO_2 being injected into the storage formation is leaking.[113] In Queensland, Victoria, and under the Commonwealth and offshore legislation, this includes a situation where there is a significant risk that CO_2 will leak from the formation.[114]

7.10 REVIEW QUESTIONS

1. Explain why carbon capture sequestration (CCS) is imperative within a carbon economy and how it differs from biological sequestration.
2. How is CCS regulated, if at all, under common law?
3. What changes have the onshore CCS statutory frameworks in Victorian and Queensland introduced?
4. What type of interest does the Crown/state acquire in subsurface geological storage space and what are some of the inherent difficulties associated with this type of ownership?
5. What sort of rights are conferred pursuant to a greenhouse gas injection licence?
6. Why do you think it is necessary to impose volume limits upon CCS processes?
7. What rights, if any, does a surface estate owner have to consult with, negotiate or object to the development of a CCS project underneath private land that they own?
8. Consider the following problem:

   The Gorgon Project in Western Australia is incorporating pre-combustion CCS technology into its newly constructed gas plant. This will result in the separation of the hydrogen and carbon prior to combustion. The carbon will then be injected into subsurface reservoirs. What are the possible difficulties that might be encountered if the reservoirs prove unsuitable and leakage of carbon causes environmental damage to the pristine habitat surrounding Barrow Island where the project is located?

---

110 *GGGSA* (Vic) ss 89–91; Petroleum and Geothermal Energy Legislation Amendment Bill 2013, cl 78(3) amending s 69(3).
111 *OPGGSA* (Cth) ss 380(1)(g), (h); *GGGSA* (Vic) ss 182(f), (g); *OPGGSA* (Vic) ss 406(1)(g), (h).
112 *OPGGSA* (Cth) s 380(1)(c); *GGSA* (Qld) s 364(2); *GGGSA* (Vic) s 182(c); *OPGGSA* (Vic) s 406(1)(c).
113 *OPGGSA* (Cth) s 379(1); *GGSA* (Qld) s 363(1); *GGGSA* (Vic) s 6; *OPGGSA* (Vic) s 405(1). Unlike the other jurisdictions, the Petroleum and Geothermal Energy Resources Bill 2013 (WA) does not propose to include directions for responding to 'serious situations'.
114 *OPGGSA* Cth s 379(1)(b); *GGSA* (Qld) s 363(1)(b); *OPGGSA* (Vic) s 405(1)(b).

9. What powers does the Minister have to intervene in a CCS project in Victoria where a 'serious situation' arises and how is a 'serious situation' defined?
10. Where a CCS storage reservoir exists underneath private land what type of compensation is the surface estate holder entitled to apply for?

7.11 FURTHER READING

O L Anderson, 'Geologic CO_2 Sequestration: Who Owns the Pore Space?' (2009) 9 *Wyoming Law Review* 97.

C Brennan, 'The Common Heritage of Mankind Principle in International Law' (1983) 21 *Columbia Journal of Transnational Law* 305.

W Buzbee, 'Recognising the Regulatory Commons: A Theory of Regulatory Gaps' (2003) 89 *Iowa Law Review* 1.

W Buzbee, 'State Greenhouse Gas Regulation, Federal Climate Change Regulation and the Preemption Sword' (2009) 1 *San Diego Journal of Climate and Energy Law* 23.

G Campbell, 'Carbon Capture and Storage: Legislative Approaches to Liability – Managing Long-Term Obligations and Liabilities' (2009) *AMPLA Yearbook* 324.

Cato 2, *The CATO-2 programme* http://www.co2-cato.org/cato-2.

Council of Australian Governments, *Communique* (7 December 2009) http://www.coag.gov.au/node/475.

A B Endres, 'Geologic Carbon Sequestration: Balancing Efficiency Concerns and Public Interests in Property Rights Allocations' (2011) *University of Illinois Law Review* 623.

European Biofuels Technology Platform, *Bio-Mass with CO_2 Capture and Storage (Bio-CCS): The way forward for Europe* (20 June 2012) http://www.biofuelstp.eu/downloads/bioccsjtf/EBTP-ZEP-Report-Bio-CCS-The-Way-Forward.pdf.

V B Flatt, 'Paving the Legal Path for Carbon Sequestration from Coal' (2009) 19 *Duke Environmental Law & Policy Forum* 211.

Global CCS Institute, *Report to the Global CCS Institute on Legal and Regulatory Developments Related to Carbon Capture Storage between November 2010 and June 2011* (13 December 2011) <http://www.globalccsinstitute.com/publications/legal-and-regulatory-developments-related-carbon-capture-and-storage-between-november-2>.

Global CCS Institute, *CO_2 Capture Technologies: Oxy Combustion with CO_2 Capture* (January 2012) http://www.globalccsinstitute.com/publications/co2-capture-technologies-oxy-combustion-co2-capture.

Global CCS Institute, *Rotterdam CCS Network* (2013) http://www.globalccsinstitute.com/projects/12661.

Global CCS Institute, *Comparing Different Approaches to Managing CO_2 Storage Resources in Mature CCS Futures* (March 2014) http://decarbonise/sites/default/files/publications/162748/comparing-different-approaches-managing-co2-storage-resources-mature-ccs-futures.pdf.

Global CCS Institute, *Moving Below Zero: Understanding Bio-energy with Carbon Capture and Storage* (9 April 2014) <http://www.globalccsinstitute.com/publications/moving-below-zero-understanding-bioenergy-carbon-capture-storage>.

Global CCS Institute, *The Global Status of CCS: Released by the Global Carbon Capture Institute* (5 November 2014) <http://www.globalccsinstitute.com/publications/global-status-ccs-2014>.

Global CCS Institute, *How CCS Works – Transport* http://www.globalccsinstitute.com/content/how-ccs-works-transport.

Global CCS Institute, S Whittiker and E Perkins, *Technical Aspects of CO_2 Enhanced Oil Recovery and Associated Carbon Storage* (2013) http://decarboni.se/sites/default/files/publications/118946/technical-aspects-co2-enhanced-oil-recovery-and-associated-carbon-sto.pdf.

B N Grave, 'Carbon Capture and Storage in South Dakota' (2010) 55 *South Dakota Law Review* 72.

K. Gray, 'Regulatory Property and the Jurisprudence of Quasi-Public Trust' (2010) 32 *Sydney Law Review* 221.

S Hazeldine, 'Geological Factors in Framing Legislation to Enable and Regulate Storage of Carbon Dioxide Deep in the Ground' in I Havercroft, R Macrory and R Steward (eds), *Carbon Capture and Storage: Emerging Legal and Regulatory Issues* (Hart Publishing, 2011).

M Heller, 'The Tragedy of the Anti-Commons: Property in the Transition from Marx to Markets' (1998) 111 *Harvard Law Review* 621.

S Hepburn, 'Native Title Rights in the Territorial Sea and Beyond: Exclusivity and Commerce in the *Akiba* Decision' (2011) 34(1) *University of New South Wales Law Journal* 159.

S Hepburn, 'Ownership Models for Geological Sequestration: A Comparison of the Emergent Regulatory Models in Australia and the United States' (2014) 44 *Environmental Law Reporter* 10310.

IEAGHG, *Deployment of CCS in the Cement Industry* (December 2013) http://decarboni.se/sites/default/files/publications/162743/deployment-ccs-cement-industry.pdf.

Intergovernmental Panel on Climate Change, *Carbon Dioxide Capture and Storage* (2005) http://www.ipcc.ch/pdf/special-reports/srccs/srccs_wholereport.pdf.

Intergovernmental Panel on Climate Change, O Edenhofer, R Pichs-Madruga, Y Sokona (et al), 'Chapter 7: Energy Systems' in *Climate Change 2014 Mitigation of Climate Change: Working Group III Contribution to the Fifth Assessment Report of the Intergovernmental Panel on Climate Change* (Cambridge University Press, 2014).

Intergovernmental Panel on Climate Change, T F Stocker, D Qin, G K Plattner, M Tignor, S K Allen, J Boschung, A Nauels, Y Xia, V Bex and P M Midgley (eds), 'Summary for Policymakers' in *Climate Change 2013 The Physical Science Basis: Working Group I Contribution to the Fifth Assessment Report of the Intergovernmental Panel on Climate Change* (Cambridge University Press, 2013).

International Energy Agency, CO_2 *Emissions from Fuel Combustions: Highlights* (2013) http://www.iea.org/publications/freepublications/publication/co2emissionsfromfuelcombustionhighlights2013.pdf.

International Energy Agency, *Technology Roadmap; Carbon Capture and Storage* (2013) http://www.iea.org/publications/freepublications/publication/technology-roadmap-carbon-capture-and-storage-2013.html.

International Energy Agency, *World Energy Outlook 2014* (13 November 2013).

S Kalen, 'Coal's Plateau and Energy's Horizon' (2013) *Public Land and Resources Law Review* 145.

T M Kerr, 'Carbon Dioxide Capture and Storage: Priorities for Development' (2008) 4 *Carbon and Climate Law Review* 335.

E Kriegler, J P Weyant, G J Blanford, V Krey, L Clarke, J Edmonds, A Fawcett, G Ludere, K Riahi, R Richels, S K Rose, M Tavoni and D. Van Vuuren, 'The role of technology for achieving climate policy objectives: overview of the EMF 27 study on global technology and climate policy strategies' (2014) 123 *Climate Change* 353.

T Lloyd, 'Carbon Capture and Sequestration: Removing The Legal and Regulatory Barriers' (2014) 4 *Climate Law* 187.

T J Logan, 'Carbon Down Under: Lessons From Australia: Two Recommendations for Clarifying Sub-Surface Property Rights to Facilitate Onshore Geologic Carbon Sequestration in the United States' (2010) 11 *San Diego International Law Journal* 561.

A Long, 'Complexity in Global-Energy Environmental Governance' (2014) 15 *Minnesota Journal of Law, Science and Technology* 1055.

Minister for State and Energy Resources, State of Victoria, 'Victoria Leads Australia with Carbon Capture and Storage Bill' (Press Release, 9 September 2008) 1 http://www.premier.vic.gov.au/component/content/article/4704/html.

Organisation for Economic Co-operation and Development, International Energy Agency, *Carbon Capture and Storage in the CDM* (November 2007) http://www.iea.org/publications/freepublications/publication/CCS_in_CDM.pdf.

Power Technology.com, 'Leading the Way: CCS Fitted Coal-Fired Stations Now a Reality' (14 July 2014) http://www.power-technology.com/features/featureleading-the-way-ccs-fitted-coal-fired-power-stations-now-a-reality-4317055/.

A W Reitze Jr, 'State and Regional Control of Geological Carbon Sequestration (Part 1)' (2011) 41 *Environmental Law Reporter* 10348.

C M Rose, 'Property in all the Wrong Places?' (2005) 114 *Yale Law Journal* 991.

S Som, 'Creating Safe and Effective Carbon Sequestration' (2008) 17 *New York University Environmental Law Journal* 961.

State Government of Victoria, Department of Energy and Earth Resources, *A Regulatory Framework For The Long-Term Underground Geological Storage Of Carbon Dioxide In Victoria* (January 2008) http://www.dpi.vic.gov.au/energy/about/legislation-and-regulation/ccs-regulations/a-regulatory-framework-for-the-long-term-underground-geological-storage-of-carbon-dioxide.

N Swayne and A Phillips, 'Legal Liability for Carbon Capture and Storage in Australia: Where should the losses fall?' (2012) 29 *Environmental Planning Law Journal* 189.

A Thorpe, 'Too Little, Too Soon: An Assessment of Australian Carbon Capture and Storage Legislation Against the New Standards set for the Clean Development Mechanism' (2012) 3 *Climate Law* 139.

K Trisolini, 'Holistic Climate Change Governance: Towards Adaptation and Mitigation Synthesis' (2014) 85 *University of Colorado Law Review* 615.

United States Environmental Protection Agency, *Standards of Performance for Greenhouse Gas Emissions From New Stationary Sources: Electric Utility Generating Units; Proposed Rule* (2014) 40 CFR pts 60, 70, 71, 79 and 98.

K Vatopoulos, E Tzimas, 'Assessment of CO_2 capture technologies in cement manufacturing process' (2012) 32 *Journal of Cleaner Production* 251.

P J Vergragt, *CCS: The Next Technological Lock-in? A Case study from The Netherlands* (Tellus Institute, Boston, 2008).

M Wang, A Lawal, P Stephenson, J Sidders, C Ramshaw and H Yeung, 'Post-Combustion CO_2 with Chemical Absorption: A State of the Art Review' (2011) 89(9) *Chemical Engineering and Design* 1609.

J B Wiener, 'Global Environmental Regulation: Instrument Choice in Legal Regulation' (1999) 108 *Yale Law Journal* 677.

Zero CO_2NO, *The Netherlands* http://www.zeroco2.no/projects/countries/the-netherlands.

8

CLIMATE CHANGE AND MINING AND ENERGY POLICY

8.1	Introduction	316
8.2	Changes in the climate system: Atmosphere, ocean, cryosphere, and sea level	318
8.3	The legal framework	320
8.4	The economics of climate change	335
8.5	The impact of climate change on Australian mining and energy industries	342
8.6	Review questions	348
8.7	Further reading	349

8.1 Introduction

Climate change is a profound concern for the modern world. Global debate regarding whether it is occurring, how it is occurring, and how it might be mitigated continues unabated. Increasingly, however, there appears to be international consensus from climate scientists, economists, and world leaders that human induced anthropogenic activities since the pre-industrial age have contributed significantly to climate change.[1] For example, the burning of fossil fuels for energy production has increased levels of greenhouse gases (GHGs), such as CO_2, being emitted into the earth's atmosphere and this has generated an unnatural 'increase in global warming'[2] 'by around 0.7 degrees Celsius'.[3] There is little doubt that the effects of this warming have contributed to detrimental changes in the earth's climate where 'global average temperatures have risen considerably since measurements began in the mid-1800s',[4] 'weather patterns have been affected'[5] together with unnatural 'changes in the oceans and sea level'.[6] Whilst the earth's climate has always been variable, with historical data indicating the occurrence of major climate shifts across periods as modest as a few decades, and an increase during the twentieth century of the earth's global average surface temperature by 0.8°C, one of the most significant and unsettled issues from a policy perspective is how climate change can be expected to behave during the next century under natural and human influences. A further and more critical issue is how ecosystems and other activities may be impacted. Our understanding of these issues will necessarily inform decisions, at both the global and the regional level, about energy and infrastructure.[7]

Decisions made regarding the infrastructure and technologies that we install today will affect emissions throughout the century.[8] The next 15 years are particularly crucial. This is because of the greenhouse gas-emitting capital stock that

---

1 Explanatory Memorandum, Carbon Pollution Reduction Scheme Bill 2009 (Cth) 3.
2 Intergovernmental Panel on Climate Change (IPCC), *Climate Change 2007 – The Physical Science Basis: Contribution of Working Group I to the Fourth Assessment Report of the IPCC* (2007). For example, see *Summary for Policymaker*.
3 Australian Government, *Productivity Commission for the Prime Ministerial Task Group on Emissions Trading* (Productivity Commission, 2007) ch 1, 2.
4 R Garnaut, *The Garnaut Climate Change Review: Final Report* (Cambridge University Press, 2008) 78.
5 Ibid 82.
6 Ibid.
7 See the discussion by S E Koonin, 'A Degree of Uncertainty', *The Wall Street Journal* (online), 23 September 2014, http://online.wsj.com/articles/climate-science-is-not-settled-1411143565?mod=WSJ_myyahoo_module.
8 See F Jotzo, 'Prerequisites and Limits for Economic Modelling of Climate Change Impacts and Adaptation', (Research Report No 1055, Australian National University, March 2010).

already exists globally and, in addition, the US$90 trillion of new infrastructure investment that is expected during this period, across the cities, land use, and energy systems where greenhouse gas emissions are concentrated. Action to reduce greenhouse gas emissions has become even more urgent because of the long lag effects – or 'lock in' – that has occurred, both in terms of atmospheric physics and human infrastructure.

To begin with, there is a long atmospheric time-lag as it can take between 25 to 30 years for CO_2 molecules to reach the upper atmosphere and cause the 'greenhouse effect' of trapping heat. Thus, moderating climate change in 2040–2050 necessarily requires cutting GHG emissions today and over the next 10 years. CO_2 remains in the atmosphere for several centuries; hence, the only way to truly limit its impact is to avoid excessive emissions in the first place. Further, GHG emissions are derived from expensive capital infrastructure that has a long life span. Once a power station, building, factory, or car has been built, it will generate emissions at about the same rate through its entire duration. This can be 40–50 years for a power station and even longer for some buildings on the assumption that there is no modification for new technology or energy efficiency. Once capital assets are built, their lifetime emissions are potentially irrevocable for decades.[9] The difficulty in encouraging an immediate response to 'lock in' climate change is that it requires a fundamental shift in human behaviour which itself depends upon acts which surpass 'universal human egotism'.[10]

Climate change has a particular cogency for mining and energy industries because the most significant problem in the climate change issue is the increasing release of carbon into the atmosphere. Much of this comes from coal-fired power stations but the use of oil and gas is also relevant. Most global statistics indicate that at least half – and probably much more – of the greenhouse gas emissions are the product of energy production and use.[11] The International Energy Agency

---

9 For a discussion of the 'lock in' effect see S J Liebowitz and E S Margolis, 'Path Dependence, Lock-In, and History' (1995) 11 (1) *Journal of Law, Economics and Organization* 205, 207–10.
10 See especially the discussion by R Epstein, 'Behavioral Economics: Human Errors and Market Corrections' (2006) 73 (1) *The University of Chicago Law Review* 111, 112 where the author argues that any social or economic framework 'has to contend with behaviour that inflicts harm on others and at the other extreme, must give breathing room for acts of selflessness that falsify the claims of universal human egotism'.
11 See the discussion by R Lyster and A Bradbrook, *Energy Law and the Environment* (Cambridge University Press, 2006) 51 where the authors note that climate change has the potential to be the most serious of all the problems of atmospheric pollution in terms of global environmental impact. See also the discussion by E Ostrom, 'Polycentric Systems for Coping with Collective Action and Global Environmental Challenge' (2010) 20 *Global Environmental Change* 550, 555 where the author notes that there are 'no panaceas for … for complex problems such as global warming'.

(IEA) estimates that even with a significant shift away from coal and oil, anticipated emissions through 2035 correspond with a long-term global average temperature rise of 3.6°C.[12]

8.2 Changes in the climate system: Atmosphere, ocean, cryosphere, and sea level

According to art 1[3] of the *United Nations Framework Convention on Climate Change (UNFCCC)*, the climate system is 'the totality of the atmosphere, hydrosphere, biosphere and geosphere and their interactions'.[13] Changes in the earth's climate system – or 'climate change' – are a 'continuing natural phenomenon' that flows from the cyclical orbiting of the earth around the sun and the natural volcanic or seismic activity.[14] Until the twentieth century, climate change had '... occurred more slowly allowing communities and ecosystems to adapt gradually or to migrate to more suitable habitats.'[15] In the last 100 years, however, there has been an enormous increase in global warming which far exceeds the natural process. The scientific evidence from climate scientists indicate that the increase in global warming cannot be fully explained by the various processes associated with the earth's radiative budget – for example, orbital variations and the radiative perturbation of mineral dust – and that human activities, particularly fossil fuel consumption, deforestation, and agriculture, have contributed to climate sensitivity.[16]

The Intergovernmental Panel on Climate Change has confirmed that human activities are 'very likely' to be responsible for the global warming that has occurred over the last 50 years and this has made global warming one of the most crucial

---

12 See International Energy Agency, *World Energy Outlook – Factsheet* (IEA 2012) 1, anticipating natural gas to overtake coal as an energy source by 2035 by which time rises in energy related greenhouse gas emissions will have resulted in a long-term average temperature increase of 3.6°C. See also the discussion by A Long, 'Complexity in Global Energy-Environment Governance' (2014) 15 *Minnesota Journal of Law Science and Technology* 1054, 1065.
13 *United Nations Framework Convention on Climate Change*, opened for signature 4 June 1992, 1771 UNTS 107 (entered into force 21 March 1994), art 1[3].
14 W Gumley, 'Legal and Economic Responses to Global Warming – An Australian Perspective' (1997) 14 *Environmental and Planning Law Journal* 341, 352.
15 Ibid.
16 See P Kohler, R Bintanja, R Rischer, H Joos, F Knutt, R Lohmann, and V Masson-Delmotte, 'What caused Earth's temperature variations during the last 800,000 Years? Data-based evidence on radiative forcing and constraints on climate sensitivity' (2010) 29 *Quaternary Science Reviews* 129, 132–3.

environmental, social, and economic challenges of recent history.[17] Consequently, the UNFCCC has recalibrated their definition of climate change to refer to change that has a human induced quality. The UNFCC states: '... change of climate which is attributed directly or indirectly to human activity that alters the composition of the global atmosphere and which is in addition to natural climate variability observed over comparable time periods'.[18]

Over the last decade, the Intergovernmental Panel on Climate Change ('IPCC') has carefully examined the connection between human induced climate change and global warming.[19] In 2007, the IPCC, in its *Working Group I Report*, commented that 'the increase of global surface temperatures, changes in precipitation patterns and observations of changes in ocean and arctic temperatures'[20] all pointed to the conclusion that '[human induced] climate change is an indisputable threat'.[21] In its *Fifth Assessment Report*, released in 2013 ('IPCC Report'), the IPCC Working Group summarised the outcomes of a range of different, independent scientific investigations that had been conducted since the release of the fourth assessment report in 2007. The IPCC Report concluded that the majority of the computer modellings conducted indicated an 'unequivocal' warming of the climate system since the 1950s, holding that such changes were 'unprecedented' over the decades leading to the millennia.[22] The warming of the oceans and the atmosphere, the shrinking of the ice-caps, and the rise in GHG in the atmosphere have – according to this report – been clearly apparent from the climate science and a warming of between 0.65–1.06°C occurred between the period of

---

17 See IPCC, above n 2, where a discussion on characteristics of climate change over the last century is made. It is noted that climate change can be 'characterised by a change in the average temperature over time or by a change in the distribution and severity of weather events ... Since modern measurements began in the late 1800s, global average surface temperature has increased by around 0.7°–0.8°C ... In 1910, the average temperature over the previous 20 year period was 13.73°C ... In 2009, it was 14.53°C'.
18 *United Nations Framework Convention on Climate Change*, opened for signature 4 June 1992, 1771 UNTS 107 (entered into force 21 March 1994), art 1[2].
19 The IPCC is a scientific international body, which operates 'under the auspices of the United Nations (UN). It reviews and assesses the most recent scientific, technical and socio-economic information produced worldwide relevant to the understanding of climate change.' See Intergovernmental Panel on Climate Change, *Organization* <http://ipcc.ch/organization/organization.shtml>.
20 Intergovernmental Panel on Climate Change, *Climate Change 2001 Synthesis Report: Summary for Policymakers* (2001) <http://www.ipcc.ch/pdf/climate-changes-2001-synthesis-spm/synthesis-spm-en.pdf>.
21 Ibid.
22 Intergovernmental Panel on Climate Change, T F Stocker, D Qin, G K Plattner, M Tignor, S K Allen, J Boschung, A Nauels, Y Xia, V Bex and P M Midgley (eds) 'Summary for Policymakers' in *Climate Change 2013: The Physical Science Basis. Contribution of Working Group I to the Fifth Assessment Report of the Intergovernmental Panel on Climate Change* (Cambridge University Press, 2013) 2.

1880–2012, and during the period 2003–2012 an accelerated warming between 0.72–0.85°C.[23]

The IPCC Report suggests that the modelling shows, inter alia, a decrease in snow cover in the Northern Hemisphere since the mid-nineteenth century, substantial Arctic warming since the mid-twentieth century, a significant drop in the Arctic summer sea ice from 1900–2012, an upward trend in the global average ocean heat content and a global mean sea level rise of approximately 0.19 m. More specifically, the IPCC Report suggests that each of the last three decades has been successively warmer at the earth's surface than any preceding decade since 1850. In the Northern Hemisphere, 1983–2012 was 'likely the warmest 30 year period of the last 1400 years'.[24] The IPCC Report also suggests that changes in many extreme weather and climate events have been observed since about 1950 and that it is 'likely that the number of cold days and nights has decreased and the number of warm days and nights has increased on the global scale'. Further, the IPCC Report specifically states that it is

> ... likely that the frequency of heat waves has increased in large parts of Europe, Asia, and Australia. There are likely more land regions where the number of heavy precipitation events has increased than where it has decreased. The frequency or intensity of heavy precipitation events has likely increased in North America and Europe.[25]

The IPCC Report also concludes that it is 'virtually certain' that the upper ocean (0–700 m) warmed from 1971–2010 and it likely warmed between the 1870s and 1971 and that Greenland and Antarctic ice sheets have been losing mass, glaciers have continued to shrink almost worldwide, and Arctic sea ice and Northern Hemisphere spring snow cover have continued to decrease.[26] Finally, the IPCC Report concluded that the rate of sea level rise since the mid-nineteenth century has been larger than the mean rate during the previous two millennia. Over the period 1901–2010, global mean sea level rose by approximately 0.19 m.

8.3 The legal framework

From the 1990s–2009 the primary focus for addressing climate change was through the development of the *United Nations Framework Convention on Climate Change*. Global climate change was an important concern at the Rio Conference[27] because of the research collected by the IPCC.

---

23 Ibid.
24 See IPCC, above n 22, B1.
25 Ibid.
26 Ibid.
27 This is also known as the United Nations Conference on Environment and Development.

8.3.1 The IPCC

The IPCC was created in 1988 by the World Meteorological Organization ('WMO') and the United Nations Environment Program ('UNEP'). The role of the IPCC is to research scientific, technical, and socio-economic information relevant to a comprehension of the scientific basis of risk of human induced climate change, its potential impacts, and the range of options available for mitigation and adaptation. The IPCC has three Working Groups and a Task Force. Working Group I assesses the 'physical science' basis of climate change. Working Group II assesses climate change impacts, adaptation, and the vulnerability of socio-economic and natural systems to climate change. This includes the consequences of climate change and options for adaptation. Working Group III assesses options for limiting greenhouse gas emissions and mitigating climate change. Working Groups meet in plenary session with the relevant government representative. The Taskforce on National Greenhouse Gas Inventories is responsible for looking after the IPCC's National Greenhouse Gas Inventories Program and the main objective is to develop and refine a methodology for the calculation and reporting of national greenhouse gas emissions and removals.[28]

8.3.2 The *United Nations Framework Convention on Climate Change*

There are two key instruments relevant to the international regulation of climate change: the *United Nations Framework Convention on Climate Change* and the *Kyoto Protocol*. The *United Nations Framework Convention on Climate Change* ('*UNFCCC*') is an international environmental treaty negotiated at the United Nations Conference on Environmental Development ('UNCED'), also known as the Earth Summit, and held in Rio de Janeiro in 1992. The Preamble to the *UNFCCC* sets out: that the earth's climate and adverse effects are a common concern for humankind; that the greenhouse effect will increase warming of the earth's surface and atmosphere and will adversely affect natural ecosystems and humankind; that there is a need for an appropriate international response in accordance with 'common but differentiated responsibilities'; that developed countries should take immediate action to develop comprehensive strategies to protect the climate system for present and future generations; and that these strategies should be coordinated with social and economic development.[29]

---

[28] This information is available at the IPCC website: See Intergovernmental Panel on Climate Change, *Structure: How does the IPCC work?* http://www.ipcc.ch/organization/organization_structure.shtml.

[29] The *United Nations Framework Convention on Climate Change*, opened for signature 9 May 1992, 177 UNTS 107 (entered into force on 21 March 1994). See Preamble. As of March 2014, the UNFCCC has 196 parties.

A full extract of the Preamble and the objective as outlined in Article 2 of the *UNFCCC* is extracted below:

The Parties to this Convention,

Acknowledging that change in the Earth's climate and its adverse effects are a common concern of humankind,

Concerned that human activities have been substantially increasing the atmospheric concentrations of greenhouse gases, that these increases enhance the natural greenhouse effect, and that this will result on average in an additional warming of the Earth's surface and atmosphere and may adversely affect natural ecosystems and humankind,

Noting that the largest share of historical and current global emissions of greenhouse gases has originated in developed countries, that per capita emissions in developing countries are still relatively low and that the share of global emissions originating in developing countries will grow to meet their social and development needs,

Aware of the role and importance in terrestrial and marine ecosystems of sinks and reservoirs of greenhouse gases,

Noting that there are many uncertainties in predictions of climate change, particularly with regard to the timing, magnitude and regional patterns thereof,

Acknowledging that the global nature of climate change calls for the widest possible cooperation by all countries and their participation in an effective and appropriate international response, in accordance with their common but differentiated responsibilities and respective capabilities and their social and economic conditions,

Recalling the pertinent provisions of the Declaration of the United Nations Conference on the Human Environment, adopted at Stockholm on 16 June 1972,

Recalling also that States have, in accordance with the Charter of the United Nations and the principles of international law, the sovereign right to exploit their own resources pursuant to their own environmental and developmental policies, and the responsibility to ensure that activities within their jurisdiction or control do not cause damage to the environment of other States or of areas beyond the limits of national jurisdiction,

Reaffirming the principle of sovereignty of States in international cooperation to address climate change,

Recognizing that States should enact effective environmental legislation, that environmental standards, management objectives and priorities should reflect the environmental and developmental context to which they apply, and that standards applied by some countries may be inappropriate and of unwarranted economic and social cost to other countries, in particular developing countries,

Recalling the provisions of General Assembly resolution 44/228 of 22 December 1989 on the United Nations Conference on Environment and Development, and resolutions 43/53 of

6 December 1988, 44/207 of 22 December 1989, 45/212 of 21 December 1990 and 46/169 of 19 December 1991 on protection of global climate for present and future generations of mankind,

Recalling also the provisions of General Assembly resolution 44/206 of 22 December 1989 on the possible adverse effects of sea-level rise on islands and coastal areas, particularly low-lying coastal areas and the pertinent provisions of General Assembly resolution 44/172 of 19 December 1989 on the implementation of the Plan of Action to Combat Desertification,

Recalling further the Vienna Convention for the Protection of the Ozone Layer, 1985, and the Montreal Protocol on Substances that Deplete the Ozone Layer, 1987, as adjusted and amended on 29 June 1990,

Noting the Ministerial Declaration of the Second World Climate Conference adopted on 7 November 1990,

Conscious of the valuable analytical work being conducted by many States on climate change and of the important contributions of the World Meteorological Organization, the United Nations Environment Program and other organs, organizations and bodies of the United Nations system, as well as other international and intergovernmental bodies, to the exchange of results of scientific research and the coordination of research,

Recognizing that steps required to understand and address climate change will be environmentally, socially and economically most effective if they are based on relevant scientific, technical and economic considerations and continually re-evaluated in the light of new findings in these areas,

Recognizing that various actions to address climate change can be justified economically in their own right and can also help in solving other environmental problems,

Recognizing also the need for developed countries to take immediate action in a flexible manner on the basis of clear priorities, as a first step towards comprehensive response strategies at the global, national and, where agreed, regional levels that take into account all greenhouse gases, with due consideration of their relative contributions to the enhancement of the greenhouse effect,

Recognizing further that low-lying and other small island countries, countries with low-lying coastal, arid and semi-arid areas or areas liable to floods, drought and desertification, and developing countries with fragile mountainous ecosystems are particularly vulnerable to the adverse effects of climate change,

Recognizing the special difficulties of those countries, especially developing countries, whose economies are particularly dependent on fossil fuel production, use and exportation, as a consequence of action taken on limiting greenhouse gas emissions,

Affirming that responses to climate change should be coordinated with social and economic development in an integrated manner with a view to avoiding adverse impacts on the latter, taking into full account the legitimate priority needs of developing countries for the achievement of sustained economic growth and the eradication of poverty,

Recognizing that all countries, especially developing countries, need access to resources required to achieve sustainable social and economic development and that, in order for developing countries to progress towards that goal, their energy consumption will need to grow taking into account the possibilities for achieving greater energy efficiency and for controlling greenhouse gas emissions in general, including through the application of new technologies on terms which make such an application economically and socially beneficial,

Determined to protect the climate system for present and future generations ...

Have agreed as follows:

...

Article 2

OBJECTIVE

The ultimate objective of this Convention and any related legal instruments that the Conference of the Parties may adopt is to achieve, in accordance with the relevant provisions of the Convention, stabilization of greenhouse gas concentrations in the atmosphere at a level that would prevent dangerous anthropogenic interference with the climate system. Such a level should be achieved within a time frame sufficient to allow ecosystems to adapt naturally to climate change, to ensure that food production is not threatened and to enable economic development to proceed in a sustainable manner.[30]

The primary objective of the *UFCCC* as outlined in art 2 is framed in traditional environmental law harm prevention; namely, to stabilise greenhouse gas emissions in the atmosphere at a level that would prevent dangerous anthropogenic interference with the climate system.[31] The means to do so are conceived in terms of quantifiable targets for the reduction of greenhouse gases.[32] The basic model was to introduce a legally authoritative institution with powers to impose sanctions for non-compliance. The focus was primarily upon developing 'flexible mechanisms' so that the overall cost of compliance could be reduced.

8.3.3 The *Kyoto Protocol*

The second relevant legal instrument is the ratified *Kyoto Protocol* ('*Protocol*').[33] The *Kyoto Protocol* is an international agreement created by signatories

---

30 See the outline at: United Nations, *United Nations Framework Convention on Climate Change*, (1992) http://unfccc.int/resource/docs/convkp/conveng.pdf.
31 *UNFCCC* art 2.
32 See the outline by E L F Schipper, 'Conceptual History of Adaptation in the UNFCCC Process' (2006) 15 *Review of European Community and International Environmental Law* 82, 86.
33 *Kyoto Protocol to the United Nations Framework Convention on Climate Change*, opened for signature (11 December 1997), 2303 UNTS 148 (entered into force 16 February, 2005).

under the *UNFCCC* in Japan in 1997. On 3 December 2007, Australia ratified the *Protocol*. This instrument was accepted on 12 December 2007. Australia's ratification came into effect on 11 March 2008. The *Protocol* establishes binding commitments to limit and reduce GHG emissions for the developed countries listed in Annex I to the *UNFCCC*, in accordance with the targets stated in Annex B to the *Protocol*. These targets are to be achieved in the commitment period of 2008–2012 and represent an average reduction of 5.2 per cent compared to 1990 levels. Pursuant to art 3.3, reductions in greenhouse gas emissions resulting from forestry activities – limited to afforestation, reforestation and deforestation since 1990 – may be used to meet the commitments of each party.[34] Further, commitments to human-induced activities relating to revegetation, forest management, cropland management, and grazing land management may be counted towards commitment in the second and subsequent periods.[35]

A second commitment period, extending from 1 January 2013 until 2020 was agreed to by the parties at the United Nations Climate Change Negotiations in Durban, South Africa, in 2011.[36] Subsequently, at the Doha Climate Change Conference in 2012, Australia agreed to join the second commitment period. This commitment is known as the Quantified Emission Limitation or Reduction Objective ('QELRO'). Pursuant to the QELRO, Australia agreed to an unconditional 2020 target of 5 per cent below 2000 levels and retained the option later to move up within its 2020 target of 5 to 15, or 25 per cent below 2000 levels. The European Union, as a component of a global and comprehensive agreement, reiterated its conditional offer to adopt a 30 per cent reduction by 2020 compared to 1990 levels; however, this was subject to the proviso that other developed countries commit themselves to comparable emission reductions in accordance with their respective responsibilities and capabilities.

Further, art 3 para 7 of the second commitment set out that those countries included in Annex I of the first commitment, for whom land-use change and forestry constituted a source of GHG in 1990 under the first commitment period, were entitled the GHG aggregate minus removals by sinks from land-use change when calculating their obligations.[37]

The principle of 'common but differentiated responsibilities' is a core assumption underpinning the *Protocol*, since no binding commitments of GHG reductions

---

34 Ibid art 3.3.
35 Ibid art 3.9.
36 Durban Climate Change Conference, November/December 2011. At this conference the parties decided to adopt a universal legal agreement on climate change no later than 2015. Details of the conference are available at: http://unfccc.int/meetings/durban_nov_2011/meeting/6245.php
37 *Doha Amendment to the Kyoto Protocol*, opened for signature 8 December 2012, COP 18 (not yet in force).

were formulated for non-Annex I countries (developing countries). In order to assist Annex I countries to fulfill their commitments, the *Protocol* includes the 'flexible mechanisms'; namely, joint implementation (JI), the clean development mechanism (CDM), and emissions trading. JI means that developed countries can invest in projects in other developed countries to acquire credits that will assist them in meeting assigned amounts. The acquisition of emission reduction units from a JI project must be 'supplemental' to domestic actions for emission reduction. The CDM entitles developed countries to invest in emission reducing projects in developing countries and to obtain certified emission reductions towards meeting assigned amounts. The CDM allows developed countries to meet their own emission reduction target and also to establish an export market for sustainable energy technologies. These innovative mechanisms are based on the assumption that, since there are no boundaries in the atmosphere, investments in carbon abatement measures may be carried out where reductions are cheaper to obtain, with a general view to lowering the costs of implementing the *Protocol*.[38]

Participation in the flexible mechanisms requires Annex I parties to have ratified the *Protocol*.[39] They must also have determined the assigned amount of emissions in terms of tonnes of CO_2 and have in place a national system for estimating emissions and removals of greenhouse gases, as well as a registry to record and track the creation and movement of emissions.[40] The signatory countries have formulated their own group of measures aimed at the fulfillment of their obligations, comprising a combination of direct regulations and economic instruments that include either carbon taxes or schemes of emissions trading.

Articles 3 and 5 of the *Protocol* impose GHG emission targets on annex countries in accordance with reduction commitments prescribed in Annex B. Article 3 entitles the parties to determine their own rules, guidelines and modalities regarding how human induced activities may be added to or subtracted from the assigned obligations of the parties. Article 5 imposes an obligation on the parties to establish a national system for the estimation of anthropogenic emissions by the parties. Articles 3.1 and 5 of the *Protocol* are extracted below.

Kyoto Protocol

Article 3

1. The Parties included in Annex I shall, individually or jointly, ensure that their aggregate anthropogenic carbon dioxide equivalent emissions of the greenhouse gases listed in Annex A do not exceed their assigned amounts, calculated pursuant to their quantified

---

38 *Protocol* arts 6, 12.
39 *Protocol* art 12. See also *Doha Amendment to the Kyoto Protocol* art 1, Amendment F (amending art 3 para 7 of the *Protocol*).
40 *Protocol* art 12.

emission limitation and reduction commitments inscribed in Annex B and in accordance with the provisions of this Article, with a view to reducing their overall emissions of such gases by at least 5 per cent below 1990 levels in the commitment period 2008 to 2012.

Article 5

1. Each Party included in Annex I shall have in place, no later than one year prior to the start of the first commitment period, a national system for the estimation of anthropogenic emissions by sources and removals by sinks of all greenhouse gases not controlled by the Montreal Protocol. Guidelines for such national systems, which shall incorporate the methodologies specified in paragraph 2 below, shall be decided upon by the Conference of the Parties serving as the meeting of the Parties to this Protocol at its first session.

2. Methodologies for estimating anthropogenic emissions by sources and removals by sinks of all greenhouse gases not controlled by the Montreal Protocol shall be those accepted by the Intergovernmental Panel on Climate Change and agreed upon by the Conference of the Parties at its third session. Where such methodologies are not used, appropriate adjustments shall be applied according to methodologies agreed upon by the Conference of the Parties serving as the meeting of the Parties to this Protocol at its first session. Based on the work of, inter alia, the Intergovernmental Panel on Climate Change and advice provided by the Subsidiary Body for Scientific and Technological Advice, the Conference of the Parties serving as the meeting of the Parties to this Protocol shall regularly review and, as appropriate, revise such methodologies and adjustments, taking fully into account any relevant decisions by the Conference of the Parties. Any revision to methodologies or adjustments shall be used only for the purposes of ascertaining compliance with commitments under Article 3 in respect of any commitment period adopted subsequent to that revision.

3. The global warming potentials used to calculate the carbon dioxide equivalence of anthropogenic emissions by sources and removals by sinks of greenhouse gases listed in Annex A shall be those accepted by the Intergovernmental Panel on Climate Change and agreed upon by the Conference of the Parties at its third session. Based on the work of, inter alia, the Intergovernmental Panel on Climate Change and advice provided by the Subsidiary Body for Scientific and Technological Advice, the Conference of the Parties serving as the meeting of the Parties to this Protocol shall regularly review and, as appropriate, revise the global warming potential of each such greenhouse gas, taking fully into account any relevant decisions by the Conference of the Parties. Any revision to a global warming potential shall apply only to commitments under Article 3 in respect of any commitment period adopted subsequent to that revision.[41]

---

41 *Kyoto Protocol* arts 3.1, 5.

The *Protocol* is insufficient, in itself, to combat the complexities of climate change as its provisions contain what have now been acknowledged to be fundamental structural flaws.[42] Attempts to impose quantifiable GHG emissions reduction targets has generated deep inequity issues between developed and developing countries; it has also required signatory countries to implement a range of core economic strategies which, despite lacking the technological capacity and expertise to support such changes, have been implemented and have impacted significantly upon the energy sector.[43] The *Protocol* was never intended to represent a final solution. Indeed, the *UNFCCC* and the *Protocol* have always been regarded as the first step in the process of addressing climate change. However, it is clear that the initiatives implemented under the *Protocol* are at a virtual standstill. In 2012, the International Energy Agency concluded:

> Taking all new developments and policies into account, the world is still failing to put the global energy system onto a more sustainable path. Global energy demand increases by over one third in the period 2035. Energy related emissions rise from an estimated 31.2 gig tons in 2011 to 37 gig tons in 2035 pointing to a long-term average temperature increase of 3.6 degrees Celsius.[44]

The 2012 United Nations Climate Change Conference and the second commitment to the *Protocol* in Doha has not generated significant results. The *Protocol* has failed to achieve its objectives for a number of reasons. First, some of the largest emitters are not constrained by the *Protocol*; it is not ratified by the United States and it does not include emission targets for some of the most rapidly-growing economies in the developing world. Second, the CDM has been criticised for crediting projects that would have occurred anyway: a concept known as 'additionality'. Third, the commitment is over a relatively short period of time when the long-term problem

---

42 See the discussion by R N Cooper, 'The Kyoto Protocol: A Flawed Concept' (2001) 31 *Environmental Law Reporter* 11,484, 11,486 where the author notes, inter alia, that one of the major problems with the *Protocol* is that effective action cannot be taken by a small group of countries alone and further, that the *Protocol* lacks comprehensive coverage because economic activities might re-locate from countries with GHG emission ceilings to those without. See also the discussion by C Hepburn, 'Regulation by Prices, Quantities or Both: A Review of Instrument Choice' (2006) 22 *Oxford Review of Economic Policy* 226, 231–3 where the author reviews issues of efficiency and uncertainty associated with climate change policy instruments.

43 See S Handke and E Hey, 'Climate Change Negotiations in a Changing Global Energy Landscape: A Wicked Problem' (2013) 2(7) *European Society of International Law* 8 arguing that climate change policies will alter fundamental economic structures but there is an absence of the technological means to support new and emergent energy markets. See also Richard Lazarus, ' Super Wicked Problems and Climate Change: Restraining the Present to Liberate the Future' (2009) 94 *Cornell Law Review* 1153.

44 International Energy Agency, *World Energy Outlook, Factsheet* (2012) http://www.worldenergyoutlook.org/publications/weo-2012/.

of climate change is considered.[45] Despite this, the *Protocol* was only ever intended to function as a starting point; it represented the catalyst for policy change, seeking to engage governments and stakeholders in discussion and cooperation on the issue of climate change. Whilst the internal mechanisms it set up have not been effective, the enduring success of the *Protocol* lies in the fact that it has informed and activated the international community of the risk of inaction and this, in turn, has stimulated many governments across the world to implement domestic legal regimes for emission control.[46]

8.3.4 The *Energy Charter* and the *Energy Charter Treaty*

Finally, the *Energy Charter* and the *Energy Charter Treaty* ('*Energy Charter*' and '*Energy Treaty*') are relevant to climate change issues because of their objectives in encouraging investment and trade in the energy markets of Eastern Europe. The *Energy Treaty* is an international agreement creating a multi-lateral framework for the implementation of cooperative energy relations in the energy sector. At its inception, the *Energy Charter* sought to bring together the energy sector in the Soviet Union and Eastern Europe following the end of the Cold War, given the growing demand for energy in Europe. The first *Energy Charter* was signed in 1991 by 47 nations including Australia, the United States, Canada, New Zealand, and Japan. This Charter was replaced by the *International Energy Charter*, which maps out common principles for international cooperation in the energy sector and which is based on the text of the original charter.[47] The *International Energy Charter* provided the groundwork for the subsequent implementation of the *Energy*

---

[45] For an extensive discussion of the concerns associated with the *Kyoto Protocol* see F A Felder, 'Climate Change Mitigation and the Global Energy System' (2014) 25 *Villanova Environmental Law Journal* 89, 90–1.

[46] See also the discussion by S Freeland, 'Kyoto Protocol: An Agreement Without a Future' (2001) 24 *University of New South Wales Law Journal* 532, 541–2.

[47] *European Energy Charter*, signed December 1991, The Hague. The text of this original Charter has been updated by the newly-developed *International Energy Charter*. Approximately 80 states took part in negotiations on the *International Energy Charter*, which were conducted throughout 2014 at the Headquarters of the Energy Charter Secretariat in Belgium. The negotiations started on the basis of the text of the *European Energy Charter* which was adopted in 1991 in The Hague. Non-signatories of the original *European Energy Charter* joined the negotiations on an equal footing with its signatories. The objective is to create a new instrument that reflects the modern energy challenges as well as the ambition of the *Energy Charter* to play a leading role in the evolving architecture of global energy governance. It is anticipated that the *International Energy Charter* will be adopted in May 2015. The newly proposed *International Energy Charter* text is available at: http://www.encharter.org/fileadmin/user_upload/document/IEC_FAQ.pdf

Treaty[48] and the *Energy Charter Protocol on Energy Efficiency and Related Environmental Aspects* ('*PEERA*').[49] To date, the *Energy Treaty* has been signed or acceded to by 52 states, the European Community and the European Atomic Energy Community (EURATOM).

The *Energy Charter* was, in essence, a declaration of political intent to promote energy cooperation. The *Energy Treaty*, however, is a legally binding multilateral instrument aimed at strengthening the rule of law on energy issues through the creation of a level playing field where regulations must be observed by all participating governments. The creation of these rules helps to mitigate the risks associated with energy-related investment and trade.[50] Both the *Energy Treaty* and the *Protocol* play a strong role in protecting international energy security. Whilst the *Energy Treaty* has been operative for over two decades now, it retains strong relevance in a world where energy and capital have been globalised and there is increasing interdependence between exporters and importers of energy. This environment makes multilateral rules providing a strong and balanced legal framework increasingly crucial.[51]

Most of the terms of the *Energy Treaty* focus upon international energy trade and investment. The primary provision relevant to climate change is art 19(1), which sets out that:

Energy Charter Treaty

Article 19(1)

each Contracting Party shall strive to minimize in an economically efficient manner harmful Environmental Impacts occurring either within or outside its Area from all operations within the Energy Cycle in its Area, taking proper account of safety.

[and] ...

each Contracting Party shall strive to take precautionary measures to prevent or minimize environmental degradation. The Contracting Parties agree that the polluter in the Areas of Contracting Parties, should, in principle, bear the cost of pollution, including transboundary pollution, with due regard to the public interest and without distorting investment in the Energy Cycle or international trade.'

---

48 *Energy Charter Treaty*, opened for signature 17 December, 1994, 2080 UNTS 95 (entered into force 16 April 1998).
49 *Energy Charter Protocol on Energy Efficiency and Related Environmental Aspects*, opened for signature 17 December 1994, 2080 UNTS 95 (entered into force 16 April 1998). Unlike the *Energy Treaty*, which was based upon the 1991 *European Energy Charter*, the *PEERA* was drawn up as a declaration of intent to promote East-West energy cooperation and deals specifically with intergovernmental cooperation in the energy sector.
50 For a discussion of the structure of the *Energy Treaty* see A Konoplyanik, 'Energy Charter Treaty and its Role in International Energy' (2006) 24 *Journal of Energy and Natural Resources Law* 523.
51 Konoplyanik, above n 50, 527.

Article 19 is aimed at improving energy efficiency in all applications and focuses upon both energy use and production; however, the provision is non-binding in a legal sense because parties need only 'strive' to minimise harmful environmental impacts and take precautionary measures. They are not legally compelled to do so. This effectively precludes any international enforcement of the different environmental efficiency aspects of art 19. It is also clear that the economic considerations take priority over the environmental obligations. In this respect, the environmental obligations for energy efficiency represent an element of the investment and trade focus of the *Energy Treaty*; however, they not constitute an independent legal mandate.[52]

The full text of art 19 of the *Energy Charter Treaty* is extracted below:

Energy Charter Treaty

Article 19 Environmental aspects

(1) In pursuit of sustainable development and taking into account its obligations under those international agreements concerning the environment to which it is party, each Contracting Party shall strive to minimize in an economically efficient manner harmful Environmental Impacts occurring either within or outside its Area from all operations within the Energy Cycle in its Area, taking proper account of safety. In doing so each Contracting Party shall act in a Cost-Effective manner. In its policies and actions each Contracting Party shall strive to take precautionary measures to prevent or minimize environmental degradation. The Contracting Parties agree that the polluter in the Areas of Contracting Parties, should, in principle, bear the cost of pollution, including transboundary pollution, with due regard to the public interest and without distorting Investment in the Energy Cycle or international trade. Contracting Parties shall accordingly:

    (a) take account of environmental considerations throughout the formulation and implementation of their energy policies;

    (b) promote market-oriented price formation and a fuller reflection of environmental costs and benefits throughout the Energy Cycle;

    (c) having regard to Article 34(4), encourage co-operation in the attainment of the environmental objectives of the Charter and co-operation in the field of international environmental standards for the Energy Cycle, taking into account differences in adverse effects and abatement costs between Contracting Parties;

---

52 See especially the discussion by E Emeseh, A Aboah and H Barmakhshad, 'Framework for Achieving Sustainability in Investment Decisions: Reflections on Rio 20' (2014) 16 *Environmental Law Review* 21, 24 noting that there are a 'plethora of bilateral investment treaties and regional treaties but as yet no international legal instrument on investment which provides a binding framework for integrating sustainable development into international investment law'. See also S P Subedi, *International Investment Law: Reconciling Policy and Principle* (Hart Publishing, Oxford, 2nd ed, 2012) 41.

(d) have particular regard to Improving Energy Efficiency, to developing and using renewable energy sources, to promoting the use of cleaner fuels and to employing technologies and technological means that reduce pollution;

(e) promote the collection and sharing among Contracting Parties of information on environmentally sound and economically efficient energy policies and Cost-Effective practices and technologies;

(f) promote public awareness of the Environmental Impacts of energy systems, of the scope for the prevention or abatement of their adverse Environmental Impacts, and of the costs associated with various prevention or abatement measures;

(g) promote and cooperate in the research, development and application of energy efficient and environmentally sound technologies, practices and processes which will minimize harmful Environmental Impacts of all aspects of the Energy Cycle in an economically efficient manner;

(h) encourage favourable conditions for the transfer and dissemination of such technologies consistent with the adequate and effective protection of Intellectual Property rights;

(i) promote the transparent assessment at an early stage and prior to decision, and subsequent monitoring, of Environmental Impacts of environmentally significant energy investment projects;

(j) promote international awareness and information exchange on Contracting Parties' relevant environmental programmes and standards and on the implementation of those programmes and standards;

(k) participate, upon request, and within their available resources, in the development and implementation of appropriate environmental programmes in the Contracting Parties.

(2) At the request of one or more Contracting Parties, disputes concerning the application or interpretation of provisions of this Article shall, to the extent that arrangements for the consideration of such disputes do not exist in other appropriate international fora, be reviewed by the Charter Conference aiming at a solution.

(3) For the purposes of this Article:

(a) "Energy Cycle" means the entire energy chain, including activities related to prospecting for, exploration, production, conversion, storage, transport, distribution and consumption of the various forms of energy, and the treatment and disposal of wastes, as well as the decommissioning, cessation or closure of these activities, minimizing harmful Environmental Impacts;

(b) "Environmental Impact" means any effect caused by a given activity on the environment, including human health and safety, flora, fauna, soil, air, water, climate, landscape and historical monuments or other physical structures or the interactions among these factors; it also includes effects on cultural heritage or socio-economic conditions resulting from alterations to those factors;

(c) "Improving Energy Efficiency" means acting to maintain the same unit of output (of a good or service) without reducing the quality or performance of the output, while reducing the amount of energy required to produce that output;
(d) "Cost-Effective" means to achieve a defined objective at the lowest cost or to achieve the greatest benefit at a given cost.[53]

Other relevant provisions of the *Energy Treaty* include art 1(2)(a), which seeks to promote 'energy efficiency policies consistent with sustainable development'; art 1(2)(b), which sets out that energy markets should be based on a 'fuller reflection of environmental costs and benefits'; and arts 3, 5 and 8, which impose both national and international obligations to support energy efficiency. The international obligations relate to assistance and cooperation, whereas the national obligations require participating countries to 'develop and implement energy efficiency policies, laws and regulations'.[54] The proposed *International Energy Charter* ('the *IEC*') is expected to come into force in May 2015. This new declaration seeks to introduce amendments which better reflect the realities of the energy sector, particularly the importance of developing countries, and emerging economies. The *IEC* is, however, a political declaration which has the explicit aim of modernising the principles relevant to energy cooperation. Unlike the *Energy Treaty*, the *IEC* it is not a legally binding document. Rather, the *IEC* explicity recognises the global challenge posed by the trilemma between energy security, economic development and environmental protection, and efforts by all countries to achieve sustainable development.[55]

8.3.5 The Harvard Project: a new proposed framework for climate change

Global demand for energy means that any disruption in the energy system will necessarily have widespread market impacts. The effectiveness of a legal framework for climate change mitigation ultimately depends upon how scientifically sound, economically rational, and politically pragmatic the framework has become.[56] Post-Kyoto

---

53 The *Energy Treaty* is taken from the Energy Charter Website where the Charter is extracted http://www.encharter.org/fileadmin/user_upload/document/EN.pdf.
54 *Energy Treaty*, pt IV, art 7 confirms sovereignty for domestic governments over energy resources. For international obligations see, for example, pt IV, arts 20, 23 and 25 which deal with: transparency, observance by sub-national authorities, and economic integration agreements. Part II, art 3 compels all trade related investment matters between parties to be consistent with the provisions of the GATT (General Agreement on Tariffs and Trade).
55 See the Preamble to the proposed text for the *International Energy Charter* at p. 2. Available at: http://www.encharter.org/fileadmin/user_upload/document/IEC/IEC_text_brochure_ENG.pdf.
56 See J E Aldy and R N Stavins, 'Designing the Post-Kyoto Climate Regime: Lessons From the Harvard Project on International Climate Agreements: An Interim Progress Report for the 14th Conference of the Parties' (Interim Progress Report, Harvard Environmental Economics Program, 24 November 2008) xi http://belfercenter.ksg.harvard.edu/files/post%20kyoto%20final%20HIGH%20QUALITY.pdf.

frameworks have been proposed. The Harvard Project for Designing the Post-Kyoto Climate Change Regime ('the Harvard Project') involved a group of researchers setting out to determine how a truly international climate change project might operate, drawing upon the experiences of the *Kyoto Protocol*. The framework for the Harvard Project proposed including the establishment of an international agreement that incorporates an emissions cap system linking national and regional caps and trade frameworks to allow for global trading. Developing countries would not be required to be subject to costly obligations in the early stages of the agreement and further, would not be expected to make sacrifices that differ in any substantial way to those experienced by developed countries. Participating countries would not be expected to accept any emission target that costs more than 1 per cent of gross domestic product or more than 5 per cent of gross domestic product over a particular year.[57]

The core elements of this post-Kyoto-international-cap-and-trade agreement would be: (i) a 'progressivity' factor which requires developed countries to make stronger emission cuts; (ii) a late-comer factor requiring nations that did not commit to the *Protocol* to make gradual emission cuts taking account of their additional emissions since 1990; and (iii) a gradual equalisation factor resulting in emissions in each country being slowly reduced to an average per capita emission output. The overall objective, as the researchers conclude, is to establish a framework where 'every country will feel that it is contributing its fair share'.[58]

This proposed framework adopts a different architecture to the *Kyoto Protocol* as it relies upon a series of linked international agreements to separately address various sectors and would establish global standards for specific industries, such as the energy sector, financial aid for developing countries, trade restrictions to enforce agreements, international agreements relating to research and development, and separate agreements to cover adaptation assistance for developing countries. Some of the key design features include burden sharing in an international climate change agreement; incentives for technology transfers; reform of CDM to ensure that reductions are offered for real, verifiable and permanent changes; and an alignment of global climate policy with global trade policy.[59]

Ultimately, a legal framework that is capable of 'catalyzing a transformation of the global energy system to dislodge fossil fuel dominance and initiate large-scale diffusion of renewable energy technologies' would appear to provide the best chance of success for climate change mitigation.[60] The existing international legal

---

57 Aldy and Stavins, above n 56, viii.
58 See Aldy and Stavins, above n 56, xii–xiii.
59 See Aldy and Stavins above n 56, xiv–xvi.
60 A Long, 'Global Energy-Environment Governance' above n 12, 1057. See also A J Wildermuth, 'The Next Step: The Integration of Energy Law and Environmental Law' (2011) 31 *Utah Environmental Law Review* 369, 373–5 arguing that energy law and environmental law are fundamentally oppositional and successful climate change mitigation depends upon creating strategies and frameworks that promote mutuality.

framework for climate change does not provide a synthetic approach to the particular difficulties and impediments that the mining and energy industry face in responding to climate change. Energy lies at the core of every human necessity and is inextricably entwined with core human values of fairness, justice, economic opportunity and environmental protection.[61] The demand flowing from climate change imperatives to transform energy into cleaner, more affordable and accessible forms has introduced opportunities and challenges within an already highly complex framework. There is a need for swift regulatory innovation in order to reinforce climate change imperatives and develop a more effective and dynamic energy framework capable of bridging different levels of governance and institutional divisions and in so doing respond more effectively to new possibilities for institutional innovation.[62]

8.4 The economics of climate change

Nicholas Stern, a former chief economist for the World Bank, in his seminal 2006 review, *The Economics of Climate Change* ('the Stern Review'), strategically addressed the economic consequences of climate change. The Stern Review examined the costs and risk of climate change in detail and the prospect of doing nothing in the face of catastrophic changes resulting from climate change were rejected. The Review stated:

> The scientific evidence is now overwhelming: climate change is a serious global threat, and it demands an urgent global response. This Review has assessed a wide range of evidence on the impacts of climate change and on the economic costs, and has used a number of different techniques to assess costs and risks. From all of these perspectives, the evidence gathered by the Review leads to a simple conclusion: the benefits of strong and early action far outweigh the economic costs of not acting. Climate change will affect the basic elements of life for people around the world – access to water, food production, health, and the environment. Hundreds of millions of people could suffer hunger, water shortages and coastal flooding as the world warms. Using the results from formal economic models, the Review estimates

---

61 See H M Osofsky, 'Learning from Environmental Justice: A New Model for International Environmental Rights' (2005) 24 *Stanford Environmental Law Journal* 71, 92–7.
62 H J Wiseman and H M Osofsky, 'Dynamic Energy Federalism' (2013) 72 *Maryland Law Review* 773, 819 where the authors propose what they describe as a 'dynamic energy federalism' model capable of integrating complex institutional dynamics of energy laws 'tripartite' structure.

that if we don't act, the overall costs and risks of climate change will be equivalent to losing at least 5% of global GDP each year, now and forever. If a wider range of risks and impacts is taken into account, the estimates of damage could rise to 20% of GDP or more.[63]

Subsequently, in 2014, Lord Nicholas Stern (as he had then become) stated that since the release of his groundbreaking review in 2006, global weather patterns had changed substantially, making it 'very unwise' to ignore the recommendations in his original review. His Lordship noted that the conclusion of the Intergovernmental Panel on Climate Change is that it is now '95% likely that most of the rise in global average temperature since the middle of the 20th century is due to emissions of greenhouse gases, deforestation and other human activities'.[64]

From an economic perspective, Stern indicated that climate change will have a major financial impact: equivalent to losing at least 'five percent of global gross domestic product (GDP) each year, now and forever in a business as usual emissions scenario'.[65] The reason for this lies in the fact that climate change creates the risk of major disruption to social and economic activities.

The Stern Review was reinforced in Australia by the Garnaut Review. According to Professor Garnaut in his 2008 Climate Change Review ('the Garnaut Review'), which was derived from data in the IPCC 2007 *Fourth Assessment Report*:[66]

> Almost 60 per cent of [greenhouse gas emissions from human activity] ... was emissions of carbon dioxide from fossil fuel combustion and other carbon dioxide-emitting industrial processes (such as cement production and natural gas flaring) ... [where] ... richer countries tend to have much higher per capita emissions than poorer countries.[67]

The 2008 Garnaut Review concluded that 'carbon dioxide emissions from fossil fuels are the largest and fastest-growing source of greenhouse gases ... [where they] have expanded by 3 per cent a year in the early 21st century'.[68] Subsequently,

---

63 N Stern, *The Economics of Climate Change: The Stern Review* (Cambridge University Press, 2007) vii. The full report is available at http://mudancasclimaticas.cptec.inpe.br/~rmclima/pdfs/destaques/sternreview_report_complete.pdf.
64 See N Stern, 'Climate change is here and it could lead to global conflict', *The Guardian* (online), 14 February 2014 http://www.theguardian.com/environment/2014/feb/13/storms-floods-climate-change-upon-us-lord-stern.
65 Stern, above n 64, xv.
66 Intergovernmental Panel on Climate Change, R K Pachauri and A Reising (eds) *Fourth Assessment Report of the Intergovernmental Panel on Climate Change: Synthesis Report* (2007).
67 The Garnaut Climate Change Review, *Final Report* (2008) http://www.garnautreview.org.au/index.htm.
68 Garnaut, above n 67, ch 3 at 55.

in his 2011 Review, Garnaut commented that '[e]missions from fossil fuels are the largest source of atmospheric carbon dioxide from human activities ... [noting that between] 2000 and 2008, the annual increase in fossil fuel emissions grew to 3.4 per cent'.[69] Professor Garnaut further predicted, consistent with the Stern Review, that the economic consequences of failing to implement climate change mitigation strategies are likely to be significant and noted particularly dire consequences for the agricultural and mining sectors in Australia:

> Without mitigation, the best estimate for the Murray-Darling Basin is that by mid-century it would lose half of its annual irrigated agricultural output. By the end of the century, it would no longer be a home to agriculture. The other sector that is hit hard is mining. Output of this sector is projected to decline by more than 13 per cent by 2100. This result is mainly driven by the deceleration of global economic growth. Most coal produced in Australia is exported. The international modelling implies that the world demand for coal falls by almost 23 per cent. Iron ore activity declines for much the same reason as for coal.[70]

The global challenges associated with climate change are enormous. Domestically, Australia faces scorching weather, severe floods, devastating bushfires, water shortages, and the dire prospect that crop production and clean water will diminish, and animal and plant species will become extinct.[71] There is 35 000 km of coastal road and rail worth $60 billion, which will all be at risk from sea level rises and storm surges.[72]

The IPCC 2013 *Fifth Assessment Report* indicates that the atmospheric concentrations of CO_2, methane, and nitrous oxide have increased to levels unprecedented in at least the last 800 000 years. Carbon dioxide concentrations have risen by 40 per cent since pre-industrial times, primarily from fossil fuel emissions and secondarily from net land-use change emissions. The ocean has absorbed about 30 per cent of the emitted anthropogenic CO_2, which has generated ocean acidification.[73]

---

69 Garnaut Climate Change Review, *Update, 2011: Australia in the Global Response to Climate Change Summary*, (31 May 20111), chs 1, 3 http://www.garnautreview.org.au/update-2011/garnaut-review-2011/summary-20June.pdf.
70 Garnaut (2011), above n 69, ch 11 at 258.
71 See the discussion by B Saul, S Sherwood, J McAdam, T Stephens and J Slezak, *Climate Change and Australia: Warming to the Global Challenge* (Federation Press, 2012) 1.
72 The Climate Institute, *Coming Ready or Not: Managing Climate Risks to Australia's Infrastructure,* (2012) 30.
73 See IPCC, *Climate Change 2013: The Physical Science Basis. Contribution of Working Group I to the Report of the Intergovernmental Panel on Climate Change* above n 22, B5. The report suggests at p. 12 that from 1750–2011, CO_2 emissions from fossil fuel combustion and cement production have released 375 [345 to 405] GtC to the atmosphere, while deforestation and other land use change are estimated to have released 180 [100 to 260] GtC. This results in cumulative anthropogenic emissions of 555 [470 to 640] GtC.

Increasingly, the physical impacts of climate change and predictions regarding future impacts have caused businesses to re-evaluate existing practices by both themselves and their investors which will make them more competitive in the short to medium-term and more likely to exist in the longer-term. For example, British Telecom is planning its infrastructure spending on the basis of recurring flooding scenarios occurring 50–100 years in the future.[74] In China, new coal-fired power plants will be banned in the region surrounding Beijing, in the Yangtze Delta region near Shanghai, and in the Pearl River Delta region of Guangdong province; whilst in the United States, emissions standards have been imposed on new coal-fired power stations which effectively preclude the construction of any new coal power station without the additional construction of carbon capture storage technology.[75]

The importance of imposing a price on carbon emissions, accelerating low-emission technology, progressing renewable energy industries, and promoting cost-effective energy-efficiency measures should not be underestimated. Economic growth needs to be decoupled from greenhouse gas emissions and this depends upon a reduction in energy consumption of around 75 to 85 per cent by 2050. This reduction needs to be combined with the implementation of a comprehensive efficiency scheme that will necessarily involve a more competent and resourceful use of fossil fuels, the growth of renewables and nuclear energy, and the capture of carbon-heavy emissions.[76] Middle-income countries – such as China and India – comprise 70 per cent of the world population and were central to the boom in developing country growth, globalisation, and urbanisation in the 2000s.[77] These countries have a large and growing middle class and account for around half of world energy consumption and carbon emissions from energy use, and these proportions are rising rapidly. These countries are often already tackling complex problems associated with structural change and institutional modernisation but are also critical to a transition to a resource-efficient, low-carbon global economy. Much will depend upon the capacity to deal with inefficiencies and dysfunctions connected with urbanisation, industrialisation, and energy use.

In Australia there has, unfortunately, been a 'fragmented and inconsistent' policy response to climate change mitigation. Since 2007, the Australian Government has put forward three different legislative regimes to reduce greenhouse gas

---

74 E Lowitt, 'How to Survive Climate Change and Still Run a Thriving Business' *Harvard Business Review* (April 2014) <https://hbr.org/2014/04/how-to-survive-climate-change-and-still-run-a-thriving-business-checklists-for-smart-leaders>.
75 G Parkinson, 'China, United States Say No To New Coal Plants', *RenewEconomy*, (16 September 2013) http://reneweconomy.com.au/2013/china-us-say-new-coal-plants-37836.
76 See J O'Neil, *The Growth Map: Economic Opportunities in the BRICs and Beyond* (Penguin Publishing, 2011). Note: 'BRICs' meaning Brazil, Russia, India and China.
77 The Global Commission on the Economy and Climate, *The New Climate Economy Report* (2014) ch 1, Available at: http://newclimateeconomy.report/.

emissions.[78] The carbon pricing mechanism introduced in 2011 was subsequently abolished in 2014 and replaced with an emission reduction fund which encourages companies to put forward emission reduction projects and allows them to register those projects with the clean energy register and, where approved, receive funding for those projects.[79] Within such a context, addressing the impact of climate change has become a strong social and political challenge.[80] Uncertainty as to the degree of impact that climate change will have upon ecosystems and human activities should not, however, justify inaction.

In Australia, the lack of bipartisan approach to climate change mitigation has impeded our ability to start transitioning as a nation towards some of the core economic transformations that have already been initiated by the rest of the world.

The ostensible argument is that the implementation of an emissions trading scheme would put Australia at a competitive disadvantage with the rest of the world because other countries have not followed suit. In the context of this argument, the existence of the European emission trading scheme appears to have been largely ignored, possibility because of its initiation difficulties and some of the problems it experienced with market progression; similarly, the extensive scope of China's five-year clean energy plan has not been properly taken into account. The political environment in Australia has reached the point where any regulation of greenhouse gas emissions are generally opposed by Australian industry groups: the Australian Petroleum Production and Exploration Association have actually requested the complete removal of greenhouse gas abatement requirements where they are imposed as a component of environmental review processes for resource titles.[81]

In reality, however, this self-focused, insular approach is fundamentally inconsistent with global developments. Across the world, emitters of greenhouse gases have been acting in response to rising greenhouse gas emissions. Europe has had an emission trading scheme (ETS) in place since 2005 and China and the United

---

78 See the discussion by M Wilder, 'The Global Economic Environment and Climate Change' in CEDA, *The Economics of Climate Change* (CEDA, June 2014) 17 http://www.ceda.com.au/research-and-policy/research/2014/06/18/climatechangeeconomics. The author notes that major policy reviews including the Energy White Paper, Agricultural Competiveness White Paper, and the Productivity Commission Inquiry into Public Infrastructure make no mention of climate change or its associated risks.
79 For full details see the Australian Government, Department of the Environment, *Emission Reduction Fund White Paper* (April 2014) http://www.environment.gov.au/climate-change/emissions-reduction-fund/publications/white-paper.
80 See especially the discussion by M Porter, 'Reforms in the Greenhouse Era: Who Pays and How? in *Committee for Economic Development of Australia Growth 61: A Taxing Debate – Climate Policy beyond Copenhagen* (2009) 20.
81 Carbon and Environment Daily, *Direct Action Sparks New Carbon Deregulation Campaign* (28 April 2014) http://adminpanel.ceda.com.au/FOLDERS/Service/Files/Documents/22090~Economics-of-Climate-Change.pdf.

States are following closely behind. Korea was one of the first countries to enshrine green growth in its national development strategy. During the global financial crisis in 2008, Korea dedicated 80 per cent of its fiscal stimulus plan to green growth projects, particularly infrastructure and transportation. Subsequently, in 2009, it announced plans to invest US$85 billion in clean energy technologies and implementing its green growth plan, estimated to create more than one million new jobs and bolster a clean-tech export industry. Korea further committed two per cent of gross domestic product to the creation of a knowledge and technological foundation to sustain a green growth economy.[82]

Over the past 30 years, China has experienced enormous growth which entailed rapid growth in capital accumulation, exports, an energy-intensive industry, and high levels of fossil fuel use, particularly coal. This produced an urban sprawl and severe local air pollution. New directions are being implemented with growth being driven by innovation, more efficient resource use, cleaner energy sources and reduced coal consumption, cleaner air to breathe, more compact and productive cities, and greater reliance on growth in domestic consumption and services. These structural changes offer what has been described as a 'no-regrets' opportunity for decarbonisation to become a firm part of China's efforts to achieve national economic and social goals.[83]

No-regrets options – also referred to as 'win win' policy measures – are by definition GHG emissions reduction options that have negative net costs, because they generate direct or indirect benefits large enough to offset the costs of implementing the options.[84] These measures encompass reform that seeks to address market failures and improve the effectiveness of existing regulation and policies by building more effective capacities to adapt to future climate change. This must necessarily include economic reform which is justifiable in its own right and which will enhance the adaptive capacity of the economy by encouraging a more efficient allocation of resources – that is, land, labour and capital – which will allow those resources to respond more effectively to future climate change. Climate change therefore strengthens the domestic and international case for 'no-regrets' reform.[85]

---

[82] See World Bank, *Korea's Global Commitment to Green Growth* (2012) http://www.worldbank.org/en/news/feature/2012/05/09/Korea-s-Global-Commitment-to-Green-Growth.

[83] See The New Climate Economy, 'Chapter 1: Climate Change and the Economy' in *The Global Commission on the Economy and Climate* (2014) http://newclimateeconomy.report/.

[84] See IPCC Report, above n 2, 7.43.4.2.

[85] See Australian Government, Productivity Commission, 'Chapter 6: No Regrets Policy' in *Climate Change Adaptation Policy* (2009) http://www.pc.gov.au/__data/assets/pdf_file/0009/119709/09-climate-change-adaptation-chapter6.pdf. See also D De Sousa and J Thwaites, 'Regulatory Responses to Facilitate Adaptation of Existing Infrastructure to Climate Change' (Paper presented at National Climate Change Adaptation Research Facility: Climate Adaptation in Action 2012 Conference, Melbourne, 26 June 2012).

In pursuit of these goals, China has implemented an aggressive investment approach under its five-year plan with the introduction of seven provincial and regional emissions trading schemes – in Beijing, Tianjin, Shanghai, Chongqing, Shenzhen, Guangdong, and Hubei – covering an estimated 1 billion metric tonnes of CO_2 as well as its strong support of the renewable energy industry.[86] The emissions trading scheme introduced in China is the second largest in the world and it deals with emission permits equivalent to 1.1 billion tonnes of CO_2 each year. China's National Development and Reform Commission recently stated that by 2020 China's national carbon market will encompass between 3 billion and 4 billion tonnes of carbon, which is twice the size of the European emissions trading scheme. The market will be worth between 60 billion and 400 billion Yuan (A$11 billion to $73 billion). This implies an average price of 20–100 Yuan – or between A$4 and $18 – per tonne of CO_2 at the current exchange rate. This means it could be as much as double the current EU trading price.[87] The impact of such a carbon price in China is likely to be profound. The reason for this is because of the emissions intensity of the Chinese economy where the ratio of CO_2 to gross domestic product is three to seven times higher than the emissions intensity in Europe.[88] Investment in greater energy productivity and economic restructuring that steers away from carbon-intensive industries is likely to generate productivity gains; and for China, will create important co-benefits for air pollution and energy security. These factors, combined with climate change mitigation and the likelihood that China's actions will influence other countries, means that the actions taken by China have a global significance.[89]

Within this global environment, Australian industries take strong risks in failing to develop law and policy to address climate change. This is particularly true for mining and energy industries. Mining and energy industries are not only direct emitters of greenhouse gases but also indirect emitters given the fact – particularly apparent with extractive industries – of the progression of energy intensive technologies. These emissions will imply direct and indirect costs for mining companies given the compliance costs of the environmental obligations plus the costs

---

86 L Bifera, 'Carbon Trading in China: Short Term Experience Long Term Wisdom' on Center for Climate and Energy Solutions Blog, (4 September 2014) http://www.c2es.org/blog/biferal/carbon-trading-china-short-term-experience-long-term-wisdom.
87 See F Jotzo, 'China Heads for Price on Carbon: Energy Overhaul is Next', *The Conversation* (19 September 2014) http://theconversation.com/china-heads-for-price-on-carbon-energy-market-overhaul-is-next-31119.
88 See F Teng and F Jotzo, 'Reaping the Economic Benefits of De-Carbonisation for China' (Working Paper, Australian National University, 2014) 1413. The authors argue that decarbonisation improves productivity and accords with multiple national policy objectives.
89 Ibid.

associated with increasing energy prices. Apart from increases in cost, international response to climate change generates changes in market demand; for example, the shift to renewable energy apparent from China's five-year plan indicates a reduction in demand for Australia's fossil fuels and commodities. Furthermore, overseas investments will become increasingly regulated through international costs being imposed upon carbon.

Coal mining and oil and gas companies in Australia are likely to be significantly impacted. Further recent measures introduced in China include limiting coal to 65 per cent of primary domestic energy consumption by 2015 and banning all new coal generation in Beijing, Shanghai, and Guangzhou.[90] Changes to the existing mining and energy sector will, however, take time to filter through. Altering energy prices and making changes to regulatory and ownership structures can have significant impacts on the profitability of established mining and energy industries. Hence, it is only where companies are able to save money and increase profits by cutting emissions that they will acquire sufficient incentive to shift to lower-carbon energy and, in so doing, improve their energy efficiency.

8.5 The impact of climate change on Australian mining and energy industries

In order to fully evaluate the impact of climate change and to evaluate the nature and scope of mitigation strategies and how they will affect Australian mining and energy industries it is appropriate to commence with some energy statistics. Australia has an estimated 46 per cent of uranium resources, 6 per cent of coal resources, and 2 per cent of natural gas resources in the world. By contrast, Australia only has 0.3 per cent of world oil reserves. Overall, Australia produces about 2.4 per cent of total world energy and is a major supplier of energy to world markets as it exports more than three-quarters of its energy output, which is worth an estimated A$80 billion.[91] Australia is the world's largest exporter of coal. Coal accounts for more than half of Australia's energy exports. Australia is also one of the world's largest exporters of uranium and is ranked sixth in terms of liquefied natural gas

---

90 The Global Commission on the Economy and Climate 2014, above n 83.
91 See Australian Bureau of Agriculture and Resource Economics and Sciences (ABARES), A Schultz and R Putney, *Australian Energy Statistics – Energy Update 2011* (2011), 3–10 http://www.agriculture.gov.au/abares/publications/display?url= http://143.188.17.20/anrdl/DAFFService/display.php?fid=pe_abares99010610_12c.xml.

(LNG) exports and this market is expanding with the development of the Gladstone Project in Queensland.[92] In contrast, more than half of Australia's liquid fuel requirements are imported.

Globally, Australia is the world's twentieth largest consumer of energy and fifteenth in terms of per capita energy use. The primary energy consumption in Australia is coal. Coal generates 75 per cent of Australia's electricity, followed by gas at 16 per cent, hydro at 5 per cent, and wind at 2 per cent. Australia has substantial resources of coal, both black and brown. The most significant black coal resources are located in the Bowen-Surat (Queensland) and Sydney Basins (New South Wales). Coal is Australia's largest commodity export with annual coal exports to Japan, India, European Union, Republic of Korea, and Taiwan amounting to more than $40 billion. Economic demonstrated resources of black coal are currently estimated to be adequate for about 90 years at current rates of production. Australia also has significant brown coal resources which are located mostly in the Gippsland Basin in Victoria, where brown coal is used to generate electricity. It is estimated that there is approximately 500 years of brown coal resources remaining.[93]

These statistics indicate that Australia continues to be heavily reliant upon carbon-intensive fossil fuels for energy production despite the imperatives associated with climate change and increased greenhouse gas emissions. Furthermore, with increases in demand from its domestic and international markets, greenhouse gas emissions are likely to continue to increase in Australia unless significant mitigation strategies – and this inevitably includes the implementation of a carbon price – are developed to provide incentives for Australian mining and energy industries to shift forward. What is urgently required is a reduction in carbon-intensive energy production and a significantly increased focus upon renewable energy and improved energy efficiency strategies.

The Emissions Reduction Fund Green Paper issued in 2013 sought to reduce Australia's greenhouse gas emissions by creating positive incentives to adopt better technologies and practices to reduce emissions.[94] This Paper acknowledged

---

92 The first LNG exports from Queensland are due by the end of 2015 from BG Group's Queensland Curtis project, with shipments to follow next year from Santos's Gladstone LNG venture and Origin Energy's Australia Pacific LNG project. Adding strength to the Queensland LNG industry is the recent joining of forces between industry lobby group, the Australian Petroleum Production and Exploration Association (APPEA), and the Queensland Government's 30-year ResourcesQ initiative. See Queensland Government, Department of Energy and Water Supply *2012 Gas Market Review* (2012) http://www.dews.qld.gov.au/__data/assets/pdf_file/0006/77775/gas-market-review-2012.pdf.

93 See Australian Government, Geoscience Australia, *Basics* http://www.ga.gov.au/scientific-topics/energy/basics.

94 See Australian Government, Department of the Environment, *Emissions Reduction Fund Green Paper* (2013) http://www.environment.gov.au/climate-change/emissions-reduction-fund/green-paper.

that the commitment of the Australian Government continues to recognise 'the science of climate change and [still] supports national and global efforts to reduce greenhouse gas emissions'.[95] However, it is apparent from this Paper that 'without positive and direct action by the Government, industry and community, Australia's national emissions will grow strongly to 2020 as the economy grows'.[96] Such emission growth must be countered by strong changes to domestic energy policy.

Achieving this shift necessarily requires complex and strategic changes to core legal, economic, and social frameworks. Government policies towards the energy sector have an enormous impact on the future global energy landscape and international climate governance. Fossil fuel dominance and the global scale of the energy sector means that inevitably, strong climate change measures will cause severe financial and economic disruptions to the energy sector, particularly given the fact that fossil-fuel infrastructure worth hundreds of billions of dollars will, eventually, turn into 'stranded assets' and significant infrastructure and technology costs will be required to escalate the renewable sectors. Climate change and the nexus between climate change and energy policy has been described as a 'wicked problem' because of the fact that it is characterised by intricate interdependencies and changing contexts.[97]

The existing legal framework for regulating climate change does not currently differentiate between the emission of greenhouse gases through consumption and the emission of greenhouse gases through energy production. Most of the greenhouse gases that contribute to human-induced climate change are emitted during the production of energy from fossil fuels for the benefit of rising global energy demands. As outlined above, combusting fossil fuels to produce energy is responsible for approximately 70 per cent of all global greenhouse gas emissions and 68 per cent of all CO_2 emissions. The average annual increase of carbon dioxide emissions connected with the burning of fossil fuel was 1 per cent in the 1990s. However, from 2000 onwards these emissions grew by an annual average of 3 per cent: from 24 150 million metric tonnes in 2000 to 32 579 million metric tonnes in 2011. The primary reason for this was carbon-intensive growth in developing countries flowing from energy intensive activities. In particular, between 1990 and 2012, China's coal consumption almost quadrupled and there was a particularly steep increase in the period between 2000 and 2010. Similarly, India's coal consumption almost tripled between 1990 and 2012. The burning of coal constitutes approximately 43 per cent of total emissions and has been the core reason underlying the increase in CO_2 emissions in the last decade. The International Energy Agency indicated that

---

95 Ibid 1.
96 Ibid.
97 See the discussion by Handke and Hey, above n 43, 8. See also R Lazarus, 'Super Wicked Problems and Climate Change: Restraining the Present to Liberate the Future' (2009) 94 *Cornell Law Review* 1153, 1155.

total emissions from coal combustion increased from 9060 million metric tonnes in 2000 to a total of 14 416 million metric tonnes in 2011.[98]

Climate change is therefore closely connected with the progression of a carbon-intensive economy. Addressing climate change will ultimately be ineffectual in the absence of strong mitigation and adaptation measures which have been comprehensively amalgamated into the global energy policy framework. Understanding the importance of the linkage between climate change mitigation strategies and global energy policy is critical. Integrating economic mechanisms into global energy policy is crucial because these mechanisms focus upon the human structures connected with climate change and therefore seek to adjust those structures in order to achieve an effective outcome.[99] Climate change is an ecological issue in that it has an impact upon natural processes; however, from a policy perspective, climate change mitigation is primarily an economic concern. This is particularly the case given the connection that exists between climate change, energy production, and energy consumption.[100] The implementation of environmental policy in the absence of economic measures is ineffective because the core focus of pure environmental regulation is primarily to address the issue of harm prevention.

Energy policy is generally derived from domestic social and economic imperatives rather than any collective global power shift producing adjustments in international energy and climate governance.[101] The regional focus of energy policy results, as Handke and Hey have outlined, in an increased number of 'influential stakeholders gaining from the unrestricted use of fossil fuels, which complicates the process of defining global climate measures'.[102] Hence, in order to achieve effective changes to energy policy, greater changes need to be implemented at an international level. A shift in global energy policy has the potential to affect a much more expansive share of total emissions than a single national domestic policy and may also target massive emission concentrations that are occurring within developing countries; global regulatory strategies have the potential to achieve transformative mitigation. Climate change is undeniably a worldwide phenomenon and reviewing energy policy on a global scale allows consideration to be given to

---

98 See International Energy Agency Statistics, *CO$_2$ Emissions from Fuel Combustion: Highlights* (2012), 8 http://www.iea.org/publications/freepublications/publication/CO2emissionfromfuelcombustionHIGHLIGHTS.pdf.
99 See generally M Faber and R Manstetten, *Philosophical Basics of Ecology and Economy* (Routledge, 2010) 16–24 where the authors outline the strong linkage between the disciplines of ecology and economics.
100 Ibid.
101 See the discussion by C R Sunstein and L A Reische, 'Automatically Green: Behavioural Economics and Environmental Protection' (2014) 38 *Harvard Environmental Law Review* 127, 130–8 where the authors argue that variables beyond strict economic incentives are also relevant in the choice consumers make regarding energy efficiency.
102 Handke and Hey, above n 43, 3.

the different components relevant to holistic climate change. This would open up an 'array of possibilities for reinventing global climate governance that cannot be seen when one focuses on the domestic level'.[103]

There are also problems with the privatisation of infrastructure assets, particularly with respect to fossil fuel industries. Energy companies around the world now function as publicly traded corporations operating within global markets. Market mechanisms dominate the global energy business and this has resulted in many developments being beyond the control of the state and, in particular, beyond the effective reach of climate change regulation. A good example is the Hazelwood Coal-fire Plant in the Latrobe Valley in Victoria. Open-cut coalmines are highly susceptible to infrastructure, operational, environmental, and safety failures. The Emergency Risks in Victoria Report calculated that the annual likelihood of a medium impact mine failure in Victoria at almost 100 per cent and these failures can have catastrophic social and environmental costs.[104] The 2014 Hazelwood coal fire caused an estimated $100 million in infrastructure damages and untold environmental and health costs to the region. The Commission into the fire concluded that the fire was the consequence of GDF Suez, a large multinational company, failing to implement adequate risk management procedures. The Commission found that the disaster could have been foreseen, but the 'foreseeable risk' was not properly managed.[105]

The regulatory failures connected with this disaster highlight the deeper concerns associated with incorporating coal-fire energy and the infrastructure that supports this into the global market. The rehabilitation bond of $15 million initially paid by GDF Suez was patently inadequate to support ongoing rehabilitation obligations and this amount was never revised despite further coal mining titles being issued. The behaviour of GDF Suez in failing to implement risk management strategies and in failing to properly comply or effectively monitor rehabilitation obligations is ultimately reflective of economic imperatives: in a context where market mechanisms rule, it was cheaper for GDF Suez to forego the bond rather than proactively instigate proper and effective remediation management processes.

A further problem associated with climate change and energy policy lies in the technological advancements, which have, effectively, allowed for an increasingly globalised approach to trade in oil, coal, and natural gas. Developments in domestic or regional economies now affect prices across the world. Relatively low prices for imported coal and LNG impose constraints upon European electricity producers

---

103 See Long, 'Complexity in Global Energy-Environment Governance' above n 12, 1072.
104 See State Government of Victoria, Department of Justice and Regulation, *Emergency Risks in Victoria* (2014), 41 <http://www.justice.vic.gov.au/home/safer+communities/emergencies/emergency+risks+in+victoria+report>.
105 Parliament of Victoria, *The Hazelwood Mine Fire Inquiry* (28 August 2014) http://report.hazelwoodinquiry.vic.gov.au/executive-summary-2/hazelwood-mine-fire.

making the additional coats associated with climate change mitigation even more economically unpalatable.[106] On the other hand, the strong export market for LNG in Australia places significant economic pressure upon domestic industries. In a situation of economic, financial, and political volatility, private energy companies are increasingly reluctant to enter into long-term gas contracts. In Australia, wholesale prices have nearly tripled from approximately $3 a gigajoule to $8 a gigajoule for natural gas as a result of rising export demand and open export policy. Australia is set to become the leading exporter of LNG by 2017, surpassing Qatar, and LNG exports from Australia to Asia will commence in 2016. Adaptation and mitigation for domestic industries exposed to rising domestic gas prices therefore becomes an additional cost, which lacks regional support.[107]

Energy policy is, however, also responsive to market and technological change. The evolving situation in Europe provides a good outline. European countries are gradually creating a system in which there are a greater range of low-carbon and intermittent energy sources: more energy suppliers; more modern power stations (replacing coal and nuclear plants); increased and better storage; and more energy traded across borders. This is all connected by the creation of 'smart grids', that inform energy consumers how much power they are using and shut off appliances when they are not needed, therefore managing energy demand far more efficiently. Within this context, the economic prevalence of fossil fuel industries is diminishing. Coal and gas electricity is increasingly regarded as the generator of last resort, utilised only when wind and solar generators are exhausted. In this context, the technological developments underpinning the renewable sector in combination with a rigorous adherence to climate change mitigation has gradually fostered market changes.

The complex interplay that exists between markets and shifting policy objectives necessitates a continuous realignment of climate and energy policies. The key to addressing climate change mitigation lies in implementing a dynamic process of feasible changes to economic strategies. It also depends upon the introduction of a regulatory and policy framework – particularly at the international level – that is capable of effectively enforcing these economic strategies with the objective of addressing existing market forces and encouraging behavioural shifts. The inevitable consequence of a market realignment and a market correction is market disruption and extensive costs associated with stranded fossil fuel infrastructure. Acknowledging and accepting these

---

106 See 'How To Lose Half a Trillion Euros' *The Economist* (online) 12 October 2013 <http://www.economist.com/news/briefing/21587782-europes-electricity-providers-face-existential-threat-how-lose-half-trillion-euros> outlining that losses to European electricity utilities is the product of increasing demand for renewable energy sources but also the availability of gas following the US shale gas revolution.
107 See J Kehoe, 'Australia Shows US How Not to Export Gas', *Australian Financial Review* (online) 30 July 2014 <http://www.afr.com/p/business/resources/energy/gas/australia_shows_us_how_not_to_export_ObsvU7S8je3WEC3T4OUTOM>.

consequences as a necessary and unavoidable externality is crucial. A further imperative lies in recognising that the existing economic and policy framework precludes the market from accurately reflecting the true cost to society of the production of fossil fuel. This uncorrected market failure produces profligacy. As outlined by Lord Nicholas Stern:

> At the heart of economic policy must be the recognition that the emission of greenhouse gases is a market failure ... the social cost of production and consumption exceeds the private cost, so that markets without policy intervention will lead to too much of such goods being produced and consumed. By producing and consuming less of these products and more of others, we create economic gains that can make everyone better off. Markets with uncorrected failures lead to inefficiency and waste.[108]

8.6 REVIEW QUESTIONS

1. Describe what is meant by 'climate change' and why this has a particular cogency for the mining and energy sector.
2. What is the significance of the Preamble to the *United Nations Framework Convention on Climate Change*?
3. What is the principle of 'common but differentiated' responsibilities and where does it come from?
4. What are some of the structural flaws in the *Kyoto Protocol* that have prevented it from achieving its objectives?
5. Do you think art 19 of the *Energy Charter Treaty* is an effective means of encouraging mining and energy industries to reduce emissions and if not, how might it be modified or improved to better align with climate mitigation imperatives?
6. Why is it important to 'decouple' economic growth from greenhouse gas emissions?
7. How does the economic approach of Australia to climate change compare with developments in the rest of the world and how is/will this affect our economic progression?
8. What is a 'no-regrets' option and how has it been incorporated into China's response to climate change?
9. Discuss how energy policy evolves and consider why a shift in 'global energy policy' is crucial for emissions reductions and how, if at all, such a shift might be best achieved?
10. What is meant by a 'stranded asset' and why is this relevant when considering a shift from fossil fuel dependence to renewable energy?

---

[108] N Stern, *A Blueprint for a Safer Planet: How We Can Save the World and Create Prosperity* (Vintage Books, 2010) ch 1 at p.14.

8.7 FURTHER READING

J E Aldy and R N Stavins, 'Designing the Post-Kyoto Climate Regime: Lessons From the Harvard Project on International Climate Agreements: An Interim Progress Report for the 14th Conference of the Parties' (Interim Progress Report, Harvard Environmental Economics Program, 24 November 2008) http://belfercenter.ksg.harvard.edu/files/post%20kyoto%20final%20HIGH%20QUALITY.pdf.

Australian Bureau of Agriculture and Resource Economics and Sciences, A Schultz and R Putney, *Australian Energy Statistics – Energy Update 2011* (2011) http://www.agriculture.gov.au/abares/publications/display?url=http://143.188.17.20/anrdl/DAFFService/display.php?fid=pe_abares99010610_12c.xml.

Australian Government, *Productivity Commission for the Prime Ministerial Task Group on Emissions Trading* (Productivity Commission, 2007).

Australian Government, Department of the Environment, *Emissions Reduction Fund Green Paper* (2013) http://www.environment.gov.au/climate-change/emissions-reduction-fund/green-paper.

Australian Government, Department of the Environment, *Emission Reduction Fund White Paper* (April 2014) http://www.environment.gov.au/climate-change/emissions-reduction-fund/publications/white-paper.

Australian Government, Geoscience Australia, *Basics* http://www.ga.gov.au/scientific-topics/energy/basics.

Australian Government, Productivity Commission, 'Chapter 6: No Regrets Policy' in *Climate Change Adaptation Policy* (2009) http://www.pc.gov.au/__data/assets/pdf_file/0009/119709/09-climate-change-adaptation-chapter6.pdf.

L Bifera, 'Carbon Trading in China: Short Term Experience Long Term Wisdom' on Center for Climate and Energy Solutions Blog, (4 September 2014) http://www.c2es.org/blog/biferal/carbon-trading-china-short-term-experience-long-term-wisdom.

S Bruce, 'Climate Change Mitigation Through Energy Efficiency Laws: From International Obligations to Domestic Regulation' (2013) 31 *Journal of Energy and Natural Resources Law* 313.

Carbon and Environment Daily, *Direct Action Sparks New Carbon Deregulation Campaign* (28 April 2014) http://adminpanel.ceda.com.au/FOLDERS/Service/Files/Documents/22090~Economics-of-Climate-Change.pdf.

C Calarne, 'Exploring Methodological Challenges within the Context of Climate Change Law and Policy' (2011) 105 *American Society of International Law* 255.

S Christensen, 'Regulation of Emissions Under the Carbon Pricing Mechanism: A Case Study of Australia's Coal Fired Electricity Sector' (2013) 15 *Asia Pacific Journal of Environmental Law* 17.

Climate Institute, *Coming Ready or Not: Managing Climate Risks to Australia's Infrastructure,* (2012).

R N Cooper, 'The Kyoto Protocol: A Flawed Concept' (2001) 31 *Environmental Law Reporter* 11,484.

D De Sousa and J Thwaites, 'Regulatory Responses to Facilitate Adaptation of Existing Infrastructure to Climate Change' (Paper presented at National Climate Change Adaptation Research Facility: Climate Adaptation in Action 2012 Conference, Melbourne, 26 June 2012).

E Emeseh, A Aboah and H Barmakhshad, 'Framework for Achieving Sustainability in Investment Decisions: Reflections on Rio 20' (2014) 16 *Environmental Law Review* 21.

R Epstein, 'Behavioural Economics: Human Errors and Market Corrections' (2006) 73(1) *The University of Chicago Law Review* 111.

M Faber and R Manstetten, *Philosophical Basics of Ecology and Economy* (Routledge, 2010).

F A Felder, 'Climate Change Mitigation and the Global Energy System' (2014) 25 *Villanova Environmental Law Journal* 89.

S Freeland, 'Kyoto Protocol: An Agreement Without a Future' (2001) 24 *University of New South Wales Law Journal* 532.

R Garnaut, *The Garnaut Climate Change Review: Final Report* (Cambridge University Press, 2008) http://www.garnautreview.org.au/index.htm.

Garnaut Climate Change Review, *Update, 2011: Australia in the Global Response to Climate Change Summary*, (31 May 2011) http://www.garnautreview.org.au/update-2011/garnaut-review-2011/summary-20June.pdf.

W Gumley, 'Legal and Economic Responses to Global Warming – An Australian Perspective' (1997) 14 *Environmental and Planning Law Journal* 341.

W Gumley 'Carbon Pricing Options for a Post-Kyoto Response to Climate Change in Australia' (2011) 39 *Federal Law Review* 131.

S Handke and E Hey, 'Climate Change Negotiations in a Changing Global Energy Landscape: A Wicked Problem' (2013) 2(7) *European Society of International Law Reflections* 8.

C Hepburn, 'Regulation by Prices, Quantities or Both: A Review of Instrument Choice' (2006) 22 *Oxford Review of Economic Policy* 226.

'How To Lose Half a Trillion Euros', *The Economist* (online) 12 October 2013 http://www.economist.com/news/briefing/21587782-europes-electricity-providers-face-existential-threat-how-lose-half-trillion-euros.

Intergovernmental Panel on Climate Change, *Climate Change 2001 Synthesis Report: Summary for Policymakers* (2001) <http://www.ipcc.ch/pdf/climate-changes-2001/synthesis-spm/synthesis-spm-en.pdf>.

Intergovernmental Panel on Climate Change, *Climate Change 2007 – The Physical Science Basis: Contribution of Working Group I to the Fourth Assessment Report of the IPCC* (2007).

Intergovernmental Panel on Climate Change, *Organization* http://ipcc.ch/organization/organization.shtml.

Intergovernmental Panel on Climate Change, *Structure: How does the IPCC work?* http://www.ipcc.ch/organization/organization_structure.shtml.

Intergovernmental Panel on Climate Change, R K Pachauri and A Reising (eds) *Fourth Assessment Report of the Intergovernmental Panel on Climate Change: Synthesis Report* (2007).

Intergovernmental Panel on Climate Change, T F Stocker, D Qin, G K Plattner, M Tignor, S K Allen, J Boschung, A Nauels, Y Xia, V Bex and P M Midgley (eds) 'Summary for Policymakers' in *Climate Change 2013: The Physical Science Basis. Contribution of Working Group I to the Fifth Assessment Report of the Intergovernmental Panel on Climate Change* (Cambridge University Press, 2013).

International Energy Agency, *World Energy Outlook – Factsheet* (2012) http://www.worldenergyoutlook.org/publications/weo-2012/.

International Energy Agency Statistics, *CO₂ Emissions from Fuel Combustion: Highlights* (2012) http://www.iea.org/publications/freepublications/publication/CO2emissionfromfuelcombustionHIGHLIGHTS.pdf.

F Jotzo, 'Prerequisites and Limits for Economic Modelling of Climate Change Impacts and Adaptation', (Research Report No 1055, Australian National University, March 2010).

F Jotzo, 'China Heads for Price on Carbon: Energy Overhaul is Next', *The Conversation* (19 September 2014) http://theconversation.com/china-heads-for-price-on-carbon-energy-market-overhaul-is-next-31119.

J Kehoe, 'Australia Shows US How Not to Export Gas', *Australian Financial Review* (online) 30 July 2014 http://www.afr.com/p/business/resources/energy/gas/australia_shows_us_how_not_to_export_ObsvU7S8je3WEC3T4OUTOM.

M D King and J Gulledge, 'The Climate Change and Energy Security Nexus' (2013) 37 *Fletcher Forum of World Affairs* 25.

P Kohler, R Bintanja, R Rischer, H Joos, F Knutt, R Lohmann, and V Masson-Delmotte, 'What Caused Earth's Temperature Variations During the Last 800,000 Years? Data-based evidence on radiative forcing and constraints on climate sensitivity'. (2010) 29 *Quartenary Science Reviews* 129.

A Konoplyanik, 'Energy Charter Treaty and its Role in International Energy' (2006) 24 *Journal of Energy and Natural Resources Law* 523.

S E Koonin, 'A Degree of Uncertainty', *The Wall Street Journal* (online), 23 September 2014, http://online.wsj.com/articles/climate-science-is-not-settled-1411143565?mod=WSJ_myyahoo_module.

R Lazarus, 'Super Wicked Problems and Climate Change: Restraining the Present to Liberate the Future' (2009) 94 *Cornell Law Review* 1153.

S J Liebowitz, and E S Margolis, 'Path Dependence, Lock-In, and History' (1995) 11(1) *Journal of Law, Economics and Organization* 205.

A Long, 'Complexity in Global Energy-Environment Governance' (2014) 15 *Minnesota Journal of Law Science and Technology* 1055.

E Lowitt, 'How to Survive Climate Change and Still Run a Thriving Business' *Harvard Business Review* (April 2014) <http://hbr.org/2014/04/how-to-survive-climate-change-and-still-run-a-thriving-business-checklists-for-smart-leaders>.

R Lyster and A Bradbrook, *Energy Law and the Environment* (Cambridge University Press, 2006).

S McInerney-Lankford, 'Climate Change and Human Rights: An Introduction to Legal Issues' (2009) 33 *Harvard Environmental Law Review* 431.

The New Climate Economy, 'Chapter 1: Climate Change and the Economy' in *The Global Commission on the Economy and Climate* (2014) http://newclimateeconomy.report/.

J O'Neil, *The Growth Map: Economic Opportunities in the BRICs and Beyond* (Penguin Publishing, 2011).

H M Osofsky, 'Learning from Environmental Justice: A New Model for International Environmental Rights' (2005) 24 *Stanford Environmental Law*.

E Ostrom, 'Polycentric Systems for Coping with Collective Action and Global Environmental Challenge' (2010) 20 *Global Environmental Change* 550.

G Parkinson, 'China, United States Say No To New Coal Plants', *RenewEconomy*, (16 September 2013) http://reneweconomy.com.au/2013/china-us-say-new-coal-plants-37836.

Parliament of Victoria, *The Hazelwood Mine Fire Inquiry* (28 August 2014) http://report.hazelwoodinquiry.vic.gov.au/executive-summary-2/hazelwood-mine-fire.

M Porter, 'Reforms in the Greenhouse Era: Who Pays and How?' in *Committee for Economic Development of Australia Growth 61: A Taxing Debate – Climate Policy beyond Copenhagen* (2009).

Queensland Government, Department of Energy and Water Supply *2012 Gas Market Review* (2012) http://www.dews.qld.gov.au/__data/assets/pdf_file/0006/77775/gas-market-review-2012.pdf.

B Saul, S Sherwood, J McAdam, T Stephens and J Slezak, *Climate Change and Australia: Warming to the Global Challenge* (Federation Press, 2012).

E L F Schipper, 'Conceptual History of Adaptation in the UNFCCC Process' (2006) 15 *Review of European Community and International Environmental Law* 82.

J Seth, 'Climate Change and Global Justice: Crafting Fair Solutions for Nations and People' (2009) 33 *Harvard Environmental Law Review* 297.

State Government of Victoria, Department of Justice and Regulation, *Emergency Risks in Victoria* (2014) http://www.justice.vic.gov.au/home/safer+communities/emergencies/emergency+risks+in+victoria+report.

N Stern, *The Economics of Climate Change: The Stern Review*, (Cambridge University Press, 2007). The full report is available at http://mudancasclimaticas.cptec.inpe.br/~rmclima/pdfs/destaques/sternreview_report_complete.pdf.

N Stern, *A Blueprint for a Safer Plant: How We Can Save the World and Create Prosperity* (Vintage Books, 2010).

N Stern, 'Climate change is here and it could lead to global conflict', *The Guardian* (online), 14 February 2014 http://www.theguardian.com/environment/2014/feb/13/storms-floods-climate-change-upon-us-lord-stern.

S P Subedi, *International Investment Law: Reconciling Policy and Principle*, (Hart Publishing, Oxford, 2nd ed, 2012).

C R Sunstein and L A Reische, 'Automatically Green: Behavioural Economics and Environmental Protection' (2014) 38 *Harvard Environmental Law Review* 127.

F Teng and F Jotzo, 'Reaping the Economic Benefits of De-Carbonisation for China' (Working Paper, Australian National University, 2014).

M Wilder, 'The Global Economic Environment and Climate Change' in CEDA, *The Economics of Climate Change* (CEDA, June 2014). http://www.ceda.com.au/research-and-policy/research/2014/06/18/climatechangeeconomics.

A J Wildermuth, 'The Next Step: The Integration of Energy Law and Environmental Law' (2011) 31 *Utah Environmental Law Review* 369.

H J Wiseman and H M Osofsky, 'Dynamic Energy Federalism' (2013) 72 *Maryland Law Review* 773.

E A Wisnosky, A Ricardo, M Ardizzone and A Frederico, 'International Energy and Natural Resources Law' (2012) 46(1) *International Lawyer* 305.

World Bank, *Korea's Global Commitment to Green Growth* (2012) http://www.worldbank.org/en/news/feature/2012/05/09/Korea-s-Global-Commitment-to-Green-Growth.

ENVIRONMENTAL REGULATION

9.1	Introduction	355
9.2	Jurisdictional framework	360
9.3	Bilateral agreements	362
9.4	Environmental impact assessment	365
9.5	Environmental assessment of onshore mining projects in Western Australia	368
9.6	Environmental assessment of onshore mining projects in Queensland	374
9.7	Environmental assessment of onshore mining projects in New South Wales	381
9.8	Commonwealth environmental legislation: The *Environment Protection and Biodiversity Conservation Act 1999*	387
9.9	Review questions	398
9.10	Further reading	399

9.1 Introduction

The environmental regulation of mineral exploration and mining activity is a relatively new area that has evolved dramatically since the early 1990s. The increasing trend to develop and enhance environmental policy and regulation for mining and energy projects has been achieved via the imposition of strict environmental evaluation, planning, mitigation, and rehabilitation requirements upon mining proponents. Mining and energy operations across the world are now routinely subjected to environmental impact assessments and environmental management conditions; however, there is considerable debate regarding the scope and implementation of this regulation. In some countries, environmental impact assessment constitutes a precondition to the issuance of an exploratory resource title; whereas in others, whilst it may be a prerequisite to the commencement of mining operations, environmental assessment does not constitute a precondition to the issuance of a resource title in the first instance. There is also considerable range regarding the rigour and focus of environmental assessment. Some countries mandate a full and comprehensive environmental impact assessment of mining projects – particularly large-scale projects – whilst others are satisfied with less onerous evaluation and this has the capacity to cause longer-term environmental degradation.[1] Many countries have also sought to incorporate, as an additional component of an environmental impact assessment, social and community concerns regarding the impact of a mining project upon local regions. Most countries now accept the importance of imposing ongoing rehabilitation obligations upon mining and energy proponents through different stages of the project but there are different perspectives regarding the best way to incentivise land rehabilitation during the exploration and production phase of a project and once it is completed. Options vary from the imposition of environmental 'bonds' to the endorsement of progressive self-financing mechanisms; whereby mining proponents guarantee the costs of site rehabilitation by setting aside prescribed amounts from revenues received from operations.[2]

This chapter examines the regulatory framework in Australia for the environmental assessment and management of resource titles issued for mining and energy projects that seek to extract and produce fossil fuels. Rigorous environmental

---

1 This is a particular concern for developing countries. See, eg, in Brazil where environmental licensing has largely fallen short of international best practice. This is discussed by L Parry, 'New Laws Could Hand Miner's 10% of Brazil's National Parks and Indigenous Lands', *The Conversation* (7 November 2014) http://theconversation.com/new-laws-could-hand-miners-10-of-brazils-national-parks-and-indigenous-lands-33912.
2 See the discussion by E Bastide, T Waelde and J Warden-Fernandez, *International and Comparative Mineral Law and Policy: Trends and Prospects* (Kluwer, 2006) 56–7.

regulation is crucially important for the mining and energy sector given the heavy interconnection between resource exploitation and environmental health and sustainability. Concern for the welfare of the environment following extractive mining, particularly with the advent of new technological advances, has a global relevance and has generally been addressed via core principles of environmental federalism to ensure that the scale of the issue is addressed at the correct level and further, to ensure that global imperatives, such as food security, ecological sustainability and climate change mitigation, are addressed at a more central level.[3] The environmental concerns that mining and energy projects generate can be organised according to the particular stage of mining production the project has reached.

The initial stage will generally involve a direct evaluation of the environmental impacts connected with the exploration, investigation and sampling of fossil fuel reservoirs. This evaluation occurs at the speculative stage of the project when an exploratory resource title is first sought. Consideration at this stage must be given to the potential for contamination or degradation flowing from the investigative process. This could include impacts upon any adjacent water resources as well as the prospect of any chemical contamination of natural vegetation – including water resources – and the probability of any land and air degradation. Once a mining project moves into the production phase, stronger environmental assessment and rehabilitation conditions become imperative, given the deeper impact that extractive processes can have upon the natural environment. Rigorous impact assessment and management conditions are needed to mitigate the dangers of environmental degradation and promote proper post-production land rehabilitation and longer-term ecological sustainability. A further crucial aspect of environmental regulation and policy lies in the need to take account of the broader environmental impacts of mining and energy projects upon fragile ecological frameworks including, for example, world heritage areas, protected species, areas of strategic biophysical relevance for agricultural production, areas of critical importance for tourism, and other established land-based industries.[4]

The environmental assessment of mining projects necessarily involves a multivariate range of factors that must be carefully balanced to determine whether the social and economic imperatives connected with the development of the mining

---

3 See M Radetski, 'Economic Growth and Environment', in P Low (ed), *International Trade and the Environment* (Vol 1), (World Bank: Discussion Papers 159, 1992). For the Australian context see R Briese, 'Climate Change Mitigation Down Under – Legislative Responses in a Federal System' (2010) 13 *Asia Pacific Journal of Environmental Law* 75, 95 where the author notes that Australia has generally adopted models of cooperative environmental federalism whereby the jurisdiction level of regulation matches the scale of the environmental problem; i.e. local environmental problems addressed at a local level and cross-boundary problems addressed at a national level.
4 This will generally involve an assessment of national matters of environmental significance outlined in the Commonwealth legislation: the *Environment Protection and Biodiversity Conservation Act 1999* (Cth). This is discussed in more detail further in this chapter.

project are not outweighed by the immediate and longer-term environmental impacts.[5] The potentially devastating impact that mining projects can have upon the natural environment means that the approval and management process is a crucial aspect of longer-term project sustainability.[6] There are two fundamentally different policy paradigms around which regulatory frameworks for environmental assessment have largely been structured. The first can be broadly described as the 'cost-benefit approach' to environmental assessment, whose objective is to optimise social welfare by 'predicting, weighing and aggregating all relevant consequences of policy proposals in order to identify choices that represent welfare maximising uses of public resources'.[7] The second is a more pragmatic and risk-averse approach known as the 'precautionary approach'. The precautionary approach to environmental regulation accords with that fundamental principle of ecological sustainable development, the 'precautionary principle', and is fundamentally grounded in minimising the risk of environmental damage. A precautionary approach to environmental regulation mandates that when an activity raises possible threats of harm to the environment, precautionary measures should be immediately adopted even in circumstances where the scientific consequences are not fully determined.[8] The debate between the cost-benefit approach (CBA) and the precautionary principle (PP) is polarising. Defenders of the PP approach focus on the potential for damage from new technological processes and ultimately prefer an ex-ante approach whenever a project has the possibility of causing harm to human health or the environment. The difficulty, however, with the PP approach is that it has been described as 'asymmetric' because it fails to take proper account of the opportunity costs that might be associated with the project and that the failure to treat opportunity costs 'pari passu with the primary risks targeted by policy measures is indefensible'.[9]

The regulatory paradigm adopted for environmental assessment is also impacted by the nature of the ownership framework in which it operates. As a general principle, public ownership frameworks for minerals and resources have stronger

---

5   The importance of balancing immediate social and economic benefits with cumulative environmental impacts is well established. See especially the discussion by R Harding (ed) *Environmental Decision Making* (Federation Press, 1998) ch 7. See also G Bates, *Environmental Law in Australia* (Lexis Nexis, 7th ed, 2010) ch 10. For a specific evaluation of the way in which this balancing process is conducted see D Kysar, 'It Might Have Been: Risk, Precaution and Opportunity Costs' (2007) 22 *Journal Of Land Use and Environmental Law* 1, 3 where the author evaluates the cost-benefit analysis against the precautionary principle approach to environmental regulation.
6   See E Bastide, T Waelde and J Warden-Fernandez, *International and Comparative Mineral Law and Policy: Trends and Prospects* (Kluwer Law, 2010) 57.
7   See Kysar, above n 5, 3–4.
8   For a discussion on this see R W Hahn and C R Sunstein, 'The Precautionary Principle as a Basis for Decision-Making' (2005) 2(2) *The Economist's Voice*, art 8.
9   Kysar, above n 5, 8.

environmental assessment processes than those where the minerals and resources are privately controlled. This is largely a consequence of the fact that the government has control of the *in situ* resources.[10] The public ownership framework fundamentally alters the underlying 'private' ownership assumptions and disaggregates the land asset from the minerals that reside within that asset. Public ownership of natural resources is based upon the assumption that the corpus of the land belongs to the sovereign and this, in turn, prompts the government to regulate in order to prevent wasteful exploitation.[11]

In Australia, the implementation of environmental regulation has traditionally been resisted by the strong and entrenched assumptions of private landowners. Traditionally, this has largely meant that environmental responsibility for natural resource management was regarded as something inextricably connected with the owner of private land rather than a public responsibility.[12] Private ownership rights create powerful incentives to preserve and enhance the value of what people own.[13] However, when the conferral of property rights generates negative externalities, particularly in terms of environmental degradation, consideration should be given as to whether the owner should be allowed to impose the harm, whether the owner should compensate for the harm, or whether the owner should be stopped from engaging in the harmful conduct.[14]

In a contemporary Australian framework where the occurrence of mining projects on private land has dramatically increased – particularly following the rapid expansion of unconventional gas in the eastern states – private ownership rights continue to retain considerable influence in the development of environmental policy. Today, the fundamental entitlement of the state and, in some instances, the Commonwealth, to regulate the environmental impact and effect of mining projects is well established. Despite this, the success of the regulatory framework is largely dependent upon their capacity to respond and

---

10 P Wieland, 'Going Beyond Panaceas: Escaping Mining Conflicts in Resource Rich Countries Through Middle-Ground Policies' (2014) 20 *New York University Environmental Law Journal* 199, 204. See also the discussion by Bates, above n 5, 46–7 where the author notes that environmental regulation cuts across common law rights and centuries of legal and cultural tradition that support the pre-eminence of private ownership in land.
11 See E Duruigbo, 'The Global Energy Challenge and Nigeria's Emergence as a Major Gas Power: Promise, Peril or Paradox of Plenty' (2009) 21 *Georgetown International Environmental Law Review* 395, 440.
12 See especially the discussion by M Crommelin, 'Resource Law and Public Policy' (1983) 15 *University of Western Australia Law Review* 1, 9 where the author notes that the implementation of public ownership of minerals and petroleum in Australia has meant that state legislatures have taken control of the development of mining projects by private companies.
13 See the discussion by T Anderson and D Leal, *Free Market Environmentalism* (Palgrave, 1991) 3.
14 See J Singer, 'How Property Norms Construct the Externalities of Ownership' in G S Alexander and E Penalver (eds), *Property and Community* (Oxford University Press, 2009).

deal with the 'cultural and political pressures exerted by the actual or perceived rights that attach to land and to the natural resources' that may reside within that land.[15]

Ensuring that natural resources, particularly fossil fuel reserves which have been forming for billions of years, are effectively regulated is increasingly important given the emergent issues associated with the exploitation of these resources which directly impacts upon the welfare and benefit of all citizens.[16] Strong and effective environmental constitutionalism is imperative to ensure that fossil fuel resources are appropriately managed within the context of climate change imperatives and within the context of dwindling supplies. Balancing these global concerns is best achieved by promoting regulatory and policy inter-linkages between the environment, culture, and the market. At the international level, environmental regulation has been significantly weakened because nation-states have refused to work together in global agreement or with institutions to achieve meaningful environmental improvement. Kysar has argued that one of the core difficulties stems from the flawed economic models that guide political reaction to environmental problems. Effective negotiations are often impeded by narrow, cost-benefit calculations conducted within the parameters of a cost-benefit regulatory model that prioritises the commercialisation of energy resources, rather than longer-term cumulative models which are more responsive to environmental science models.[17]

In Australia, mining proponents who seek to exploit natural resources that have been vested in the state – whether on private or Crown land – are always subject to environmental review although the scope and focus of that review and the way in which it is conducted can vary considerably depending upon whether it is conducted at the state or federal level. Understanding how environmental regulation has impacted mining projects in Australia should necessarily be prefaced by a brief overview of the underlying jurisdictional framework and an evaluation of the relationship between the Commonwealth and the states regarding natural resource management.

---

15 Bates, above n 5, 48.
16 P Wieland, 'Going Beyond Panaceas: Escaping Mining Conflicts in Resource Rich Countries Through Middle-Ground Policies' (2014) 20 *New York University Environmental Law Journal* 199, 204.
17 See D Kysar, 'Executive Summary: Professor Douglas Kysar's Analysis of Flaws in Predictive International Climate Policy Models' (2013) 40 *British Columbia Environmental Affairs Law Review* 409, 410 where the author argues that the fundamental pursuit of 'maximised social welfare is an unsustainable and flawed goal'. Policy decisions that support this goal, despite appearing 'suboptimal and inefficient' must be adopted because the global environmental challenge of climate change requires us to act in a manner which is 'contrary to assumptions that now prevail in climate change policies and politics'.

9.2 Jurisdictional framework

The environmental regulation of the mining and energy sector has a legislative rather than a common law foundation. The environmental approval and management of mining projects is controlled by the specific legislative regimes that have been implemented within each state and territory. Consequently, the nature and scope of the environmental regulation that a mining proponent is subject to depends largely upon the state in which the project is located.

The Commonwealth Government has never played a significant role in the environmental regulation of mining projects in Australia. This stems largely from the fact that the *Commonwealth Constitution* does not directly grant to the Commonwealth any specific power to make laws with respect to the environment. As such, the environment has always been regarded as a local issue and the environmental regulation of mining projects proposed within the jurisdiction of the state is subject to state law as they retain ownership and control over their natural resources. It is only where a particular project triggers a specific matter of national environmental significance that assessment at the Commonwealth level becomes necessary.[18]

State-based environmental frameworks can vary considerably and the environmental review process is itself invariably shaped by the nature and scope of the resource sector that characterises the state. For example, Western Australia is the most resource-intensive state in Australia. The primary mining focus for Western Australia is iron ore as it has the world's largest economic reserves for iron ore. In 2011, the iron ore output for Western Australia was 474 million tonnes and in the 2011–2012 fiscal year this generated in excess of $3.9 billion in royalties for the state.[19] In light of this, the framework for environmental assessment is relevant not only to the environmental sector but also to the industry sector. The two relevant government departments are the Department of Industry and Resources (DoIR) and the Department of Environment (DoE), which includes the Environmental Protection Authority. The DoIR explicitly seeks to develop the industry and resources for the benefit of Western Australians to ensure sustainable prosperity and

---

18 For a detailed discussion of this see M Crommelin, 'Commonwealth Involvement in Environmental Policy: Past, Present and Future' (1987) 4 *Environmental Planning Law Journal* 101; L Godden and J Peel, 'The Environment Protection and Biodiversity Conservation Act 1999 (Cth): Dark Sides of Virtue' (2007) 31 *Melbourne University Law Review* 262. More recently, see the discussion by J M Rieder, 'Evaluation of Two Environmental Acts: The National Environmental Policy Act and the Environment Protection and Biodiversity Conservation Act' (2011) 14 *Asia Pacific Environmental Law Journal* 105.
19 See the Government of Western Australia, Department of Mines and Petroleum, *Western Australian Mineral and Petroleum Statistics Digest 2008-09* (2009) http://www.dmp.wa.gov.au/documents/Statsdigest09web.pdf.

a better quality of life. The DoE, on the other hand, focuses upon sustaining a healthy environment through the sustainable use of natural resources. The tension in Western Australia between focused and rigorous environmental assessment and the commercial imperatives associated with industry development is clear and that generates a strong potential for the approval process to be compromised by vested commercial incentives to progress iron ore mining projects.

In addition to the mining and petroleum legislation that exists in each state to regulate the licensing framework for mining and energy projects, the environmental legislation that impacts upon the assessment of such titles and the ongoing environmental management of such projects is extensive. It includes the following:

- **Western Australia**: *Environmental Protection Act 1986*; *Planning and Development Act 2005* and associated regulations
- **Victoria**: *Environment Effects Act 1978*; *Environment Protection Act 1970*; *Planning and Environment Act 1987* and associated regulations
- **New South Wales**: *Environmental Planning and Assessment Act 1979*; *Protection of the Environment Operations Act 1997* and associated regulations
- **Queensland**: *Environmental Protection Act 1994*; *Energy and Water Ombudsman Act 2006*; *Environment Protection (Greentape Reduction) and Other Legislation Amendment Act 2012* and associated regulations
- **South Australia**: *Environment Protection Act 1993* and associated regulations
- **Tasmania**: *Environmental Management and Pollution Control Act 1994* and associated regulations
- **Northern Territory**: *Environmental Assessment Act; Planning Act* and associated regulations
- **Australian Capital Territory**: *Environment Protection Act 1997*; *Planning and Development Act 2007* and associated regulations.

The *Commonwealth Constitution* contains no specific power regarding environmental regulation because this was not a strong concern when the *Constitution Act* was passed in 1900. States are entitled to legislate on all matters not specifically referred to in the Commonwealth by the *Constitution*. The Commonwealth only has exclusive jurisdiction over matters outlined in s 52 – which includes Commonwealth places and territories and the Commonwealth public service – and s 90 – which is the power to levy duties of customs and excise. Hence, from a jurisdictional perspective, it is possible for environmental issues relevant to particular mining and energy projects to be dealt with by either the Commonwealth or the state government. The national Commonwealth legislation is the *Environment Protection and Biodiversity Conservation Act 1999* (Cth) ('the *EPBCA*'). The existence of a dual jurisdictional structure in Australia is consistent with the federal Constitutional framework; however, it also means that mining and energy projects that affect matters of national environmental significance and thereby trigger the *EPBCA* will be covered by both national and state environmental regimes.[20]

In order to avoid duplication and reduce overlap, the Commonwealth may delegate responsibility to a state to undertake environmental assessment on behalf of the Commonwealth. The *EPBCA* explicitly enables the Commonwealth to provide for accreditation of state or territory processes for environmental assessment. These processes are known as 'assessment bilaterals'. The core aim of this process is to streamline the environmental assessment between the Commonwealth and the state, particularly with respect to matters of national environmental significance ('MNES'). The Commonwealth Government has not yet formalised the capacity of the state government to approve proposals that have a national environmental significance; however, these changes have been proposed.

In Queensland, the idea of a 'one stop shop' whereby the state has the capacity to evaluate all matters of national environmental significance has generated significant concern for stakeholders who feel that this should only occur where the environmental impact assessment relevant to the Queensland legislation enshrines best practice environmental standards as well as the core objectives of the *EPBCA*. Where a mining development constitutes a conflict of interest because, for example, the state has a significant financial stake in the project, any assessment of a matter of national environmental significance that the project might raise should, it is argued, be excluded from bilateral agreements and remain within the province of the Commonwealth.[21]

9.3 Bilateral agreements

Bilateral agreements may be entered into between the Commonwealth and state and territory governments in circumstances where concurrent environmental assessment and approval processes exist. The underlying rationale of the bilateral agreement is to ensure a more streamlined and efficient approach to environmental assessment and approval that avoids the overlap and duplication that may occur when both state and Commonwealth agencies have jurisdiction. There are two core types of bilateral agreement under the *EPBCA*: (i) assessment bilateral agreements; and (ii) approval bilateral agreements. An assessment bilateral agreement allows the Commonwealth to accredit a particular state/territory assessment and approval

---

20 See J Crawford, 'The Constitution and the Environment' (1991) 13 *Sydney Law Review* 11, 12.
21 See the detailed discussion on this by the Queensland Environmental Defenders Office in their discussion on the *Draft Approval Bilateral Agreement Made pursuant to Sections 45 and 46 of the EPBC Act* (2014): EDO NSW, *Draft Approval Bilateral Agreements: NSW & Qld* (2014) http://www.edonsw.org.au/draft_approval_bilateral_agreements_for_nsw_and_qld; See also Environmental Defenders Office Qld, *Say no to Commonwealth delegating environmental approval powers to Queensland* (12 June 2014) http://www.edoqld.org.au/news/say-no-to-commonwealth-delegating-environmental-approval-powers-to-queensland/.

process. Assessment bilateral agreements may be entered into between the Commonwealth and the relevant state/territory for any proposed mining project in order to determine core environmental assessment processes. Where a proposed project is covered by an assessment bilateral agreement, then it will be assessed pursuant to the accredited state or territory process. Following assessment, the proposed action will require approval from the Commonwealth Environment Minister in accordance with the *EPBCA*. Assessment bilateral agreements may be made pursuant to s 47 of the *EPBCA* which enables the Commonwealth minister to determine whether to approve of an action that is based upon the accredited environmental assessment process by a state or territory. An approved bilateral agreement on the other hand will be fully assessed and approved by the relevant state or territory. Where an approved bilateral agreement exists, no further approval is required from the Commonwealth Environment Minister pursuant to the *EPBCA*. Approved bilateral agreements may be issued pursuant to s 46 of the *EPBCA*, which obviates the need for Commonwealth approval where a state or territory has approved of the action in accordance with an accredited approvals process.

A good example of a s 47 assessment bilateral agreement is that entered into between the state and the Commonwealth to accredit the assessment and approval process undertaken by the relevant state or territory Environmental Protection Agency for the purpose of assessing matters of national environmental significance under the *EPBCA*.[22]

Section 44 of the *EPBCA* sets out the object for providing bilateral agreements between the Commonwealth and states/territories. The Commonwealth Environment Minister may enter into a bilateral agreement only in circumstances where he or she is satisfied that it is consistent with the objects of the *EPBCA* and meets the specific requirements. Section 45 of the *EPBCA* confers power upon the Minister to enter into a bilateral agreement and outlines the criteria for a bilateral agreement.

Section 44 and 45 of the *EPBCA* are extracted below:

Environment Protection and Biodiversity Conservation Act 1999 (Cth)

44 Object of this Part

The object of this Part is to provide for agreements between the Commonwealth and a State or self-governing Territory that:

(a) protect the environment; and
(b) promote the conservation and ecologically sustainable use of natural resources; and

---

22 This type of assessment bilateral agreement was entered into between the Western Australian Government and the Commonwealth Government in June 2014 which thereby allowed the Commonwealth Government to rely upon assessments made by the Western Australian Environmental Protection Agency when assessing matters referred to it which are deemed to be of national environmental significance.

(c) ensure an efficient, timely and effective process for environmental assessment and approval of actions; and

(d) minimise duplication in the environmental assessment and approval process through Commonwealth accreditation of the processes of the State or Territory (and vice versa).

45 Minister may make agreement

Making bilateral agreement

(1) On behalf of the Commonwealth, the Minister may enter into a bilateral agreement.

Note 1: A bilateral agreement can detail the level of Commonwealth accreditation of State practices, procedures, processes, systems, management plans and other approaches to environmental protection.

Note 2: Subdivision B sets out some prerequisites for entering into bilateral agreements.

What is a **bilateral agreement**?

(2) A **bilateral agreement** is a written agreement between the Commonwealth and a State or a self-governing Territory that:
   (a) provides for one or more of the following:
      (i) protecting the environment;
      (ii) promoting the conservation and ecologically sustainable use of natural resources;
      (iii) ensuring an efficient, timely and effective process for environmental assessment and approval of actions;
      (iv) minimising duplication in the environmental assessment and approval process through Commonwealth accreditation of the processes of the State or Territory (or vice versa); and
   (b) is expressed to be a bilateral agreement.

Publishing notice of intention to enter into agreement

(3) As soon as practicable after starting the process of developing a draft bilateral agreement with a State or self-governing Territory, the Minister must publish, in accordance with the regulations (if any), notice of his or her intention to develop a draft bilateral agreement with the State or Territory.

Publishing bilateral agreements and related material

(4) As soon as practicable after entering into a bilateral agreement, the Minister must publish in accordance with the regulations:
   (a) the agreement; and
   (b) a statement of the Minister's reasons for entering into the agreement; and
   (c) a report on the comments (if any) received on the draft of the agreement published under Subdivision B.

The effect of entering into an assessment bilateral agreement is that the assessment or approval process relevant to the state or territory will be accredited to the Commonwealth which means that the Commonwealth is entitled to take that assessment or approval process into consideration in making its own determination. Entering into the agreement does not mean that the relevant state or territory acquires full jurisdiction over the assessment and approval process but rather, that the assessment and approval process undertaken by the state may be taken into account by the Commonwealth which may obviate the need for the Commonwealth to undertake any further assessment or approval.

9.4 Environmental impact assessment

Environmental impact assessment (EIA) is the formal process of examining and evaluating the full range of environmental impacts that proposed activities are likely to have upon the environment. The EIA is a crucial component of environmental assessment for mining and energy projects, particularly where it is mandated as a precondition to the approval and issuance of a resource title. The significance of the EIA lies in the fact that it has the strategic capacity to not only identify significant environmental impacts but, armed with that knowledge, to preclude a project from proceeding or, alternatively, to ensure that the mitigation of those impacts is a condition to approval.[23] In this respect, the EIA is a universally recognised strategy for sustainable development.[24] EIA is a proven technique that is utilised to avoid or minimise unanticipated adverse environmental effects. In this respect, it represents a process for institutionalising foresight and broadly accords with the precautionary approach to environmental regulation.[25] The structure of the EIA is essentially the same worldwide, although every jurisdiction has tailored the EIA to meet specific geographical and environmental needs and to identify specific socio-economic conditions. The core flexibility of the EIA as an instrument has allowed it to function effectively across a range of cultural, political, and socio-economic conditions which makes it extremely effective in the context of mining projects in developed and undeveloped countries.

---

[23] See especially the discussion by M Raff, 'Ten Principles of Quality in Environmental Impact Assessment' (1997) 14 *Environmental Planning Law Journal* 207; I Thomas and P Murfitt, *Environmental Management Processes and Practices for Australia* (Federation Press, 2005) pt II: Environmental Management Processes and Systems; G Bates, *Environmental Law in Australia* (Lexis Nexis, 7th edition, 2010) ch 10. For an international perspective on environmental impact assessment, see A Sifakis, 'Precaution, Prevention and the Environmental Impact Assessment Directive' (1998) 7 *European Environmental Law Review* 349; N A Robinson, 'International Trends in Environmental Impact Assessment' (1992) 19 *British Columbia Environmental Affairs Law Review* 591, 595.

[24] See, eg, Johannesburg Plan of Implementation of the World Summit on Sustainable Development III 19(e) where the environmental impact assessment is listed as an essential practice for implementing 'sustainable development'.

[25] See Robinson, above n 23, 591.

The purpose of the EIA is to generate an informative statement regarding predicted environmental outcomes and how those outcomes might be managed. The EIA has a common application to projects requiring development approval because, in such instances, the possibility of environmental harm is strong. The EIA is generally mandated within specific legislation, usually environmental planning legislation.[26] It may also be required within environmental protection legislation.[27] In Victoria, the *Environmental Effects Act 1978* (Vic) deals specifically with environmental evaluation and mandates the requirement for an 'environmental effects statement', which differs completely from an EIA. An environmental effects statement (EES) may be required in circumstances where the project could have a significant effect on the environment in Victoria which includes major effects on landscape values of regional importance, extensive or major effects on land stability, beneficial uses of water bodies, and social or economic well-being due to direct or indirect displacement of non-residential land use activities.[28] It is generally accepted that the EES is not as rigorous as the EIA because it is issued at the discretion of the Minister and usually only applied at the early stages of assessment, prior to formal licensing and approval.[29]

Where an EIA is required, it will ordinarily involve the following minimal requirements:

- a description of the proposed activity
- a description of the potentially affected environment, including specific information necessary for identifying and assessing the environmental effects of the proposed activity
- a description of practical alternatives as appropriate
- an assessment of the likely or potential environmental impacts of the proposed activity and alternatives, including the direct, indirect, cumulative, short-term and long-term effects
- an identification and description of measures available to mitigate adverse environmental impacts of the proposed activity and alternatives, and an assessment of those measures

---

26 See, eg, *Environmental Planning and Assessment Act 1979* (NSW); *Sustainable Planning Act 2009* (Qld).

27 See, eg, *Environmental Protection Act 1994* (Qld); *Environmental Protection Act 1986* (WA); *Environmental Management and Pollution Control Act 1994* (Tas).

28 See S Rao, 'Reforming the Environmental Assessment Process in Victoria' (2010) 1 *National Environmental Law Review* 34, 37–43 where the author notes that an EES is rarely prepared for mining projects as it is only required where the Minister for Planning requires it, which is seldom.

29 See Victorian Planning and Environmental Law Associate, J Power, *Review of the Environment Effects Act (1978)* (Inquiry into the Environmental Effects Statement Process in Victoria, Submission no 55, 2002), 4–5 http://www.parliament.vic.gov.au/images/stories/committees/enrc/environmental_effects/submissions/55_Victorian_Planning__Environmental_Law_Association.pdf.

- an indication of gaps in knowledge and uncertainties which may be encountered in compiling the required information
- an indication of whether the environment of any other state or areas beyond national jurisdiction are likely to be affected by the proposed activity, and possible alternatives
- a brief non-technical summary of the information provided under the above headings.

In assessing the likely or potential environmental impacts of a proposed activity, the EIA is usually subject to the guidelines or terms of reference that specify the scope of the assessment. In most jurisdictions, the legislation specifically articulates broad guidelines for the relevant content of the EIA. These guidelines help to identify the affected environment, the anticipated effects on the environment, the proposed standards and safeguards, and any monitoring or management programs. In the context of mining projects, anticipated effects can be broad-ranging but generally include direct impacts upon ecological systems, waterways and water resources, air pollution, chemical contamination, surface-estate industries, and the likely success of land rehabilitation and remediation. For example, an EIA relevant to a CSG mining proposal would generally consider the impact of water-pumping extraction processes on groundwater systems. It would also consider the impact of the local redistribution of water in the alluvium in response to water table drawdown which may compromise water quality in some water bores; the redistribution of water with potential quality changes; the capacity of a groundwater aquifer to recharge; the impact of any discharge of associated water into rivers or streams; and, during water table drawdown, water in the alluvium may be redistributed so that in some cases low quality water may flow to areas where water quality was previously high. Such a change to water quality may affect a number of bores even though the effect or magnitude is hard to predict.[30]

In most jurisdictions in Australia, an EIA may only be issued where there has been a prior referral of a proposed mining project or activity to the relevant assessing authority. Depending on the jurisdiction, the referral process may be made by the mining proponent or the relevant responsible department and it may be voluntary or legislatively mandated. For example, under s 38(5) of the *Environmental Protection Act 1986* (WA), the Environmental Protection Authority may require a mining proponent or a 'relevant decision making authority' to refer a proposal which the authority has received notice of but which has not been voluntarily referred.

---

30 These were some of the predicted impacts outlined in the EIA prepared pursuant to the *EPBCA*: See C Moran and S Vink, 'Assessment of impacts of the proposed coal seam gas operations on surface and groundwater systems in the Murray-Darling Basin' (Assessment report, University of Queensland, 29 November 2010) http://www.environment.gov.au/epbc/notices/assessments/pubs/coal-seam-gas-operations-impacts.pdf.

All states in Australia provide for an EIA; however, the framework in each state does not necessarily result in this type of environmental review, particularly for the issuance of a resource title. This effectively means that the stricter environmental assessment requirements associated with the EIA – including such requirements as public consultation – may not be incorporated into the approval process for mining licences.[31] The different environmental review frameworks regulating the issuance and management of resource titles for onshore fossil fuel mining projects in Queensland, New South Wales, and Western Australia are summarised below. These state frameworks illustrate the regulatory variances that exist between the most resource intensive states in Australia with respect to environmental governance.

9.5 Environmental assessment of onshore mining projects in Western Australia

In Western Australia, the Department of Mines and Petroleum (DMP) is responsible for the environmental review of mineral, petroleum and geothermal exploration. The authorisation of these activities pursuant to specific resource titles may only be approved where it is clear that it is consistent with principles of responsible and ecologically sustainable exploration and development. Environmental approvals are issued in accordance with the *Mining Act 1978* (WA); the *Petroleum and Geothermal Energy and Resources (Environment) Regulations 2012* (WA); *Petroleum Pipelines (Environment) Regulations 2012* (WA); and the *Petroleum Submerged Lands (Environment) Regulations 2012* (WA).

All applications for a mining lease in Western Australia must either be accompanied by a mining proposal or by a mineralisation report and a supporting statement. A mining proposal must be submitted to the DMP for the purpose of assessing the environmental impacts of a mining proposal and – as discussed below – will include a range of issues relevant to the environmental impact of the area which is subject to the mining proposal. A statement in support of a mineralisation report is a much briefer document than a full EIA and it need only set out when mining is likely to commence, the anticipated method by which mining will be conducted, and the location and area of the land required for the mining project.

---

31 This is discussed by G Bates, *Environmental Law in Australia* (LexisNexis, 7th edition, 2010) 300.

In addition to the assessment of a mining proposal, the DMP may need to consult with other departments regarding the environmental impacts of the project. Where a mining proposal does not trigger referral to the Environmental Protection Authority (EPA) and is approved by the DMP, the supporting statement will constitute the sole document that determines the environmental conditions to be subsequently imposed on a mining lease.

A mining proposal will trigger referral to the EPA where it is likely to have a significant impact on the environment. In such circumstances, referral is required pursuant to pt IV of the *Environmental Protection Act 1986* (WA). Where such a referral occurs and the EPA requires a formal EIA, the approval process will be much longer, with time frames varying between 6–18 months.[32]

Assessment regarding whether a project is deemed too high risk by the EPA involves an assessment as to whether the proposal is likely to have a significant effect on the environment. This determination is based upon the significance test which is explicitly outlined in s 7 of the *Environmental Impact Assessment (Part IV Divisions 1 and 2) Administrative Procedures 2012*. Section 7 is extracted below:

Environmental Impact Assessment (Part IV Divisions 1 and 2) Administrative Procedures 2012 (Cth)

7 Significance Test

The EPA makes a decision about whether a proposal is likely to have a significant effect on the environment using professional judgement, which is gained through knowledge and experience in the application of EIA. In determining whether a proposal is likely to have a significant effect on the environment, whether the proposal would meet the EPA's objectives for environmental factors and consequently whether or not a referred proposal should be assessed, some of the matters to which the EPA may have regard to include –

(a) values, sensitivity and quality of the environment which is likely to be impacted;
(b) extent (intensity, duration, magnitude and geographic footprint) of the likely impacts;
(c) consequence of the likely impacts (or change);
(d) resilience of the environment to cope with the impacts or change;
(e) cumulative impact with other projects;
(f) level of confidence in the prediction of impacts and the success of proposed mitigation;
(g) objects of the Act, policies, guidelines, procedures and standards against which a proposal can be assessed;
(h) presence of strategic planning policy framework;

---

32 See Government of Western Australia, Department of Mines and Petroleum, *Guidelines for Mining Proposals in Western Australia* (2006), 10 http://www.dmp.wa.gov.au/documents/ENV-MEB-200.pdf.

(i) presence of other statutory decision-making processes which regulate the mitigation of the potential effects on the environment to meet the EPA's objectives and principles for EIA; and

(j) public concern about the likely effect of the proposal, if implemented, on the environment.

The significance test will include a consideration not only of the environmental values impacted and the intensity and duration of the impact but also the public concern about the impact and the extent to which other statutory decision-making processes meet the EPA's objectives as well as the principles of EIA. The DMP will generally refer a proposal to the EPA if an activity is proposed in or within 500 m of an environmentally sensitive area. The DMP will also liaise with the EPA if a proposed activity is within 2 km of a town site, the coastline, or likely to impact a water resource area, including a water reserve, water catchment and groundwater protection area, and declared or proposed water supply catchment area. In respect of unconventional gas projects, the EPA has adopted the approach of determining whether to assess projects that incorporate hydraulic fracturing extraction techniques on an individual basis, in the same manner as is the case with other petroleum and mining proposals.

All mining proposals in Western Australia are required to outline the following issues:

- a brief discussion of the geology of the area including the minerals to be mined and the status, category, and estimates of the resource
- the potential for acid rock drainage must also be ascertained through a detailed assessment of mining wastes and tailings and, where it is deemed to be significant, particularised management strategies should be outlined within the proposal
- an analysis of topsoil and subsoil layers should be conducted to carefully identify soil deficiencies and possible adverse parameters
- a description of the hydrology of the area, with a particular focus on surface and subsurface water resources and the potential, if any, for flooding
- the possible impact of the mining proposal upon any nearby water bodies, wetlands or groundwater dependent eco-systems potentially adversely affected by any change or alteration in hydrology
- where relevant, a flora and fauna survey
- any significant climatic impacts likely to affect the project
- the social environment in which the project is located including sites of state, national, or Aboriginal heritage
- a preliminary mine closure plan, detailing the methodology relevant to a mining closure.

Since 1 July 2011, a Mine Closure Plan must accompany all new applications for mining tenements and should be prepared in accordance with DMP Guidelines. Effective rehabilitation strategies are a crucial component of mine closure, particularly in circumstances where a tenement holder is directly responsible for cleaning contaminated sites following the cessation of the tenement. A new Mining Rehabilitation Fund has also been introduced in Western Australia to address the state's unfunded liability for abandoned mine site rehabilitation.

Effective from 1 July 2013, most tenement holders under the *Mining Act 1978* (WA) will be required to contribute an annual levy to the Fund which is to be calculated as 1 per cent of the estimated total mine closure cost.[33] A Mine Closure Plan must be prepared where mine closure planning is specifically required as part of the EPA's assessment of the project pursuant to pt IV of the *Environmental Protection Act 1986* (WA). Assessment by the EPA of mine closure will only occur in circumstances that give rise to high environmental risk.[34] Where the EPA does assess a mine closure, a condition will normally be applied requiring a Mine Closure Plan to be prepared in accordance with the DMP Guidelines. Compliance monitoring of these conditions may also be delegated by the EPA to the DMP.

A mine rehabilitation plan should also be included within a mining proposal which carefully outlines and addresses the site-specific issues in a durable and effective manner. The mine rehabilitation plan should concentrate on every aspect of the planned mining site and should also set out rehabilitation completion criteria which seek to incorporate a rehabilitation endpoint. Particular sites may require an initial rehabilitation program which may then need to be complemented by ongoing environmental augmentation programs, with rehabilitation criteria and outcomes being updated and rearticulated within the Annual Environmental Report. Progressive rehabilitation obligations and outcomes should also be detailed within the Annual Environmental Report. The DMP Guidelines specifically recommend the immediate commencement of rehabilitation field trials to determine the suitability of rehabilitation methodologies.[35]

The main objective of the assessment of the mining proposal by the DMP is to identify all possible environmental impacts and to ascertain whether significant impacts require the implementation of special management procedures. Where this is the case, specific procedures include commitments by the

---

33 See *Mining Rehabilitation Fund Act* 2012 (WA) s 9A.
34 See the discussion of 'high risk' in Government of Western Australia, *Guidelines for Mining Proposals*, above n 32, 8. The Western Australian EPA will, however, continue to assess mine closure for mining projects not subject to the *Mining Act 1978* (WA), such as pre-1899 titles, minerals-to-owner tenure, Hampton locations, or State Agreement Act projects.
35 Government of Western Australia, *Guidelines for Mining Proposals*, above n 32, 12.

mining proponent to minimise, control, ameliorate, and rehabilitate the land in accordance with a structured environmental reporting program approved by the DMP.

Once the formal environmental approval process is completed, any mining project that involves ground disturbance will be required to submit an environmental bond prior to approval. The bond will be calculated according to the type and area of disturbance anticipated. The obligation to lodge an environmental bond is a condition on title and any approval to commence the authorised mining activity will be subject to the lodgement of the bond. The aim of the environmental bond is to give the state recourse to specifically designated funds in circumstances where it is clear that the tenement holder has failed to meet environmental commitments. Additional environmental conditions may be imposed pursuant to ss 46A, 63AA, 84 and 89 of the *Mining Act 1978* (WA). For example, all mining proposals are subject to a condition that the proponent submits an Annual Environmental Report (AER) outlining the activities that have been conducted in the previous 12 months as well as the activities proposed for the following 12 months. Once the conditions are imposed, the Mineral Titles Services Division will update the tenement register, which is publicly accessible. An environmental commitment set out within a mining proposal will become a legally binding obligation once the proposal is imposed as a tenement condition in accordance with s 84.

Section 84 of the *Mining Act 1978* (WA) is extracted below:

Mining Act 1978 (WA)

84 Conditions for prevention or reduction of injury to land

(1) On the granting of a mining lease, or at any subsequent time, the Minister may impose on the lessee reasonable conditions for the purpose of preventing or reducing, or making good, injury to the land in respect of which the lease is sought or was granted, or injury to anything on or below the natural surface of that land or consequential damage to any other land.

(2) Without limiting the generality of subsection (1) the Minister may, on the granting of the mining lease or at any subsequent time, if it is reasonable in all the circumstances so to do, impose on the lessee a condition that mining operations shall not be carried out within such distance of the natural surface of the land in respect of which the lease is sought or was granted, as the Minister may specify.

(3) Any condition imposed under this section may at any time be cancelled by the Minister or from time to time varied by him.

(4) A condition imposed in relation to a lease under this section –
    (a) may, either in full or with sufficient particularity as to identify the recommendation or other source from which it derives, be endorsed on the original and the duplicate

of the lease, for which purpose the lessee shall produce the duplicate of the lease on demand; and

(b) whether or not so endorsed, on notice of the imposition of the condition being given in writing to the lessee shall for all purposes have effect as a condition to which the lease is subject; and

(c) where it is set out or otherwise sufficiently identified in the notification of the grant of the lease, shall have effect as though the lease had been issued duly endorsed as to the terms of that condition.

The environmental approval process relevant to the EPA in Western Australia was the subject of a major review in 2009.[36] Specific recommendations in that review included mandating a referral of all major projects – including major mining projects to both the Department of Environmental Conservation and the EPA at the outline and developmental phase of the proposed project. This would enable consideration and resolution of major/strategic issues early in the development process and detailed design issues could then be deferred to the subdivision phase. Further recommendations include: reducing the duplication between agencies and providing a clearer framework by clarifying the roles and responsibilities of relevant agencies, reforming the EPA by reducing ministerial discretions and providing improved transparency for assessment and decision-making regarding proposals affecting wetlands, and improving the transparency and clarity of assessment criteria. The Western Australian Government adopted the recommendations of the Review and the EPA has been implementing the reforms. The reform program includes 47 Review recommendations with the overarching objective of improving the timeliness and effectiveness of the EPA's functions through:

- the gradual introduction of outcome based conditions
- the use of risk-based assessment where applicable
- improved project tracking in the Office of the EPA
- improved rigour and consistency in the scoping phase
- an improved focus on timelines
- the provision of greater guidance to proponents with the aim of improving certainty, clarity and consistency
- the creation of the new Office of the EPA to more effectively support the work of the EPA.

---

36 See Government of Western Australia, Environmental Protection Authority, *Review of Environmental Impact Assessment process* (2009) <http://www.epa.wa.gov.au/abouttheepa/eiareview/Pages/default.aspx>.

9.6 Environmental assessment of onshore mining projects in Queensland

Like Western Australia, Queensland is a strong resource state. Queensland has more than 30 billion tonnes of coal deposits along with metals, phosphate rock, oil shale, and minerals, and is in the world's top six regions for the production of lead, zinc, bauxite, and silver. It also has a petroleum industry, predominantly in coal seam gas, that exceeds A$1 billion in production value. In 2012–2013 the mining and petroleum industry in Queensland generated A$25.6 billion (or 8.8 per cent) of gross state product and represented 60 per cent of all state exports. The sector attracted 68 per cent of all capital investment in Queensland and exploration expenditure of over A$1.3 billion.[37]

In Queensland, all mining and petroleum/gas projects require a resource title issued from the Department of Natural Resources and Mines in accordance with the relevant requirements in the legislation: the *Mineral Resources Act 1989* or the *Petroleum and Gas (Production and Safety) Act 2004*. As discussed in Chapter 2, this title confers upon the holder the right to access the land and to undertake exploration, resource assessment, feasibility studies, prospecting, or production. Further, the environmental regulation framework in Queensland requires prior environmental review before resources titles may be authorised. All environmental authority (EA) applications for resource activities will be categorised in one of three ways: a standard application, a variation application, or a site-specific application. Significantly, however, it will only be site-specific applications for new resource activities ('greenfield' sites) or the amendment of existing EAs ('brownfield' sites) that will be subject to environmental impact statement (EIS) in accordance with the provisions under the *Environmental Protection Act 1994* (Qld).

An EA for a resource activity must also be obtained from the Department of Environment and Heritage Protection (DEHP) in order to undertake an environmentally relevant activity. Mining and resource projects come within the definition of an 'environmentally relevant activity' because of their potential to release contaminants into the environment. The EA includes a range of information that may also assist in the imposition of ongoing environmental management conditions. Before a mining or petroleum/gas project may apply for an environmental authority, the *Environmental Protection Act 1994* (Qld) ('the *EPA*')

---

37 These statistics are found at Queensland Government, Department of Natural Resources and Mines, *Queensland Mining Update,* (19 January 2015) <https//: mines.industry.qld.gov.au/mining/queensland-mining-update.htm>.

will require the project's likely environmental impacts to be assessed and measures to be proposed for the avoidance or minimisation of any adverse impacts. A low-impact mining project will be subject to the relevant *Code of Environmental Compliance* and will attract the minimum level of environmental assessment and generally will only need to apply for a standard environmental authority which is the relevant environmental authority for low-risk activities.[38]

On the other hand, a mining proposal that is not low-impact may need to submit an environmental management plan ('EM plan') that formally identifies what the *EPA* describes as the 'environmental values' that either will or have the potential to be affected by the project and which seeks to assess the potential adverse and beneficial impacts of the mining proposal upon those values. In addition to outlining impacts, the EM plan must also propose environmental protection commitments to safeguard and/or enhance the impacted environmental values. The EM plan assists the DEHP's capacity to develop specific conditions for an environmental authority.

An environmental value is defined in s 9 of the *EPA* to refer to: (a) a quality or physical characteristic of the environment that is conducive to ecological health or public amenity or safety; or (b) another quality of the environment identified and declared to be an environmental value under an environmental protection policy or regulation. An environmental value is therefore conceptual rather than physical and refers to emergent but ultimately indeterminate environmental notions, such as ecological integrity, biological diversity, and conservation value.[39]

Hence, whilst water itself cannot come within the definition of an environmental value, the benefit of clean drinking water for public health is a recognisable

---

38 A standard range of conditions for environmentally relevant activities are set out in specific Codes of Environmental Compliance. These Codes are now described as eligibility criteria and standard conditions. Any new operation commencing from 31 March 2013 that meets the eligibility criteria for an activity and all of the standard conditions can make a standard application for an environmental authority to carry out this activity. The conditions that apply to the EA will be the standard conditions. See Queensland Government, Department of Environment and Heritage Protection, *Eligibility criteria and standard conditions* (18 December 2013) <http://www.ehp.qld.gov.au/licences-permits/compliance-codes/>. The general requirements for environmental authorities are set out in the *EPA* (Qld) s 125. Section 125(j) specifically sets out that if the application is a standard or variation application it should include a declaration that each relevant activity complies with the eligibility criteria.

39 The concept of an environmental value and its expansion within the context of environmental law was discussed by D B Spence, 'Paradox Lost: Logic, Morality and the Foundations of Environmental Law in the 21st Century' (1995) 20 *Columbia Journal of Environmental Law* 145, 158–62 where the author argues that the emergence of environmental values has transformed the economics and politics of environmental regulation.

value. The breadth and scope of the concept of 'environmental value' inevitably allows regulators in the mining and energy sector to draw upon underlying conceptions of ecological economics in ascertaining sustainability principles. Within this paradigm, environmental resources and environmental protection are attributed an ascertainable value in order to ensure that the full environmental costs and benefits of resource allocation are factored into policy and decision-making.[40] In accordance with this model, if it is assumed that the optimal scale of resource exploitation is a sustainable one, then the normative consequence must be that governments regulate the scale of the economy so that 'it requires no more resources or produces no more wastes than can be regenerated or absorbed respectively by the environment'.[41] In this way, the rationale underpinning the explicit regulatory articulation of environmental values lies in the assumption that regulators, in referencing such values, will have a greater capacity to comprehensively evaluate the most effective way to sustain an ecological context for future generations. This is particularly crucial for the mining and energy sector, given the size and scope of the sector. Within this sector, issues of equity and fairness in resource allocation between different groups and across different generations is a strong concern given the fact that the ecosystem is, ultimately, finite.[42]

The framework for environmental regulation in Queensland bifurcates the concept of an environmental value and the broad notion of environmental harm. It articulates an 'environmental harm' as any action that produces an adverse impact upon an environmental value. This includes pollution, chemical contaminants, land clearing, soil erosion, and water depletion.[43] Material environmental harm is harm that is not trivial or negligible and serious environmental harm is harm that causes actual or potential irreversible damage to environmental values which is widespread or which has a wide impact.[44]

---

40 See the discussion by L Godden and J Peel, *Environmental Law: Scientific, Policy and Regulatory Dimensions* (Oxford University Press, 2010) 29–30 where the authors argue that environmental 'existence' values are difficult to determine as there is no available economic proxy from which to derive a value; for example, how do we 'value' the possibility of a species becoming extinct?

41 D A Kysar, 'Law, the Environment and Vision' (2003) 97 *Northwestern University Law Review* 675, 683.

42 The capacity of markets to circumvent environmental regulation was outlined by R M Solow, 'The Economics of Resources or The Resources of Economics' (1974) 64 *American Economic Review* 1, 8–9.

43 In *Cougar Energy Ltd v Debbie Best, Chief Executive Under the EPA* [2011] QPEC 150, [43] the Court held that environmental harm existed because there was a risk that 'contaminants in the ground water will travel horizontally through the kunioon coal seam aquifer until a fracture, or other preferential pathways encountered and contaminates be transported vertically into the upper aquifer'.

44 See *EPA* (Qld) ss 9, 14, 16 and 17.

The *EPA* makes it clear that a person is not entitled to carry out an activity that causes environmental harm unless reasonable or practical measures are taken to prevent or minimise the harm. Section 319 states the general environmental duty and is extracted below:

Environmental Protection Act 1994 (Qld)

319 General environmental duty

(1) A person must not carry out any activity that causes, or is likely to cause, environmental harm unless the person takes all reasonable and practicable measures to prevent or minimise the harm (the general environmental duty).
(2) In deciding the measures required to be taken under subsection (1), regard must be had to, for example-
   (a) the nature of the harm or potential harm; and
   (b) the sensitivity of the receiving environment; and
   (c) the current state of technical knowledge for the activity; and
   (d) the likelihood of successful application of the different measures that might be taken; and
   (e) the financial implications of the different measures as they would relate to the type of activity.

Section 493A then provides that an activity that causes serious or material environmental harm or environmental nuisance is unlawful unless it is approved under the *EPA* or the general environmental duty is complied with.

Large-scale impacts associated with a resource project may trigger an environmental impact assessment by the DEHP. This will be determined according to the scale, intensity, and duration of an impact. Most coal seam gas projects in the Surat and Bowen Basins will be treated as site-specific and will attract environmental impact assessment. Further, many large-scale resource projects may be characterised as 'coordinated projects' pursuant to the *State Development and Public Works Organisation Act 1972* (Qld) ('the *SDPWOA*') which confers overall supervision by the State Coordinator-General who may then refer the project to the DEHP for an EIS.

The factors considered by the Coordinator-General in determining whether a declaration under the *SDPWOA* should be made include the strategic significance of the project and the likelihood of significant environmental effects and significant infrastructure requirements.[45]

Section 143 of the *EPA* describes the circumstances pursuant to which a resource activity must or may be subject to EIA and is extracted below:

---

45 *SDPWO* (Qld) s 27.

Environmental Protection Act 1994 (Qld)

143 EIS may be required

(1) This section applies for a site-specific application for a resource activity if –
   (a) the application does not relate to a coordinated project; and
   (b) an EIS relating to the activity has not been submitted under chapter 3, part 1.

(2) Without limiting section 140(1), the administering authority may include in an information request a requirement that the applicant provide an EIS for the application.

(3) In deciding whether an EIS is required for an application, the administering authority must consider the standard criteria.

(4) A requirement under subsection (2) ceases to have effect if a relevant activity or tenure for the application is, or is included in, a coordinated project.

The criteria for s 143 relevant to both new (greenfield) projects and the expansion of existing (brownfield) projects is set out in Appendix A of the Guidelines and includes:

Guideline: Triggers for environmental impact statements (EIS) under the Environmental Protection Act for mining and petroleum activities

Appendix A – EIS triggers under section 143 of the EP Act

1. Mining activities – triggers new applications
    An EIS is required for mining activities that would result in:
    - the removal of two million tonnes/year (t/y) or more of run-of-mine (ROM) ore or coal
    - the removal of one million t/y or more of ROM ore or coal on or under a floodplain or in a coastal hazard area
    - the introduction of a novel or unproven resource extraction process, technology or activity.

2. Mining activities – triggers for major amendment applications
    An EIS is required for a proposed major amendment of mining activities as a result of one or more of the following:
    - for existing mines extracting between 2–10 million tonnes/year (t/y) ROM ore or coal, an increase in annual extraction of more than of 100% or 5 megatonnes/y (Mt/y) (whichever is the lesser)
    - for existing mines extracting over 10 million t/y ROM ore or coal, an increase in annual extraction of more than 50% or 10 Mt/year (whichever is the lesser)
    - for existing mines extracting more than 20 million t/y ROM ore or coal extraction, an increase in annual extraction greater than 25%

- proposed activities in a Category A or B environmentally sensitive area, unless previously authorised under Queensland legislation
- a substantial change in mining operations, e.g. from underground to open cut, or (for underground mining), or a change from minor subsidence to potentially substantial subsidence
- the introduction of a novel or unproven resource extraction process, technology or activity.[46]

A decision may, however, be made requiring an EIS application, even where no EIS criteria have been triggered. This will occur in circumstances where the DEHP or the Minister for Environment and Heritage Protection determines that the project applied for would involve a significant environmental impact or a high level of uncertainty about potential impacts, or would involve a high level of public interest. In determining whether an EIS is required in this context the standard criteria articulated in Appendix B must be considered. This is extracted below:

Guideline: Triggers for environmental impact statements (EIS) under the Environmental Protection Act for mining and petroleum activities

Appendix B – Standard criteria

...

Standard criteria means –

(a) the following principles of environmental policy as set out in the Intergovernmental Agreement on the Environment –
   (i) the precautionary principle;
   (ii) intergenerational equity;
   (iii) conservation of biological diversity and ecological integrity; and
(b) any Commonwealth or State government plans, standards, agreements or requirements about environmental protection or ecologically sustainable development; and
(c) any relevant wild river declaration; and
(d) any relevant environmental impact study, assessment or report; and
(e) the character, resilience and values of the receiving environment; and
(f) all submissions made by the applicant and submitters; and
(g) the best practice environmental management for activities under any relevant instrument, or proposed instrument, as follows –

---

46 See Queensland Government, Department of Environmental and Heritage Protection, *Guideline: Environmental Impact Statements: Triggers for environmental impact statements under the Environmental Protection Act 1994 for mining and petroleum activities* (2014), 5 http://www.ehp.qld.gov.au/management/impact-assessment/pdf/eis-guideline-trigger-criteria.pdf.

(i) an environmental authority;
(ii) a transitional environmental program;
(iii) an environmental protection order;
(iv) a disposal permit;
(v) a development approval; and

(h) the financial implications of the requirements under an instrument, or proposed instrument, mentioned in paragraph (g) as they would relate to the type of activity or industry carried out, or proposed to be carried out, under the instrument; and

(i) the public interest; and
(j) any relevant site management plan; and
(k) any relevant integrated environmental management system or proposed integrated environmental management system; and
(l) any other matter prescribed under a regulation.[47]

The EIS process will generally identify similar forms of environmental values as articulated within an EM plan although, given the range and detail of environmental impact review, the assessment under an EIS will be far more wide-ranging. One crucially important aspect of the environmental impact assessment for mining and energy projects in Queensland lies in the capacity of the assessment to incorporate a greater level of public scrutiny. The draft terms of reference for an EIS are made publicly available for a minimum period of 30 business days to allow stakeholders and members of the public to read and review the assessment and comment, where deemed appropriate, upon the values and commitments that it outlines. The commencement of a review process is set out on the website of the DEHP as well as in local and circulating newspapers. Owners of land which is directly relevant to, or whose land is adjacent to, areas relevant to the environmental impact assessment may also be notified specifically by mail.[48]

Once environmental assessment of a mining or energy project has been completed and a resource title is issued, that title may be subject to a range of environmental conditions that are intended to specifically address the identifiable environmental risks or harms outlined within the EA or the EIS. A mining proponent may appeal a decision to refuse a mining project or, alternatively, may appeal a decision to impose extensive environmental conditions upon a project. The appeal may be heard by the Land Appeal Court of Queensland and leave may subsequently be granted to appeal to the Queensland Court of Appeal.[49]

---

47 See Queensland Government, Department of Environment and Heritage Protection, above n 46, 7.
48 *EPA* (Qld) ss 41–3.
49 *EPA* (Qld) ss 523, 539.

9.7 Environmental assessment of onshore mining projects in New South Wales

News South Wales is another strong resource state with significant supplies of coal and natural gas, particularly coal seam gas. As in Queensland and Western Australia, the NSW regulatory framework contains, comparatively, some of the most detailed environmental regulation in the country. As outlined in Chapter 5, NSW is currently in the process of strategically mapping each and every part of its domain. It has also recently introduced a new gas policy which involves undertaking a strategic review to determine whether new exploration licences should be issued based upon an assessment of economic, environmental, and social factors.[50] Areas identified as biophysical, strategic agricultural land (BSAL) will be subject to a more focused environmental evaluation – known as a 'gateway assessment' – where the project involves state significant mining or CSG.[51] Areas may be classified as BSAL in circumstances where it is likely that the project will reduce the agricultural productivity of the land as a result of surface areas disturbance, impacts on the soil profile, fertility, and salinity. Pursuant to this framework, a development application for a coal seam gas project, or a state significant mining project – and this applies to either a stand alone ('greenfield') or an extension ('brownfield') project, and to new or extended mining leases – cannot be lodged unless a gateway certificate has been issued or the land has been verified as not containing BSAL pursuant to a strategic agricultural land verification application.

The gateway panel will take account of all advice received from the Commonwealth Independent Expert Scientific Commission on CSG and Large Coal Mining Development (IESC) and as well as factors relevant to the Aquifer Interference Policy. Bio-regional assessments of the ecology, hydrology, and geology of all BSAL areas are being conducted and this is particularly relevant for the environmental evaluation of CSG proposals.

A mining proposal that is approved by the gateway panel will receive an unconditional certificate and the development application will then proceed to a full merit evaluation.[52] Where, however, the proposal is refused, a conditional

---

50 The changes proposed by the NSW gas policy are discussed in Chapter 5.
51 See, eg, the NSW Government, Department of Planning and Infrastructure, *Strategic Regional Use Plan: Upper Hunter*, (September 2012) ch 11 pp. 21–2 <http://www.nsw.gov.au/sites/default/files/initiatives/upperhunterslup_sd_v01.pdf>. The definition of 'strategic agricultural land' covers 'biophysical, strategic, agricultural land' and 'critical industry clusters'. These definitions include land with high rainfall, land used for agricultural product that provides significant employment, land used for equine activities, and land used for viticulture.
52 See NSW Government, above 51, 80.

certificate may be issued requiring the applicant to address the conditions outlined in any subsequent development application. Under the proposed gas policy, the strategic release framework will control the release of areas within NSW that are amenable to gas exploration; however, exclusion zones in areas close to the Sydney water catchment area would remain in place.[53]

Areas which are not yet subject to strategic regional land use mapping will continue to be assessed, in accordance with established environmental assessment regime established under the *Environmental Planning and Assessment Act 1979* ('the *EPAA*'), the *Environmental Planning and Assessment Regulations 2000* (NSW) and the relevant environmental planning instrument, whether that be a state environmental planning policy (SEPP) or a local environmental policy (LEP). Environmental planning instruments provide for the protection of the environment, including the preservation of trees, native flora, fauna and animals and the preservation of vulnerable ecological communities. These instruments also prescribe the circumstances where a 'development consent' for a particular project is required.[54]

The broad approach of the *EPAA* is to conduct a comprehensive and merit-based review, which may or may not incorporate environmental impact assessment, but which does cover a wide range of environmental issues including biodiversity, water management, alternative uses of the land, and an assessment of whether the proposed development is in the best interests of the state. Broader assessment themes associated with environmental review under the *EPAA* include the ecological sustainability of the project, concepts of inter-generational equity, and general cost-benefit considerations of social and economic gains.

The explicit incorporation of ecological sustainability into the regulatory framework of the *EPAA* is significant as it highlights the domestic incorporation of the core notions of ecologically sustainable development (ESD), as initially elaborated internationally by the Brundtland Commission, which were subsequently endorsed at the United Nations Conference on Environment and Development (UNCED) held in Rio de Janeiro in 1992.[55] These discussions formed the basis for the subsequent

---

[53] This is discussed in Chapter 5. See also the exclusion zones for critical industry clusters which has been given legal effect via an amendment to the *State Environmental Planning Policy (Mining, Petroleum Production and Extractive Industries) 2007* (NSW) ('Mining SEPP').

[54] *EPAA* (NSW) s 26.

[55] The Brundtland Commission refers to the World Commission on Environment and Development, convened in 1983 by the United Nations General Assembly and headed up by the former Norwegian Prime Minister, Gro Harlem Brundtland. The 1987 Brundtland Commissions Report, *Our Common Future*, provided a crucial outline for future global environments. The Brundtland Commission Report led to the convening of a United Nations Conference on Environment and Development (UNCED) (also called the Earth Summit) in Rio de Janeiro in 1992. This summit produced two principal instruments relevant to ecological sustainability: (i) a declaration of principles (the 'Rio Declaration'); and (ii) a detailed action plan. UNCED reconvened in 2002 for the World Summit on Sustainable Development.

emergence of concepts of ecologically sustainable development that seek to ensure a more holistic focus upon ecological, economic, and utilitarian ethics and which reflect an inter-generational concern for 'ensuring the wise use of scarce natural resources'.[56]

For most large mining and energy projects, the mining proponent must submit an application for development consent to the NSW Department of Planning and Infrastructure and must also prepare an EIS. By contrast, smaller projects or exploratory activities will only need to submit a Review of Environmental Factors ('REF'). A REF is mandated under pt 5 of the *EPAA*. Unlike the EIS, the REF will address all potential environmental impacts connected with the mining proposal, particularly those relevant to the land, water resources, and the community. Approval of an REF will be subject to the determination that the environmental impacts of the specific project are acceptable.

In essence, the REF seeks to determine the significance of the likely environmental impacts of a proposal and the measures required to mitigate any adverse impacts upon the environment. It serves two core purposes: first, to assist and document the authority's determination of whether an activity should be approved, taking into account all possible matters affecting or likely to affect the environment;[57] second, to evaluate the likely impact of the proposal on the environment or upon significantly threatened species, populations or ecological communities or their habitats. An REF will not be required where it is determined that a significant environmental impact is likely, and an EIS is appropriate prior to the issuance of a resource title.[58] Factors relevant to the REF include broader and, by contrast with the EIS, more generalised landscape and pollution assessments including: the impact of the project on a particular ecosystem, how the project will transform the aesthetic, recreational or environmental value of an area, and long-term environmental degradation, pollution, and environmental problems associated with the disposal of waste.[59]

In 2011, the *EPAA* introduced two new categories of development relevant to mining projects: state significant developments (SSD) and state significant

---

[56] See the discussion by L Godden and J Peel, *Environmental Law: Scientific Policy and Regulatory Dimensions* (Oxford University Press, 2010), 135.

[57] *EPAA* (NSW) s 111.

[58] *EPAA* (NSW) s 112. See the discussion in NSW Department of Environment, Climate Change and Water, *Guidelines for the Review of Environmental Factors* (2011) http://www.environment.nsw.gov.au/resources/protectedareas/110028REFProGde.pdf.

[59] *EPAA* (NSW) s 111. The factors relevant to an REF are set out in the *Environmental Planning and Assessment Regulation 2000* (NSW) reg 228. The Department of Industry and Investment has also issued guidelines on how to prepare a Review of an Environmental Factor which specifically require the potential impacts on surface and groundwater to be evaluated.

infrastructure (SSI).[60] Neither the SSD nor the SSI are specifically defined and will only exist where so declared by a state environmental planning policy.[61] SSD applications are assessed by the NSW Department of Planning and Infrastructure. Development and consent for these projects may only be granted by the Minister pursuant to the provisions of div 4.1 and 5.1 of the *EPAA*.[62]

SSD applications specifically include: development for the purpose of petroleum production; drilling or operating petroleum exploration wells in an environmentally sensitive area of state significance; and development for the purpose of petroleum related works which is ancillary to another state significant development or with a capital investment value of more than $30 million.[63] All SSD projects must be reviewed pursuant to an EIS.[64]

As outlined in Chapter 5, neither an SSD nor an SSI will be exempt from the need to obtain an aquifer interference approval under the *Water Management Act 2000* (NSW).[65] Section 89J(1)(g) of the *EPAA* (NSW) confirms the need for a mining proponent to undergo a separate aquifer interference approval where the activities of the project involve interference, penetration, or obstruction of water from an aquifer.[66]

Where a project is approved, environmental management conditions may be imposed to minimise potential environmental impacts. These conditions include rehabilitation and environmental performance conditions for titles issued under the both the *Mining Act 1992* (NSW) and the *Petroleum (Onshore) Act 1991* (NSW).

---

60 The categories of state significant development and state significant infrastructure are set out in the *State Environmental Planning Policy (State and Regional Development) 2011* (NSW). State Significant Development is assessed and determined under div 4.1 of pt 4 of the *EPAA* (NSW) and State Significant Infrastructure is assessed and determined under pt 5.1 of the *EPAA* (NSW).
61 See the definition set out in the *State Environmental Planning Policy (State and Regional Development) 2011* (NSW) schs 1, 2 and 3 for a detailed outline.
62 *EPAA* (NSW) s 89C.
63 *State Environmental Planning Policy (State and Regional Development) 2011* (NSW) cl 5 of sch 1 and sch 3.
64 *EPAA* s 89E. See also *Environmental Planning and Assessment Regulation 2000* (NSW) sch 2, reg 3(4A)(a) and (b) which sets out that: '(a) if a gateway certificate has been issued in relation to a State significant development to which an application for environmental assessment requirements relates, the Director-General, in preparing the requirements, must address any recommendations of the Gateway Panel which are set out in the certificate, and (b) if a gateway certificate ... in relation to the State significant development to which an application for environmental assessment requirements relates, the Director-General, in preparing the [EIA] requirements, must consult with the Gateway Panel and have regard to the need for the requirements to assess any key issues raised by that Panel.'
65 *EPAA* (NSW) s 89J(1)(g) which sets out that activities under s 91 of the *Water Management Act 2000* (NSW) – apart from the aquifer interference approval – will be exempted where a development consent for a state significant development exists. For state significant infrastructure a similar provision exists in s 115G(1).
66 This is discussed more extensively in Chapter 5.

Rehabilitation conditions are generally imposed with the aim of improving environmental management and rehabilitation outcomes through the imposition of focused, remedial obligations. These obligations are progressive and will generally endure for the duration of the mining project. Environmental management conditions may also be directed to mining closure obligations. The conditions will include a requirement to submit a Mining Operations Plan (MOP) prior to the commencement of any mining operation and also, to submit Annual Environmental Management Reports (AEMRs). Both the MOP and AEMR constitute the Mining, Rehabilitation and Environmental Management Process (MREMP).

The ability to impose environmental management conditions upon mining projects coming within the application of the *Mining Act 1992* (NSW) is outlined in s 239 which is extracted below:

Mining Act 1992 (NSW)

239 Rehabilitation etc of area damaged by mining

(1) The conditions subject to which an authority or mineral claim is granted or renewed may include such conditions relating to:

   (a) the rehabilitation, levelling, regrassing, reforesting or contouring of such part of the land over which the authority or claim has effect as may have been damaged or adversely affected by prospecting operations or mining operations, and

   (b) the filling in, sealing or fencing off of excavations, shafts and tunnels,

   as may be prescribed by the regulations or as the Minister or mining registrar may, in any particular case, determine.

(1A) The Minister or mining registrar may, in any particular case, determine that an authority or mineral claim be granted or renewed subject to conditions relating to the afforestation (including for carbon sequestration within the meaning of section 87A of the Conveyancing Act 1919 and related environmental purposes) of such part of the land over which the authority or claim has effect as may have been damaged or adversely affected by prospecting operations or mining operations.

(1B) However, a condition referred to in subsection (1A) may only be imposed at the request of the applicant for, or holder of, the authority or claim.

(2) The Minister or mining registrar may amend an authority or mineral claim:

   (a) that does not contain conditions of the kind that may be imposed under this Division, or

   (b) that does contain such conditions, being conditions that the Minister or mining registrar considers are inadequate,

   so as to include conditions or further conditions of that kind or so as to alter any such conditions.

(3) Any conditions of the kind referred to in subsection (1)(a) or (1A) are to be in a form approved by the Commissioner of the Soil Conservation Service and after consultation with the Director of National Parks and Wildlife.

(4) An amendment takes effect on the date on which notice of the amendment is served on the holder of the authority or mineral claim or on such later date as may be specified in the notice.

(5) This section has effect despite anything to the contrary in section 93 of the *Environmental Planning and Assessment Act 1979*.

Rehabilitation conditions are attached to mining leases in order to regulate environmental management and rehabilitation. The conditions include requirements to submit a MOP prior to the commencement of any operations and also a requirement to submit AEMRS.

Rehabilitation conditions function as a management tool for operations within the mine as well as a means by which the environmental performance of mining operations may be identified and addressed. The nature and scope of the environmental management conditions may also be utilised to estimate the amount of security deposit that a mining proponent must pay to the NSW Trade and Investment Division of the Department of Resources and Energy. Payment of a security deposit functions as an insurance measure in the event of a default on rehabilitation obligations. A security deposit covering the full rehabilitation costs is required on all mining authorisations. This requirement effectively ensures that the state does not incur financial liabilities for any default. The mining proponent must provide an estimate of the rehabilitation costs when determining the amount of security deposit. The security deposit will be only be released once the Department is satisfied that the rehabilitation obligations have been satisfied.[67]

Rehabilitation conditions are crucial components for the ongoing environmental management of productive mining operations. In NSW, as in other states, the incentive to comply with rehabilitation obligations lies in the financial impost of forfeiting the security deposit. In light of this, it is crucial to ensure that the security deposit is sufficient to maximise the prospect of compliance. Non-compliance is always a danger where the security deposit or bond is insufficient. This was a strong issue in the Victorian framework where the Hazelwood Mine Fire Commission Report indicated that strong rehabilitation obligations which proponents are incentivised to comply with play a clear role in eliminating or reducing the risk of a coal fire.[68] The report concluded that once a worked out batter in a coalmine has been properly rehabilitated, coal will no longer be exposed. It therefore follows that the rehabilitated land bears no greater fire risk than any other part of the rural landscape. Conversely,

---

[67] *Mining Act 1992* (NSW) pt 11, div 3. See also NSW Government, Department of Trade and Investment, ESG3: Mining Operations Plan (MOP) Guidelines, September 2013 (9 September 2013) 44 http://www.resourcesandenergy.nsw.gov.au/__data/assets/pdf_file/0007/527911/ESG3-Mining-Operations-Plan-MOP-Guidelines-September-2013.pdf.

[68] Parliament of Victoria, Hazelwood Mine Fire Inquiry, *Report: Part 3: Fire Risk Management* (August 2014) http://report.hazelwoodinquiry.vic.gov.au/microsoft-word-document-version-text-size-18pt.

where rehabilitation obligations are not properly and adequately complied with, the risk of catastrophic environmental danger increases exponentially. The NSW mining rehabilitation framework seeks to ensure that parties are adequately encouraged to comply with progressive environmental management conditions. The MREMP framework attempts to facilitate the development of mining in NSW by requiring all mining operations to be safe, and by requiring, as far as possible, that the environment is protected through strong rehabilitation objectives. The extent to which these objectives may be achieved depends upon the capacity of the regulatory framework to properly monitor breaches of progressive rehabilitation obligations. If the sanctions imposed for such breaches do not act as a sufficient deterrent, the effectiveness of a progressive environmental risk management strategy is undermined. Through the imposition of stronger and higher sanctions and rigorous environmental enforcement procedures, regulatory frameworks increase the cost of engaging in non-rehabilitative and environmentally damaging behaviour. The core premise is that costly external sanctions promote compliance with progressive environmental management obligations because compliance is the most cost effective way of conducting operations. Importantly, however, sanctions should only be imposed where it is clear that a breach has occurred and that the legislation authorises the imposition of the sanction. For example, in *Metgasco Limited v Minister for Energy and Resources*,[69] Button J concluded that Metgasco did not break consultation obligations and that such obligations could not support the imposition of a suspension of title because this was not authorised under the relevant Act.

9.8 Commonwealth environmental legislation: *Environment Protection and Biodiversity Conservation Act 1999*

The dual jurisdictional framework for environmental assessment of mining and energy projects has been outlined above. Commonwealth involvement in the environmental assessment of a mining or energy project will depend upon whether the triggers contained within the *Environment Protection and Biodiversity Conservation Act 1999* (Cth) ('the *EPBCA*') are activated. The fundamental

---

69 [2015] NSWSC 453.

framework of the *EPBCA* is structured around the premise that if a mining project involves an action that will have – or is likely to have – an impact on a matter of national environmental significance (MNES), the activity may not proceed without approval from the relevant Commonwealth minister.

The *EPBCA* replaced its predecessor, the *Environmental Protection (Impact of Proposals) Act 1974* (Cth), which was perceived to be too reliant upon ministerial discretion because environmental impact assessment was not automatic and could only be triggered in circumstances where the relevant 'action Minister' determined that a particular activity was likely to affect the environment to a 'significant extent'.[70]

The *EPBCA* was introduced with the aim of providing a substantially revised environmental impact assessment process although there are concerns that the number of specifically listed MNES within the *EPBCA* are too few.[71] The current *EPBCA* incorporates nine MNES:

1. World heritage properties (set out by UNESCO in accordance with art 8 of the *World Heritage Convention*. Examples include the Sydney Opera House, the Melbourne Exhibition Building, and the Great Barrier Reef[72].
2. National heritage properties (listed by the Australian World Heritage Advisory Committee. Sites on the national list include Cockatoo Island, Bondi Beach, and Fraser Island and amount to approximately 8 million hectares across Australia.
3. Wetlands of international importance (relevant wetlands are those designated under the Ramsar Wetlands List)[73]
4. Listed threatened species (six categories are included and the list is extensive)[74]
5. Listed migratory species protected under international agreements
6. Commonwealth Marine Areas
7. The Great Barrier Reef Marine Park

---

70 See *Environmental Protection (Impact of Proposals) Act 1974* (Cth) s 5. See also the discussion by J Peel and L Godden, 'The Environment Protection Biodiversity and Conservation Act 1999 (Cth): Dark Sides of Virtue' (2007) 31 *Melbourne University Law Review* 106, 111 where the authors describe the environmental assessment process of the *EPBCA* as 'narrow and outdated'.
71 See especially the discussion by L Ogle, 'The Environment Protection Biodiversity Conservation Act 1999 (Cth): How Workable Is It?' (2000) 17 *Environmental and Planning Law Journal* 468, 470 where the author notes that there are no direct triggers which provide the Commonwealth government with clear power to control things such as new projects emitting greenhouse gases, land clearing of native vegetation, or significant water allocation decisions.
72 *World Heritage Convention*, opened for signature 23 November 1972, 1037 UNTS 151 (entered into force 15 December 1975).
73 The Ramsar Wetlands List is set out in the *Ramsar Convention*, opened for signature 2 February 1971, 996 UNTS 245 (entered into force 21 December 1975). Wetlands are the only habitat that has specific a international convention dedicated to their protection and preservation.
74 *EPBCA* (Cth) s 178(1).

8. Nuclear actions – including uranium mines ('nuclear action' is defined, inter alia, to include the establishment or modification of a nuclear installation and the transportation of spent nuclear fuel or radioactive waste products)[75]
9. Water resources from coal seam gas development and large coal mining development.

The *EPBCA* defines 'action' broadly to include a project, development, undertaking, activity or series of activities, or an alteration or modification to any project, development, undertaking, or existing infrastructure.[76] This definition is clearly broad enough to incorporate mining and energy projects. Whilst the *EPBCA* acknowledges that 'actions' can have both beneficial and adverse impacts upon the environment, the primary focus of the *EPBCA* approval process is upon adverse impacts. There are two components to the *EPBCA* approval process: (i) referral; and (ii) assessment. The referral stage occurs where a proponent or government body 'refers' a proposal to the Commonwealth Minister in circumstances where they believe a project is likely to have an impact on a matter of national environmental significance. In determining whether the action will have a significant impact the proponent must consider the 'sensitivity, value and quality of the environment which is impacted as well as the intensity, duration, magnitude and geographic extent of the impacts'.[77] A significant impact is defined as an impact which is 'important, notable or of consequence, having regard to its context or intensity' and also an impact which has a 'real chance' of occurring.[78] Assessment under the *EPBCA* can range from preliminary documentation to a public environment report or a full-scale public inquiry. Most assessments will, however, require production of an environmental impact assessment and a period for public comment.[79]

In the context of mining and energy projects, the *EPBCA* has been utilised infrequently because there is not a specific and discrete trigger impacted automatically by the initiation of a mining and/or energy project. There are, however, a range of circumstances where environmental assessment under the *EPBCA* may become necessary; for example, if exploratory underground drilling, mine sampling, or mine construction impacts on an endangered or critically endangered species because it is likely to damage habitat critical to the survival of the species or disrupt the breeding cycle of a population of the species, it will constitute a matter of national environmental significance. This type of action may also have a

---

75 *EPBCA* (Cth) s 22.
76 *EPBCA* (Cth) div 1, subdiv A, s 523.
77 Australian Government, Department of the Environment, *Significant Impact Guidelines 1.1: Matters of National Environmental Significance* (2013) 3 <http://www.environment.gov.au/epbc/publications/significant-impact-guidelines-11-matters-national-environmental-significance>.
78 Ibid.
79 *EPBCA* (Cth) pt 8, div 3.

significant impact on listed threatened ecological communities where, for example, it adversely impacts on their habitat.

Mining projects may also have a significant impact upon a matter of national environmental significance where the project is carried out within a national heritage area. For example, if a project will disturb Indigenous burial grounds or artefacts with national heritage values it is likely to trigger the application of the *EPBCA*. It is also important to consider the Ramsar criteria if the exploratory drilling is to occur in or immediately adjacent to a Ramsar wetland.[80]

The recent implementation of the 'water resources from coal seam gas and large coal mining development' trigger in the *EPBCA* is the most direct and explicit MNES for mining and energy projects. This is because it has an automatic application to projects that involve the large-scale removal of water, such as coal seam gas extraction.[81] The core objective of this new trigger is to address concerns relating to the depletion and contamination of water resources as a result of coal seam gas and coal mining projects. The trigger does not have an application to shale gas mining and therefore does not address the issue of hydraulic fracturing.

As discussed in Chapter 5, extracting large volumes of low-quality water has the capacity to impact upon connected surface and groundwater systems, some of which may already be fully or over-allocated; this includes the Great Artesian Basin and the Murray-Darling Basin. Coal seam gas mining can cause a dramatic depressurisation in the coal seam that can lead to changes in pressures of adjacent aquifers with consequential changes in water availability and reductions in surface water flows for connected systems. Coal seam gas mining can also generate land subsidence, which may impact adversely upon surface water systems, ecosystems, irrigation, and grazing lands. Further, the production of large volumes of treated waste water has the strong capacity to alter natural flow patterns of surface water resources, as well as impact deleteriously on water quality, river, and wetland health. Hydraulic fracturing when utilised in coal seam gas mining has the potential to induce connection and cross-contamination between aquifers and this can also dramatically impact upon groundwater quality. Finally, the reinjection of treated waste water into other aquifers and water resources has the potential to change the beneficial use characteristics of those aquifers.[82]

In accordance with the Commonwealth focus of the *EPBCA* as well as its constitutional limitations, the trigger will only apply to coal seam gas development

---

80 See *Ramsar Convention*, above n 72.
81 This trigger passed the Parliament on 19 June 2013 and came into effect on 2 June 2013. The amendment is now set out in the *EPBCA* (Cth) s 24D. Section 24D was introduced because of increasing concerns regarding the impact of CSG extraction upon water resources and the perceived need for a national environmental focus.
82 For an outline of the range of possible impacts upon water resources that CSG mining can have upon the environment see Australian Government, National Water Commission, *Coal seam gas* (June 2012) http://www.nwc.gov.au/nwi/position-statements/coal-seam-gas.

activities conducted by a constitutional corporation, the Commonwealth, a Commonwealth agency, or a person conducting the action for the purpose of domestic (state or territory jurisdictions) or international trade.

The trigger covering coal seam gas development is defined broadly to include any activity involving CSG extraction that has, or is likely to have, a significant impact on water resources, including any direct or cumulative impacts connected with salinity and salt production.[83] A water resource is also defined broadly to include surface or ground water, a watercourse, lake, wetland, or aquifer and will include water, organisms, and other components and ecosystems that contribute to the water resource.[84]

The *EPBCA* does not explicitly outline what constitutes a 'significant impact' on a water resource although impacts have been held to include changes in the quantity, quality, or availability of surface or ground water, alteration in ground pressure, alteration in drainage patters, or any substantial reduction in the availability of water for the environment.[85] A significant impact will include both direct and indirect impacts.

Importantly, the development of associated infrastructure that is not a component of the actual extraction process for CSG mining will not be included within the specific definition of 'CSG development' or 'large coal mining development' under section 24D of the *EPBCA*. This means that pipelines and road or rail development associated with coal seam gas or large coal mine projects will not automatically trigger the application of *EPBCA*. The development of infrastructure is not regarded as sufficiently proximate to the extraction of CSG or coal because the action causing the impact needs to be a constituent of an entire project rather than a separate and isolated component.

The *Significant Impact Guidelines 1.3: Coal seam gas and large coal mining developments – impacts on water resources* provides the following examples as illustrative of the type of actions that will not come within the application of the coal seam gas and large coal mining water resource trigger:

Example 1 – Associated infrastructure that would not have the water trigger applied
Processing Company is proposing to construct a new CSG processing facility and associated transport infrastructure. The processing facility will receive gas from *CSG Fields A and B*, which are already approved and operational. The project will include the processing facility, underground pipelines, a haul road, and accommodation facilities. As there is no

---

83 *EPBCA* (Cth) s 528.
84 Ibid.
85 These impacts were outlined by the Australian Government, Department of the Environment, *Water resources – 2013 EPBC Act amendment – Water trigger* http://www.environment.gov.au/epbc/what-is-protected/water-resources. See also Australian Government, Department of the Environment, *Significant Impact Guidelines 1.3: Coal seam gas and large coal mining developments – impacts on water resources* (December 2013) http://www.environment.gov.au/system/files/resources/d078caf3-3923-4416-a743-0988ac3f1ee1/files/sig-water-resources.pdf.

extraction of CSG as part of *Processing Company's* proposed action, the proposal is not a 'CSG development' or 'large coal mining development' for the purpose of the water trigger.

Example 2 – Associated infrastructure that would not have the water trigger applied
Coal Mining Company (CMC) is operating existing *Coal Mine A*, which was approved under the *EPBC Act* before the commencement of the Amendment Act on 22 June 2013. *CMC* now wants to construct new accommodation facilities, which will involve new sewage treatment arrangements and clearing of additional land. As there is no extraction of coal as part of *CMC*'s proposed action, the proposal is not a 'CSG development' or 'large coal mining development' for the purpose of the water trigger.

Example 3 – Associated infrastructure that would have the water trigger applied
Coal Mining Company (CMC) is developing a new coal mine (*Coal Mine B*), which will include an open-cut mine, a holding dam for water generated through mine de-watering, a coal washing facility, a rail line and some office buildings. In assessing whether the water trigger applies to the action, *CMC* should consider the impacts on water from the mine void, the holding dam and the coal washing facility. If a referral is made, incorporating all aspects of the operations, and the action is determined to be a controlled action (i.e. a significant impact on a water resource is likely), the Department will assess the impacts on water resources of the whole of the action, including the rail line and office buildings.

Example 4 – Associated infrastructure that would not have the water trigger applied
CSG Company is proposing to conduct well testing and pilot studies prior to seeking approvals for commercial operations. This testing will consist of a small number of wells that will operate for only a short period of time. To do this, *CSG Company* will establish accommodation facilities and access roads. The environmental assessment undertaken by *CSG Company* provides sufficient information to demonstrate that there is unlikely to be a significant impact on a water resource as a result of the exploratory CSG extraction. However, *CSG Company* also needs to construct a flood mitigation dam to protect the accommodation facilities. This dam may have an impact on catchment drainage patterns. The exploration activities may be within the definition of CSG development because there will be extraction of CSG. However, even if the flood mitigation dam could be considered to have a significant impact on a water resource, the water trigger will not be applied. This is because the extractive process itself will not have a significant impact on a water resource, taking into account cumulative and indirect impacts. However, if the entire project is referred as a single action, and the extractive activities come within the definition of CSG development, then all of the action's impacts on water resources will be considered.[86]

---

86 Australian Government, Department of the Environment, *Significant Impact Guidelines 1.3:Coal seam gas and large coal mining developments – impact on water resources* (December 2013) 9–10. http://www.environment.gov.au/system/files/resources/d078caf3-3923-4416-a743-0988ac3f1ee1/files/sig-water-resources.pdf.

Most coal seam gas projects will automatically attract the trigger and will therefore need to be referred to the Federal Minister for approval and, subject to a determination that the project amounts to a controlled action, the project will then require environmental assessment under the *EPBCA*. The possibility of overlap and inconsistency between state and Commonwealth legislation is mitigated, as discussed earlier in this Chapter, by the existence of bilateral agreements, which may be entered into between the Commonwealth and the state.[87] A key function of a bilateral agreement, is to reduce duplication of environmental assessment and regulation between the Commonwealth, states, and territories. These agreements allow the Commonwealth to 'accredit' particular state or territory assessment processes and approval decisions.

Unlike most state environmental assessment frameworks, the *EPBCA* explicitly articulates the fundamental and internationally accepted tenets of ecologically sustainable development (ESD) and incorporates those tenets into particular aspects of the assessment process. The principles of ESD are explicitly defined within s 3A of the *EPBCA*, which is extracted below:

Environment Protection and Biodiversity Conservation Act 1999 (Cth)

3A Principles of ecologically sustainable development

The following principles are principles of ecologically sustainable development:

(a) decision-making processes should effectively integrate both long-term and short-term economic, environmental, social and equitable considerations;

(b) if there are threats of serious or irreversible environmental damage, lack of full scientific certainty should not be used as a reason for postponing measures to prevent environmental degradation;

(c) the principle of inter-generational equity – that the present generation should ensure that the health, diversity and productivity of the environment is maintained or enhanced for the benefit of future generations;

(d) the conservation of biological diversity and ecological integrity should be a fundamental consideration in decision-making;

(e) improved valuation, pricing and incentive mechanisms should be promoted.

The mandatory considerations which must be taken into account in environmental assessment under the *EPBCA* are set out in s 136(1) and include: (a) matters relevant to the matter protected and (b) social and economic matters. In considering these matters, the Minister, in accordance with s 136(2)(a), must take into account the principles of ESD. The specific incorporation of internationally recognised principles such as the precautionary principle and the principle of inter-generational

---

87 *EPBCA* (Cth) ch 3, pt 5.

equity ensures that the environmental assessment process coheres with international expectations. Indeed, the precautionary principle, in particular, has evolved to such an extent in the international arena that it is tantamount to a norm of customary international law.[88] The precautionary principle has been utilised in an enormous variety of intergovernmental declarations, resolutions, and action programs and in more than sixty multilateral treaties involving a vast range of environmental concerns.[89] Whilst its inclusion in the *EPBCA* is primarily aspirational, as s 136(2)(a) lists the principles of ESD as a 'factor to take into account', it nevertheless provides a strong point of difference from state-based environmental assessment processes.

The precautionary principle, as outlined earlier, effectively sets out that whenever there are irreversible or serious threats to the environment or human health, decision-makers cannot use lack of scientific certainty as a reason to postpone preventative measures. In other words, the Minister under the *EPBCA* should, where assessing a mining project that triggers a matter of national environmental significance, err on the side of caution by avoiding the risk of irreversible environmental harm that a particular project may cause, even if the potential for such harm occurring is not fully established or supported by scientific certainty.[90]

The precautionary principle has been criticised for its lack of clarity and for its status as a 'law of fear'; the concern being that the principle is so vague that it provides very little guidance, but rather has the capacity to exacerbate irrational public perceptions of risk which then promote reactive processes including loss aversion, the myth of a benevolent nature, the heuristic of non-availability, and the assumption of system neglect.[91] Despite the flaws underlying the precautionary principle, its inclusion within the *EPBCA* has strong strategic significance, especially when the environmental impact of many new and emergent technologies associated with resource extraction remain unclear. The precautionary principle is directly responsive to this because it acknowledges the vulnerability of

---

[88] See especially the discussion by J E Hickey Jr and V R Walker, 'Refining the Precautionary Principle in International Environmental Law' (1995) 14 *Virginia Environmental Law Journal* 423, 425 discussing the widespread inclusion of the precautionary concept in international environmental treaties.

[89] See, eg, United Nations, *Report of the World Commission on Environment and Development: Our Common Future* (1987) <http://www.un-documents.net/our-common-future.pdf>.

[90] The precautionary principle was formulated by the *Rio Declaration on Environment and Development* (Rio De Janeiro, 3–14 June 1992) Principle 15 http://www.un.org/documents/ga/conf151/aconf15126-1annex1.htm

[91] See especially C S Sunstein, *Laws of Fear: Beyond the Precautionary Principle*, (Cambridge University Press, 2005) 61. See also A Trouwborst, 'Prevention, Precaution, Logic and the Law – The Relationship Between the Precautionary Principle and the Preventative Principle in International Law and Associated Questions' (2009) 2 *Erasmus Law Review* 105, 108 where the author discusses the legal status of the precautionary principle as a matter of general international law.

the environment and the fact that anthropogenic impacts are generally long-term and irreversible. The complexity and variability of natural systems and processes and the catastrophic impact that mining and extraction processes may have on matters of national environmental significance provide a strong foundation for the implementation of a clear risk-averse approach to environmental assessment.

The other well established principle of ESD, also incorporated indirectly into the *EPBCA*, is that of inter-generational equity. Inter-generational equity has two distinct elements in terms of its application to the utilisation of natural resources. The first is the 'inter' fairness requirement which seeks balance and proportion in the manner in which resources are utilised between past, present, and future generations of humans.[92] This requires the consumptive demands of contemporary society to be countered by the need to ensure adequate resources are available for successive generations. The second is an 'intra' fairness requirement whereby natural resources should be shared fairly amongst all humans in order to address fundamental socio-economic asymmetry in resource usage between nations and individuals.[93] Recent climate change jurisprudence has argued that the explicit incorporation of inter-generational equity into the *EPBCA* requires the Minister to pay regard to the issue of climate mitigation and consider how the greenhouse gas implications of approving a licence for fossil fuel mining projects.

In the 2006 Anvil Hill decision of *Gray v The Minister for Planning*, Pain J, in the New South Wales Land and Environment Court, held that the impact of greenhouse gas emissions from burning coal was a relevant consideration for the national environmental legislation and should therefore be taken into account when determining whether, on the specific facts, the overall environmental impact associated with an extension of a pre-existing coal licence.[94] His Honour held that whilst greenhouse gas emissions may be difficult to measure and climate change is a global concern, generated by a range of different contributors and flowing from the voluntary actions associated with increased energy consumption, it would be naïve not to adopt a whole of project assessment when determining the environmental impact of a coal licence extension. Climate change is directly linked to the production of energy from fossil fuel sources. Where an important downstream greenhouse gas emission is omitted from the Minister's assessment it becomes more difficult for the

---

92 See O Schachter, *Sharing the World's Resources* (Columbia University Press, 1977) 11–12. See also E B Weiss, 'Inter-Generational Equity in International Law' (1987) 1 *American Society of International Law Proceedings* 127, 128.
93 See also the discussion by G F Maggio, 'Inter/Intra Generational Equity: Current Applications Under International Law For Promoting the Sustainable Development of Natural Resources' (1997) 4 *Buffalo Environmental Law Journal* 161, 163. See also L M Collins, 'Revisiting the Doctrine of Intergenerational Equity in Global Environmental Governance' (2007) 30 *Dalhousie Law Journal* 79, 82.
94 *Gray v The Minister for Planning* [2006] NSWLEC 720.

final decision maker to be fully cognisant of all matters relevant to the permitting process. His Honour stated:

> The coal intended to be mined is clearly a potential major single contributor to GHG emissions deriving from NSW given the large size of the proposed mine. That the impact from burning the coal will be experienced globally as well as in NSW, but in a way that is currently not able to be accurately measured, does not suggest that the link to causation of an environmental impact is insufficient.[95]

His Honour concluded that the principles of ESD, particularly the principle of inter-generational equity, was legally required to be taken account of by the Minister when determining the environmental impact of a coal licence under the *EPBCA*. This issue has been raised more recently in a claim brought by the NSW EDO against a licence issued for the development of the new Carmichael coal mine in Queensland.[96]

Unfortunately, not all ESD principles are incorporated into the *EPBCA*. Notable exclusions from the *EPBCA* framework are the public trust doctrine (which imposes a duty to hold the environment in trust for the benefit of the public), the subsidiarity principle (where decisions involve the community most closely affected by them), and the polluter and user pays principles (which makes users accountable for the costs of using environmental resources).[97]

The explicit 'mandatory' considerations that have been articulated within s 136 of the *EPBCA* (and which are extracted below) require consideration not only of environmental impact issues relevant to the specific action but also consideration of broader economic and social factors that may be relevant to the environmental approval. The inclusion of this criteria lacks a clear focus and scope in this context, although it has been utilised to promote a range of fiscal and community advantages such as the economic impact of a mining project closing down upon a local community and the capacity, in remote areas, for a mining project to stimulate employment and activity within Indigenous communities. The breadth of the criteria also allows the Commonwealth to take into account the environmental history of an applicant when considering whether to grant an approval. The type of information relevant to the person's environmental history will be that which indicates whether a person is likely to comply with the conditions of an approval. This will

---

95 Ibid [98].
96 For a discussion of this see S Hepburn, 'Court Challenge will Test Coal Mining's Climate Culpability', *The Conversation*, 16 January 2015 https://theconversation.com/court-challenge-will-test-coal-minings-climate-culpability-36285.
97 This is discussed by S Marsden, 'Strategic Environmental Assessment in Australia – An Evaluation of Section 146 of the Environment Protection Biodiversity Conservation Act 1999 (Cth)' (1999) 8 *Griffith Law Review* 394, 395–8.

generally include details of any previous *EPBCA* approvals and permits held and the level to which the proponent has complied with those approval and permit conditions. It may also entitle the Minister to consider details of any environmental audits to which they may have been subjected.[98]

Environment Protection and Biodiversity Conservation Act 1999 (Cth)

136 General considerations

Mandatory considerations

(1) In deciding whether or not to approve the taking of an action, and what conditions to attach to an approval, the Minister must consider the following, so far as they are not inconsistent with any other requirement of this Subdivision:
 (a) matters relevant to any matter protected by a provision of Part 3 that the Minister has decided is a controlling provision for the action;
 (b) economic and social matters.

Factors to be taken into account

(2) In considering those matters, the Minister must take into account:
 (a) the principles of ecologically sustainable development; and
 (b) the assessment report (if any) relating to the action; and
 (ba) if Division 3A of Part 8 (assessment on referral information) applies to the action – the finalised recommendation report relating to the action given to the Minister under subsection 93(5); and
 (bc) if Division 4 of Part 8 (assessment on preliminary documentation) applies to the action:
  (i) the documents given to the Minister under subsection 95B(1), or the statement given to the Minister under subsection 95B(3), as the case requires, relating to the action; and
  (ii) the recommendation report relating to the action given to the Minister under section 95C; and
 (c) if Division 5 (public environment reports) of Part 8 applies to the action:
  (i) the finalised public environment report relating to the action given to the Minister under section 99; and
  (ii) the recommendation report relating to the action given to the Minister under section 100; and

---

98 See Australian Government, Department of Sustainability, Environment, Water, Population and Communities *Environment Protection and Biodiversity Conservation Act 1999 (Cth): Policy Statement: Consideration of a Person's Environmental History when making Decisions under the EPBC Act* (2013) http://www.environment.gov.au/system/files/resources/57853bb3-a51b-4aed-8565-f603fa1868d2/files/epbc-act-policy-environmental-history.pdf.

(ca) if Division 6 (environmental impact statements) of Part 8 applies to the action:
  (i) the finalised environmental impact statement relating to the action given to the Minister under section 104; and
  (ii) the recommendation report relating to the action given to the Minister under section 105; and
(d) if an inquiry was conducted under Division 7 of Part 8 in relation to the action – the report of the commissioners; and
(e) any other information the Minister has on the relevant impacts of the action (including information in a report on the impacts of actions taken under a policy, plan or program under which the action is to be taken that was given to the Minister under an agreement under Part 10 (about strategic assessments)); and
(f) any relevant comments given to the Minister in accordance with an invitation under section 131 or 131A; and
(fa) any relevant advice obtained by the Minister from the Independent Expert Scientific Committee on Coal Seam Gas and Large Coal Mining Development in accordance with section 131AB; and
(g) if a notice relating to the action was given to the Minister under subsection 132A(3) – the information in the notice.

Note: The Minister must also take into account any relevant comments given to the Minister in response to an invitation under paragraph 131AA(1)(b). See subsection 131AA(6).

Person's environmental history

(4) In deciding whether or not to approve the taking of an action by a person, and what conditions to attach to an approval, the Minister may consider whether the person is a suitable person to be granted an approval, having regard to:
  (a) the person's history in relation to environmental matters; and
  (b) if the person is a body corporate – the history of its executive officers in relation to environmental matters; and
  (c) if the person is a body corporate that is a subsidiary of another body or company (the *parent body*) – the history in relation to environmental matters of the parent body and its executive officers.

Minister not to consider other matters

(5) In deciding whether or not to approve the taking of an action, and what conditions to attach to an approval, the Minister must not consider any matters that the Minister is not required or permitted by this Division to consider.

9.9 REVIEW QUESTIONS

1. Describe the two policy paradigms around which all environmental policy frameworks for the regulation of mining and energy projects are structured.

2. Are all mining proponents who seek to develop a mining or energy project in Australia automatically subject to environmental regulation?

3. Outline the jurisdictional basis for environmental regulation in Australia and explain how it relates to the Constitutional framework? In what circumstances, if any, will the Commonwealth Minister become involved in the environmental assessment of a proposed mining project?

4. What is a 'bilateral agreement' and in what circumstances may it be entered into?

5. Explain the pros and cons of an environmental impact assessment in the context of a proposal to develop a mining project such as, for example, (i) a coal seam gas project and (ii) a coal-fired plant.

6. Assume the following facts:

> Blue Energy is involved in setting up operations for its coal seam gas plant in Queensland. The relevant planning scheme requires a full environmental impact assessment to be carried out before any licence is to be issued. In preparing its environmental impact assessment, Blue Energy fails to include accurate details regarding the volume of subsurface water to be removed. Blue Energy relies on approximate figures that turn out to be less than half the volume required for a production site of its size. Will the proposal be rejected as a result of this error? Advise Blue Energy.

7. How does the EPA determine whether a mining proposal will have a 'significant effect on the environment' in Western Australia? What factors does it take into account?

8. Explain the concept of an 'environmental value' in the *Environmental Protection Act 1994* (Qld) and how it relates to the environmental assessment process.

9. Outline the purpose of the new Gateway Panel under the New South Wales environmental assessment process for mining and energy projects.

10. Consider the following problem:

> Santos is proposing to conduct well testing and pilot studies prior to seeking an exploration licence for a coal seam gas project. This testing will consist of a small number of wells that will operate for only a short period of time. The environmental assessment indicates that there is unlikely to be a significant impact on a water resource as a result of the exploratory CSG extraction. Santos must also construct a flood mitigation dam to protect the accommodation facilities which may have an impact on catchment drainage patterns. Will this attract the coal seam gas trigger under the *Environment Protection Biodiversity Conservation Act 1999* (Cth)?

9.10 FURTHER READING

T Anderson and D Leal, *Free Market Environmentalism* (Palgrave, 1991).

Australian Government, Department of the Environment, *Significant Impact Guidelines 1.1: Matters of National Environmental Significance* (2013) <http://www.environment.gov.au/

epbc/publications/significant-impact-guidelines-11-matters-national-environmental-significance>.

Australian Government, Department of the Environment, *Significant Impact Guidelines 1.3: Coal seam gas and large coal mining developments – impact on water resources* (December 2013) http://www.environment.gov.au/system/files/resources/d078caf3-3923-4416-a743-0988ac3f1ee1/files/sig-water-resources.pdf.

Australian Government, Department of the Environment, *Water resources – 2013 EPBC Act amendment – Water trigger*. http://www.environment.gov.au/epbc/what-is-protected/water-resources.

Australian Government, Department of Sustainability, Environment, Water, Population and Communities, *Environment Protection and Biodiversity Conservation Act 1999 (Cth): Policy Statement: Consideration of a Person's Environmental History when making Decisions under the EPBC Act* (2013) http://www.environment.gov.au/system/files/resources/57853bb3-a51b-4aed-8565-f603fa1868d2/files/epbc-act-policy-environmental-history.pdf.

Australian Government, National Water Commission, *Coal seam gas* (June 2012) http://www.nwc.gov.au/nwi/position-statements/coal-seam-gas.

E Bastide, T Waelde and J Warden-Fernandez, *International and Comparative Mineral Law and Policy: Trends and Prospects* (Kluwer, 2006).

C J Bateman and J T B Tripp, 'Towards Greener FERC Regulation of the Power Industry' (2014) *Harvard Environmental Law Review* 275.

G Bates, *Environmental Law in Australia* (LexisNexis, 7th ed, 2010).

R Briese, 'Climate Change Mitigation Down Under – Legislative Responses in a Federal System' (2010) 13 *Asia Pacific Journal of Environmental Law* 75.

J F Castrilli, 'Environmental Regulation of the Mining Industry in Canada: An Update of Legal and Regulatory Requirements' (2001) 34 *University of British Columbia Law Review* 91.

L M Collins, 'Revisiting the Doctrine of Intergenerational Equity in Global Environmental Governance' (2007) 30 *Dalhousie Law Journal* 79.

J Crawford, 'The Constitution and the Environment' (1991) 13 *Sydney Law Review* 11.

M Crommelin, 'Resource Law and Public Policy' (1983) 15 *University of Western Australia Law Review* 1.

M Crommelin, 'Commonwealth Involvement in Environmental Policy: Past, Present and Future' (1987) 4 *Environmental Planning Law Journal* 101.

E Duruigbo, 'The Global Energy Challenge and Nigeria's Emergence as a Major Gas Power: Promise, Peril or Paradox of Plenty' (2009) 21 *Georgetown International Environmental Law Review* 395.

EDO NSW, *Draft Approval Bilateral Agreements: NSW & Qld* (2014) http://www.edonsw.org.au/draft_approval_bilateral_agreements_for_nsw_and_qld.

Environmental Defenders Office Qld, *Say no to Commonwealth delegating environmental approval powers to Queensland* (12 June 2014) http://www.edoqld.org.au/news/say-no-to-commonwealth-delegating-environmental-approval-powers-to-queensland/.

L Godden and J Peel, 'The Environment Protection and Biodiversity Conservation Act 1999 (Cth): Dark Sides of Virtue' (2007) 31 *Melbourne University Law Review* 262.

L Godden and J Peel, *Environmental Law: Scientific, Policy and Regulatory Dimensions* (Oxford University Press, 2010).

Government of Western Australia, Department of Mines and Petroleum, *Guidelines for Mining Proposals in Western Australia* (2006) http://www.dmp.wa.gov.au/documents/ENV-MEB-200.pdf.

Government of Western Australia, Department of Mines and Petroleum, *Western Australian Mineral and Petroleum Statistics Digest 2008–09* (2009) http://www.dmp.wa.gov.au/documents/Statsdigest09web.pdf.

R W Hahn and C R Sunstein, 'The Precautionary Principle as a Basis for Decision-Making' (2005) 2(2) *The Economist's Voice*, art 8.

R Harding (ed), *Environmental Decision Making* (Federation Press, 1998).

S Hepburn, 'Court Challenge will Test Coal Mining's Climate Culpability', *The Conversation*, 16 January 2015 http://theconversation.com/court-challenge-will-test-coal-minings-climate-culpability-36285.

V Heyvaert, J Thornton and R Drabble, 'With Reference to the Environment: The Preliminary Reference Procedure, Environmental Decisions and the Domestic Judiciary' (2014) 130 *Law Quarterly Review* 413.

J E Hickey Jr and V R Walker, 'Refining the Precautionary Principle in International Environmental Law' (1995) 14 *Virginia Environmental Law Journal* 423.

A Ingelson and C Nwapi, 'Environmental Impact Assessment for Oil, Gas and Mining Projects in Nigeria: A Critical Analysis' (2014) 10 *Law, Environment and Development Journal* 1.

D Kysar, 'Law, the Environment and Vision' (2003) 97 *Northwestern University Law Review* 675.

D Kysar, 'It Might Have Been: Risk, Precaution and Opportunity Costs' (2007) 22 *Journal Of Land Use and Environmental Law* 1.

D Kysar 'Executive Summary: Professor Douglas Kysar's Analysis of Flaws in Predictive International Climate Policy Models' (2013) 40 *British Columbia Environmental Affairs Law Review* 409.

G F Maggio, 'Inter/Intra Generational Equity: Current Applications Under International Law For Promoting the Sustainable Development of Natural Resources' (1997) 4 *Buffalo Environmental Law Journal* 161.

S Marsden, 'Strategic Environmental Assessment in Australia – An Evaluation of Section 146 of the Environment Protection Biodiversity Conservation Act 1999 (Cth)' (1999) 8 *Griffith Law Review* 394.

C Moran and S Vink, 'Assessment of impacts of the proposed coal seam gas operations on surface and groundwater systems in the Murray-Darling Basin' (Assessment report, University of Queensland, 29 November 2010) http://www.environment.gov.au/epbc/notices/assessments/pubs/coal-seam-gas-operations-impacts.pdf.

S O Nliam, 'International Oil and Gas Environmental Legal Framework and the Precautionary Principle: The Implications for the Niger Delta' (2014) 22 *African Journal of International and Comparative Law* 22.

NSW Department of Environment, Climate Change and Water, *Guidelines for the Review of Environmental Factors* (2011) http://www.environment.nsw.gov.au/resources/protectedareas/110028REFProGde.pdf.

NSW Government, Department of Planning and Infrastructure, *Strategic Regional Use Plan: Upper Hunter* (September 2012) <http://www.nsw.gov.au/sites/default/files/initiatives/upperhunterslup_sd_v01.pdf>.

NSW Government, Department of Trade and Investment, *ESG3: Mining Operations Plan (MOP) Guidelines, September 2013* (9 September 2013) http://www.resourcesandenergy.nsw.gov.au/__data/assets/pdf_file/0007/527911/ESG3-Mining-Operations-Plan-MOP-Guidelines-September-2013.pdf.

L Ogle, 'The Environment Protection Biodiversity Conservation Act 1999 (Cth): How Workable Is It?' (2000) 17 *Environmental and Planning Law Journal* 468.

K Palmer and D Grinlinton, 'Developments in Renewable Energy Law and Policy in New Zealand' (2014) 32 *Journal of Energy and Natural Resources Law* 245.

Parliament of Victoria, Hazelwood Mine Fire Inquiry, *Report: Part 3: Fire Risk Management* (August 2014) http://report.hazelwoodinquiry.vic.gov.au/microsoft-word-document-version-text-size-18pt.

L Parry, 'New Laws Could Hand Miner's 10% of Brazil's National Parks and Indigenous Lands', *The Conversation* (7 November 2014) http://theconversation.com/new-laws-could-hand-miners-10-of-brazils-national-parks-and-indigenous-lands-33912.

Queensland Government, Department of Environment and Heritage Protection, *Eligibility criteria and standard conditions* (18 December 2013) http://www.ehp.qld.gov.au/licences-permits/compliance-codes/.

Queensland Government, Department of Environmental and Heritage Protection, *Guideline: Environmental Impact Statements: Triggers for environmental impact statements under the Environmental Protection Act 1994 for mining and petroleum activities* (2014) http://www.ehp.qld.gov.au/management/impact-assessment/pdf/eis-guideline-trigger-criteria.pdf.

Queensland Government, Department of Natural Resources and Mines, *Queensland Mining Update*, (19 January 2015) https//:mines.industry.qld.gov.au/mining/queensland-mining-update.htm.

M Radetski, 'Economic Growth and Environment', in P Low (ed), *International Trade and the Environment* (Vol 1), (World Bank: Discussion Papers 159, 1992).

M Raff, 'Ten Principles of Quality in Environmental Impact Assessment' (1997) 14 *Environmental Planning Law Journal* 207.

S Rao, 'Reforming the Environmental Assessment Process in Victoria' (2010) 1 *National Environmental Law Review* 34.

J M Rieder, 'Evaluation of Two Environmental Acts: The National Environmental Policy Act and the Environment Protection and Biodiversity Conservation Act' (2011) 14 *Asia Pacific Environmental Law Journal* 105.

N A Robinson, 'International Trends in Environmental Impact Assessment' (1992) 19 *British Columbia Environmental Affairs Law Review* 591.

O Schachter, *Sharing the World's Resources* (Columbia University Press, 1977).

A Sifakis, 'Precaution, Prevention and the Environmental Impact Assessment Directive' (1998) 7 *European Environmental Law Review* 349.

J Singer, 'How Property Norms Construct the Externalities of Ownership' in G S Alexander and E Penalver (eds), *Property and Community* (Oxford University Press, 2009).

R M Solow, 'The Economics of Resources or The Resources of Economics' (1974) 64 *American Economic Review* 1.

D B Spence, 'Paradox Lost: Logic, Morality and the Foundations of Environmental Law in the 21st Century (1995) 20 *Columbia Journal of Environmental Law* 145.

C S Sunstein, *Laws of Fear: Beyond the Precautionary Principle*, (Cambridge University Press, 2005).

M W Tabb, 'An Environmental Conversation' (2014) 50 *Natural Resources Journal* 143.

I Thomas and P Murfitt, *Environmental Management Processes and Practices for Australia*, (Federation Press, 2005).

A Trouwborst, 'Prevention, Precaution, Logic and the Law – The Relationship Between the Precautionary Principal and the Preventative Principle in International Law and Associated Questions' (2009) 2 *Erasmus Law Review* 105.

United Nations, *Report of the World Commission on Environment and Development: Our Common Future* (1987) <http://www.un-documents.net/our-common-future.pdf>.

Victorian Planning and Environmental Law Association, J Power, *Review of the Environment Effects Act (1978)* (Inquiry into the Environmental Effects Statement Process in Victoria, Submission no 55, 2002) http://www.parliament.vic.gov.au/images/stories/committees/enrc/environmental_effects/submissions/55_Victorian_Planning__Environmental_Law_Association.pdf.

E B Weiss, 'Inter-Generational Equity in International Law' (1987) 1 *American Society of International Law Proceedings* 127.

P Wieland, 'Going Beyond Panaceas: Escaping Mining Conflicts in Resource Rich Countries Through Middle-Ground Policies' (2014) 20 *New York University Environmental Law Journal* 199.

10

MINING AGREEMENTS AND REVENUE FRAMEWORKS

10.1	Introduction	406
10.2	Mining agreements	407
10.3	The core elements of a mining agreement	417
10.4	Revenue frameworks	422
10.5	Review questions	437
10.6	Further reading	338

10.1 Introduction

Mining and energy projects have different socio-economic impacts upon economies compared with other sectors, such as agricultural, manufacturing, or service. There are many different reasons for this but primarily, because large mining projects operate with unusually large economies of scale, they have the capacity to generate significant investment and employment. In developing countries, mining projects can have such a significant localised impact that they overwhelm adjacent local economies and affect the social and economic foundations of those communities. The environmental impacts increase in both geographical scale and economic intensity as a project progresses through the exploration and extraction phases.[1] Air pollution, water pollution, tailings, and waste are all natural consequences of a large-scale mining project and the localised effect of these environmental impacts can be quite profound in poorer countries.[2]

Furthermore, mining and energy projects have the capacity to generate suboptimal political and economic outcomes, particularly within developing countries where, in a weak institutional environment, the dependence on natural resource wealth exacerbates governance problems.[3]

The domestic economic impact of mining depends heavily upon the way in which mining revenue is generated by the government through taxes, royalties, and profit sharing.[4] The distribution of mining revenue has a crucial impact upon economic development. With large-scale petroleum mining projects, revenue can constitute one-fifth of the GDP. Mining revenue can, however, be notoriously difficult to manage. This is particularly true for natural gas in Australia where the revenue is volatile and high, relative to the domestic GDP.

Mining projects also reflect the inherent tension that exists between sovereign control and private development. Mineral resources that reside within a state or territory belong to that state and the state retains the power to regulate those resources. However, the successful development of those minerals depends largely upon private investment and initiative given the high-risk nature

---

1 See the discussion by R Auty, 'Mining Enclave to Economic Catalyst: Large Mineral Projects in Developing Countries' (2007) 13 *Brown Journal of World Affairs* 135, 138.
2 See the discussion by T Walde, 'Environmental Policies in Developing Countries (1992) 10 *Journal of Energy and Natural Resources Law* 327, 331.
3 This is discussed by A Gillies and A Heuty, 'Does Transparency Work – The Challenges of Measurement and Effectiveness in Resource Rich Countries' (2011) 6 *Yale Journal of International Affairs* 25, 27.
4 See especially K Naito, F Remy and J Williams, *Review of Legal and Fiscal Frameworks for Exploration and Mining* (World Bank Group Mining Department, 2001) ch 1. See also E Blanco and J Razzaque, *Globalisation and Natural Resource Law: Challenges, Key Issues and Perspectives* (Edward Elgar, 2011).

of many mining projects in terms of exploration, extraction, processing, transportation, and commercialisation.[5]

The degree of state intervention in mining agreements will vary according to the domestic context. If the domestic sector lacks the capital to progress mining projects much of the impetus must come from foreign investment. This will generally result in improved state participation to ensure effective management of the foreign operator.[6] The resource market may be international in scope; however, most mining projects are, necessarily, grounded in the localised circumstances relevant to the particular region or area in which the project is situated. This means that in many contexts, terms and conditions relevant to the mining agreement may be subject to the requirements and expectations of affected local communities.[7] The method by which governments impose control in the articulation of mining agreements is largely a product of the 'maturity and stability of the country's legal framework for mining'.[8]

10.2 Mining agreements

Mining agreements entered into between governments and mining proponents are commonplace because they provide the parties with greater clarity regarding the terms and conditions of exploration or extraction and provide the mining proponent with greater assurance regarding key aspects of their investment over, often lengthy, periods of time.[9] These agreements are of particular importance because they outline the financial benefits and risks connected with the project and structure the way in which the benefits and risks are to be allocated. Mining agreements can provide greater certainty and protection for all stakeholders and this is important for generating foreign investment in the mining sector.

Mining agreements generally operate in tandem with existing general, domestic regulation and, in this respect, their existence provides foreign investors with greater confidence and assurance, particularly where the domestic regulatory framework

---

5 See Naito, Remy and Williams, above n 4, 4–5. See also E Bastida, T Walde and J Warden-Fernandez, *International and Comparative Mineral Law And Policy – Trends and Prospects* (Kluwer Law International, 2005) ch 2.
6 See Naito, Remy and Williams above n 4, 136.
7 See the discussion by J P Williams, 'Global Trends and Tribulations in Mining Regulations' (2012) 30 *Journal of Energy and Natural Resources Law* 391, 392. For a more expansive discussion of the impact of mining agreements upon third world regions, see M Radetzki, 'State Ownership in Developing Country Mineral Industries' in *The Use of State Enterprises in the Solid Minerals Industry in Developing Countries*, Proceedings of the Interregional Seminar held in Budapest, 5–10 October 1987 (United Nations Department of Technical Co-operation for Development, TCD/SEM.88/5, INT-87-R37, 1989) 13, 15.
8 See J P Williams, 'Global Trends and Tribulations in Mining Regulations' above n 7, 395.
9 The term 'mining agreement' is used here interchangeably with other like terms such as 'franchise agreement' or 'indenture'.

is deficient. Mining agreements will generally outline the way in which the revenue generated from mineral extraction is to be shared between governments and mining proponents. Long-term mining projects can provide governments with extensive sources of revenue and, for this reason, governments often seek to encourage investment through subsidies, discounts, and tax exemptions. Benefits and conditions may all be carefully articulated through individualised arrangements with mining proponents that cater to the specific requirements of the project, the resource, the economic conditions and the jurisdiction.

The widespread use of mining agreements between the project developer and the state – which are subsequently ratified by the legislature – is a definitive feature of Australian mining law, particularly following the rapid expansion of the mining sector in the 1950s.[10] There has also been a rise – particularly with respect to large mining projects – of broader alliance frameworks whereby contractual relationships between parties involved in a mining project are either open-ended or endure for a term of years not defined by a specific project.[11] There is strong strategic utility in developing a coalition of stakeholders, particularly in areas where profit margins are tight: the industry is relatively mature and future prospects are unclear. An alliance can ensure that factors relevant to risk, funding, and escalating costs are appropriately hedged.[12]

There are essentially four different forms of mining agreements that operate globally: the concession agreement, the profit sharing agreement, the risk service contract, and the joint venture agreement. The form of agreement adopted will generally determine the terms and conditions of a long-term mining project and also allocate the method by which mining revenue is to be shared. Whilst each arrangement may achieve the same objectives, they are conceptually different in focus. Each form of agreement provides for different levels of control by the mining proponent, different revenue and compensation arrangements, and differing levels of government involvement.[13]

---

10 See A Fitzgerald, *Mining Agreements: Negotiated Frameworks in the Australian Minerals Sector* (Lexis Nexis, 2001) 2.
11 See National Economic Development Office, *Partnering: Contracting Without Conflict*, (National Economic Development Office, London, 1991).
12 See the discussion by J Lacey, 'Partnering and Alliancing: Back to the Future' (2007) 26 *Australian Resources and Energy Law Journal* 69, 81 discussing the legal frameworks for alliancing including joint ventures, partnerships, and unit trusts. The author notes that project alliancing was created in an attempt to design a project delivery strategy which would see project capital costs and operating costs restrained to ensure projects remained economically feasible.
13 See the discussion by P McNamara, 'The Enforceability of Mineral Development Agreements to which The Crown in the Right of the State is a Party' (1982) 5 *University of New South Wales Law Journal* 263, 266 noting the restrictive rule which sets out that where a Minister or Director of Mines has a statutory discretion to grant or withhold a mining lease or other production tenement they cannot, without further statutory authority, enter a binding agreement for the granting of a statutory tenement. See also *Cudgen Rutile (No 2) Pty Ltd v Chalk PC* [1975] AC 520, 533–4.

10.2.1 Concession agreements

The concession agreement is essentially a form of public-private arrangement whereby the government grants a private party exclusive licensing rights to operate, maintain, and manage a mining project for an extended period of time, in exchange for either a premium and/or revenue payments.[14] The concession agreement effectively constitutes a negotiated contract between the mining proponent and the state, allowing the mining proponent to develop a mining project in accordance with terms and conditions that have been agreed upon in return for the issuance of licensing entitlements. The payment received by the state is generally based upon a royalty system.

The concession agreement presumes that a licence must be issued by the state to the mining proponent, subject to the imposition of specific terms and conditions, that may be fixed by legislation and modified on a case-by-case basis. A crucial element of the concession agreement lies in the fact that the state retains the power to modify, at any point, terms and conditions that have been fixed by legislation.[15]

Offshore petroleum resources in Australia are generally subject to a licensing concession agreement.[16] The reason for this is the state is generally unwilling to invest large amounts of public capital into high-risk exploratory ventures and therefore relies heavily upon a public-private partnership agreement. Each party is mutually dependant upon the other: the private mining proponent needs the state, because the state is the owner of the resource, and the state needs the private mining proponent as they bring capacity, technology, and financial investment. The concession agreement functions as an effective interface in a framework where the ownership of the resource resides in the state because this provides a core incentive for public and private parties to negotiate as the project is unable to proceed in the absence of legally enforceable rights held by each party.[17] It should be borne

---

14 See the outline by W Onorato, 'World Petroleum Legislation: Frameworks that Foster Oil and Gas Development' (Policy Research Working Paper No 2, World Bank, 1995) 5.4.5.
15 See the discussion by T Hunter, 'Comparative Law as an Instrument in Transnational Law: The Example of Petroleum Regulation' (2009) 21 *Bond Law Review* 42, 45. See also T Daintith, 'Evaluation of the Petroleum (Submerged Lands) Act as a Regulatory Regime' (2000) *Australian Mining and Petroleum Law Association Yearbook* 91, 91–2.
16 See Hunter, above n 15, 50–2 where the author discusses the prevalence of the licensing concession agreement in the offshore petroleum framework of both Australia and Norway.
17 See especially the discussion by D Custos and J Reitz, 'Public-Private Partnerships' (2010) 58 *American Journal of Comparative Law* 555, 562 where the authors discuss the reciprocal advantages relevant to particular sectors. See also K Hogan, 'Protecting the Public in Public-Private Partnerships: Strategies for Ensuring Adaptability in Concession Contracts' (2014) 2 *Columbia Business Law Review* 420, 423 discussing the difficulties with concession agreements because of the perceived trade-off between short-term economic gain and long-term public interest where the agreements effectively confer upon a private company ownership and control over an infrastructure asset for many decades.

in mind, however, that despite the conferral of rights, the resource itself continues to be owned by the state until it is actually produced or sold.[18] Hence, the concession agreement confers upon the mining proponent a right to extract and exploit minerals that belong to the state; however, the conferral of those rights do not in themselves constitute a transfer of ownership.[19]

The exploitation right is a vital criterion that distinguishes the concession agreement from other forms of public contract. The creation of a right to exploit requires the concessionaire to make a payment to those who are permitting use of the resource or structure over a given period of time. It also results in the transfer of responsibilities to the concessionaire, who assumes the inherent risk associated with any construction and utilisation of the facilities. The operational risks connected with the concession agreement are imposed upon the concessionaire at a simultaneous point to the creation of the exploitation right.[20]

10.2.2 Profit sharing contracts

Profit sharing contracts ('PSCs') are commonly utilised in the petroleum industry, but not so often in the mining industry. The core elements of the PSC are straightforward: the mining proponent bears all the financial cost of the mining operation which it then deducts from the gross value of the mineral produced to arrive at the net profit. The net profit is then shared between the mining proponent and the government, in accordance with pre-determined proportions stipulated in the terms and conditions of the PSC. This effectively means that the mining proponent's share of production should not be subject to any royalty, provided that the proportion received by the government reflects the absence of royalty payment. Many PSCs do not completely remove royalty provisions as they are often levied on the value of the mining proponent's share of resource prior to calculating the net profit.

A PSC is entered into between the state and a mining proponent and, unlike the concession agreement, its primary assumption is that the extracted petroleum will be 'shared' between the parties rather than the petroleum proponent acquiring a licensing entitlement allowing it to exclusively exploit the resource. In this regard,

---

18 See Hogan, above n 16, 427. See also the outline of the new concession framework in Brazil by S Sewalk, 'Brazil's Energy and Policy Regulation' (2014) 25 *Fordham Environmental Law Review* 652, 687.
19 See especially the discussion by T Daintith in *Finders Keeper's: How the Rule of Capture Shaped the World Oil Industry* (RFF Press, 2010) 367 where the author notes that concession 'is the traditional instrument through which states have invited oil companies to contribute their capital and expertise to the search for oil, in return for the right to treat any oil they can find and recover as their own'.
20 C Georgeta, 'The Oil and Mining Concession in Perspective' (2013) 2 *Perspectives of Business Law Journal* 87, 93.

the PSC involves the grant, by the state, of a contractual entitlement upon the petroleum proponent to explore and commercialise an area for petroleum production and then subsequently divide both the produced oil and the profit oil between the state and the petroleum proponent in accordance with agreed percentages. The percentage awarded to the state effectively replaces government taxes, duties, and other miscellaneous charges. Further, it is clear that the 'the investor can count on the fact that any subsequent changes in the rate of taxation or in the accounting of the taxable base are not going to influence the economy of the project, because they will be absorbed by the government take.'[21]

The PSC was defined by Professor Omorogbe in the following terms:

> Arrangements where the foreign firm and the government share the output of the operation in predetermined propositions. This new form has been regarded as being a substantial departure from the old concessions in that the host state is theoretically the undisputed owner of the petroleum, with the foreign corporations being engaged as contractors to perform certain specified tasks in return for a fee in kind.[22]

Under the terms of a PSC, a petroleum proponent is given the right to explore and produce a resource. The petroleum proponent bears the mineral and financial risk.[23] The PSC is not used extensively in Australia given that most of Australia's petroleum is imported; however, PSCs are utilised extensively in the Middle East and central Asia where large oilfields are operative.

The PSC is a successful contractual model for oil extraction because of the large capital costs associated with developing a project. The PSC entitles a successful petroleum proponent to utilise money from the 'produced' oil to recover capital and expenditures. The 'profit oil' is divided between the government and the petroleum proponent and, typically, the government receives a much higher proportion of 'profit oil' than the petroleum proponent.[24] The PSC will, however, generally mean that a petroleum proponent has reduced discretion over the way

---

21 G Cordero-Moss, 'Contract or Licence – Regulation of Petroleum Investment in Russia and Foreign Legal Advice' (2003) 13 *Transnational Law and Contemporary Problems* 519, 527.
22 Y Omorogbe, 'The Legal Framework for the Production of Petroleum in Nigeria' (1987) 5 *Journal of Energy and Natural Resources Law* 273, 279.
23 See the discussion by S Adepetun, 'Production Sharing Contracts – The Nigerian Experience' (1995) 13 *Journal of Energy and Natural Resources Law* 21, 21–2.
24 4 See Adepetun, above n 23, 25 where the author notes that generally the proportion by which profit oil is shared differs from one type of contract area to the other (i.e. onshore, shallow offshore and deep offshore). See also D K Espinosa, 'Environmental Regulation of Russia's Offshore Oil (and Gas) Industry and its Implication for the International Petroleum Market' (1997) 6 *Pacific Rim Law and Policy Journal* 647, 648 noting the problem of overlapping production sharing agreements in Russia which has introduced law on PSA pursuant to the *Law on Production Agreements* No 225–3, Economic Law of Russia.

in which operations are conducted than it would have under a concession agreement. One of the principal rationales for using the PSC is that it involves no surrender of the host country's sovereignty because title to the petroleum does not pass to a foreign mining company allowing the host country to retain a degree of autonomy and control.[25]

This type of arrangement is particularly popular in an environment where heightened global interest in resource endowments means that many countries are seeking to maximise their economic participation in the wealth generated by the petroleum resource.[26] Unlike the royalty framework, which is the revenue arrangement that generally applies to a concession agreement, changes in international resource prices or production rates will affect the company's share of production within a PSC and this is something that all of the parties and the stakeholder's may need to anticipate.

The PSC is essentially an agreed program for the extraction of mineral resources that is to be carried out by the petroleum proponent (acting as investor) in favour of the government. The architecture of the PSC means that the government is regarded as having hired the mining proponent as a contractor to perform the work envisioned by the program. Payment is not made in money, but rather, through the conferral of a portion of the produced product. This is the core element underlying the 'profit sharing' arrangement because it is premised upon the sharing of the work performed by the petroleum proponent rather than the licensing out of rights to exploit. The state, without investing its own funds into the prospecting, exploration, and extraction of mineral resources and without bearing any commercial risks receives a substantial part of the product produced by the petroleum proponent.[27] Sometimes, to encourage exploration and development in uncertain and unproved areas, the government may agree to a much smaller proportion of the profit.

The petroleum proponent, as investor, must perform the activities specified in the agreement – that is, exploration, extraction and infrastructure creation – at its own expense and risk. Hence, if the project turns out to be economically unprofitable, the underlying legal assumption of the PSC is that expended funds may not be refunded. To determine the volume of the extracted resources and to carry out production sharing, the concept of the 'point of measurement' is used. This represents an arbitrary point that relates to the movement of extracted resources – that is, the mouth of the shaft or the delivery point. At the point of

---

25 E E Smith, 'From Concessions to Service Contracts' (1992) 27 *Tulsa Law Journal* 493, 515–19.
26 See the discussion by J P Williams, 'Global Trends and Tribulations in Mining Regulation' above n 7, 421 noting that national governments have become more active in 'directing the way in which mining companies affect the local economy of the host country'.
27 See Adepetun, above n 23, 25.

measurement, all extracted resources will continue to remain the property of the state.[28]

The expenditures of petroleum proponents may be taken from the product that is produced (cost recovery product). The remainder of the produced product (profit product) is divided between the government and the petroleum proponent in a proportion consistent with the terms provided for in the PSC. The petroleum proponent acquires ownership rights to the cost-recovery product and its part of the profit production at the point of measurement. The state acquires rights to the product, but the petroleum proponent may, in its place, transfer value.

An important requirement in the PSC is that the government must transfer access to the subsoil in the area specified in the agreement for use by the mining proponent. The issuance of an exclusive right of usage means that the government must abstain from any activity that interferes with the mining project. It does not, however, mean that ownership of the subsoil or ownership of the extracted mineral is transferred over to the mining proponent.[29]

Whilst the government is a party to a PSC, it retains its usual prerogatives and therefore must continue to carry out its public functions. Hence, outside the framework of the PSC, the government will continue to make decisions concerning subsoil usage as an authoritative and sovereign regulator.[30]

For the duration of the PSC, the existing tax framework is replaced by the terms and conditions relating to profit sharing that are set out in the PSC. The idea is that the PSC encourages overall 'production sharing' and that the shared profits represents an alternative revenue framework to the conventional royalty system that applies to the concession agreement. The replacement may be partial or complete in the sense that taxes will be fully replaced by profit product. In Libya, for example, all taxes are met out of the petroleum proponent's share of production and there are no withholding taxes on the distribution of profits.[31] The replacement may also be partial in the sense that some taxes may be levied during the currency of the PSC. In Indonesia, for example, income tax and dividend tax are imposed in addition to the profit sharing provisions of the PSC arrangement.

---

28 For a discussion on the nature of the point of measurement see H Devold *Oil and Gas Production Handbook* (ABB Oil and Gas, 2010) 2.2.5.
29 See Adepetun above n 23, 26. See also R Fabrikant, 'Production Sharing Agreements in the Indonesian Petroleum Industry' (1975) 16 *Harvard International Law Journal* 303, 336–7 where the author describes the fact that the sovereignty of the government does not preclude the assumption that contractors assume they will not be dislodged as long as they faithfully perform their agreed contractual duties.
30 See E E Smith, 'From Concessions to Service Contracts' (1992) 27 *Tulsa Law Journal* 493, 515.
31 See International Tax & Investment Center, *Petroleum Taxation in Libya* (May 2005) http://www.iticnet.org/file/document/watch/1632.

10.2.3 Risk service contracts

As with the profit sharing contract, the underlying impetus for entering into a risk service contract is to generate participation by the host country in the profit flowing from the exploitation of resources contained within its jurisdiction. A further similarity that the risk service contract shares with the profit sharing contract is that it is generally favoured by petroleum proponents. These similarities aside, the structure and fiscal arrangements of the risk service contract differ substantially from the PSC. A risk service contract involves an international oil company (petroleum proponent) supplying services and expertise – for example, technical, financial, managerial or commercial services – to the state, for all stages of the project ranging from exploration through to production and, sometimes, marketing phases. This service and expertise is provided in direct exchange for an agreed fixed fee or some other form of compensation.

The petroleum proponent who enters the risk service contract bears all of the financial costs associated with exploration. In return, where exploration efforts are successful, the government allows the contractor to recover incurred costs through the sale of the oil or gas and will pay the petroleum proponent an agreed fee which is calculated on a percentage of the remaining revenues. The fee which is payable may be subject to taxes. Hence, this contractual arrangement means that the petroleum proponent bears all the risks – especially exploration risks – and is compensated when a commercial discovery is made. The petroleum proponent is entitled to a share of the profits but not a share of the production.

Ownership of the oil produced – that is, ownership of the petroleum – and major installations (except where leased) remain with the state. In some cases, the petroleum proponent may negotiate an option to buy oil back at world prices. Payments to the petroleum proponent under a risk service contract will generally be made in oil. As such the risk service contract is often a more appropriate contractual arrangement in marginal oil fields because the focus is upon the maximisation of oil production and recovery rates and the revenue framework is consistent with this focus. The contracting petroleum proponent will generally receive a fee payment commencing from first production and continuing throughout the duration of the contract. One of the major difficulties with the risk service contract is that upfront capital investment in infrastructure may not be as extensive, and this can generate expensive operational limitations.

10.2.4 Joint venture agreements

Parties involved in a major and collaborative mining or energy project may enter into a mutual and reciprocal agreement in order to cooperate and share the financial responsibilities associated with the exploration and production of resources.

The most common legal arrangement for such a collaborative alliance is a joint venture. In this context, the joint venture is a contractual, unincorporated, non-partnership arrangement.[32]

A mining agreement that amounts to a joint venture differs fundamentally from a standard concession agreement because it results in the state being involved in a common undertaking with the developers to produce minerals or energy. Entering into a concession agreement will not automatically give rise to a partnership between the state and the mining proponents as this type of legal relationship must be agreed upon by an express term within the contract. A joint venture agreement will generally only have an application to a specific project or development and the parties will enter into a joint operating agreement.[33] A joint venture agreement, like any mining partnership, will require cooperation amongst the parties. Usually, an operator – a joint venture partner who is engaged in the business of extracting the resource – will invest funds, actively develop market production, and partner with a non-operator. A non-operator is a joint venture partner who is not actively engaged in the business of extracting the resource and is purely an equity partner and therefore only contributes money and acquires a proportionate share in the outcome.[34]

The definition of a joint venture was articulated by the High Court in *United Dominions Corporation v Brian* where Mason and Brennan JJ stated:

> The term 'joint venture' is not a technical one with a defined common law meaning. As a matter of ordinary language, it connotes an association of persons for the purposes of a particular trading, commercial, mining or other financial undertaking or endeavor with a view to mutual profit, with each participant usually, but not necessarily, contributing money, property or skill.[35]

The joint venture relationship will be subject to the terms and conditions contained in the mining agreement. Generally, a joint venture arrangement will be conducive in a mining context where the parties require an additional investor to provide capital and the pooling of funds, property, and expertise helps to ensure the success of the project. A joint venture agreement can also help to minimise a range of risks and costs associated with large mining projects because they can facilitate collaboration between those parties with specific mining expertise and those parties

---

32 See Fitzgerald above n 10, 20. See also J D Merrells, 'Mining and Petroleum Joint Ventures in Australia: Some Basic Legal Concepts' (1981) 3 *Australian Mining and Petroleum Journal* 1, 2–3.
33 See the discussion by E Kuntz, *A Treatise on the Law of Oil and Gas* (Lexis Nexis, 1989) §19A.8.
34 See the outline by C J Meyers, *Manual of Oil and Gas Terms: Annotated Manual of Legal Engineering Tax Words and Phrases* (Lexis Nexis, 15th ed, 2012) 662. See also L E Schroeder, *Oil and Gas Law: A Legal Research Guide* (W S Hein, 2012) ch 5.
35 (1985) 157 CLR 1, [5] (Mason and Brennan JJ).

whose only contribution is funding. The risks associated with a large mining project are spread and joint operations help to generate good economies of scale. State participation with foreign investors can be advantageous where the government is stable and capable of providing strong financial backing. There are, however, some disadvantages associated with a joint venture between state and private parties. These stem largely from the potential incongruity that can arise between the public goals of the state and the profit making interests of the foreign investor.[36]

A joint venture mining agreement will generally attract fiduciary responsibilities between the parties, including between operating and non-operating parties, but only to the extent that the parties are entrusted with the power and authority to act on behalf of the other joint venture partners. As outlined by Beech J in *Red Hill Iron Ltd v API Management Pty Ltd*:

> The relationship will be fiduciary to the extent, and only to the extent, that the fiduciary has agreed or undertaken to exercise powers or discretions for the principal, or, in the case of a horizontal relationship, for the parties jointly.[37]

The operator's responsibility to take appropriate care in the management and operation of the project is crucial because of the dependency of the non-operating partner. Non-operating parties will often contribute enormous funding but must be fully prepared to put their trust in the skill and expertise of the operating party to properly develop the mining project. The economic framework for mining development relies heavily upon this reciprocal trust because non-operating parties will only be prepared to enter into complex and difficult to monitor joint venture arrangements in the mining and energy sector where they have full knowledge and expectation of the reliable performance of complex expert tasks. The steadfast and dependable implementation of reciprocal obligations of trust and responsibility, through the mechanism of a contractual arrangement, is a crucial component of a successful mining project.[38] The additional application of equitable fiduciary obligations is also important in this context, not just because it provides a panoply of remedies that extend well beyond compensatory damages for breach of contract to include a disgorgement

---

36 See M Wallace, 'Joint Venture as an Institution for Development: Legislative History' (1978) *Arizona State Law Journal* 173, 176.
37 [2012] WASC 323, [365]–[380]. See also *Hospital Products v United States Surgical Corporation* (1986) 156 CLR 41, 96–7; *Pilmer v The Duke Group Ltd* (2001) 207 CLR 165, 196–7 and *Grimaldi v Chameleon Mining NL (No 2)* [2012] FCAFC 6, [179].
38 See M Friedman, *Capitalism and Freedom* (University of Chicago Press, 1962), 13–15. See also the discussion by K N Llewellyn, 'The Effect of Legal Institutions on Economics' (1925) 15 *American Economic Review* 665, 678–80. More recently, see J Blocher, 'Institutions in the Marketplace of Ideas' (2008) 57 *Duke Law Journal* 821, 830 outlining the shortcomings of an idealised view of an uninhibited, costless and perfectly efficient free market.

of profit for a conflict of interest, but more fundamentally because the fiduciary principle seeks to hold an operating party to a high duty of responsibility and disclosure.

The operating party must make sure that they make full disclosure and provide the non-operating party (principal) with all relevant information that may affect their interests.[39] One of the core fiduciary duties that applies to mining and energy projects is the duty of loyalty regarding the use of joint information within the geographic area covered by the joint agreement. The parties come together to jointly develop a mining project for their mutual benefit. The operator is in a strong position because it is generally the recipient of detailed geologic information regarding the character and size of potential resources and the nature and dimensions of the site. It is imperative that the operating party be subject to strict duties that preclude them exploiting this ascendant position. As classically stated by Justice Cardozo in this context, the fiduciary 'rule of undivided loyalty is relentless and supreme'.[40]

It may often be necessary for joint venture partners to enter into a further confidentiality agreement in circumstances where the prospect of a disclosure may jeopardise the commercial viability of a project. Given the risks involved, particularly for a non-operating party, where a joint venture is proposed it is usually appropriate to require parties to enter into a confidentiality agreement to ensure that all information acquired during the course of the venture is treated as confidential and may therefore only be disclosed in accordance with permitted uses that have been mutually agreed upon by the parties.[41]

10.3 The core elements of a mining agreement

A mining agreement grants exclusive rights to the mining or petroleum proponent to explore, search and drill, produce, and commercialise the resource for a specified

---

39 See the discussion by L P Hendrix and S L T Golding, 'The Standard of Care in the Operation of Oil and Gas Properties: Does the Operator Owe a Fiduciary Duty to the Non-Operators?' (1993) 44 *Institute of Oil And Gas Law and Taxation* 10–1.
40 *Meinhard v Salmon* (1928) 164 NE 545, 548. See also H R Williams and C J Meyers in P H Martin and B M Kramer (eds) *Oil and Gas Law* (Lexis Nexis, 2004) §437.1.
41 See the discussion by W Kirk, 'Resource Contracts – Confidentiality Agreements' (1993) *Australian Mining and Petroleum Law Association Yearbook* 194, 195. See also P Armitage, 'Confidentiality Agreements: A Necessary Inconvenience?' [1998] *AMPLA Yearbook* 100, 115–17; D A McPherson, 'Confidentiality and the Public Domain – a Lesson for the Canadian Mining Industry' (1995) 13(2) *Journal of Energy Natural Resources Law* 61; P W Machin, 'Areas of Mutual Interest' [1993] *AMPLA Yearbook* 221.

period of years. The essential feature of a mining agreement in Australia is that it involves an undertaking by the state to grant rights to exploit the resource in return for the commitment by a mining or petroleum proponent to undertake and finance the project and to pay royalties.

Under a concession agreement, a contract is entered into in addition to the issuance of a licence or lease (the different forms having been outlined in Chapter 2). The contract will articulate the rights and obligations of both parties throughout the life of a mining project although many of these rights and obligations may be contingent upon a mining or petroleum proponent demonstrating a capacity to finance a project and also to submit a logical and coherent development proposal. This will generally require a mining or petroleum proponent to:

- prepare a feasibility study to define the project (and provide a detailed outline of issues that need to be covered in a development proposal)
- address relevant environmental and Indigenous issues
- clearly identify the mining proponent and any relevant stakeholders.

Once negotiated and settled, the contract between the government and the mining or petroleum proponent must be ratified by parliament. This ratification is carried out to ensure that the terms of the contract (also known as a 'State Agreement') override any inconsistent provisions that may exist in any other law. In addition to conferring a clear right to mine, the Agreement ensures that all of the requirements of a major mining project may be properly managed within a single enactment.

Whilst a resource title may be issued to correspond with the specific phase of a mining project – that is, exploration, retention, or production – a State Agreement may range across the project holistically, covering all aspects of the project from exploration to production. In this respect, a State Agreement allows mining/petroleum proponents and the government to create a specific legal regime for each particular resource project and, in so doing, overcome difficulties or impediments that may exist in the general regulation.

A State Agreement has the benefit of providing the parties with an integrated framework to manage and monitor the coordination of a mining project. Once executed and ratified, a State Agreement may only be altered by mutual agreement by both the government and the mining/petroleum proponent. The ratification of a State Agreement has the effect of formalising the allocation of responsibilities, providing for certainty and stability of tenure, and precluding any possibility of the government resuming any land that may be subject to the agreement. This type of security is particularly crucial for mining projects as they involve a significant investment in capital expenditure.[42]

---

42 See Fitzgerald, above n 10, ch 9 where the author notes at p.160 that the agreements provide a coherent basis for articulating the relationship between the mining proponent and the state.

State Agreements are highly effective at encouraging economic development and facilitating expenditure on socially beneficial infrastructure that may support a project, such as public roads, improved water, or healthcare facilities. Australia has one of the world's largest economic resources of gold, iron ore, lead, rutile, zircon, nickel, uranium, and zinc. The country also ranks among the top six worldwide for known resources of bauxite, black coal, recoverable brown coal, cobalt, copper, ilmenite, lithium, magnesite, manganese ore, niobium, silver, tantalum, tungsten, and vanadium. The volume of Australia's mineral resources will allow the mining sector to continue as the most important export earning sector of the Australian economy in the foreseeable future.[43] Indeed, Australia derives a greater portion of its wealth from mineral and energy resources than any other country except for Norway.[44] Seen in this context, the economic benefits of ensuring that the parties enter into strong, articulated, and focused State Agreements is unequivocal.

A state concession agreement will generally include a range of indicative terms which include:

- an outline of the mining area as specified by attachment of a geographical map
- ratification and operation of the project upon signing despite any contrary legal enactment
- obligations of the mining proponent. These may include requirements to undertake feasibility studies, environmental assessments, development proposals, compliance with relevant native title laws, provision to the government of evidence of financial capability for the commercialisation of extracted minerals, the obligation of the state to issue a mining licence or lease to support mining operations, and a general commitment by the government to make land available for infrastructure support facilities
- a regime of royalty payments by the mining proponent to the government with the methodology for calculating and paying such royalties clearly detailed. The inclusion of fiscal arrangements within a state agreement is particularly important because provisions dealing with the collection of rental, royalty, and stamp duty can vary markedly
- termination rights and obligations on the occurrence of certain events
- dispute resolution procedures
- term; for example, the State Agreement operates for 50 years from the commencement date

---

43 See the report by Australian Government, Geoscience Australia, *Australia's Identified Mineral Resources 2013* (2013) 3 http://www.ga.gov.au/corporate_data/78988/78988_AIMR_2013.pdf.
44 See the factsheet from Australian Government, Geoscience Australia, *Applying geoscience to Australia's most important challenges* <http://www.ga.gov.au/scientific-topics/minerals/mineral-resources/aimr>. See also A Stoeckel, *Minerals: Our Wealth Down Under* (Centre for International Economics, Canberra, 1999) 2–5.

- any state benefits granted in conjunction with the project; for example, any subsidy, discount, or exemption of the project from stamp duty
- any other applicable state taxes.

One of the fundamental requirements for a State Agreement – prior to any agreement execution, registration or legislative ratification – is the submission by the mining proponent of a development proposal around which the terms and conditions of the State Agreement will be structured. The development proposal functions as a negotiation instrument which, following approval, will form the structural foundation for the subsequent State Agreement. A development proposal, submitted to the relevant minister for a mining project, will generally include details of the arrangements for implementing the project. These might include:

- how the minerals will be mined and recovered
- any physical separation or beneficiation process that the minerals are to be subjected to
- the method by which an extracted resource is to be transported
- how the employees are to be accommodated or any ancillary issues relevant to the operation of the workforce
- utilities supply
- use of local labour, professional services, manufacturers and suppliers
- the method by which residue is to be disposed of
- how a mining proponent will finance a project
- any anticipated further applications for mining leases which may be required for future proposed operations.

The statutory ratification of the mining agreement is an important component of the overall process. State Agreements generally represent long-term resource development contracts. The agreement is signed off by the project developer and the minister responsible for resource development or, alternatively, the Premier on behalf of the state.[45] Following this, the agreement is given legislative ratification as a Schedule to an Act of Parliament. This effectively means that the body of the Agreement is contained in the Schedule rather than the Act, which is itself very short and contains only a few provisions.

Legislative ratification is vital for State Agreements as it ensures that the terms and conditions of the agreement are valid and enforceable and this, in turn, provides certainty for the parties involved.[46] This is crucial for both investor and public

---

45 Fitzgerald, above n 10, 6. The author notes that legislatively ratified agreements are used across a range of resource sectors but their most extensive application applies with respect to mining.
46 See N Seddon, *Government Contracts — Federal, State and Local*, (Federation Press, 4th ed, 2009) [2.2].

confidence. Doubt regarding the power of the state to enter into the contract is removed once the agreement is legislatively ratified. Generally, state governments are capable of entering into contracts in the same way as any private individuals, subject to any constitutional limitations which may exist. As outlined by Dixon J in the High Court in *NSW v Bardolph*, '... the principles of responsible government do not disable the executive from acting without the prior approval of Parliament'.[47] However, because the state is regarded as the custodian of public resources, any commercial activity regarding those resources should always be carefully examined.[48] The ratification of the State Agreement gives it a higher level of public scrutability because it ensures that the agreement is accessible as a schedule to a legislative act.

Commonwealth governments have, however, had their private contracting capacity articulated differently. In *Williams v Commonwealth of Australia*, French CJ confirmed that 'The Commonwealth is not just another legal person like a private corporation or a natural person with contractual capacity'.[49] His Honour held that the Commonwealth was restricted in its capacity to enter into private contracts and that the only areas that were substantively valid were: (i) contracts arising out of a constitutionally valid statute; (ii) contracts arising out of the administration of a department of State, as per s 64 of the *Commonwealth Constitution*; (iii) contracts arising out of the existence of an existing prerogative power; (iv) contracts arising from the exercise of a statutory power, or from executive action to give effect to a statute; and (v) contracts arising out of the implied nationhood power, where such contracts involve a national endeavour and concern the Commonwealth as the national government.[50] Following the decision in *Williams*, the Commonwealth government introduced the amendments to the *Financial Management and Accountability Act 1997* (Cth) which now make it clear that the Commonwealth has the authority to spend, where no other legislative authority exists.[51]

The risk of a State Agreement, which is not authorised by statute, becoming invalid is borne by the private mining or petroleum proponent. As such, the developer and its financiers need to be prudent and should ensure any possible

---

47 (1934) 52 CLR 455, 509 (Dixon J). In *Williams v Commonwealth of Australia* (2012) 248 CLR 156, [79] ('*Williams*') where French CJ noted, however, that the conclusions in *Bardolph* did not involve a consideration of the powers of the executive government of the Commonwealth acting under ss 61 and 64 of the *Constitution*.
48 See N Seddon above n 46, 3.
49 (2012) 248 CLR 156, [38].
50 These categories were outlined by R McLean, 'The Power of the Commonwealth Executive to Contract Post-Williams v Commonwealth of Australia' (2012) 11(2) *Canberra Law Review* 150, 157–8.
51 Schedule 1 of the Bill inserts a new s 32B into the *Financial Management and Accountability Act 1997* (Cth) to provide the requisite statutory authority for Commonwealth spending, where no other legislative authority exists. See McLean, above n 50 158.

issue of invalidity is removed through the process of ratification. Where specific provisions in the state agreement are found to be inconsistent with existing legislation, the terms and conditions of the ratified agreement will override that inconsistency.[52]

10.4 Revenue frameworks

10.4.1 Royalties

A common, although not universal, form of revenue that underpins the concession licence agreement is the royalty. As the state began to assume control of the ownership of mineral resources, mineral and petroleum royalties were introduced and, over time, these royalties became incorporated into the general fiscal regime.[53] It has been argued that the evolution of the mining royalty is symbolic of the willingness of mining and petroleum proponents to 'pay for risk reduction'.[54] The payment of mineral royalties provides the government with a flexible fiscal tool with the capacity to be adapted to meet changing circumstances. Royalty payments have a greater capacity to be channelled into targeted distribution than other revenue structures. This has prompted them to be recalibrated, not merely as fiscal instruments, but as fiscal instruments with an inherent capacity to promote social and strategic change.[55] Royalties are appealing to governments because the revenue may be received once production commences and they are much easier to administer than other more complex fiscal instruments because they ensure that mining and petroleum companies make, at least, a minimum payment in return for extraction and production.[56]

A royalty represents a payment to the owners of a resource – and in Australia that is the state – for the right to sell, dispose of, or use the resources. A royalty is

---

[52] See the discussion by M Crommelin, 'State Agreements: Australian Trends and Experience' [1996] *AMPLA Yearbook* 328, 330. See also M Hunt and M Lewis, *Mining Law in Western Australia* (Federation Press, 1993) 2.8.

[53] See generally J Otto, C Andrews, F Cawood, M Doggett, P Guj, F Stermole, J Stermole and J Tilton, *Mining Royalties: A Global Study Of Their Impact on Investors, Government and Civil Societies* (The World Bank, 2006) 16 noting that the exploitation of a non-renewable resource justifies a royalty, which is described as akin to an 'ownership transfer tax'.

[54] J Otto and J Cordes, *The Regulation of Mineral Enterprises: A Global Perspective on Economics, Law and Policy* (Mineral Law Foundation, 2002) 52–8.

[55] See generally E Bastida, T W Vaelda, J Warden-Fernandez (eds) *International and Comparative Law and Policy: Trends and Prospects* (Kluwer, 2005).

[56] See the discussion by E M Sunley, T Baunsgard, and D Simard, 'Revenue from the Oil and Gas Sector: Issues and Country Experience' (Paper presented at IMF Conference on Fiscal Policy Formulation and Implementation in Oil Producing Countries, Washington, 5–6 June 2002) http://siteresources.worldbank.org/INTTPA/Resources/SunleyPaper.pdf.

payable on the basis that the state retains ownership in all minerals located on or below the surface of land and all petroleum produced to the surface of land or in a natural underground reservoir. The intent of the royalty is to charge the producer of the mineral or petroleum for the right to mine the minerals produced. Mining and petroleum royalties represent revenue that is a direct component of the state agreement and therefore should be distinguished from taxes that are paid on the profits which have been earned from those minerals.[57]

Typically, a royalty will entitle the government to a specified fraction of gross production as soon as production commences. Where the government wishes to encourage exploration and development in an uncertain and unproved area, the government take may be relatively modest. The most common form of royalty is a variable royalty. The variable royalty is usually set as a sliding scale, based on levels of production. The initial royalty may be based on a low percentage of production in order to encourage interest in areas and spur on development. The royalty percentage will then cumulatively increase as production and commercialisation evolves. This system is generally preferable to a fixed royalty that may act as a disincentive to the development of the mining project in the early stages. Sliding scales can also be based on factors other than levels of production. For example, in some countries, the royalty rate can be decreased in circumstances where exploration and development encounters exceptional difficulty.[58]

The collection of mining royalties is based on 'self assessment'. This means it is the responsibility of the mining lease-holder to calculate and promptly lodge all royalty returns. To ensure royalties are calculated and lodged correctly, audits are conducted annually.

One of the key features connected with a royalty framework is to determine whether the system will discriminate between the different types of minerals that exist. Hence, it is necessary to determine whether the royalties that are payable will be uniform or, alternatively, whether the royalties that are payable will depend upon the type of mineral involved. The benefit of specificity lies in the fact that it can be individualised to suit the marketing, physical properties and profitability of the mineral.[59]

In Australia, each state or territory has its own royalty framework and the rate applied will therefore depend upon the mineral involved. In the offshore context,

---

57 Ibid 505–16.
58 The Algerian mineral law regime fixes a base royalty rate of 20 per cent, but reduces the percentage to 16.25 per cent and 12.5 per cent in areas which present exceptional difficulties for exploration and development. See Abu Dhabi Concession, art 13 (1980). This is discussed by N E Terki, 'Comment: The Algerian Act of 1986 and the Encouragement of Foreign Investment in the Area of Hydrocarbons' (1988) 3 *Oil & Gas Law Taxation Revenue* 80.
59 J Otto, C Andrews, F Cawood, M Doggett, P Guj, F Stermole, J Stermole and J Tilton, above n 53, 43.

royalties are shared between the Commonwealth and state governments. Oil production had been taxed on the basis of a value-based excise although this framework was largely replaced with the implementation of a petroleum resource rent tax (PRRT) which is imposed at the rate of 40 per cent of profit and will only arise when the oil producer has recouped all its exploration and development costs and has started to obtain an economic return on their investment.[60]

Most mining royalties are imposed by reference to a fixed amount per tonne or a percentage of the value of the mineral production. In Australia, mining royalties are payable on minerals and the rate and method of determining the royalty will therefore necessarily depend upon the category of mineral to which they apply. There are three primary categories:

- non-coal royalties
- coal royalties
- petroleum royalties.

There are two primary methods for determining how mining royalties are to be calculated. The first method is known as the 'quantum' method. Pursuant to this method the royalty is levied at a flat rate per unit of quantity. Hence, the amount of royalty that is paid depends entirely on the amount of mineral that is extracted. A fixed quantum royalty is applied to a physical rather than a financial base and provisions may be incorporated for progressive adjustments of the fixed royalty rate in accordance with inflation or changes in commodity prices. This type of royalty generates a stable revenue income and it is administratively efficient and easy to audit. However, it can also be economically inefficient and distortionary especially in circumstances where the value of the resource increases over time. For this reason, a quantum royalty is generally only applied to low-value to volume resources such as gypsum, limestone and clays.[61]

The second method of calculating a royalty is known as the 'ad valorem' method. Ad valorem royalties are levied as a percentage of the total value of minerals recovered or the 'ex-mine' value. The ex-mine value refers to the value of the mineral once it is mined and brought to the surface. In its simplest form, an ad valorem royalty consists of a uniform percentage (the rate) of the value (the base) of the mineral(s) in the products sold by the miner. An ad valorem royalty is applied to high-value to volume minerals.[62]

---

60 See the discussion by P Broek, 'Australia Prospects for Tax-Efficient Mineral Wealth' (1992) *International Tax Review* 13.
61 See the outline by J Otto, 'Foreword' in *The Taxation of Mineral Enterprises* (Kluwer, London, 1995) xiii.
62 See J Otto et al, *Mining Royalties: A Global Study of Their Impact on Investors, Government, and Civil Society* (The World Bank, 2006) 11.

Ad valorem royalties can be levied on two possible bases:

(i) **Realised value of sales**: This represents the value set out upon the actual invoice; that is, the actual sale price. The realised value is clearly articulated upon the invoice making all subsequent audits much easier to conduct, reducing administrative costs, and minimising disputes. Gains and losses associated with extraction and production will be automatically incorporated into the realised value.

(ii) **Gross value**: The gross value of the resource is obtained by determining the value of the resource by reference to a quoted market price on the day of the sale and then multiplying this by reference to the amount involved. Auditing of this type of royalty assessment can be complex as it requires verification of specific quantities and grades of product sold.[63]

The ad valorem royalty is the most common form of royalty imposed in the concession licensing agreement. The primary reason for this lies in its consistency and simplicity because, whilst the mine is operative, the royalty is payable. The ad valorem royalty applies to the value base regardless of how that base is defined and irrespective of the nature of the product sold. This approach is inequitable where products have had value added and creates a disincentive to invest in downstream processing.[64] Hence, many jurisdictions apply progressively lower royalty rates as the product evolves; for example, from crude ore to metal.

10.4.2 Royalty rates and mining taxes in the Australian context

The royalty and taxation framework in Australia is structured to respond directly to the nature, scope, and economic potential of our natural resources. Australia is a resource rich country. This is particularly true for its minerals. As previously outlined, Australia has large reserves of brown coal, mineral sands (rutile and zircon), nickel, silver, uranium, zinc, and lead. It also has significant reserves of bauxite, copper, gold, iron ore, niobium, tantalum, and manganese ore, as well as black coal, industrial diamonds, ilmenite, lithium, vanadium, and antimony.[65] Australia is

---

63 See the discussion by P Guj, *Mineral Royalties and other Mining-Specific Taxes* (International Mining for Development Centre, University of Western Australia and the University of Queensland, 2012) 4 http://im4dc.org/wp-content/uploads/2012/01/UWA_1698_Paper-01_-Mineral-royalties-other-mining-specific-taxes1.pdf.
64 Ibid.
65 See Australian Government, Geoscience Australia, *Australia's Identified Mineral Resources* (2013) 4–8 http://www.ga.gov.au/corporate_data/78988/78988_AIMR_2013.pdf. The report notes at p.4 that 'Australia's continuing position as a premier mineral producer is dependent on continuing investment in exploration to locate high quality resources and/or upgrade known deposits to make them competitive on the world market, as well as investment in beneficiation processes to improve metallurgical recoveries'.

also the third largest producer of uranium, although production has experienced a slight downturn with the increasing costs associated with mining and ore drilling.[66]

Iron ore and coal account for approximately 56 per cent of total mineral production. Western Australia produces approximately 97 per cent of the iron ore output, while Queensland and New South Wales jointly account, in roughly equal proportions, for 97 per cent of coal production. In addition, WA is a large producer of gold, nickel, alumina, and mineral sands; Queensland of alumina, gold, and base metals; South Australia of copper, gold, and uranium; and the Northern Territory of manganese, alumina, and uranium.[67]

Australia's energy industry is also in the initial stages of a new era, with multibillion dollar investments being directed at the development of coal seam gas into LNG. The Gorgon Project in Western Australia is one of the world's largest natural gas projects and the largest single resource development in Australia's history. It is located within the Greater Gorgon area, between 130 and 220 km off the northwest coast of Western Australia. It includes the construction of a 15.6 million tonne per annum (MTPA) liquefied natural gas (LNG) plant on Barrow Island and a domestic gas plant with the capacity to supply 300 terajoules of gas per day to WA.[68]

The Gorgon Project is also investing approximately $2 billion in the design and construction of the world's largest commercial-scale CO_2 injection facility to reduce the project's overall greenhouse gas emissions by between 3.4 and 4.1 million tonnes per year. The Australian Government has committed $60 million to the Gorgon Carbon Dioxide Injection Project as part of the Low Emissions Technology Demonstration Fund.[69]

---

[66] Ibid 143 where the report notes that in 2013 there were 435 operable commercial nuclear power reactors in 31 countries that produced approximately 11 per cent of the world's electricity in 2012. Additionally, there are a total of 167 reactors in the planned category, including 65 under construction. In 2012, China resumed approval of new reactor constructions after completion of safety reviews. India, the Russian Federation, and South Korea are expanding their nuclear generating capacity and several non-nuclear countries are moving ahead with reactor construction programs or are considering building nuclear power plants in the future. The United Arab Emirates recently commenced its first nuclear power plant. Saudi Arabia, Vietnam, Bangladesh, Poland, Turkey, and Belarus are proposing to proceed with nuclear power development. Hence, whilst the Fukushima incident has affected nuclear power projects and policies in some countries, nuclear power remains a key part of the global electricity mix.

[67] Ibid.

[68] For a full discussion on the scope and nature of the Gorgon Project on Barrow Island see Chevron Australia, *Gorgon Project* http://www.gorgon.com.au. For a detailed outline of the progress of the Gorgon Project, see Chevron, *Australia: Project Progress* (December 2014) http://www.chevron.com/countries/australia/businessportfolio/projectprogress/.

[69] For a full outline of the CCS capabilities of the Gorgon Project see Global CCS Institute, *The Global Status of CCS* (November 2014) http://decarboni.se/publications/global-status-ccs-2014/33-geographical-trends-large-scale-ccs-project. The report notes that it is expected that over 100 million tonnes of CO_2 will be injected over the life of the project.

To date, iron ore and coal have dominated Australia's exports. Given the high value of these commodities they have attracted focused revenue regulation. It is estimated that in 2010–2011 total Australian mining taxation amounted to A$23.4 billion, and this includes AUD$14.6 billion in corporate income tax and A$8.8 billion from mineral royalties, most of which is attributable to iron ore and coal.[70] In the future, a greater percentage of revenue is likely to stem from LNG production and export with export revenues predicted to rise by 60 per cent over the next five years and grow at an annual rate of 7 per cent to reach a predicted A$284 billion by the end of 2017.[71]

The framework for mining royalties in Australia is sourced in our core ownership structure. The vesting of resources in the state and territory governments allows the government to license out the right to extract and produce those resources to third party mining and petroleum proponents in return for payment. Royalties are payments made to the relevant state or territory government, as the owner of the mineral, in consideration of a right granted to extract and remove minerals. Royalties are calculated in respect of the profit derived from the minerals or petroleum taken or produced as a result of the permissions which have been granted. Royalties are the most intuitive form of revenue because of the incremental nature of the extraction process associated with fossil fuel projects. Royalties must be distinguished from taxes which are payments made on revenue generated from business activities and are not based upon any permissory exchange.[72]

In most instances, royalties will be payable on either an ad valorem or a quantum basis. This will depend upon the particular state or territory and the nature of the mineral. In the Northern Territory, a profit-based royalty regime applies. According to this framework, the net value of a mine's production is used to calculate the royalty that is payable. Many small mining operations are effectively exempted from this regime because no liability applies to the first $50 000 of net value. The advantage of a profit-based royalty framework is that it has increased allocative efficiency. However, this type of royalty regime may also create uncertainties in revenue stream and further, the cost of compliance for both the mining proponent and for government can be quite high.[73]

---

70 See Deloitte Access Economics, 'Estimated Companies tax, MRRT and Royalties expenses for the Mineral Sector' (Report Prepared for the Mineral Council, 30 July 2014) 2, 6 http://www.minerals.org.au/file_upload/files/reports/DAE__MCA_Royalty_and_company_tax_estimates_30_July_%281%29.pdf. The report notes at p.6 that estimated iron ore royalties have been reduced in 2013–14 by some $200 million below the projections in the WA budget. This latter adjustment has been adopted in light of the recent sharp falls in spot iron ore prices.
71 Bureau of Resources and Energy Economics, *Australian Energy Resource Assessment* (2nd ed, 2014) 83.
72 See the definition of a royalty in *Mineral Royalty Act 1982* (NT) s 4.
73 See Guj, above n 63, 4.

10.4.2.1 Coal royalty rates

In New South Wales, the coal ad valorem royalty rates are 6.2 per cent for deep underground mines (coal extracted below 400 m), 7.2 per cent for underground mines and 8.2 per cent for open cut mines. In Queensland, the ad valorem royalty rates are 7 per cent where the value of the coal produced does not exceed $100/tonne and 10 per cent on the value of the coal where it does exceed $100/tonne. In South Australia, a royalty rate of 3.5 per cent applies to all mineral leases, except those with a new mine status. Newly approved mining leases can apply for new mine status and this entitles them to a reduced royalty rate of 1.5 per cent for five years. In Victoria, brown coal royalties are calculated on a quantum rate at $0.0588 per GJ, adjusted in accordance with the consumer price index. In Western Australia, if the coal is exported the rate is 7.5 per cent ad valorem. If it is not exported, a quantum $1/tonne, adjusted each year at 30 June is applied in accordance with comparative price increases.[74]

10.4.2.2 Iron ore royalty rates

In Western Australia, beneficiated iron ore – that is, refined iron ore where the waste materials are separated out – is 5 per cent ad valorem. Fine ore is 5.625 per cent and lump ore is 7.5 per cent. In Queensland, iron ore royalties are 2.7 per cent ad valorem with a discount of 20 per cent if the iron ore is processed in Queensland and the metal content is at least 95 per cent. In New South Wales, the royalty rate is 4 per cent, in Victoria 2.75 per cent, and in South Australia it is 3.5 per cent.[75]

10.4.2.3 Petroleum royalty rates

In Western Australia, onshore petroleum is 5–10 per cent of the well-head value for primary licences and 10–12.5 per cent for secondary licences. Offshore petroleum is 11–12.5 per cent of well-head value. In New South Wales, onshore petroleum is nil for the first 5 years and it increases to 10 per cent of well-head value at the

---

74 See *Mineral Resources (Sustainable Development) Act 1990* (Vic) s 12; *Mineral Resources (Sustainable Development) (Minerals Industries) Regulations 2013* (Vic) ss 6–10; *Mining Act 1992* (NSW) ss 291, 291A, 292; *Mining Regulation 2010* (NSW) regs 61A–65; *Mineral Resources Act 1989* (Qld) ss 320–5; *Mineral Resources Regulation* pt 3, div 1; *Mining Act 1978* (WA) ss 109, 109A; *Mining Regulations 1981*, div 5; *Mineral Resources Development Act 1995* (Tas), div 5; *Mineral Resources Regulations 2006* pt 3; *Mining Act 1971* (SA) pt 3; *Mining Regulations 2011* pt 2; *Mineral Royalty Act 1982* (NT) pt II.

75 See *Mineral Resources (Sustainable Development) Act 1990* (Vic) s 12; *Mineral Resources (Sustainable Development) (Minerals Industries) Regulations 2013* regs 6–10; *Mining Act 1992* (NSW) ss 291, 291A, 292; *Mining Regulation 2010* (NSW) regs 61A–65; *Mineral Resources Act 1989* (Qld) ss 320–5; *Mineral Resources Regulation* pt 3, div 1; *Mining Act 1978* (WA) ss 109, 109A; *Mining Regulations 1981*, div 5; *Mining Act 1971* (SA) pt 3; *Mining Regulations 2011* pt 2.

end of the tenth year in relation to offshore petroleum. In Queensland, Victoria, and South Australia it is 10 per cent of well-head value. Coal seam gas is charged at the same royalty rate as petroleum in Queensland, New South Wales, Western Australia, and South Australia. In Victoria, it is 2.75 per cent ad valorem. In Queensland, exemptions apply to flared or vented coal seam gas, incidental coal seam gas mined under a mining lease, and coal seam gas mined under a mineral hydrocarbon mining lease.[76]

Offshore petroleum production is also regulated by the Commonwealth in circumstances where the offshore production project exists beyond the territorial seas and within areas of Commonwealth jurisdiction (this is outlined in Chapter 3). Royalties for petroleum production are governed by the provisions set out in the *Offshore Petroleum Greenhouse Storage Act 2006* and the *Offshore Petroleum (Royalty) Act 2006* (Cth).[77] The provisions in the *Offshore Petroleum (Royalty) Act 2006* (Cth) make it clear that offshore petroleum royalty rates must be determined according to the type of licence that is issued. The royalty rate for a petroleum production licence is 10 per cent with a similar rate applicable to renewed licences.[78] The royalty rate for a petroleum exploration licence and a petroleum retention licence is 10 per cent of the value at the well-head.[79] There is provision for a reduction in royalties payable in circumstances where the rate of recovery from a well has become so reduced that further recovery would be uneconomic.[80] There are also provisions for exemptions. Royalties are not payable where petroleum is unavoidably lost before it is ascertained, or in circumstances where it is used for petroleum exploration activities – for example, flared or vented – nor is it payable if returned to a natural reservoir.[81]

10.4.2.4 Privately owned minerals

The public ownership system for minerals and resources justifies the imposition of royalties and revenue payments for minerals that reside in private land because those minerals no longer belong to the surface estate owner. The rationale being that the state and territories own the resources for the benefit of the public at large

---

76 *Petroleum and Geothermal Energy Act 2000* (SA) s 43; *Petroleum Act 1998* (Vic) ss 149–50; *Petroleum (Onshore) Act 1991* (NSW) s 85(2); *Petroleum (Onshore) Regulation 2007* (NSW) reg 23; *Mineral Resources Development Act 1995* (Tas) s 102; *Mineral Resources Regulations 2006* (Tas) reg 7, sch 1 ('well head'); *Petroleum Act 1923* (Qld) s 49 ('wellhead'); *Petroleum and Gas (Production and Safety) Act 2004* (Qld) s 155(1), ch 6; *Petroleum and Geothermal Energy Resources Act 1967* (WA) ss 52, 142; *Offshore Petroleum (Royalty) Act 2006* (Cth) s 5.
77 See *Offshore Petroleum And Greenhouse Storage Act 2006* (Cth) ss 631–5; *Offshore Petroleum (Royalty) Act 2006* (Cth) ss 6–10
78 See *Offshore Petroleum (Royalty) Act 2006* (Cth) s 6.
79 Ibid ss 7, 8.
80 Ibid s 9.
81 Ibid s 10.

and therefore retain the capacity to exploit those resources for public benefit. Public ownership of natural resources represents a fundamental shift in cultural and historical perspectives toward resource management. Historically, landowners in Australia have enjoyed the unlimited benefits of mineral ownership. This freedom has, however, been gradually eroded over time, with the state vesting retrospective ownership in minerals in order to 'chip away at the full panoply of such power'.[82]

In New South Wales, private ownership of minerals will only exist if the Crown grant is dated prior to the *Crown Lands Act 1884* (NSW) ('the *1884 Act*') and minerals were not expressly reserved prior to that date by the terms of that express grant. A general policy in favour of mineral reservation to the Crown was introduced by the *1884 Act*. This Act states that all grants of land issued pursuant to its terms must contain a reservation of all minerals. The *1884 Act* has since been replaced by the *Crown Lands Consolidation Act 1913* (NSW), which continues the prospective policy in favour of a reservation of all minerals to the Crown. There is, however, no future capacity to extend private ownership of minerals. In other Australian jurisdictions, legislation has been passed conferring ownership of minerals in private land to the state and deeming that ownership to operate retrospectively.[83]

In all states, the holder of a mining lease is liable to pay a royalty to the Crown on privately owned minerals recovered from the mining area as if those minerals were publicly owned. Where a royalty is collected on privately owned minerals, legislation in all states require that seven-eights or 87.5 per cent of the royalty collected be paid to the private mineral owner, with the rest being paid to the state.[84] Mineral ownership may be determined by a search of the mineral reservation which is shown on the original land grant of the particular land title at the Land Titles Office. It may also be ascertained from any subsequent interpretation of the information which may be gained from that search.

10.4.2.5 Meaning of a well-head

The assessment of petroleum royalties is determined by reference to the value at the 'well-head'. A well-head is, essentially, the valve which exists at the surface of

---

[82] See the discussion by P Babie, 'Sovereignty as Governance: An Organising Theme for Australian Property Law' (2013) 36 *University of New South Wales Law Journal* 1075, 1103. See also A Cox, 'Land Access for Mineral Development in Australia' in R G Eggert (ed), *Mining and the Environment: International Perspectives on Public Policy* (Resources for the Future, Washington DC, 1994) 21.

[83] In South Australia, all privately owned minerals reverted to the Crown on 3 July 1972 by virtue of the *Mining Act 1971* (SA) s 16. In Victoria, all privately owned minerals reverted to the Crown on 30 October 1985 by virtue of the *Mines (Amendment) Act 1983* (Vic); now see *Mineral Resources (Sustainable Development) Act 1990* (Vic) s 9. In the Northern Territory, all rights to minerals were vested in the Crown by virtue of the *Minerals (Acquisition) Act* (NT) s 3; now see the *Northern Territory (Self-Government) Act 1978* (Cth) s 69(4).

[84] See *Mining Act 1992* (NSW) s 284.

an oil or gas well which provides the structure and pressure interface for any drilling or production equipment. The meaning of a 'well-head' has, however, been the subject of some judicial deliberation and was considered by the High Court in *BHP Petroleum Pty Ltd v Balfour* (1987).[85]

> **Facts:** This case concerned the extent of the liability of BHP Petroleum Pty Ltd and its other joint venture partner to pay a royalty for petroleum production from their Cobia No 2 Well in Bass Strait. That well, at seabed level, was fitted with what was known as a 'Christmas tree' assembly and which was designed to control the production flow from the well. Flowlines were connected to this assembly in order to deliver oil 4 km to the joint venturers' Mackeral A platform where it was then to be mixed with oil from other wells and pumped to the mainland for processing. The royalty for the petroleum was calculated by reference to what was then s 5 of the *Petroleum (Submerged Lands) (Royalty) Act 1967* (Cth). Section 5, when read together with s 42 of the *Petroleum (Submerged Lands) Act 1967* (Cth) effectively set out that the value at the well-head of any petroleum was the amount agreed upon between the licensee and the Designated Authority, or, in default of agreement, the amount determined by the Designated Authority as being that value. The quantity of petroleum recovered during a period was the quantity measured during that period by the measuring device which was approved by the Designated Authority and installed at the well-head. In the absence of agreement as to which 'valve station' is to be taken as the well-head, the Designated Authority has the power to determine the issue. A 'well' was defined to be the hole drilled in the seabed rather than any of the equipment affixed to that hole for the purposed of extracting the petroleum. A 'valve station' was defined to mean 'equipment for regulating the flow of petroleum'.
>
> At all relevant times, the Designated Authority in Victoria was the Victorian Minister for Mines. The joint venturers and the Minister for Mines could not agree on the appropriate location of the well-head of the particular well in issue. In January 1980, when no agreement had been reached, the Designated Authority, purporting to exercise his power under s 8 of the Royalty Act, determined certain valves on the Mackeral A platform be the well-head for the purposes of the royalty calculations. This decision resulted in the joint venturers being liable to pay more in royalties than they felt was required under the legislation. The reason for this was because the value of the petroleum at the valve station selected by the Designated Authority was higher than its value at the valve station on the seabed. The primary reason for this increase was because the maintenance costs of the pipeline to the Mackeral A platform added to that value.
>
> **Held:** The plaintiffs were successful at first instance but an appeal to the Full Court of the Supreme Court of Victoria was allowed. Nicholson J, with whom Murray J agreed, held that the purpose of the legislation was to 'provide for the payment of royalties to the Crown for

---

85 180 CLR 474 (*'Balfour'*).

> the right to extract and take possession' of the petroleum and therefore all that the Royalty Act determines is the point at which the mineral should be treated as being 'recovered' for the purpose of assessing the royalty. His Honour therefore felt that there was nothing wrong with legislation conferring a right upon the Designated Authority to determine the final point at which recovery occurred for the purpose of such assessment. Nicholson J agreed with the argument put forward by the Designated Authority which held that the word 'well-head' was, in this context, to be determined entirely from the legislation, in particular s 8 of the Royalty Act which made it clear that in the absence of any agreement the Designated Authority had the power to determine which valves would constitute the well-head.

On appeal, the High Court unanimously reinstated the decision of the trial judge. Mason CJ, Brennan, Deane, Toohey and Gaudron JJ held that the consequence of upholding the interpretation of the legislative framework put forward by the Designated Authority was to confer upon the Designated Authority an arbitrary power of determination regarding the well-head. The Court held that this could not have been the intention of parliament because it was clear from the legislative framework that the royalty was to be determined objectively. The sections raised by the Designated Authority were purely designed to permit agreement if it could be reached and, in the absence of agreement, to provide the means whereby the Designated Authority was permitted 'to fix upon a valve station which fairly accords with the description of well-head'.[86] Their Honours felt that this had not occurred on the facts because the valve station did not fairly come within the description of a well-head. Determining a valve for the purpose of calculating the royalty was not the way to determine which valve constituted the well-head. Their Honours held that the proper principle governing this situation was articulated by the dictum of Lord Diplock in *In re Racal Communications Ltd* which set out the following:

> ... where Parliament confers on an administrative tribunal or authority, as distinct from a court of law, power to decide particular questions defined by the Act conferring the power, Parliament intends to confine that power to answering the matter for courts of law to resolve in fulfillment of their constitutional role as interpreters of the written law and expounders of the common law and rules of equity. So if the administrative tribunal or authority have asked themselves the wrong question and answered that, they have done something that the Act does not empower them to do and their decision is a nullity.[87]

---

86 (1994) 180 CLR 474, [15].
87 [1981] 1 AC 374, 382–3.

The High Court held that the relevant legislation conferred no discretion at all on the Designated Authority.[88] Their Honours felt that the imposition of a royalty and the determination of the amount of royalty needed to be ascertained objectively and should not be open to manipulation by the Designated Authority by imputing legislative discretion. Moloney made the following conclusions on the *Balfour* decision:

> The High Court's reasoning suggests that where the administrative power is central to or involved directly in, the statutory calculation of the amount of a person's 'tax' liability, it is only if the legislation is expressed in clear and unambiguous language that the court will construe that power as confemng an arbitrary discretion. To approach the interpretation of a 'taxing' provision which confers such a power on an administrative official so as to preserve for him a measure of discretion is to stray from the correct path, more particularly where the relevant statutory provision is not cast in discretionary terms.[89]

To avoid this type of difficulty, the definition of a well-head was subsequently explicitly defined in the *Offshore Petroleum (Royalty) Act 2006* (Cth) as follows:

Offshore Petroleum (Royalty) Act 2006 (Cth)

11 Meaning of wellhead

For the purposes of this Act, the wellhead, in relation to any petroleum, is:

(a) such valve station as is agreed between:
   (i) the registered holder of the petroleum exploration permit, petroleum retention lease or petroleum production licence; and
   (ii) the State Minister; or
(b) if there is no agreement within such period as the State Minister allows – such valve station as the State Minister determines to be that wellhead.

The well-head value is derived by taking the gross value of the petroleum recovered and deducting all costs incurred between a defined valve on the Christmas tree and the point of sale. The well-head value system ensures that the petroleum proponent is able to deduct amounts from the gross value, associated with the cost of related well-head production. These deductible costs are generally restricted to expenses leading up to the date of sale that are associated with the processing, storage, and transport of the recovered petroleum. Exploration costs, drilling, recovery, and other inter-related expenses – including any expenses that may accrue from an abandoned drill – will not be deductible. In contrast to the facts of the

---

[88] (1994) 180 CLR 474, [16]–[18].
[89] See G Moloney, 'Case Notes: BHP Petroleum v Balfour' (1987) 16 *Melbourne University Law Review* 436, 440.

Balfour case, in a situation where the parties do enter into a specific agreement regarding the defined location and identity of the well-head, the methodology for calculating well-head value is to be included within a specific royalty schedule that is particular to every individual agreement.[90]

10.4.2.6 Mining taxes in Australia

All taxation frameworks for mining and energy industries must be cognisant of the effect that taxes can have upon the economics of a mining project and, in particular, upon future investment. The immediate fiscal rewards to be gained from the imposition of high levels of tax need to be offset against the benefits to be gained from developing a sustainable long-term mining industry. Overall, the optimum goal is for the revenues generated by royalty and taxation frameworks to contribute to economic growth. This is particularly important in developing world countries where a lack of transparency and accountability can generate corruption and conflict.[91] From the perspective of macro-economic governance, the objective should be to maximise the present value of the social benefits that flow from the mineral sector in order to sustain long-term development. If the taxation framework is too high, investors will shift focus; whereas if the taxation framework is too low, the state will loss the opportunity to gather important revenue that may be used to further develop public welfare.[92]

In Australia, the PRRT, like the MRRT, is a resource-based tax. These taxes seek to tax the profits (also called rents or economic rents) where the sale of resources exceeds the cost of exploration and extraction. Such taxation is, in many ways, a rational form of revenue in a public ownership framework, particularly where monopolistic industries operate which have the capacity to generate super profits. This type of tax also accords with the fundamental requirements of a good and equitable (vertically and horizontally) taxation framework in accordance with the

---

90 See the discussion by Government of Western Australia, Department of Mines and Petroleum, *Petroleum Royalties* (2009) <http://www.dmp.wa.gov.au/documents/acreage_releases/201305/PETROLEUM_ROYALTIES_13MAR.pdf>. This outline notes that petroleum royalty rates between 10 per cent and 12.5 per cent of well-head value generally apply. The rate applied is dependant upon the type of licence from which the petroleum is produced. Production from primary production licences attracts a rate of between 5 per cent and 10 per cent. However, in almost all cases, a rate of 10 per cent applies to a primary licence and a rate of 12.5 per cent applies once a secondary licence is taken up.

91 See the discussion by P Wieland, 'Going Beyond Panaceas: Escaping Mining Conflicts in Resource-Rich Countries Through Middle-Ground Policies' (2014) 20 *New York University Environmental Law Journal* 199, 201 where the author notes that many developing countries are resource rich, however social inequality, weak government institutions, and growing social unrest can impede the transformation of natural endowment into positive economic benefit.

92 See Otto, Andrews, Cawood, Doggett, Guj, Stermole, Stermole and Tilton, above n 53, 5.

Henry Tax Review.[93] The Henry Tax Review described the rationale underpinning the application of such a taxation framework to the mining and petroleum sector as follows:

> The finite supply of non-renewable resources allows their owners to earn above-normal profits (economic rents) from exploitation. Rents exist where the proceeds from the sale of resources exceed the cost of exploration and extraction, including a required rate of return to compensate factors of production (labour and capital). In most other sectors of the economy, the existence of economic rents would attract new firms, increasing supply and decreasing prices and reducing the value of the rent. However, economic rents can persist in the resource sector because of the finite supply of non-renewable resources. These rents are referred to as resource rent.[94]

10.4.2.7 Minerals resource rent tax

Generally, mining revenue refers to that part of an entity's sale proceeds that are attributable to the taxable resource prior to the resource undergoing what is known as 'beneficiation'. Beneficiation refers to the refinement or separation of the valuable component of iron ore from the waste material. The sale proceeds that may be attributable prior to beneficiation are known as the 'valuation point'. Mining revenue may also include revenue generated from the supply, export, or use of something produced using the taxable resource; for example, the use of electricity produced from the burning of coal that is utilised for carrying on other mining activities.

Mining operations that occur prior to the valuation point are described as 'upstream' mining operations. Mining operations that occur after the valuation point are known as 'downstream' mining operations. The distinction between upstream and downstream mining operations is important because the Minerals Resource Rent Tax (MRRT) that exists in Australia applies at the valuation point; that is, on upstream mining operations. This effectively means that there is a fundamental separation between activities and operations that occur upstream and those that occur downstream. The MRRT will only tax the value of the extracted resources and not the value that is added in the downstream activities.[95]

---

[93] Australia's future tax system, *Papers: Henry Final Report Detailed Analysis Chapter C: Land and resources taxes*, C1-1: Charging for non-renewable resources http://taxreview.treasury.gov.au/content/content.aspx%3F;doc=html/pubs_reports.htm. See also R Garnaut and A Clunies Ross, *Taxation of Mineral Rents* (Clarendon Press, 1983) 33.
[94] See Australia's future tax system, above n 93, 217.
[95] See Australian Tax Office, *Mineral resources rent tax (MRRT)* (24 September 2014) https://www.ato.gov.au/Business/Minerals-resource-rent-tax/.

In broad terms, the MRRT is a tax on profits generated from the exploitation of non-renewable fossil resources in Australia. It replaces the proposed Resource Super Profit Tax (RSPT). The *Minerals Resource Rent Tax Repeal and Other Measures Act 2014* (Cth) received royal assent on 5 September 2014. However, Schedule 1 to this Act, which repeals the MRRT law, commenced on 30 September 2014.[96]

The objects of the *Minerals Resource Rent Tax Act 2012* (Cth) are set out in s 1.10 as follows:

Minerals Resource Rent Tax Act 2012 (Cth)

1.10 Object of this Act

The object of this Act is to ensure that the Australian community receives an adequate return for its taxable resources, having regard to:

(a) the inherent value of the resources; and
(b) the non-renewable nature of the resources; and
(c) the extent to which the resources are subject to Commonwealth, State and Territory royalties.

The MRRT applies a tax rate of 30 per cent on above-normal profits. Highly profitable mines will, however, pay 50 per cent but that will only be where they are deemed to be making extraordinary profits. Above-normal profits are to be calculated as above the 10-year bond rate (currently 5 per cent) plus 7 per cent. In combination with company tax, mining companies in Australia will pay no more than a combined 50 per cent tax on their above-normal profits but the taxation on many mining projects is likely to be significantly lower. A further 25 per cent extraction allowance exists to reduce profits subject to MRRT. Significantly, the MRRT will only apply to iron ore and coal: two of Australia's most significant resources. It will have no application to petroleum products, including natural or unconventional gas. Instead, the existing Petroleum Resource Rent Tax is to be extended to cover both onshore and offshore oil and gas projects. In 2014, the number of companies affected by MRRT is likely to be around 320 and transitional arrangements have been implemented to ensure that companies have time to adjust to the new tax.[97]

10.4.2.8 Petroleum resource rent tax

The Petroleum Resource Rent Tax (PRRT) is a tax on profits generated from the sale of marketable petroleum commodities. On 1 July 2012, the PRRT regime was extended to onshore petroleum projects – including coal seam gas, tight gas and oil shale projects – and the North West Shelf project (but not to the Joint Petroleum

---

96 Ibid.
97 Ibid.

Development Area in the Timor Sea). Some alterations have been made to the PRRT framework given the unique nature of onshore petroleum projects. In particular, resources subject to the MRRT are not taxable under PRRT. This exclusion includes coal seam gas extracted as a necessary incident of mining coal or collected from the controlled burning of underground coal.

PRRT affects any entity that has an interest in an offshore or onshore exploration permit, retention lease, or production licence, or an interest in the North West Shelf project or the Bass Strait project. Joint venture partners have individual responsibilities and contractors normally do not have PRRT responsibilities. An entity's PRRT liability is calculated on the taxable profit the entity makes from an interest in a petroleum project in a year of tax. Taxable profit for PRRT is calculated by subtracting certain deductible expenditure and transferred exploration expenditure from the assessable receipts derived from the project interest. An entity's PRRT liability is levied at 40 per cent of the taxable profit made from its interest in the project. Where an entity holds an interest in an exploration permit or retention lease, it will not have a liability to pay PRRT until a production licence is derived from that interest and commercial production commences.[98]

10.5 REVIEW QUESTIONS

1. Why are mining agreements between a mining/petroleum proponent and the government subsequently ratified by legislation in Australia? How does this legislative ratification alter the mining/petroleum agreement and why is it necessary?

2. Why are offshore petroleum resources generally regulated by a concession agreement?

3. How does a 'profit sharing contract' (PSC) differ from a concession agreement and in what context is the PSC generally used and why?

4. Why are joint ventures popular for mining/petroleum projects and how do the parties protect themselves against the prospect that one joint venturer may misuse confidentially acquired information?

5. What relevance does a development proposal have during the negotiation phase of a mining agreement?

6. Consider the following problem:

    BHP Billiton conducts exploration in areas in which it holds exploration licences for natural gas exploration. An adjacent and much smaller gas company, Natural Gas Co, holds an exploration licence over an adjacent area of land. Based upon its

---

[98] See Australian Tax Office, *Petroleum resource rent tax: Work out PRRT* (22 January 2013) www.ato.gov.au/Business/Petroleum-resource-rent-tax/In-detail/PRRT-in-detail/Work-out-PRRT/.

exploration results, BHP believes that gas may extend into the area of the licence owned by Natural Gas Co. BHP approaches Natural Gas Co and proposes that it be allowed to explore into Natural Gas Co's licence area in exchange for a 50 per cent beneficial interest in the licence. BHP and National Gas Co set out the terms of BHP's expenditure on Natural Gas Co's exploration licence over an agreed period. BHP complies with this obligation and exercises its option to acquire a 50 per cent interest in Natural Gas Co's exploration licence. A joint venture is established between the companies in 2014. This gives both parties respective and mutual interests. Natural Gas Co is privy to the exploratory results from BHP Billiton's activities. The results are positive. Natural Gas Co wants to use the information it has acquired from BHP to sell off its remaining interest in the exploration licence for a strong price. Advise BHP Billiton of the most appropriate course of action.

7. How does a royalty differ from a tax and what is the justification for imposing both forms of revenue in the mining and energy sector?
8. Explain the difference between the 'quantum' method of calculating the royalty rate and the 'ad valorem' method.
9. What is a 'well-head' and why is it relevant to the assessment of petroleum royalties?
10. What is the difference between the Petroleum Resource Rent Tax and the Minerals Resource Rent Tax and what are the inherent justifications associated with the implementation of resource and economic rent taxes in a resource rich country?

10.6 FURTHER READING

S Adepetun, 'Production Sharing Contracts – The Nigerian Experience' (1995) 13 *Journal of Energy and Natural Resources Law* 21.

P Armitage, 'Confidentiality Agreements: A Necessary Inconvenience?' (1998) *AMPLA Yearbook* 100.

Australia's future tax system, *Papers: Henry Final Report Detailed Analysis Chapter C: Land and resources taxes*, C1-1: Charging for non-renewable resources http://taxreview.treasury.gov.au/content/content.aspx?doc=html/pubs_reports.htm.

Australian Government, Geoscience Australia, *Applying geoscience to Australia's most important challenges* <http://www.ga.gov.au/scientific-topics/minerals/mineral-resources/aimr.

Australian Government, Geoscience Australia, *Australia's Identified Mineral Resources 2013* (2013) http://www.ga.gov.au/corporate_data/78988/78988_AIMR_2013.pdf.

Australian Tax Office, *Petroleum resource rent tax: Work out PRRT* (22 January 2013) https://www.ato.gov.au/Business/Petroleum-resource-rent-tax/In-detail/PRRT-in-detail/Work-out-PRRT/.

Australian Tax Office, *Mineral resources rent tax (MRRT)* (24 September 2014) https://www.ato.gov.au/Business/Minerals-resource-rent-tax/.

R Auty, 'Mining Enclave to Economic Catalyst: Large Mineral Projects in Developing Countries' (2007) 13 *Brown Journal of World Affairs* 135.

P Babie, 'Sovereignty as Governance: An Organising Theme for Australian Property Law' (2013) 36 *University of New South Wales Law Journal* 1075.

C Barnett, 'State Agreements' (1996) *AMPLA Yearbook* 314.

E Bastida, T W Vaelda, J Warden-Fernandez, *International and Comparative Law and Policy: Trends and Prospects* (ed) (Kluwer, 2005).

E Blanco and J Razzaque, *Globalisation and Natural Resource Law: Challenges, Key Issues and Perspectives* (Edward Elgar, 2011).

J Blocher, 'Institutions in the Marketplace of Ideas' (2008) 57 *Duke Law Journal* 821.

P Broek, 'Australia Prospects for Tax-Efficient Mineral Wealth' (1992) *International Tax Review* 13.

J Chandler, 'Shale Gas and Government Agreements in Western Australia' (2014) 33 *Australian Resources and Energy Law Journal* 44.

Chevron, *Australia: Project Progress* (December 2014) http://www.chevron.com/countries/australia/businessportfolio/projectprogress/.

Chevron Australia, *Gorgon Project* http://www.gorgon.com.au.

G Cordero-Moss, 'Contract or Licence – Regulation of Petroleum Investment in Russia and Foreign Legal Advice' (2003) 13 *Transnational Law and Contemporary Problems* 519.

A Cox, 'Land Access for Mineral Development in Australia' in R G Eggert (ed), *Mining and the Environment: International Perspectives on Public Policy* (Resources for the Future, Washington DC, 1994).

M Crommelin, 'Federal-State Cooperation on Natural Resources: The Australian Experience' in O Saunders (ed), *Managing Natural Resources in a Federal State* (Carswell, Toronto, 1980).

M Crommelin, 'State Agreements: Australian Trends and Experience' (1996) *AMPLA Yearbook* 328.

D Custos and J Reitz, 'Public-Private Partnerships' (2010) 58 *American Journal of Comparative Law* 555.

T Daintith, 'Evaluation of the Petroleum (Submerged Lands) Act as a Regulatory Regime' (2000) *Australian Mining and Petroleum Law Association Yearbook* 91.

T Daintith, *Finders Keeper's: How the Rule of Capture Shaped the World Oil Industry* (RFF Press, 2010).

Deloitte Access Economics, 'Estimated Companies tax, MRRT and Royalties expenses for the Mineral Sector' (Report Prepared for the Mineral Council, 30 July 2014) http://

www.minerals.org.au/file_upload/files/reports/DAE__MCA_Royalty_and_company_tax_estimates_30_July_%281%29.pdf.

H Devold, *Oil and Gas Production Handbook* (ABB Oil and Gas, 2010).

D K Espinosa, 'Environmental Regulation of Russia's Offshore Oil (and Gas) Industry and its Implication for the International Petroleum Market' (1997) 6 *Pacific Rim Law and Policy Journal* 647.

R Fabrikant, 'Production Sharing Agreements in the Indonesian Petroleum Industry' (1975) 16 *Harvard International Law Journal* 303.

A Fitzgerald, *Mining Agreements: Negotiated Frameworks in the Australian Minerals Sector* (LexisNexis, 2001).

G K Foster, 'Foreign Investment and Indigenous Peoples: Options for Promoting Equilibrium between Economic Development and Indigenous Rights' (2012) 33 *Michigan Journal of International Law* 627.

G K Foster, 'Investors, States and Stakeholders: Power Asymmetries in International Investment and the Stabilizing Potential of Investment Treaties' (2013) 17 *Lewis and Clark Law Review* 361.

M Friedman, *Capitalism and Freedom* (University of Chicago Press, 1962).

R Garnaut and A Clunies Ross, *Taxation of Mineral Rents* (Clarendon Press, 1983).

R Garnaut and M Crommelin, 'Mineral Resource Rent Tax – Will it Work?' (Presented in combination with the Melbourne Energy Institute and the Grattan Institute, 30 May 2013).

C Georgeta, 'The Oil and Mining Concession in Perspective' (2013) 2 *Perspectives of Business Law Journal* 87

A Gillies and A Heuty, 'Does Transparency Work – The Challenges of Measurement and Effectiveness in Resource Rich Countries' (2011) 6 *Yale Journal of International Affairs* 25.

Global CCS Institute, *The Global Status of CCS* (November 2014) http://decarboni.se/publications/global-status-ccs-2014/33-geographical-trends-large-scale-ccs-project.

Government of Western Australia, Department of Mines and Petroleum, *Petroleum Royalties* (2009) <http://www.dmp.wa.gov.au/documents/acreage_releases/201305/PETROLEUM_ROYALTIES_13MAR.pdf>.

D P Grinlinton, *Legal Aspects of Large Scale Mineral Developments: A Case Study of Western Australia* (University of Dundee Centre for Petroleum and Mineral Law Studies, 1988).

P Guj *Mineral Royalties and other Mining-Specific Taxes* (International Mining for Development Centre, University of Western Australia and the University of Queensland, 2012) http://im4dc.org/wp-content/uploads/2012/01/UWA_1698_Paper-01_-Mineral-royalties-other-mining-specific-taxes1.pdf.

W Hauser, 'An International Fiscal Regime for Deep Seabed Mining: Comparisons to Land Based Mining' (1978) 19 *Harvard International Law Journal* 759.

J R Henderson, 'Drafting Mining Agreements: Royalties and Related Drafting Issues' (1989) 3 *Natural Resources and Environment* 12.

L P Hendrix and S L T Golding, 'The Standard of Care in the Operation of Oil and Gas Properties: Does the Operator Owe a Fiduciary Duty to the Non-Operators?' (1993) 44 *Institute of Oil And Gas Law and Taxation* 10–1.

K Hogan, 'Protecting the Public in Public-Private Partnerships: Strategies for Ensuring Adaptability in Concession Contracts' (2014) 2 *Columbia Business Law Review* 420.

M Hunt and M Lewis, *Mining Law in Western Australia* (Federation Press, 1993).

T Hunter, 'Comparative Law as an Instrument in Transnational Law: The Example of Petroleum Regulation' (2009) 21 *Bond Law Review* 42.

International Tax & Investment Center *Petroleum Taxation in Libya* (May 2005) http://www.iticnet.org/file/document/watch/1632.

W Kirk, 'Resource Contracts – Confidentiality Agreements' (1993) *Australian Mining and Petroleum Law Association Yearbook* 194.

E Kuntz, *A Treatise on the Law of Oil and Gas* (LexisNexis, 1989).

J Lacey, 'Partnering and Alliancing: Back to the Future' (2007) 26 *Australian Resources and Energy Law Journal* 69.

M Langton and M Odett, 'Poverty in the Midst of Plenty: Aboriginal People, the Resource Curse and Australia's Mining Boom' (2008) 26 *Journal of Energy and Natural Resources Law* 31.

K N Llewellyn, 'The Effect of Legal Institutions on Economics' (1925) 15 *American Economic Review* 665.

A Loewinger, 'Multilateral Funding for Mineral Exploration in the Third World' (1980) 14 *Journal of World Trade Law* 469.

P W Machin, 'Areas of Mutual Interest' (1993) *AMPLA Yearbook* 221.

D W Maloney, 'Commentary: Natural Gas: Canning Basin Joint Venture Agreement Act' (2013) 32 *Australian Resources and Energy Law Journal* 129.

R McEvoy, 'Can PRRT Be Avoided Under Section 114 of the Constitution: A Question of Substance over Form, or No Form at All? (2014) 33 *Australian Resources and Energy Law Journal* 70.

J McLaren, 'Petroleum and Mineral Resource Rent Taxes: Could These Taxation Principles have a Wider Application?' (2012) 10 *Macquarie Law Journal* 43.

R McLean, 'The Power of the Commonwealth Executive to Contract Post-Williams v Commonwealth of Australia' (2012) 11(2) *Canberra Law Review* 150.

P McNamara, 'The Enforceability of Mineral Development Agreements to which The Crown in the Right of the State is a Party' (1982) 5 *University of New South Wales Law Journal* 263.

D A McPherson, 'Confidentiality and the Public Domain – a Lesson for the Canadian Mining Industry' (1995) 13(2) *Journal of Energy Natural Resources Law* 61.

J D Merrells, 'Mining and Petroleum Joint Ventures in Australia: Some Basic Legal Concepts' (1981) 3 *Australian Mining and Petroleum Journal* 1.

C J Meyers, *Manual of Oil and Gas Terms: Annotated Manual of Legal Engineering Tax Words and Phrases*, (Lexis Nexis, 15th ed, 2012).

G Moloney, 'Case Notes: BHP Petroleum v Balfour' (1987) 16 *Melbourne University Law Review* 436.

A Morris, "Commonwealth of Australia" in A Shah(ed) *The Practice of Fiscal Federalism: Comparative Perspectives* (McGill-Queen's University Press, Montreal, 2007).

K Naito, F Remy and J Williams, *Review of Legal and Fiscal Frameworks for Exploration and Mining* (World Bank Group Mining Department, 2001).

National Economic Development Office, *Partnering: Contracting Without Conflict* (National Economic Development Office, London, 1991).

Y Omorogbe, 'The Legal Framework for the Production of Petroleum in Nigeria' (1987) 5 *Journal of Energy and Natural Resources Law* 273.

W Onorato, 'World Petroleum Legislation: Frameworks that Foster Oil and Gas Development' (Policy Research Working Paper No 2, World Bank, 1995).

J Otto, 'Legal Approaches to Assessing Mining Royalties' in *The Taxation of Mineral Enterprises* (Kluwer, London, 1995).

J Otto, C Andrews, F Cawood, M Doggett, P Guj, F Stermole, J Stermole and J Tilton, *Mining Royalties: A Global Study Of Their Impact on Investors, Government and Civil Societies* (The World Bank, 2006).

J Otto and C Cordes, *The Regulation of Mineral Enterprises: A Global Perspective on Economics, Law and Policy* (Mineral Law Foundation, 2002).

M Radetzki, 'State Ownership in Developing Country Mineral Industries' in *The Use of State Enterprises in the Solid Minerals Industry in Developing Countries*, Proceedings of the Interregional Seminar held in Budapest, 5–10 October 1987 (United Nations Department of Technical Co-operation for Development, TCD/SEM.88/5, INT-87-R37, 1989).

E Schanze, 'Mining Agreements in Developing Countries' (1978) 12 *Journal of World Trade Law* 135.

L E Schroeder, *Oil and Gas Law: A Legal Research Guide*, (W S Hein, 2012).

S L Seck, 'Transnational Business and Environmental Harm: A TWAIL Analysis of Home State Obligations' (2011) 3 *Trade Law and Development* 164.

N Seddon, *Government Contracts — Federal, State and Local*, (Federation Press, 4th ed, 2009).

S Sewalk, 'Brazil's Energy and Policy Regulation' (2014) 25 *Fordham Environmental Law Review* 652.

E E Smith, 'From Concessions to Service Contracts' (1992) 27 *Tulsa Law Journal* 493.

A Stoeckel, *Minerals: Our Wealth Down Under* (Centre for International Economics, Canberra, 1999).

E M Sunley, T Baunsgard, and D Simard, 'Revenue from the Oil and Gas Sector: Issues and Country Experience' (Paper presented at IMF Conference on Fiscal Policy Formulation and Implementation in Oil Producing Countries, Washington, 5–6 June 2002) http://siteresources.worldbank.org/INTTPA/Resources/SunleyPaper.pdf.

N E Terki, 'Comment: The Algerian Act of 1986 and the Encouragement of Foreign Investment in the Area of Hydrocarbons' (1988) 3 *Oil & Gas Law Taxation Revenue* 80.

T Walde, 'Environmental Policies in Developing Countries' (1992) 10 *Journal of Energy and Natural Resources Law* 327.

M Wallace, 'Joint Venture as an Institution for Development: Legislative History' (1978) *Arizona State Law Journal* 173.

L Warnick, 'State Agreements' (1988) 62 *Australian Law Journal* 878.

P Wieland, 'Going Beyond Panaceas: Escaping Mining Conflicts in Resource-Rich Countries Through Middle-Ground Policies' (2014) 20 *New York University Environmental Law Journal* 199.

H R Williams and C J Meyers in P H Martin and B M Kramer (eds), *Oil and Gas Law* (Lexis Nexis, 2004).

R Williams, 'Resource and title security through State Agreements – is it still working?' in *ABARE Outlook '97 – Minerals and Energy: Vol 3* (ABARE, 1997).

J P Williams, 'Global Trends and Tribulations in Mining Regulation' (2012) 30 *Journal of Energy and Natural Resources Law* 391.

INDEX

access agreements 198–9
access entitlements, for solar and wind power 28–9, 243
accession, doctrine of 12
ad valorem royalties 424–5
agreements *see* mining agreements
aquifer interference approvals 228–30
aquifers
 harm minimisation strategy 229–30
 impact of CSG mining 184–5
Arctic waters, marine environment protection 116
assessment entitlements 52
assessment leases
 in New South Wales 77–8, 209
 see also retention licences
Australasian Code for Reporting of Exploration Results, Mineral Resources and Ore Resources (JORC Code) 78
Australian Competition and Consumer Commission (ACCC) 148
Australian Energy Market Agreement 143
Australian Energy Market Commission (AEMC)
 amendment of National Gas Rules 138
 establishment 148
 on incentivising renewable energy 245
 powers and functions 152–5, 158
 recommendations for reform 170
Australian Energy Market Operator (AEMO) 155–8, 260–3
Australian Energy Regulator (AER) 143, 148–52,
158, 161, 168, 170, 260–3
Australian National Carbon Storage Council 294
Australian National Registry of Emissions 253
Australian Petroleum Production and Exploration Association 339
Australian Renewable Energy Agency (ARENA) 252–3
authority to prospect 190–1, 193
Autoridade Nacional do Petróleo (ANP) 128

Bass Strait oil and gas fields 95–6
Bayu-Undan oil field 128
biophysical, strategic, agricultural land (BSAL) areas 220–1, 381
Brazil, removal of minerals and resources on Indigenous lands 12
Brundtland Commission 382
BTEX chemical compounds 217

Callide Oxyfuel Project, Queensland 281
Canada
 carbon capture and storage 283–4
 marine environment protection in Arctic waters 116
carbon capture and storage (CCS)
 aim 282
 benefits 284–6
 capture processes 288–91
 in cement industry 290–1
 current Australian projects 280–1
 development in Australia 294–6
 enhanced oil recovery operations 290
international projects 292–3
legal liability
 environmental concerns 307–8
 regulatory standards 309–10
 tortious actions 308–9
nature of 282–6
need for 280, 286–8
offshore storage regulation
 Commonwealth 294–5, 296, 305–7
 interface with petroleum licensing framework 306–7
 and international law 304
 legal liability 309–10
 in Victoria 296, 307
onshore storage regulation
 legal liability 309–10
 Queensland 296, 301–3
 states 295–6
 Victoria 296, 297–301
oxyfuel combustion 289
post-combustion carbon capture 289
pre-combustion carbon capture 289
process 280, 284–5
storing CO_2 292
strategic initiatives in Australia 280
transporting CO_2 291–2
carbon dioxide 140, 282
carbon pricing 252, 339
CarbonNet Project, Victoria 281
cash bidding 121, 191–2
CATO-2 293
China
 decarbonisation policy 340–1, 342
 National Development and Reform Commission 341
Clean Energy Council 248, 263
Clean Energy Finance Corporation (CEFC) 253–4

444

Clean Energy Regulator
(CER) 249, 253, 255,
256
clean energy sector, growth
of 240
climate change
economics of 335–42
and energy policy 344–8
and global warming 316,
318–20
and human activities
316–19
impact on mining and
energy industries
341–8
climate change mitigation
Harvard Project 333–5
and investment and trade in
energy markets
329–33
Kyoto Protocol 324–9
policy response in
Australia 338–9
role of IPCC 321
UN Framework
Convention on Climate
Change 321–4
coal bed methane *see* coal
seam gas (CSG)
coal royalty rates 428
coal seam gas (CSG)
environmental concerns 180
nature of 179–80
produced water 180, 183
water pumping for
extraction 182–3
coal seam gas (CSG) codes of
practice 216–18
coal seam gas (CSG) industry
activity in NSW 219–20
development in
Queensland 192–3
environmental protection
licences 226–7
freeze on new petroleum
titles in NSW 203
NSW Chief Scientist's review
of 202–3
prohibitions on development
in NSW 221

coal seam gas (CSG)
regulation *see*
unconventional gas
regulation
coal seam gas (CSG)
resources 141–3
coal seam gas (CSG)
statements 190, 193
coastal waters, definition
117
common heritage
principle 113–14
Community Benefits Fund
215
Community Engagement
Guidelines for the
Australian Wind
Industry 266–70
compensatable effects 199
compensation
for compulsory land
purchases 8
conduct and compensation
agreements 68, 197, 199
for interest holders affected
by mining 214–15
for landholders affected by
mining 37–40
compulsory land purchases,
compensation for 8
compulsory pilotage, in
sensitive sea areas 116
concession agreements 409–10
concession system, validity
of 12
condensates 139
conduct and compensation
agreements 68, 197, 199
Constitution
just terms provision 12
legal status of offshore
exploration permits
70–5
contiguous zone,
regulation 107
continental shelf
and exclusive economic
zone 110
expansion of outer
limit 111–12

jurisdiction 110
maritime zone 110
payments to International
Seabed Authority 113
Council of Australian
Governments (COAG)
Environment Protection
and Heritage Council
(EPHC) 266
on fair and free trade in
natural gas 145
Ministerial Council on
Energy 146, 147, 155
Standing Council on
Energy and Resources
(SCER) 155–6, 170
CO_2 CRC Otway Project,
Victoria 281
Crown prerogative, in relation
to royal minerals 18–21
cujus est solum, ejus est
usque ad coelum et ad
inferos 5–8, 9, 12
cultural heritage rights 45–6

Darcys (units of
permeability) 178
decarbonisation, impact on
fossil fuel industry 242
Declared Transmission System
(DTS) (Vic) 157
Declared Transmission System
Service Provider
(DTSSP) 157
Deep Water Horizon oil
spill 96
Department of Environment
(DoE) (WA) 360–1,
373
Department of Environment
and Heritage Protection
(DEHP) (Qld) 374, 375,
377
Department of Industry and
Resources (DoIR)
(WA) 360–1
Department of Mines and
Petroleum (DMP)
(WA) 368–9, 370,
371–2

Department of Natural Resources and Mines (Qld) 374
Department of Planning and Infrastructure (NSW) 383, 384
Department of Resources and Energy (NSW) 386
Doha Climate Change Conference 325, 328
domanial system 11
downstream mining operations 435

East Timor *see* Timor-Leste
ecologically sustainable development 382–3, 393–6
Economic Regulation Authority (ERA) (WA) 148
electricity systems 245
Emergency Risks in Victoria Report 346
emission reduction fund 252, 339
Emissions Reduction Fund Green Paper 343–4
emissions trading schemes 339, 341
energy market
  focus on fossil fuels 2–3
  international investment and trade in 329–33
  renewable energy and property rights 30–1
  restructuring of 241
energy market regulation
  Australian Energy Market Agreement 143
  bifurcation of framework 4–5
  key legislation 143–4
Energy Modeling Forum (EMF) 287–8
'energy paradox' 30
energy policy, and climate change 344–8
energy prices, and globalisation 346–7

energy resources
  consumption 343
  nature, scope and economic potential 2–5, 342–3, 425–7
  see also mineral resources
entry notices 198
environmental bonds 372
environmental damage, liability for CCS 307–8
environmental effects statements (EES) 366
environmental impact assessment
  differences between states 368
  minimal requirements 366–7
  nature of 365
  onshore mining projects
    New South Wales 381–7
    Queensland 362, 374–80
    Western Australia 360–1, 368–73
  and ownership frameworks 357–9
  purpose 366
  referral process 367
  rigour and focus of 355
  scope 367
  and social and economic imperatives 356–7, 360–1
  structure 365
  wind farms 269
Environmental Impact Statements (EIS) 225, 383
environmental impacts, and stages of mining 356, 406
environmental planning instruments 178
Environmental Protection Authority (EPA) 209, 226–7, 369, 370, 373

environmental protection licences (EPLs), for CSG operations in NSW 226–7
environmental regulation of mining and energy sector
  at international level 359
  bilateral agreements 362–5, 393
  Commonwealth powers and legislation 361–2, 387–98
  and ecologically sustainable development 382–3, 393–6
  and inter-generational equity 395–6
  mandatory considerations 396, 397–8
  and matters of national environmental significance 230–2, 388–90
  person's environmental history 396–7, 398
  precautionary principle 393–5
  water resources trigger 390–3
  evolution 355
  jurisdictional framework 360–2
  need for rigorous regulation for mining and energy sector 355–6, 359
  state legislation 361
    New South Wales 381–7
    Queensland 362, 374–80
    Western Australia 360–1, 368–73
  see also environmental impact assessment
European Atomic Energy Community 330

European Commission,
    marine environment
    protection 116
exclusive economic zone
    (EEZ)
  overlap with continental
    shelf 110
  regulation 107–10
exploration
  definition 57
  nature of 53
exploration entitlements 52
exploration licences/permits
  access to land 59
  approval process 54–7,
    68–9
  character of 57–9
  conduct and compensation
    agreements 68
  entitlements 52, 57–8
  environmental protection
    conditions 58–9
  legal status of offshore
    permits 70–5
  legislative provisions 59–69
    New South Wales 59–62
    Queensland 64–9
    Western Australia 62–3
  offshore mining exploration
    licences 131
  petroleum exploration
    permits 120
  prioritising overlapping
    entitlements 68
  as *profit a prendre* 69–70
  proprietary status 69–70
exploration process, stages
    and techniques 53–4

fixed quantum royalties 424
fixtures, doctrine of, and
    laying of pipelines 32–4
fossil fuel industry, impact of
    decarbonisation 242
fossil fuel resources, extent
    of 244
fossil fuels, and current energy
    market 2–3
fracking *see* hydraulic
    fracturing

fracture stimulation
    management plans
    (FMSPs) 178–80
freedom of navigation
    principle 114

Garnaut Review 2008 336
Garnaut Review 2011 245,
    337
gas market, in Australia 140–3
gateway certificates 220
gateway process 219–22, 228
geothermal energy
  ownership 22–3
  and water licensing 27
Gladstone LNG (GLNG)
    project 186, 188, 306
Global Carbon Capture and
    Storage Institute 280
gold, treatment under common
    law 18–21
good oilfield practice 119
Gorgan Project, Western
    Australia 2–3, 281
Greater Sunrise gas and
    condensate fields 127,
    128–30
greenhouse gas activities 295
greenhouse gas assessment
    permits 295
greenhouse gas emission
    reduction 338
  in Australia 343–4
  and *Kyoto Protocol* 324–9
greenhouse gas emissions,
    sources 344–5
greenhouse gas holding
    leases 295, 306
greenhouse gas injection
    licences 295, 307
greenhouse gas licences 118
greenhouse gas safety
    zones 116
groundwater, impact of CSG
    mining 184–5

harm minimisation strategy,
    towards aquifer
    interference 229–30
Harvard Project 333–5

Hazelwood coal fire
    disaster 346, 386
Henry Tax Review 435
high seas region
  mineral reserves 97
  regulation 113–17
Hilmer Report 145
horizontal drilling 181–2
hydraulic fracturing ('fracking')
  environmental damage
    176–7
  health hazards 177
  process 181–2
hydrocarbon resources,
    location and regulatory
    framework 4–5
hydrocarbons 139, 178
hydro-electric projects
  ownership 22
  water licences 26–7

Independent Commission
    Against Corruption
    (ICAC) 205
Independent Expert Scientific
    Commission on
    CSG and Large Coal
    Mining Development
    (IESC) 221, 381
Independent Pricing
    and Regulatory
    Tribunal 215–16
Indigenous heritage
    protection 45–6
Industry Commission, Report
    on Energy Generation
    and Distribution 137
infrastructure licences 123
innocent passage, principle
    of 114–16
inter-generational equity
    395–6
Intergovernmental Panel
    on Climate Change
    (IPCC) 337
  on CCS 282, 286
  on global warming and
    human activity 318,
    319–20
  structure and role 321

International Energy Agency
   (IEA) 176, 285–6, 287,
   317, 328
International Maritime
   Organization (IMO)
   116
International Monetary
   Fund 259
International Seabed Authority,
   payments from resource
   title holders 113,
   117
iron ore royalty rates 428

Joint Petroleum Development
   Area (JPDA) 127–8
joint venture agreements
   414–17
JORC Code (Australasian
   Code for Reporting of
   Exploration Results,
   Mineral Resources and
   Ore Resources) 78

Kitan oil field 128
Korea, national development
   strategy 340

land, horizontal and vertical
   divisions 9
land access agreements, in
   Qld 198–9
Land Access Arbitration
   Framework Review
   (Walker Review)
   215–16
land access arrangements, in
   NSW 212–14
land access disputes, in
   NSW 215–16
land access rights
   exploration licence/permit
     holders 59
   for licence holders 34–7
   Queensland Land Access
     Code 64, 196, 199–201
   and statutory compensation
     provisions 37–40,
     214–15
Land Appeal Court (Qld) 380

Land Court (Qld) 195
Land and Environment Court
   (NSW) 211, 225–6
Large-scale Generation
   Certificates (LGCs)
   254–5, 256
Large-scale Renewable Energy
   Certificates (LRECs) 250
Large-scale Renewable Energy
   Target (LRET) 250, 251,
   254–6
Law Reform Commission
   (Vic) 244
licences, petroleum
   exploration,
   production and storage
   licences 34–7
liquefied petroleum gas
   (LPG) 139
liquid natural gas (LNG) 139
Lord Howe Rise 112

Mandatory Renewable Energy
   Target (MRET) 250–1
marine environment protection
   in Canadian-claimed Arctic
     waters 116
   jurisdiction over transiting
     ships 116
methane 139, 178
Mine Closure Plans 371
mine rehabilitation plans
   371
mineral development licences
   in Queensland 78
   see also assessment leases;
     retention licences
mineral resources, location
   and regulatory
   framework 4–5
Mineral Titles Online
   website 55
minerals, definition in Qld
   legislation 4
Minerals Resource Rent Tax
   (MRRT) 435–6
mining agreements
   concession agreements
     409–10
   core elements 417–22

joint venture
   agreements 414–17
   nature of 407–8
   profit sharing contracts
     410–13
   risk service contracts 414
   royalties 422
   State Agreements 418–22
   state intervention 406–7
   types 408
mining approval process,
   phases and related
   entitlements 52–4
mining leases
   general entitlements 84–90
   general terms and
     conditions 83–4
   land access entitlements
     34–7
   nature, purpose and role
     80
   in New South Wales 82–3,
     86–7, 209–10
   petroleum production
     licences 121–3
   in Queensland 84–6
   statutory character 80–3
   in Victoria 83, 87
   in Western Australia 84,
     88–9
Mining and Petroleum
   Legislation Amendment
   Bill 2014 (NSW)
   205–7
Mining Registrar, Warden's
   Court (WA) 56
Mining Rehabilitation
   Fund 371
mining royalties see royalties;
   royalty rates
mining taxes
   in Australia 434–7
   in Australian context
     425–7
   Minerals Resource Rent Tax
     (MRRT) 435–6
   Petroleum Resource Rent
     Tax (PRRT) 424, 436–7
   Resource Super Profit Tax
     (RSPT) 436

mining tenements, proprietary status of 14–17
Mining Tribunal (Tas) 77
Montara Commission of Inquiry 125–6
Montara oil and gas spill 125
MyMinesOnline 41–2

Narrabri CSG exploration program 187
National Centre for Cooperation on Environment and Development 232
National Competition Council (NCC) 160
National Competition Policy Review 138
National Electricity Market (NEM) 260–2
National Electricity Regulator 261–3
National Energy Market 245
national environmental significance, matters of 388–90
National Gas Law
　mirror legislation 138
　purpose 138
National Gas Rules, amendment 139
National Gas (South Australia) Bill 2008 160
National Gas Strategy 137
national heritage protection 4
National Institute for Occupational Safety (US) 177
National Native Title Tribunal (NNTT)
　appeals against determinations 45
　engagement with the state 44
　inquiries and determinations 44
　overruling of determinations 45
　role 40

National Offshore Petroleum Safety and Environment Management Authority (NOPSEMA) 119, 125–7
National Offshore Petroleum Titles Administrator (NOPTA) 119, 123, 124, 126, 127
National Water Initiative (NWI) 27
native title processes
　expedited procedure 41
　range of 41
　right to negotiate 42–5
Native Title Protection Conditions (NTPCs) 41–2
native title rights
　definition 40
　extinguishment 40
　intersection with mining rights 40–5
natural gas
　nature of 139–40
　value as a commodity 140, 176
natural gas industry, restructure 137
natural gas liquids (NGL) 139
natural gas market
　in Australia 140–3
　international demand 176
　short-term trading market (STTM) 158
　wholesale market 158
natural gas pipelines
　access arrangement information 167–71
　classification 159–64
　full access requirements 169–70
　full regulation 164–6, 167–71
　light regulation 164–6, 167–71
　limited access requirements 168–9
　network 155–6
natural gas regulation

access arrangement information 167–71
access determination disputes 166–7
and energy market framework 143–4
former natural gas access code 144–7
introduction of National Gas Law 147–8
introduction of National Gas Rules 148
light and fully regulated pipelines 164–6
objective of National Gas Law 158–9
pipeline classification 159–64
principal regulatory bodies 148
reform of National Gas Law 170–1
role of AEMC 152–5, 158
role of AEMO 155–8
role of AER 148–52, 158
role of ERA 148
natural gas resources
　conventional gas resources 140, 141
　locations 140, 141
　unconventional gas resources 140, 141–3
natural gas transportation industry, regulatory certainty and clarity 137
natural resource interests, open nature of 10–11
Netherlands, CCS projects 292–3
New South Wales
　Annual Environmental Management Reports 385, 386
　aquifer interference approvals 228–30
　assessment leases 77–8, 209
　BSAL areas 220–1, 381
　CGS mining locations 219–20

New South Wales (*cont.*)
  community concerns over CSG mining 202
  compensation 214–15
  ecological sustainability 382–3
  environmental impact assessment, onshore mining projects 381–7
  Environmental Impact Statements 225
  environmental management conditions 384–7
  environmental management obligations 210–11
  environmental planning and assessment 223–7, 382–4
  environmental protection licences 226–7
  exploration licences
    assessment 210
    community consultation 187–8
    exclusion zones 203, 221
    extinguishment of existing applications 203–4
    provisions 59–62
  gateway assessment 381–2
  gateway process 219–22, 228
  land access arrangements 212–14
  Mining Operations Plan 385, 386
  Mining, Rehabilitation and Environment Management Process 385, 387
  mining/production leases 82–3, 86–7, 209–10
  petroleum titles
    environmental management obligations 210–12
    'fit and proper' person test 204–7
    freeze on titles for CSG 203
    high and low-impact titles 210
    ownership framework 207–8
    'public interest' test 204
  planning framework reform 219
  prohibitions on CSG development 221
  prospecting titles 210
  rehabilitation conditions 385, 386–7
  Review of Environmental Factors (REF) 226
  special prospecting authorities 209, 210
  state significant developments (SSD) 223–4, 383–4
  state significant infrastructure (SSI) 223–4, 383–4
  strategic regional land use policies (SRLUPs) 219–21
  strategic release framework for CSG exploration 204
  unconventional gas regulatory framework 202
  water access licences 228
  water licensing laws 227–30
  wind power regulation 263–4, 271
New South Wales Chief Scientist, review of CSG industry 202–3
non-derogation from grant, doctrine of 16–17
Northern Territory, mining leases 83
NSW Parliament General Standing Committee, Report on CSG 213

Office of Coal Seam Gas (NSW) (OCSG) 207, 216–8
offshore areas, definition 118
Offshore Constitutional Settlement 117–18
offshore exploration, release of offshore acreages 113
offshore mineral exploration licences 131
offshore mining
  evolution of 96–7
  in high seas region 97
  regulation 130–1
  size of industry 131
offshore petroleum/gas drilling in Australia 95–6
  environmental implications 96
  nature of 95
offshore regulation
  collaboration with Timor-Leste 127–30
  complementary state and federal legislation 97–103
  contiguous zone 107
  continental shelf 110
  exclusive economic zone (EEZ) 107–10
  good oilfield practice 119
  high seas 113–17
  Joint Petroleum Development Area 127–8
  jurisdictional framework 98–103
  mining 130–1
  petroleum safety 125–7
  primary Commonwealth legislation
    key definitions, provisions and application 113–14
    resource title framework 119–24
  safety zones 118
  sea installations 124–5
  territorial sea
    Commonwealth jurisdiction 103–6
    state jurisdiction 106–7
  and United Nations Convention on Law of the Sea 103–6

450 Index

oil spills 96, 125
onshore mining projects
   New South Wales 381–7
   Queensland 374–80
   Western Australia 368–73
overlapping entitlements
   and different forms of ownership 31–4
   prioritising exploration titles 68
overlapping tenements
   greenhouse gas and petroleum interests 306–7
   in Queensland 193–5
   safety management plans 193–5
ownership
   of minerals and energy resources 4–5
   overlapping entitlements 31–4
   public ownership of minerals and petroleum 5–9
   of renewable resources 21–2
   of royal minerals 18–21
   of subsurface strata 5–9
   of water 23–7
oxyfuel combustion 289

Papua New Guinea, maritime boundaries 129–30
Partial Exemption Certificates (PECs) 256
personal property, mining tenements as 14–16
petroleum
   definition in Cth legislation 118
   definition in Qld legislation 4–5, 189
petroleum exploration licences
   land access entitlements 34–7
   New South Wales
     assessment 210
     community consultation 187–8

     for CSG 209
     exclusion zones 203
     extinguishment of applications 203–4
     provisions 59–62
   Queensland
     provisions 64–9
     public consultation 187
petroleum exploration permits
   approval process in Qld 68–9
   entitlements and requirements 120
petroleum leases
   in Queensland 189–90, 191–2
   underground water rights 192
petroleum licences
   'fit and proper' person test 204–7
   'public interest' test 204
Petroleum (Onshore) Amendment Bill 2013 (NSW) 212–14
petroleum production licences, entitlements and requirements 121–3
Petroleum Resource Rent Tax (PRRT) 424, 436–7
petroleum resources, location and regulatory framework 4–5
petroleum retention leases
   entitlements and requirements 120–1
   land access entitlements 34–7
petroleum royalty rates 428–9
petroleum safety, regulation 125–7
petroleum safety zones 118
petroleum titles
   'fit and proper' person test 204–7
   ownership framework in NSW 207–8
pipeline licences 124

pipelines, and doctrine of fixtures 32–4
Planning Assessment Commission (NSW) 210, 224
'polluter must pay' principle 248
precautionary principle 393–5
produced water 180
production entitlements 53
production leases *see* mining leases
production phase of mining/ resource projects 54
Productivity Commission 146
profit sharing contracts (PSCs) 410–13
prospecting titles, in NSW 210
public ownership, of minerals and petroleum 10–14

Quantified Emission Limitation or Reduction Objective (QELRO) 325
Queensland
   access agreements 198–9
   access and compensation framework 196–201
   authority to prospect 190–1, 193
   conduct and compensation agreements 68, 197, 199
   consultation with landowners re access 187
   CSG industry 192–3
   CSG statements 190, 193
   entry notices 198
   environmental assessment 362
   onshore mining projects 374–80
   environmental authority for resource activities 374–5
   environmental harm 376–7
   environmental impact assessment

Queensland (*cont.*)
  public scrutiny 380
  triggers 377–80
  environmental management
    (EM) plans 375
  environmental values
    375–6
  exploration licence
    provisions 64–9
  exploration permit approval
    process 68–9
  general environmental
    duty 377
  Land Access Code 64, 196,
    199–201
  mineral development
    licences 78
  mining leases 84–6
  onshore CO_2 storage
    regulation 296, 301–3,
    309–10
  overlapping tenements
    193–5
  petroleum leases 189–90,
    191–2
  resource title
    requirements 188
  safety management
    plans 193–5
  unconventional
    gas regulatory
    framework 188
  underground water
    rights 192
Queensland Gas Scheme 188

REC Registry 249, 253, 254,
  256
regalian system 11
Register of Native Title
  Claims 266
regulatory framework,
  bifurcation 4–5
rehabilitation bonds 58–9
renewable energy
  and Australian energy
    market 3
  definition 241
  economics of 258–60
  nature of 241–4

sources of 241
statutory regulation 246–50
Renewable Energy Certificates
  (RECs)
  creation and
    management 249–50
  large-scale generation
    certificates (LRECs) 250
  nature of 257
  small-scale generation
    certificates (SRECs)
    250
renewable energy market
  in Australia 244–6
  market progression 30–1,
    245–6
  and property rights 30–1
renewable energy resources
  differences from non-
    renewable energy
    resources 243
  environmental impacts 241
  future role in Australia's
    energy mix 242
  ownership of 21–2, 243
renewable energy shortfall
  charge 252, 256
Renewable Energy Target
  (RET) scheme
  aim 248–50
  background 250–2
  calculation of RET 251
  and infrastructure
    investment 247, 248
  introduction 246
  regulation 249–50
  review of 247–8
  and wind energy 263–6
Renewable Power Percentage
  (RPP) 250, 256
res communes principle 113
reservoir rock 178–9
Resource Super Profit Tax
  (RSPT) 436
resource titles
  issuing and management
    of 117
  legal character of 89–90
  regulatory framework
    119–24

retention licences
  petroleum retention
    leases 120–1
  purpose and character of 75
  in Tasmania 77
  in Victoria 78–80
  in Western Australia 76–7
  see also assessment leases;
    mineral development
    licences
Review of Environmental
  Factors (REF) 226, 383
Rio Conference (Earth
  Summit) 320, 321, 382
riparian water rights 24, 25
risk management, outer
  continental shelf
  projects 117
risk service contracts 414
Rotterdam Climate Initiative
  (RCI) 293
royal minerals, ownership
  of 18–21
royalties 422
royalty rates
  in Australian context 425–7
  coal 428
  iron ore 428
  petroleum 428–9
  privately owned
    minerals 429–30
  value at well-head 430–4

safety management plans,
  for overlapping
  tenements 193–5
sea installations,
  regulation 124–5
severance, doctrine of 13
shale gas 140, 141, 143,
  178–9, 219
silver, treatment under
  common law 18–21
small-scale generation
  certificates (SRECs) 250
Small-scale Renewable Energy
  Scheme (SRES) 250,
  251, 257–8
Small-scale Technology
  Certificates (STCs) 257

452 Index

socio-economic impacts
    of mining and energy
    projects 406
    see also compensation
Solar Credits Scheme 251,
    257–8
solar energy sector 261–3
solar power, access
    entitlements 28–9, 243
South West Hub Project,
    Western Australia 281
Southern Carnarvon Basin 112
special prospecting
    authorities 209, 210
State Agreements 418–22
statutory water
    entitlements 25–7
STC Clearing House 257–8
Stern Review (Economics of
    Climate Change) 258,
    335–6
subsurface strata, ownership at
    common law 5–9

Tasmania, retention
    licences 77
taxes *see* mining taxes
territorial sea
    Commonwealth
        jurisdiction 103–6
    state jurisdiction 106–7
tight gas 140, 141, 143, 178–9
Timor Gap 127, 130
Timor-Leste
    collaboration over resource
        development 127–30
    dispute over Timor Gap 130
    maritime boundaries 127–8
Timor Sea 127
Timor Trough 130
Torres Strait, compulsory
    pilotage 116
Torres Strait Protected Zone
    (TSPZ) 129
tortious actions, liability for
    CCS 308–9

unconventional gas extraction
    community consultation
        187

consultation with land
    owners 187
environmental issues 184–5
processes 180
social impact 185–6
unconventional gas industry
    environmental
        assessment 177
    expansion 176
unconventional gas regulation:
    Commonwealth
    environmental protection
        provisions 230–2
    water resources trigger 390–3
unconventional gas regulation:
    New South Wales
    codes of practice for
        CSG 216–18
    community concerns
        over 202
    compensation 214–15
    core regulatory
        framework 202
    environmental planning and
        assessment 223–7
    exclusion zones 203, 221
    gateway process 219–22,
        228
    land access
        arrangements 212–14
    petroleum licences and
        conditions 208–12
    petroleum titles
        'fit and proper' person
            test 204–7
        ownership
            framework 207–8
        strategic release
            framework 204
    water licensing laws 227–30
unconventional gas regulation:
    Queensland
    access and compensation
        framework 196–201
    resource title
        requirements 188
unconventional gas resources
    location and extent of 140,
        141–3
    nature and types 178–80

underground water rights
    192
United Nations Commission
    on the Limits of the
    Continental Shelf
    (CLCS) 111–12
United Nations Conference
    on Environmental
    Development (UNCED)
    (Earth Summit) 320,
    321, 328, 382
United Nations Environment
    Program (UNEP) 321
United States
    carbon capture and
        storage 283, 284
    doctrine of severance 13
    solar access rights 29
    unitisation 13
unitisation 13
upstream mining
    operations 435
usufractuary water
    entitlements 24, 25

Varanus Island gas pipeline
    explosion 125
vesting provisions
    assumptions underlying
        12
    constitutional legitimacy
        of 11–13
Victoria
    environmental effects
        statements (EES) 366
    mining leases 83, 87
    offshore CO_2 storage
        regulation 296, 307,
        309–10
    onshore CO_2 storage
        regulation 295–6,
        309–10
    rehabilitation bonds for
        mining projects 58–9
    renewable energy
        projects approvals
        processes 264
    retention licences 78–80
    wind power regulation 264,
        270

Index 453

Victorian Civil and
    Administrative Tribunal
    (VCAT), health impact
    of wind farms 271–3

Walker Review 215–16
Wallaby Plateau 112
Warburton Review 247–8
Warden's Court (WA), role of
    Mining Registrar 56
water, ownership of 23–7
water access licences
    (WALs) 24, 26, 227–30
water allocation grants 26
water entitlements 227–8
water permits 24
water resources trigger 390–3
well-head, meaning of value
    at 430–4
Western Australia
  environmental assessment
    and commercial
    imperatives 360–1

of onshore mining
    projects 368–73
environmental bonds
    372
exploration licence
    applications 55–7
exploration licence
    provisions 62–3
Mine Closure Plans 371
mine rehabilitation
    plans 371
mining leases 84, 88–9
mining proposals
    Annual Environmental
      Reports 371, 372
    environmental
      conditions 372–3
    requirements 370
Mining Rehabilitation
    Fund 371
retention licences 76–7
significance test 369–70
wind energy industry 260–3

wind farms
    access entitlements 28–9,
      243
    best practice guidelines
      266
    community and stakeholder
      consultation 266
    environmental impact
      assessment 269
    health impacts 270–3
    and national environmental
      significance 264–5
    and native title claims
      265–6
    nature of 260–3
    regulation 263–6, 270, 271
work program bidding
    system 113
World Bank 259
World Economic Forum 259
World Meteorological
    Organization
    (WMO) 321